EXPLORATION AND PRODUCTION GEOSCIENCE

Comprehensive Skills Acquisition for an Evolving Industry

Basics of Petroleum Geoscience

Exploration and Production Disciplines

Well Geology and Operations

Formation Evaluation and Reservoir Management

Livinus Nosike

EXPLORATION AND PRODUCTION GEOSCIENCE

Comprehensive Skills Acquisition for an Evolving Industry

ISBN: 978-2-36523-089-6

Publication of Extracts for Online Training, 2017, 2018, 2019

Integrated Elvee Services Limited (iESog)
contact@iesog.com / www.iesog.com

Delizon books and other media products are available at special quantity discounts, for use as premiums and sales promotion, or in corporate training programs. For more information, please send your email to the Director of Sales, sales@delizonpublishers.com, contact your local book-store or write to:

Europe
Delizon Publications
19 rue des Alliés,
64000 Pau, France.
europe@delizonpublishers.com

Africa
Delizon Publishers Limited
146 Ihe Street, Oneday Layout, Enugu,
Enugu State, Nigeria.
africa@delizonpublishers.com

Cover design by
Delizon International Book Publishers
www.delizonpublishers.com

Disclaimer:
The content of this work is entirely for educational purposes, including the maps, figures, formulas, methods and texts, and should not be taken as professional advice. The author and publishers assume no liability whatsoever for any loss or damage that results from the use of this book. Use of the material in this book is solely at the risk of the user.

Europe . Africa . North America . Asia

DEDICATION

To my children:

Great, Mighty and **Awesome**

PREFACE

Seemingly, technical books are written for those who already know the subject. Most academic books teach basics while professional books are presented with technicalities and advanced images to beginners. One may wonder if it were possible to, on one hand, present exploration and production geoscience in such a way as not to make the content cumbersome and yet, on the other hand, cover the most if not the overall topics. This is a far-fetched objective, yet the intent of this work.

The book has tried to pass on only what has been understood by the author and to do so in a simple language, framing each sentence so that the content may be easily learned, especially in the early chapters. It presents complex topics pedagogically with simple sketches. As the reader works from the simplicity of cartoon sketches in the first few chapters, the reader goes on to confront the more advanced contents of the later chapters, in the same context.

EXPLORATION AND PRODUCTION GEOSCIENCE aims to cover the full value chain of petroleum exploration and production, as may be suitable for both students and industry professionals. The work emphasises the theoretical and practical aspects and it is hoped that the reader will understand the content first as a general knowledge, and then secondly as a discipline integrating the right terminologies, principles and practical applications. The book will therefore be useful as a complementary training aid, to fully equip the geoscientist.

In this time of energy transition, when the petroleum industry is doing a final mop-up of resources and evolving fast across different skills, there is need to cover the wide expanse of exploration and production, and undertake training courses that cut across the various domains and departments in the oil industry. As the industry delves into artificial intelligence, digitalization, big data and newer technologies, there is no longer room for professionals or researchers to be oblivious of the core fields. Each work candidate must now be able to cooperate with others in the different fields of geoscience and understand what is done by every team at all stages of the upstream activities.

It is hoped that the uniqueness of this book is its inclusion of four parts that touch on the basics of petroleum geoscience, the disciplines in the upstream petroleum sector, well geology and technical well operations, formation evaluation / petrophysics and reservoir or field management, all in one volume. New ideas, findings and approaches have been introduced through this work, building on basic geological concepts and research results. The reader will no doubt see new details on geoscience operations and discover the depth as well as learn the skills required for the geoscientist's tasks.

The work starts from the core geoscience disciplines – petroleum geology, stratigraphy, sedimentary and structural geology – to the licensing and mining of oil acreages. It tackles the industry best practices in prospect evaluation, seal analyses, pore pressure prediction; it integrates the roles of well site and operations geology, overviews the geoscience and engineering aspects of drilling, well tests, reservoir and field management, emphasising how to work in a team to achieve exploration and production objectives.

Effort is made to place schematic figures of geologic bodies in 3D (and in some cases 4D), as representation has long moved from 2D seismic that formed the bases of most figures found in textbooks. It is earnestly hoped that by going through this book, the reader begins to understand what goes on through the value chain of petroleum exploration and production, thereby building individual and collective capacity, especially in developing oil host nations. The challenge will be the ability of such a book to be both versatile and in-depth, which is a limitation to overcome in subsequent communications. Such a book as this requires revisions leading to improvements in subsequent editions.

Dr. Livinus Nosike
MD & Chief Consultant, iESog
Enugu, January 2020

ACKNOWLEDGEMENTS

I remember the late Dr. Kingsley Ojoh, former Director at Total E & P Nigeria Limited, who gave me the opportunity, following his initial interview, in the process of selecting candidates for an international postgraduate merit scholarship, co-sponsored between then TotalFinaElf (Operator of the NNPC Joint Venture) and the French Government, to pursue my understanding of exploration and production geoscience further after my first degree.

Thanks to Prof. C. S. Nwajide, who at the demise of my mentor, Dr Kingsley Ojoh, continues to play that role regarding my geologic pursuits. Not that I have not grown up, but it is good to have somebody over my shoulder who says, hey, not that way, or yes, ride on. Thanks to my former lecturers: Prof. A. W. Mode, Dr. C. Ugbor, and others from the University of Nigeria, Nsukka, with whom I still share geoscience times and go for field work every year.

I owe my sincere gratitude to peers who reviewed this book, including: Dr. Aaron Auduson, Dr. Ifeanyichukwu Obi, Dr. Okwudiri Anyiam, Dr. Obinna Chudi, Dr. Dan Amogu, Mr. Lekan Oletoke, Mr. Nworie Ejiuwa and again Prof. C. S. Nwajide, who did more than reviewing by also offering useful editorial advice. I also thank the staff of Delizon Publishers and Integrated Elvee Services Limited, notably Mr Arthur Oghuma and Mr Kenny Egwu, for their reviews, suggestions and general contribution to the editorial process.

Many thanks to my Thesis Supervisor, Dr. Christopher Wibberley, who helped to shape my understanding of the core principles of structural geology and geopressures while I was a PhD student and later while we both worked at Jean Feger's Technical and Scientific Centre (CSTJF) in Pau, Southern France. As colleagues too in Total E & P, we still share useful technical and personal times together. I would like to mention other colleagues at Total E & P, particularly those with whom I have co-authored scientific papers, including: Jean-Loup Montenat, Yves Pacalin, Christophe Jardiné, Arnaud Delpeint, Gabriel Uwerikowe, Victor Biu, Adeoye Adeyemi and Musa Usman; then my first manager, a wonderful boss and *patron*, Monsieur Claude Guyot.

Other persons with whom I have had useful conversation and who were instrumental directly or indirectly to my reaching this point where I share the knowledge of geoscience to others, including members of the jury during my PhD defence at University of Nice, Sophia Antipolis (UNSA), are: Prof. Michel Popoff, Dr. Kingsley Ojoh's PhD supervisor, with whom I also spent time while he reviewed my work before my PhD defence; Prof. Jean Mascle, Director of the Research in the Geoazur Observatory in Villefranche; Dr. Christophe Larroque, Researcher at the Geoazur Laboratory in Sophia Antipolis; Prof. Peter Cobbold, Director of Research at the University of Rennes; Dr. Damien Dhont, Former lecturer and Assistant Professor at my former postgraduate school, University of Pau (later a colleague at CSTJF); Prof. Robert Butler, Professor of Tectonics at the University of Aberdeen; Dr. Bertrand Gauthier, Engineer and expert at Total E&P headquarters in France.

It is good to remember where we started, with whom we shared that space, and thank them for being part of the journey so far. Space will not allow me to individually thank the many other colleagues, partners and friends, with whom I have worked in both the private and public sectors of the industry, in the field and at base / office, during well operations with the subsurface team, as synthesis geologist in the exploration department, as development and asset geologist, monitoring wells and managing fields in the asset subdivision, including representatives of the regulatory authorities from National Petroleum Investment Management Services (NAPIMS) and Department of Petroleum Resources (DPR). I appreciate the wonderful times we spent together amidst operational challenges.

Thanks to my family and loved ones for being there for me during all the pursuits. A special one to my lovely wife, Ify, who alone can understand the passion, the patience, the challenges and the grace it took to come this far. A thank you to my children, who would run around, come over to my personal computer and try to help. They would ask: *Daddy, when will you be through with this?* I am now available to spend all the time with you!

FOREWORD

With the oil industry serving as the cash cow of several countries, cartels and companies, the need persists for modern, updated, and state-of-the-art geoscientific literature covering the entire gamut of the upstream or part thereof. The Delizon Publishers have taken a major stride towards the filling of that need. The result of that effort is the volume in your hands. The publication draws attention to the expanse of the realm of petroleum exploration and production geoscience both in its theoretical aspects and in the application of the principles to the wide range of activities spanning the entire scope of oil finding to production. Implicit in this enormous coverage is the obligation of practicing geoscientists to be easily adaptable or even aspire to some degree of versatility.

From the core areas of sedimentology/stratigraphy, their activities in the industry dovetail into physics, chemistry, and geodesy. Ancillary to these are the burgeoning areas of artificial intelligence and software engineering; so much to cope with. Well operations, involving electrical/electronic and mechanical engineering and material science, which are getting ever more sophisticated in response to the needs of designer oil wells, should not be strange to geoscientists. This is even more so as they form parts of various teams within which discipline integration is the way forward for modern–day operations. If this wide spectrum draws attention to the critical role of geoscientists in the economy of oil producing outfits, that is as well.

With such a large encompass in mind, Dr. Nosike, a practicing petroleum geoscientist, set out to cover the entire ground both in all its large encompass and great depth with surprisingly telling effect. As can be gleaned from the subheads in the contents, as well as the chapter summaries, the book gets down to not just the practical aspects of exploration and production, but where necessary, has gone further than the routine book style and delved into the details of procedures. Such "factory floor" routines, reminiscent of manuals of procedure, have been reserved for the later parts of the book, standing on the simple theoretical and easy introduction in the earlier parts – the first half. For most texts, this level of detailing would suffice. But not for Dr. Nosike, who devoted a substantial number of pages (as appendices) to a following stream of activities, duties and procedures on the well site, a hypothetical well, and down to contracts, audits and logistics.

With the advanced student, researcher or even the consultant in mind, the author has distilled out questions from each chapter. A tip to the student is to start the study of each chapter by first carefully reading the questions. This approach is sure to raise curiosity which reading the text may then satisfy, allowing the various points, however intricate, to stick to memory. The end of each chapter is also marked with references of the widest vintage to enable not only the authentication of information but also the furtherance of research. For the "coal face" practitioner, the critical areas of health, safety, environment and security have been stridently raised, reiterated and emphasized almost ad nauseam for obvious reasons: nothing works in the absence of these elements of personal and material preservation.

In all, the book points up the versatility of geoscientists, their vital place and role in national and global economies, and their obligation in environmental conservation. The energy industry in Nigeria, Africa and far beyond has had the fortune of benefitting from the author's eruditism, direct oil industry experience, as well as his passion and communicative acumen.

Professor C.S. Nwajide,
FNMGS, FNAPE, Aret Adams Laureate,
Consulting Geologist
Awka, January 2020

CONTENTS

INTRODUCTION

Exploration and Production

One of the major natural resources explored and produced globally is petroleum, such that the terms **exploration** and **production** are often used to refer to petroleum exploration and production. This book deals essentially with petroleum exploration and production aspects of geoscience. Seismic data has become one of the most relevant for exploration and production of crude oil. **Petroleum Exploration** involves the use of mainly seismic mapping and comparison of well logs exploration targets with those of well logs from offset (nearby) reservoirs, to evaluate the potential of new prospects.

Petroleum Production starts after developing (involving planning and drilling wells) the oil and gas field. When production starts, follow up seismic (4D seismic) provides information that indicates the state of the remaining reserve of hydrocarbons. The resources explored and produced are either **conventional** (in-situ oil and gas) or **unconventional** hydrocarbons (heavy oil, oil shale, tar sand, shale gas and possibly gas hydrates). The extraction of heavy oil, tar sand and shale gas proves to be more challenging with respect to hazards to the environment.

Petroleum Geoscience

Petroleum geoscience is the study of petroleum using all sciences related to its nature, search, recovery from the subsurface, and management. It is a field that covers the various areas of petroleum exploration and production, from its generation through migration and accumulation. The core fields are within geology, such as sedimentary and structural geology, stratigraphy and their application to petroleum studies, plus related courses in geophysics, geochemistry, petrophysics, as well as mainstream petroleum geology.

Petroleum geology as a course focuses on the basic materials from which petroleum is formed, and on how they generate and expel hydrocarbons. The principal material is **kerogen**, formed in source rocks. Mature source rocks expulse their hydrocarbons, which move through carrier strata to accumulate in reservoirs, when trapped under impermeable rocks known as seals. The configuration that makes this possible, in a sedimentary basin, is known as a petroleum system.

The Petroleum Geoscientist

The geoscientist should be vast in both geology and geophysics, and in the other fields of natural science related to the planet Earth. The petroleum geoscientist evaluates different kinds of source rocks, their content, kerogen types, and their maturity. The geoscientist also evaluates the presence and properties of reservoirs, such as sandstone and carbonate rocks, including their porosity and permeability. Sandstone reservoirs undergo diagenesis with burial and with aging, whereas carbonate reservoirs may undergo dissolution with burial and aging.

The geoscientist looks out for a petroleum system, with potentials for economic quantities of oil and gas in suitable traps. Traps are created by a combination of structural events and stratigraphic layering, in the presence of a seal rock. Some suitable seals for petroleum trapping are mudrocks, evaporite seals and pressure seals. The geoscientist evaluates the intrinsic nature of seals, such as the roles of stress and pressure. Efficient seals with structural integrity or lithological efficiency will retain petroleum pools in traps for millions of years in geologic timeline, such that they are available for drilling today.

The geoscience and reservoir engineering departments in oil companies carry out hydrocarbon prospect evaluation, reservoir seal assessment and formation evaluation, all of which are covered in subsequent chapters. Other important aspects of the evaluation process are checking the drillability of the well to target the resources, and the management of the petroleum field in the post-drilling phase.

History of Petroleum Geoscience

Without going into the precise chronology of petroleum exploration, it may be noted that the use of oil dates back to many years BC or BCE. Initial explorationists were field workers who searched for oil seepages, drained them, and with time dug holes to exploit the resource.

With the advent of modern geology, the use of field mapping techniques did better to highlight suitable structures for the accumulation of hydrocarbons in petroleum prone areas. Available well digging technologically at the time, and eventually drilling techniques, evolved and become the solution for extracting oil.

The development of electrical logs and drilling techniques launched petroleum exploration and production as a major operation and a scientific industry. The application of related disciplines (chemistry, biology, physics) to Earth science enlarged the scope of petroleum geoscience, with the various fields of geology such as structural, sedimentary, exploration, serving its core. In the petroleum industry, these disciplines are grouped to establish objective based departments or teams, such as exploration, prospect evaluation and seal evaluation teams.

The Occurrence of Petroleum

The very nature of petroleum makes its exploration and production a complex business. It occurs as natural gases and as crude oil, which are regarded as *conventional sources*. Other forms of petroleum are extracted from *unconventional sources* such as shale with high organic content, tar sand and, possibly, from gas hydrates. Unconventional petroleum requires advanced technology to produce properly. Care has to be taken to ensure that their production would have no adverse effects on the environment.

When occurring as gases, drilling into shallow hydrocarbon traps could let gases gush into the drilled well. This influx has to be avoided or controlled. The inflow may contain non-hydrocarbon gases, necessitating separation for pure hydrocarbon gas. If occurring as oil, it perches at the upper boundaries of the trap, underlain and mixed up with water during production. It also has to be controlled if highly pressured.

Sedimentary Basins

Basins are formed by major regional tectonics involving movement of land masses, subsidence and any process that provides lowland for sediment deposition. They occur when crustal or oceanic plates move against or away from each other, or slide past each other. The basins provide the conditions for both the generation and the accumulation of petroleum. The major types of basins are *compressive, strike-slip* and *extensional*, which are discussed in details in the appropriate chapters.

Both conventional and some unconventional petroleum (such as oil shale and tar sand) occur in sedimentary basins. This is the reason their understanding is of utmost importance in petroleum geoscience. There are many petroleum basins in the world. The type of basin, and the mechanisms that are at play in it, determine the nature of hydrocarbon reserve it will have.

Unconventional hydrocarbons are becoming a major part of oil and gas explored and produced in Canada, USA and in few other countries. This is similar to the exploration of coal in some cases, where hydrocarbons occur in solid state. Examples of these unconventional hydrocarbons are oil shale, tar sand and shale gases. Like coal beds, they have to be quarried, or cored and then heated in some ways to expel their hydrocarbons. Coal beds often have gases, which are liberated when the unit is fractured.

Surface Features and Subsurface Structures

Geology involves the study of the rocks that make up the Earth. While studies start from the surface by observing outcrops in sedimentary basins, its objective is often to understand the subsurface. In the petroleum industry, since oil and gas do not occur as surface running water, every concern eventually culminates in the subsurface. Some of the subsurface parameters of interest in petroleum geoscience include:

1. Geometries (dips and strikes of anticlines and other trap types)
2. Pore fluid – water, oil and gas (inactive or active or when flowing)
3. Subsurface temperatures and pressures (geothermal and geopressure gradients)
4. Static and dynamic fluid behaviours (involving Pressure-Volume-Temperature – PVT)
5. Basin principal stresses (acting vertically and horizontally perpendicular to each other)

The role of each of these and the factors that control them are discussed in details in the different parts of this book. How to evaluate them starts from observing surface geomorphological features, measuring orientations and dips using compass/clinometers. Seismic mapping methods are used to study both surface and subsurface features. Most petroleum basins in the world have been studied; the few remaining frontier zones are being subdivided and explored with both remote and direct mapping and exploration techniques. There are offshore zones far off the coasts and reserved land areas that are considered potential exploration frontiers, but currently controversial exploration zones.

Petroleum Geoscience Activities

Oil and gas are produced from where they occur in the subsurface, under the basic depositional and organization pattern in geology. This pattern of rock layers is known as **strata** (or **stratum** in singular form), with smaller units known as **beds** and larger units known as **formations**. With seismic images, and stratigraphy (example: seismic **sequence stratigraphy**), geoscientists are able to establish the configuration of rock units and their history, necessary for the exploration and production of hydrocarbons. They carry out basin analysis to assess the age and maturity of deposited sediments within the basin.

At the basinal exploration stage, the geoscientists want to find out factors such as:

- Suitable petroleum systems within a given basin
- Pressure and temperature evolution within the basin
- Fluid types and flow directions
- Rock properties and average petrophysical parameters
- Geomechanics interplay of stress and strain

Further studies of a petroleum field would include the assessment of:

- Compaction of sand, shale and sandstone/carbonates
- Folded structures that form traps such as anticlines and faults
- Migration of hydrocarbons and where they are eventually accumulated
- Amplitude extraction to reveal the bright and flat spots (hydrocarbon indicators)

One of the tools, seismic mapping, provide the major part of the solutions in petroleum exploration. Seismic exploration methods involve an initial **acquisition stage** where workers cut their way through lands and bushes, and use heavy vibrating trucks. These seismic vibrators, also known as **vibroseis,** are used to explore the subsurface by emitting and receiving seismic signals. Special ships are used offshore to achieve the same purpose of mapping the sea floor. They use **emitters (seismic sources)** to send sound waves below the seabed and record the returning signal on **geophones** (**hydrophones** are used for seismic acquisition at sea).

Petroleum geoscience activities are carried out using several tools, at local and basinal scales; they require complex logistics, especially once drilling starts. The ultimate purpose is to find, drill and produce hydrocarbons. The produced resources are piped and channelled to sales, refining or utility centres.

Upstream Sector of the Petroleum Industry

The upstream sector does the petroleum mining operations by drilling wells after investigations to confirm the presence of hydrocarbons, to produce confirmed reserves. Activities in a given petroleum field last for many years, from the time of initiation of the project to the *first oil*. The full value chain of the upstream sector in oil and gas business involves several stages. Below are the different methods and phases involved in the upstream sector.

Exploration and Appraisal

The aim of exploration and appraisal is to establish subsurface models, starting from geological cross-sections and horizon maps, to amplitude extraction from layers, and to 3D blocks with well paths represented on them. Eventually, a reservoir model is built to estimate potential reserves and define extraction or recovery processes.

Remote Sensing, Aerial Photo, Radar and Scanner

A first necessary stage of exploration, especially in frontier zones, is to capture the overview of the landscapes and geometries. Remote sensing refers to graphics and signals obtained from a distance. Aerial photos are taken from aircraft and images captured from satellites which, when interpreted, provide a qualitative and quantitative geomorphology of the landscape or basin of interest. Modern radar and spectral scan provide useful graphics as well for understanding the subsurface.

Geophysical Methods

The vital tool today in the exploration and appraisal phases of petroleum is geophysical. This ranges from basic methods such as magnetic and gravimetric surveying to advanced geophysical methods such as seismic imaging (2D, 3D and 4D), including reflection and refraction technology and Vertical Seismic Profiling (VSP). An advanced method, known as seismic while drilling or real-time seismic monitoring, is used to predict pore pressures and provide time-depth matching solutions while drilling.

Well Data, Field Sampling, and Modelling

The appraisal phase of petroleum geoscience uses well data from offset wells, and from initial exploration wells, to make prognoses of the reservoir properties and quantify the fluid contents. Field sampling for different parameters and their use for static modelling and dynamic simulation contribute to the appraisal of prospective oil fields. The idea is that if wells close together have one of them producing in a given type of reservoir, then the other wells would likely be positive under the same conditions. They may share similar pressures and other reservoir properties, when a given interval is correlated across the few existing wells.

Development and Drilling

Drilling and Completion

After the initial stages of exploration, the field is developed by putting in place the drilling equipment and installations. During the drilling phase, a geoscientist works with drillers to optimize formation evaluation till successful completion of the well. There are different types of drilling equipment, ranging from old *cable-tool* through *percussion drilling* to present day *rotary drilling bit* attached in long pipes whose ensemble is known as **Bottom Hole Assembly** or BHA. **Well completion** involves lower and upper installations of production equipment. The well is said to be completed when holes have been pierced across the protective **casing** or **liner** downhole or **screens** securely installed, so that hydrocarbon fluid can flow from formation into the well; then the fluid flows from the well, up through the tubings and treelike production structure with valves (known as Christmas tree), to storage tanks at the surface or export pipes.

The Formation Evaluation Process

Among the first useful parameters for formation evaluation are the drilling parameters, such as Rate of Penetration (ROP) and Torque (TRQ). They give an indication of the rock type and how they react to intrusion, essentially for drilling safety. After this, the main formation evaluation involves the use of electrical, radioactive and other types of logs to measure rock and fluid properties either during (logging while drilling) or after drilling the hole (wireline). Some of the logs used for formation evaluation are:

- Gamma ray
- Resistivity
- Sonic or acoustic
- Neutron-Porosity
- Bulk density

They help the geoscientist to determine the rock type (or lithology), and the fluid types, and how they behave downhole. They are also important for estimating the hydrocarbon reserve and choice of completion procedures.

Production and Management

Flowing the Well

This starts when the well has been completed – flow is either natural or induced. Pressure measuring equipment is installed for flow monitoring and provides information for understanding the formation and for managing production throughout the life of the well. There are different production methods, and well architectures are designed to meet their demands. A major process in the initial flowing of a well is *well test*, which provides the production and injectivity index. It also helps in confirming the pre-drill models, and in refining the other parameters whose understanding is necessary for the production and management of the field.

Enhanced Recovery

Right from the onset of production, but more frequently after an initial natural upsurge of hydrocarbon fluids, there is need to maintain subsurface pressure and drive the production by an artificial means in order to enhance recovery. Some of the Enhanced Oil Recovery (EOR) or drive processes are:

- Water Drive (water is injected for pressure support and better hydrocarbon sweep)
- Gas Drive (known as *Cap drive* when injected from the cap or anticlinal crest and as *Dissolution drive* where it is the modification of fluid properties due to gas injection that enhances flow)
- Acid cracking (dissolves acid soluble bonds, thereby reducing the viscosity of fluid and enhancing flow)
- Borehole pressure reduction (a reduction in pressure on the upper part of the well will lead to vertical flow due to differential pressure)
- Fracking (inducing micro-fractures in the formation to increase fluid flow to the well)

Field, Well and Reservoir Management

Petroleum well is managed right from the onset, and improved along the line to continue tapping un-swept hydrocarbons in the reservoir, based on flow monitoring results. The *ramp up*, *plateau* and *decline* are monitored on a curve and the objective is to maintain the production at plateau as long as possible. Then afterward manage decline to prolong the life of the field. A field may undergo processes known as *reserve revision* and *addition*. By properly managing the well and the reservoir, a field is produced optimally before abandonment.

All through the various stages of exploration and production, geologists, geophysicists, engineers, and other professionals such as accountants and medical doctors, work together to achieve success in the petroleum industry. This synergy and the various technical aspects, highlighted in this introduction, are what this book covers.

Implications of Petroleum Geoscience Operations

There is a discussion as to whether seismic acquisition has a major impact on marine life, in addition to physical displacement and impact on seabed, where seafloor profiling is involved. After the exploration, the actual drilling and production could sometimes be hazardous to the environment. This is why the oil and gas business is riddled with major environmental concerns. There is a Health, Safety and Environment (HSE) department that is charged with ensuring that exploration and production are carried out in the safest conditions possible.

Onshore, it is important to have a good relationship with the community and take absolute care not to cause oil pollution. Offshore, the geoscientist extracts amplitude shallow signals on seismic blocks, which are indicative of seafloor and shallow formation anomalies, to ensure that wells are not located on hazards such as gas pockets and gas hydrate mounds, whose dissolution could destabilize oil field installations. In all cases, during petroleum exploration and production, efforts are put in place to avoid leakages, water pollution, explosions and other risks to personnel (which could range from minor to major injuries or even fatalities).

Geoscience personnel working on rigs, in flow stations and at refineries, undergo necessary risk assessment and prevention training. Those working offshore undergo the *Helicopter Underwater Escape Technique (HUET)*, which is also a major part of other certifications needed to work in the industry. First aid and survival training is provided for those working in remote fields onshore, where health personnel form part of the response team in case of accidents.

The geoscientist therefore has a major responsibility in protecting self and in respecting the environment. While the energy provided is invaluable, Health Safety / Security and Environment (HSE/HSSE) measures remain a priority for everyone involved with the oil and gas industry in today's society.

Role of Geoscientists in Machine Learning and Automation

A new trend in the world, in the geoscience community and in the petroleum industry, is in *Big Data (BD)*, *Machine Learning (ML)* and *Artificial Intelligence (AI)*. Major corporate organisations and international oil companies (IOCs) are collaborating with Google, Microsoft,

IBM, HP, Oracle and other big data companies in order to harness the benefits of training computer systems to learn from the vast amount of data in the industry. The idea is to use parallel computing and large data 'warehouses' for the processing, storage and decentralization of oil and gas industry data. The interpretation of data would be based on intelligent designs and automation, following machine learning, sometimes with intermittent human intervention. This has been applied in well log interpretation, fault seal analysis, modelling of rock elastic properties, digital core property simulation, geostreering, intelligent well completion and enhanced oil recovery, among others.

Geoscience modelling precepts for subsurface information remain within the purview of geoscientists. They would have to provide the structure and principles for automating geoscience interpretation and modelling using AI. This is possible with the establishment of architectural frameworks, with the segmentation and parameterisation of structural and sedimentary profiles. It requires the integration of qualitative and quantitative approaches in the assessment and representation of geologic information.

Three major types of modelling provided by petroleum geoscientists are:

- Conceptual model (based on observations and experience)
- Basin modelling (aimed at providing the temperature and pressure evolutions of a sedimentary basin, their variations and consequences on the petroleum system)
- Structural modelling (for the understanding, schematic reproduction and representation of tectonic events)

The application of intelligent solutions will be useful for the petroleum industry's main interests in basin modelling, static modelling and dynamic simulations, and in the estimation of hydrocarbon in place. Examples given in this work on structures for modelling include fault-bend fold and fault-propagation fold. *Decompaction* and *backstripping* have been given as examples in structural modelling. A way of defining geologic bodies for machine learning has been exemplified with qualitative and quantitative assessments of fault zones. Methods for reservoir flow simulation and anisotropic flow models in fault seal evaluation were used to show how the computation process might be employed recursively by AI.

An Evolving Industry

The petroleum industry is evolving, using new languages, adopting new technologies, but still producing oil and gas. The argument is that it will take several decades for large manufacturing companies, big ships, the aviation industry and developing countries to switch to sufficient renewable energy solutions. Renewable energy is costly and sometimes requires non-renewable energy or fossil fuel to produce. Some concerns of pollution from fossil fuel may as well exist for renewable energy solutions. In the future, renewable energy wastes will be a concern.

With increasing world energy demand, the geoscientist's role is in energy mix, in decarbonisation, in subsurface storage, in the exploration and production of not only hydrocarbons, but also cleaner resources (such as natural hydrogen gas) and minerals (some of which are used in battery cells and solar plates or photovoltaic cell technology). Geoscientists would play a role in large installation sites study and the geoscientist's modelling capacity and understanding of geobodies will always remain an advantage. The petroleum industry understands the challenges with HSE and big projects, and has the capacity to use their heritage from fossil fuel production to invest in renewable energy solutions. This transition period should then be targeted towards optimising the full energy value chain, reducing the carbon footprint (greenhouse gases released by human activities) while harnessing new technologies in an ever-expanding world in need of energy.

Few examples of the geoscientist's role in discretisation, in the application of the principles of modelling and the representation of geobodies that can be input into intelligent designs, are mentioned in different places in this book. To participate actively in the current innovations and remain relevant in the near future, the geoscientist would have to internalize a wide range of topics and cover the scope of the core geoscience disciplines. The mastery of these basics is the focus in the next chapters.

PART ONE

BASICS OF PETROLEUM GEOSCIENCE

This part starts from the question of what is petroleum, to the origins of hydrocarbons and how source rocks mature, expulse oil and gas which migrate to be trapped in reservoirs. Migration of hydrocarbons could be primary, secondary or tertiary. Reservoirs often occur with sealing lithologies above them, in reservoir-seal pairs or sequences. Seal and trapping systems are discussed, alongside how they occur in sedimentary basins. Sedimentary processes lead to different depositional environments. Basin history, temperature and pressure evolutions play a role in ascertaining the viability of petroleum fields. In this part, how basins are formed and how to analyse them for presence of petroleum are discussed.

The concepts of stratigraphy and how it relates to other geoscience disciplines are discussed. The different types of stratigraphy: litho- chrono-bio-seismic- and sequence-stratigraphy are explained with their applications, before integrating the role of structural geology in deformation, plate movement and formation of basinal structures. Basin filling, structuration and faulting result in reservoir compartmentalisation. A vital factor in reservoir evaluation, apart from the *intrinsic* properties mentioned, is homogeneity and continuity. Overly compartmentalized reservoirs are hard to produce, while a laterally continuous and vertically thick reservoirs form very extensive reserves.

The methods used for understanding structural history, including restoration and decompaction are also presented. This gives more insight on structural traps, how they are different from stratigraphic types and how combined traps lead to complex petroleum prospects. It is important to relate the timing of trap formation to the charging history of hydrocarbons, as well as consider the impact of burial, which is integrated in basin modelling to understand the role of temperature and pressure in petroleum systems.

CHAPTER ONE

PETROLEUM GEOLOGY

1.0 Petroleum Resource

Petroleum is one of the resources that have benefited mankind and provided energy for centuries. In recent times, it tops the list of energy solutions. Despite this advantage, there has also been resentment to petroleum due to the impact of exploration and production on the environment and ecosystems. Its by-products, however, continue to be very useful and the exploration and production technology is applicable to other fields of human endeavour.

This chapter looks at what actually is petroleum, also known as hydrocarbons. The chapter will cover:

- Types of hydrocarbons and their origins
- The make-up of petroleum source rock
- Generation and migration of petroleum
- Traps: reservoir - seal rocks and accumulation
- Petroleum system and the role of burial

There are discussions as to the processes involved in petroleum generation, preservation, and transport in the subsurface, and on the exact temperatures of maturation and migration of hydrocarbons. This will be examined in this chapter. It will also discuss the difference between conventional and unconventional resources.

1.1 What is Petroleum?

Petroleum, or hydrocarbon, is a mixture of diverse hydrogen and carbon molecules, with possible traces of other organic molecules such as oxygen, nitrogen or sulphur. The word **'petroleum'** comes from the Greek word Petra (rock) and Oleum (oil), and is used to refer to natural materials mainly made up of compounds of hydrogen and carbon (hence the term **hydrocarbons**). They may however constitute traces of other organic molecules such as oxygen, nitrogen and sulphur, which can be referred to as impurities.

Petroleum originates in nature mainly from organic matter, in closed reducing biogenic environment. It can also be generated in the laboratory. Some of the generated petroleum is accumulated in *reservoir rocks* (and in hard rock fractures) with suitable *seal*. Petroleum can occur in **gaseous**, **liquid** or **solid** states under surface temperature and pressure conditions, depending on the number of its carbon molecules.

Gaseous forms of hydrocarbons include the lighter methane, ethane, propane and butane, having carbon atoms of one, two, three and four respectively; pentane and other heavier hydrocarbons exist in liquid and solid forms. The solid heavier crude will need to be broken down by a process known as **cracking** (breaking down complex heavier organic molecules into light hydrocarbons) before they can be used as a fuel product.

There are essentially four groups of hydrocarbons as described below.

Alkanes (Paraffins)

The *alkanes*, also known as **paraffins**, have a chemical formula, C_nH_{2n+2}, where n is the number of carbon atoms in their make up. This means that for example, for the first in the series, you will have $C_1H_{(2*1)+1}$ which will give CH_4. This is methane. Ethane, propane and butane are the second, third and fourth in the series respectively. Pentane (Figure 1.1) is a *straight chain* **saturated** hydrocarbon, meaning it has only *single bonds* (an attraction between the atoms that enables them form chemical compounds) between its hydrogen and carbon atoms existing in *linear* form as seen in its structure. It is liquid at room temperature and pressure.

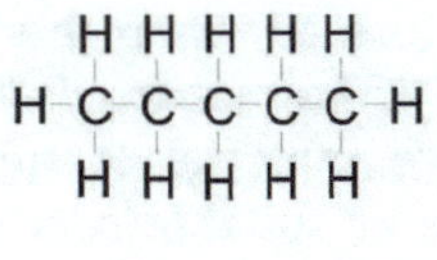

Figure 1.1: Structure of Pentane, fifth in the Alkane series

Naphthenes

Naphthenes or *Cycloalkanes* are hydrocarbons in which the carbon atoms are linked in a *ring*, and having the chemical formula is C_nH_{2n}. A simple example is cyclopropane (C_3H_6) with the structure shown below. Cyclopropane (Figure 1.2) forms a *cyclic chain* **saturated** hydrocarbon, meaning it has only single bonds and one or more carbon rings to which hydrogen atoms are attached as seen in its structure.

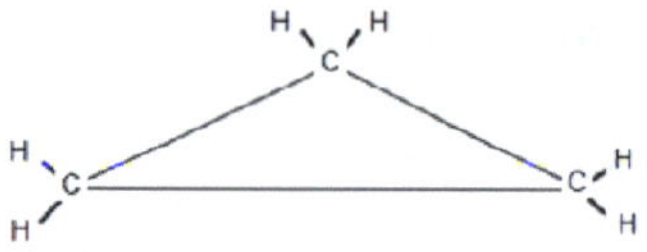

Figure 1.2: Cyclopropane (C_3H_6)

Aromatics

Aromatics are **unsaturated** (more than one single bond) hydrocarbon rings, of which the simplest example is benzene, with the formula C_nH_n. Benzene has six carbon and six hydrogen atoms in a ring, C_6H_6, represented as seen in Figure 1.3. The basic aromatics have pleasant aroma, or odours, from where they obtained their name. The structure of benzene (Figure 1.3) is a *cyclic chain* **unsaturated** hydrocarbon, meaning it has more than one single bond of carbon rings to which hydrogen atoms are attached as seen in its structure.

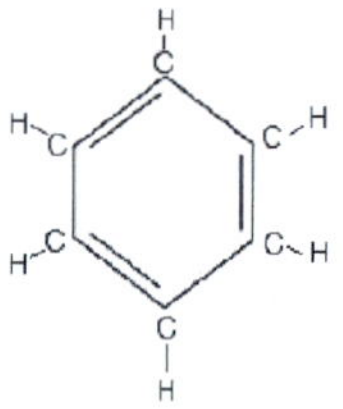

Figure 1.3: Benzene ring (C_6H_6)

Asphaltenes

Asphaltenes are the much heavier forms of hydrocarbon, which exist in solid forms, especially at surface temperature and pressure. They are formed when the hydrocarbon to be generated is **immature** (where the hydrocarbon source material is not under the most favourable or suitable conditions) to form and accumulate naturally; biologic activities can also alter the processes that may lead to the formation of petroleum. This makes them difficult to produce, hence they require cracking, or a reduction in the pressure of the reservoir to lighten them before they can be produced and used as petroleum products. They contain more traces of impurities of nitrogen, sulphur, oxygen and other molecules compared to the lighter hydrocarbons.

1.2 Origins of Hydrocarbons

Living organisms reproducing very actively in stagnant water with low-porosity sediments form hydrocarbons. Hydrocarbons are produced in environments such as:

- Delta swamps
- Coastal interlocks
- Shallow platforms
- Lakes and lagoons

For hydrocarbons to form, the living organisms, such as **algae** and **planktons** reproducing in a pool of stagnant water, need to be rapidly buried for millions of years by **sediments** (fragments or very tiny particles of rock). These sediments are essentially **shale** (fine-grained sedimentary rock composed of mud/clay and other minerals) or **marl** (lime-rich mud with variable amounts of clay and silt). Figure 1.4 shows different environments of oil and gas generation, from land through distributary channels to the ocean.

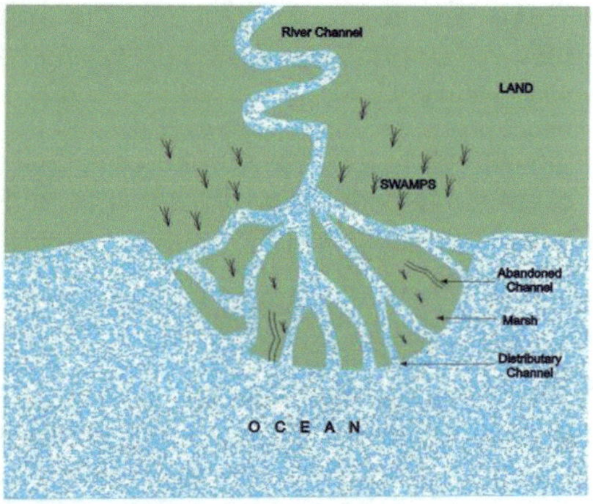

Figure 1.4: River channelling through deltaic swamps and marsh to the ocean

A river channel creates pathways where sediments are transported from the land to the sea, and allows for the burial of organic remains of living things, in different environments on land and in the ocean. Over time, millions of geologic years, oil and gas will be generated from such rocks with high organic matter, *under suitable conditions.* Without quick burial and putrefaction by reducing agents, organic matter will be oxidized and destroyed before they form petroleum products. When a sedimentary deposit has high organic content capable of yielding hydrocarbons, it is known as a **source rock**.

1.3 Petroleum Source Rocks

Petroleum source rocks are usually *shales* or *marls* (with sufficient organic content to generate hydrocarbons). The organic content of source rocks, known as **Kerogen**, is evaluated by heating it until it emits the hydrocarbons in form of gas or liquid, a process known as **pyrolysis.** *Pyrolysis* is the action of heat on the source material until it begins to emit hydrocarbons in gas or liquid form.

Kerogen and Bitumen

Before a rock rich in organic matter forms petroleum, they pass into a stage where they are soluble in organic solvent (known as kerogen) or insoluble in organic solvent (bitumen). The aromatic and asphaltic hydrocarbon extracts are of bitumen origin. On the other hand, kerogens produce the most common hydrocarbons with less molecular content.

There are different types of kerogen, depending on the *origin of the organic matter* they contain, and these include:

- *Type I – Liptinite:* Oil prone with high hydrogen per carbon but low oxygen per carbon contents. It can yield up to 80% oil.
- *Type II – Exinite:* Oil and gas prone and moderate hydrogen per carbon and oxygen per carbon contents. It can yield more than 40% oil and gas.
- *Type III – Vitrinite:* It is gas prone and has low hydrogen per carbon and high oxygen per carbon content. Its yield is often less than 30% gas.
- *Type IV – Inertinite.* It is rarely oil or gas prone, with a high carbon per hydrogen content.

In essence, each of the above kerogen can generate some form of hydrocarbon, excluding the Type IV (Inertinite) kerogen, which rarely generates oil or gas. Type I (Liptinite) kerogen generates more of oil, Type II (Exinite) kerogen generates more of oil and gas, while Type III (Vitrinite) kerogen generates more of gas. Each of the above kerogen types could also be classed according to their origin. They could be either **continental** or **marine**, with those of marine origin believed to form better hydrocarbons.

The two main controlling factors for hydrocarbon generation from source materials, whether in nature or in the laboratory, are **temperature** and **pressure** (Evamy et al., 1978). Oil generation is only possible between average ranges of *65 to 150°C.* Below this temperature, a type of gas known as **biogenic gas** will be formed due to the action of living organisms or biological processes acting on the organic matter. Above this temperature, a type of gas known as **thermal gas** will be formed, as temperatures are so high that they "bake" the organic matter responsible for generating oil.

At given temperature and pressure in the above range, hydrocarbons in source rocks will be released after their formation. The stage where hydrocarbon is released is termed **maturation**. Hydrocarbon may be expelled with non-hydrocarbon gases such as nitrogen, carbon dioxide and water, leaving residues. Such residues and gases are associated with hydrocarbon in petroleum fields.

For source rocks to generate hydrocarbon, the original organic matter (Total Organic Carbon – TOC) must be:

- rich enough (% TOC)
- isolated in an oxygen deficient environment (reducing)
- be subjected to the right temperature (for maturation)

The conducive geological environments where temperature plays its role to generate hydrocarbons is known as the "**oil kitchen**", while the depth range at which oil generation is possible is known as the "**oil window**", which varies among basins depending on *temperature gradient*, but usually at a minimum depth of about two kilometre. Source materials at such depths are subjected to high pressures due to burial load or the weight of the sediments on top of it. Time is as well important, as it determines the effectiveness with which the source rock generates hydrocarbons.

A *continental margin* (shallow water area found close to the continent that marks the transition to the ocean) is shown in Figure 1.5, where sediments are transported from swamps through beaches and finally offshore because of wave action of moving water in these environments. Water movement allows sediments to be transported and deposited at various times, depending on the energy of the waves. The deposition of sediments, under suitable conditions, leads to the formation of source rocks that generate and expel hydrocarbons when mature.

The algae and plankton prone organic shale is buried and, with compaction and increasing temperature, they mature as source rocks. The oil migrate both vertically and upward the strata (horizontally or diagonally in tilted beds, as shown in Figure 1.5, where deposits run from onshore to offshore) to become trapped as present day hydrocarbons.

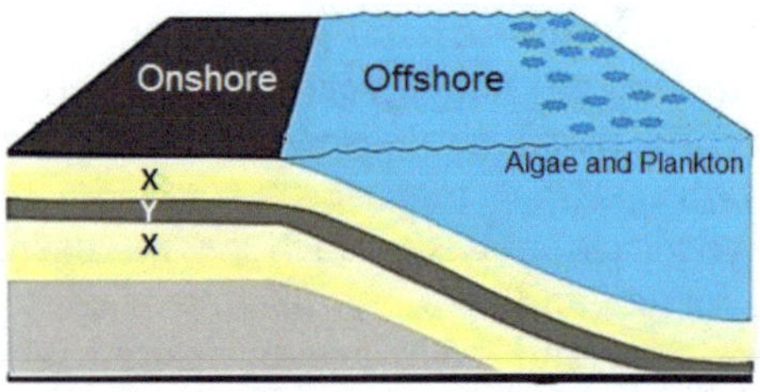

Figures 1.5: Closed petroleum generation deltaic systems (X and Y are reservoir and seal rock, respectively)

Another important condition associated to source rock formation and maturity is an enclosed oxygen deficient environment, which is common in tight shaly lithology or fined grained sediments and carbonate. Closed basins are good places for the deposition and generation of source rocks. Oil and gas continue to evolve even after accumulation in a reservoir, and eventually altered in geologic time except if produced or leaked away.

1.4 Maturation and Migration

Migration is the process that begins when generated hydrocarbons come out from the source rock, then move through carrier strata (porous media) to a place where they are accumulated (in reservoirs), and may eventually leak away from that accumulation. When a source rock becomes mature and expels hydrocarbons, the hydrocarbons may be transported away from the source rock into adjacent beds where they get to structures called traps that can allow hydrocarbons to accumulate (Hunt, 1996, 1990; Moretti, 1998). Sometimes they move away from such structures to another area where they either begin to accumulate again, or get degraded, or leak such that they are seen on the earth surface.

There are different types of migration, discussed below.

1. Primary Migration (out of the source rock)
2. Secondary Migration (through a porous carrier strata to a reservoir)
3. Tertiary Migration (out of the reservoir to leak away or to another accumulation)

Primary migration starts as soon as the source rock begins to expel the hydrocarbons. The process of **secondary migration** takes place within rocks that allow the hydrocarbons to travel until they get to an area where they can begin to accumulate. If the hydrocarbons migrate away from the accumulation then **tertiary migration** is said to take place. This can happen when the trap is full or the rock responsible for the accumulation is compromised.

The major controlling factor that determines whether migration will take place is **pressure**, or rather the difference in pressure. Usually, hydrocarbons will migrate from an area of higher pressure to that of lower pressure in order to reach equilibrium. Other important factors that facilitate the process of migration are increase in **volume**, **capillarity** and **buoyancy**. Increase in the volume of hydrocarbon fluids within the source rock will cause pressure build-up, which will force the fluid out of the rock. Where there is low capillary entry pressure, that is reduced surface tension, which allows fluids to flow through the rock grains, then migration will take place. Also, the higher the buoyancy (tendency to float) of the fluid, the easier the fluid will flow, hence migration.

Figure 1.6 depicts *primary*, *secondary* and *tertiary* migration pathways for hydrocarbons. In the presence of traps, hydrocarbons will accumulate, and in its absence they migrate and appear on the surface as oil seep. Hydrocarbons will leak away if not trapped.

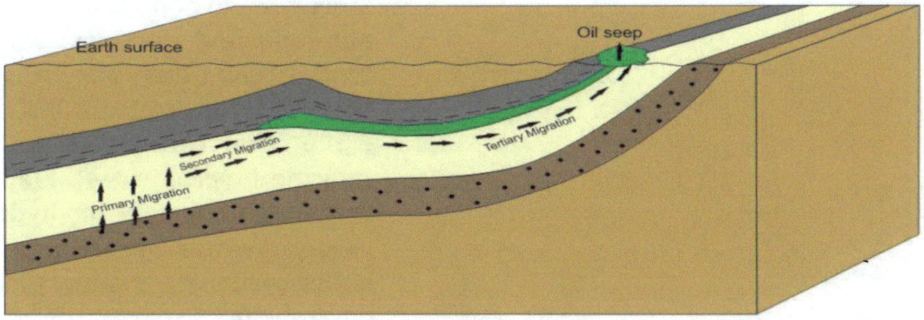

Figure 1.6: Migration of hydrocarbons with water until accumulation below an anticlinal trap and eventual leakage

1.5 Reservoir and Seal Rocks

Reservoir

The rocks where hydrocarbons expelled from the source rock flows (sometimes through an *intermediate porous rock* or *carrier strata*) to a **reservoir** rock. The major property that makes a reservoir is pore spaces and fluid flow possibilities in the form of fractures. Two properties make a rock a good reservoir rock – **porosity** and **permeability**. Porosity refers to the holes or pore spaces in rocks available for hydrocarbons to be stored, effective porosity refers to how connected these pore spaces are, which will determine permeability (i.e. hydrocarbon flow through such rocks). Porosity and permeability are studied using petrophysical methods.

Porosity is determined by *porosity logs* (record of measurements at depths) or measured directly from *cores* (cylindrical rock samples). Permeability is measured by carrying out flow test in the subsurface or in the laboratory. Reservoir lithologies include sandstones and carbonates (limestone or dolomite) and then other fractured rocks. Carbonates tend to be better reservoirs with age than sandstones because as they are buried, with time their porosities increase due to extensive crumbling of the rock material, a process known as dissolution. However, if any of the two has undergone fracturation or *secondary tectonics* (processes that change the structure of the rock after its formation), it could increase their capacity for fluid flow.

Figure 1.7 shows well sorted and poorly sorted samples from a reservoir unit. A rock unit is well sorted if it has uniformity in grain size, while a poorly sorted rock unit has varying grain sizes making up the rock matrix. Well-sorted rocks usually have higher porosity and permeability, because available spaces in poorly sorted rocks become clogged by smaller grains within the larger ones.

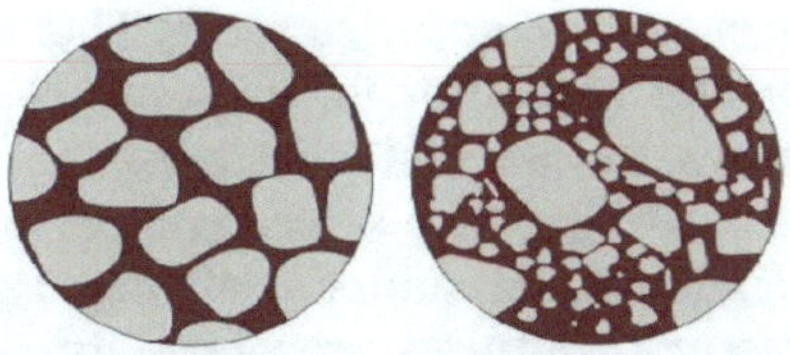

a. Well-sorted b. Poorly sorted

Figure 1.7: Well sorted and poorly sorted samples from a reservoir unit

During reservoir evaluation, when studies are carried out to check available *intrinsic properties* (character of the rock present during its formation, not *induced* after it was formed) of the reservoir such as porosity and permeability, among factors to be determined include **homogeneity** and **continuity**. A reservoir will be good enough if it is homogeneous, meaning that it has uniform property or characteristic across its entirety. Such reservoirs will be even better if it is continuous without breakage – if it is not overly compartmentalized vertically by **lithological seals (stratigraphic barriers)** and laterally by **fault (structural barriers)**.

Compartmentalized reservoirs are harder to produce, but the continuous and vertically thicker reservoirs will have more hydrocarbons stored in them (Dandekar, 2006). Following a detailed reservoir characterization, the amount of oil that can be stored in the reservoir, the *oil in place*, is calculated depending on what part of the reservoir has hydrocarbons, as the reservoir may not be *filled to spill*. Results can be enhanced when more data from studies in the area are obtained.

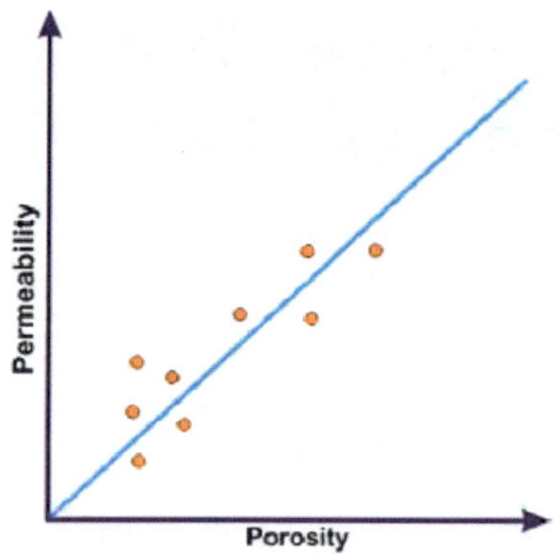

Figure 1.8: Plot of porosity versus permeability showing their relationship

Figure 1.8 is a plot showing the relationship between porosity and permeability. Generally, the higher the pore spaces, the higher the permeability, and vice versa. This is sometimes not the case though, as some relatively high porous mud rocks (like siltstone) have lower permeability; they are capable of storing fluids, but such fluids cannot easily flow through them. Fracture induced permeability is also not related to porosity.

Seal

Any rock that constitutes a blockade to flow is a **seal**. Seals are either very fined-grained or mineralized in such a way that they act as barrier to fluid flow. Such a barrier is achieved by either very high capillary entry pressure (explained in subsequent chapters) or high physical strength of the rock, that is, resistance to failure due to stress. For a rock to be a good seal, it must be able to resist any flow through it. Seals should therefore be highly *imporous* and *impermeable* (or have mechanical seal capacity).

For sandstone reservoirs, they are better when they are younger before **cementation** (hardening of sediments) and other forms of alteration occur with aging of the sediment. On the other hand, limestone and other carbonates undergo dissolution and could fracture with time, making them better reservoirs with age. Sandy reservoirs are deposited in alluvial fans, fluvial system, turbidites and more. In the marine environment for example, after being deposited by *turbiditic* current during low sea level, channel sands are covered by *hemipelagic* shales during high sea level rise.

Limestone is non-clastic, meaning that they are not deposited by movement of grains from one place to the other, but precipitated or built in-situ. Microorganisms build carbonate **reefs**, while other carbonates are formed by **corals** (marine invertebrates). Marine carbonates are the most important for hydrocarbons accumulation. Marine carbonates are produced in either carbonate shallow platforms or ramps in open shelfs.

Seals are deposited wherever very fine grain sediments occur in large quantity. Non-fractured carbonate could serve as seal, likewise salt deposits such as *halite*. However, the most common seal in most petroleum basin is **shale** or **claystone**. Figure 1.9 shows an oil block with seals (x and y) and an oil-bearing reservoir (z). The figure also illustrates a case of multiple reservoirs-seal pairs, where multiple reservoir units are overlain by multiple seal units (and underlain by seals, which could serve as source material if they have organic content buried under suitable conditions).

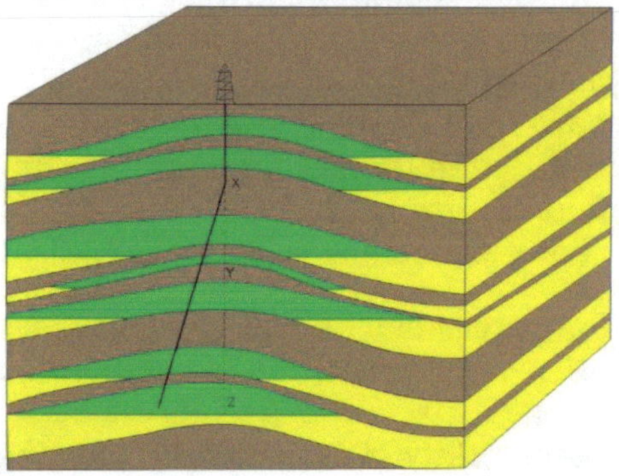

Figure 1.9: Oil block showing seals (x and y) and oil bearing reservoir (z)

Reservoir seal-pairs are commonly formed at various times of sea level rise and fall, leading to the deposition of seal lithology during high sea level and reservoir lithology during low sea level respectively. Multiple oil pay zones are therefore formed. Multiple reservoir-seal pairs often lead to multiple oil pay zones.

1.6 Traps and Accumulation

Traps

Traps are features that hold petroleum in place so that it is accumulated over a geological timescale; the process of holding the petroleum is known as *trapping*. A trap usually has the oil container known as reservoir and the seal. The reservoir is porous (like sand) while the seal is tight (like mud). Other seals are low permeability cap rock, lignite and bituminous caps and *tar* mats. Pressure seals and hydrodynamic conditions have also created an oil pool in unexpected traps (Hubbert, 1953). Hydrocarbons migrate within porous (reservoir rock) which has to be bounded by some form of seal to move within a concentrated path to its destination. It accumulates when a trap is found (Schowalter, 1979; Weber, 1978).

In Figure 1.10, the passing ground water serves as a control for hydrocarbon flow, allowing it to float and accumulate within a structure known as an anticline. An anticline is a type of structural trap, however when water pressures aid accumulation across the flanks it is said to be a *hydrodynamic trap*.

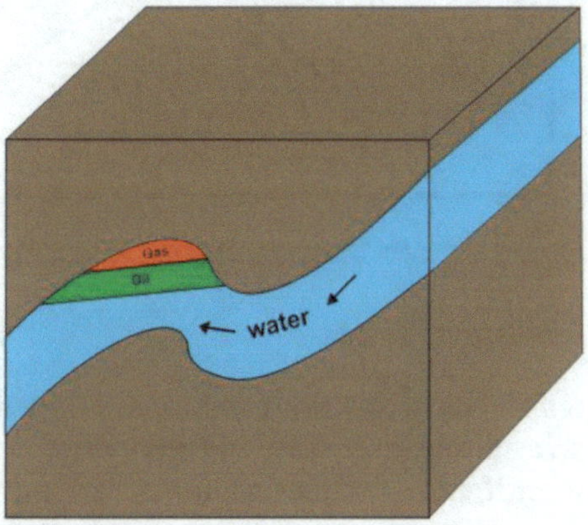

Figure 1.10: Trap showing oil migrating in water and trapped under an anticline

Accumulation

Migrating oil will need to find a place to settle, which is the process of accumulation to a place where it (is trapped) and cannot migrate further except some conditions change. In petroleum evaluation, it is important to ensure that the trap has been formed before the arrival of the hydrocarbons, or else the trap will be empty. In addition, any reactivation, redeformation of a trap, after accumulation could lead to the trapped petroleum being lost. Timing of migration versus trap formation is a standard procedure in oil evaluation of a field or a basin (England et al., 1987).

There are different kinds of traps, which are described below.

Structural Traps

These kinds of trap are emplaced by structural mechanism, created by *tectonic processes.* They could include anticlines, domes, fault assisted closures and other structures harbouring petroleum (Aydin, 2000). Figure 1.11 shows typical structural (anticlinal and fault) traps. Anticlinal are formed by tectonic processes, usually compressional forces that deform the rock, allowing subsurface fluids to be trapped below it afterwards. Where all subsurface fluids are present, they separate based on their different densities with gas (red colour) at the crest of the anticline, followed by oil (green colour), and then water (blue colour) at the base. The boundary between each fluid is termed a **contact**.

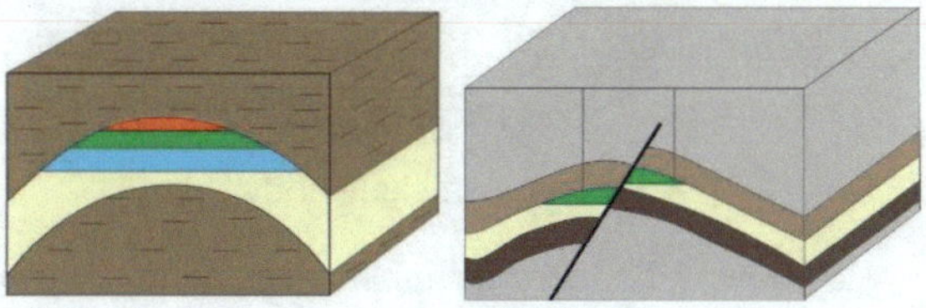

a. Anticlinal trap b. Fault trap

Figure 1.11: Typical structural trap, with a conventional anticlinal closure, showing (a.) gas-oil-water layers and (b.) a fault trap

Gas-oil contact and water-oil contact (sometimes gas-water contact in the absence of oil) are shown in the Figure 1.11 a. *(Gas Oil Contact is abbreviated as GOC and Water-Oil Contact may be abbreviated as WOC or OWC to mean Oil-Water Contact).* Fault traps (as shown in Figure 1.11 b.) are also formed by deformational processes, when there is a fracture in the rock and displacement of one of the adjacent blocks. The fault trap so formed allows oil to accumulate by acting as a barrier to flow.

Stratigraphic Traps

These kinds of trap are created by the *depositional process* itself; it is the resulting architecture after deposition of different layers of sediment that create the suitable feature for hydrocarbons to be accumulated. Examples include stratigraphic unconformities, shalying out of sand body, isolated sand lenses, pinch out and diagenetic barriers. Oil is trapped at the tipping end of the sandstone englobed by shale seal.

Figure 1.12 illustrates a stratigraphic pinchout, which describes the termination by thin out of a sandstone reservoir rock against a non-porous shale seal rock.

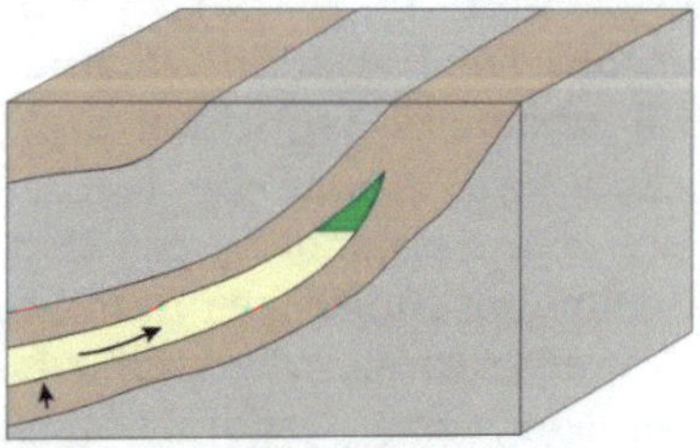

a. Pinch within a bedding unit

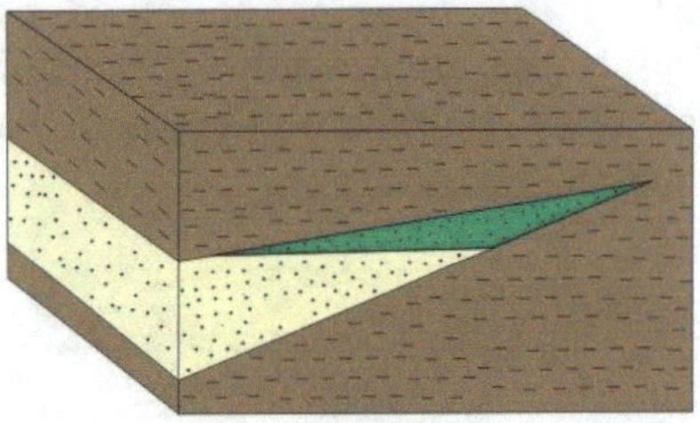

b. Pinch into a shale body

Figure 1.12: Stratigraphic trap, with a pinchout, where an intercalated sandstone stratum thins out into a shale unit

The seal rock creates a favourable geometry that serves to trap hydrocarbons after their generation and migration. Multiple pinchouts within the same stratigraphic unit is referred to as *interfingering* (Figure 1.13). This leads to the formation of multiple hydrocarbon traps.

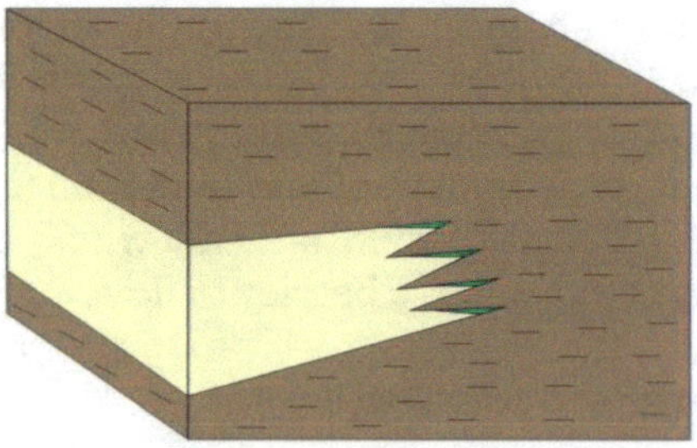

Figure 1.13: Interfingering: multiple pinchout leading to several oil traps within the same stratigraphic unit

Hydrodynamic Traps

These are traps achieved by water movement in a given direction. Usually the water movement carries the hydrocarbons to a structural or stratigraphic barrier and keeps them in place. Tilted water contact is one of the ways to identify hydrodynamic traps. They are a result of strong pressure/water head differential between two points.

A hydrodynamic trap (Figure 1.14) is made possible by differential pressure, where pressures on both ends of the trap are different, resulting in flow inhibition of the hydrocarbons. The water pressure Pw1, being greater than the hydrocarbon pressure Pw2, pushes the migrating hydrocarbons and traps them in the anticlinal structure. The tilted oil-water contact demonstrates the pressure difference and direction of flow; the tilt is lower in the flank with the lower water pressure.

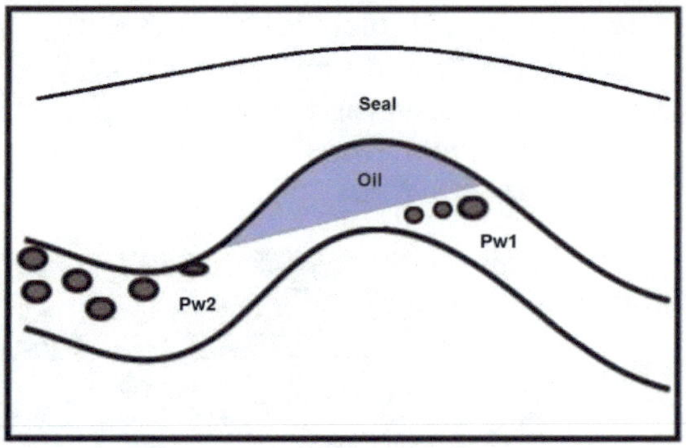

Figure 1.14: Hydrodynamic trap, where Pw1 > Pw2

Combined Traps

Combined traps are a mixture of structural and stratigraphic traps. Traps may simply be identified as structural and stratigraphic, giving examples as anticlines and pinchouts, but in reality most traps occur as combined traps. This is because it is impossible to have a structural trap without associated stratigraphic processes, for example, an anticline without associated stratigraphic layering. The factor mostly at play is used to define a trap as structural, stratigraphic, hydrodynamic or combined.

An example of combined trap is *diapiric traps* or *diapirs*. Some authors classify these as a group of trap on their own, but it is considered here as combined because most diapirs and *salt domes* have related pinching and sedimentary sag or folds associated, even sometimes faults. Figure 1.15 shows combined traps: **salt dome** and **unconformity**. A salt dome is formed when a thick bed of evaporite minerals like salt found at depths intrudes vertically into overlying rock layers. An unconformity is a period of non-deposition or period when erosion was prevalent before subsequent deposition of sediments. They inhibit hydrocarbon flow and make good traps.

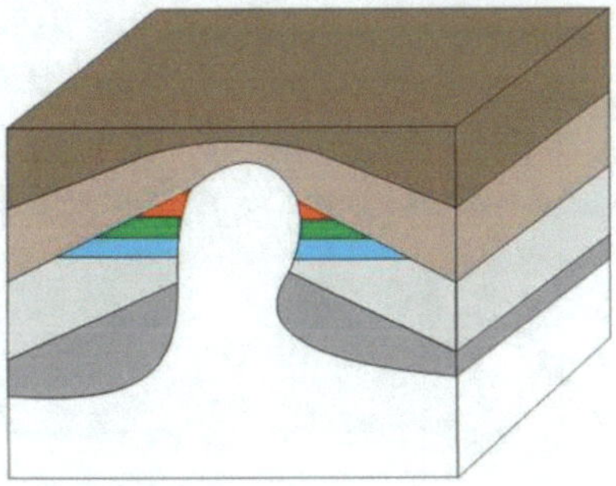

a. Salt dome

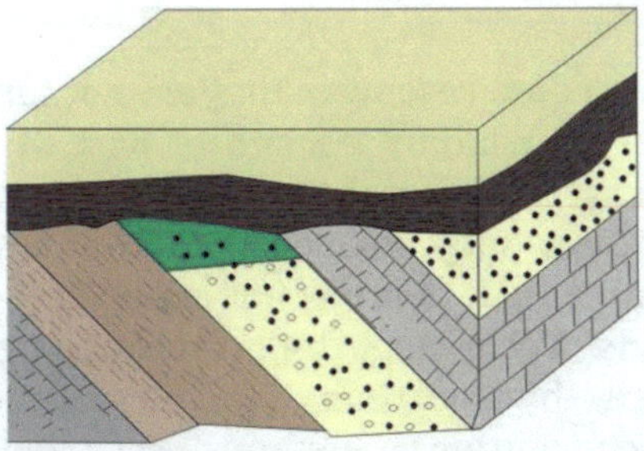

b. Unconformity

Figure 1.15: A combined trap: (a) a salt dome with oil and gas accumulations on the flanks of the folded units (b) an unconformity with oil trapped up-dip in the strata below the unconformable layer

In the angular unconformity shown in Figure 1.15, older rocks formed were tilted and partly eroded before subsequent deposition. This tilt serves as excellent traps where hydrocarbons accumulate, especially when the overlying rock is a good seal. Sometimes salt domes and unconformity related traps are also classified as structural traps.

1.7 Reservoir-Seal Pairs

The term "**reservoir-seal pair**" is used where a reservoir is directly overlain by a seal lithology, often forming multiple stacks and accumulating different pools of hydrocarbon vertically as seen in Figure 1.16. A reservoir and a seal is always necessary for a hydrocarbon trap, and form pairs which may not be adjacent or vertically overlying each other. When logged using a radioactive method, they give different responses as shown in the figure below.

A log is a record of any rock property measured at a certain depth. Figure 1.16 is a gamma ray record of the different rock types at a certain depth. Sandstones are known to show lower gamma ray responses than shales due to their lower radioactive property, hence low responses (L) at sandstone intervals. With shales having higher radioactive property, they show higher gamma responses (H) at shale intervals. There are special cases where sandstone are radioactive due to high feldspar content or other radioactive minerals present in its matrix.

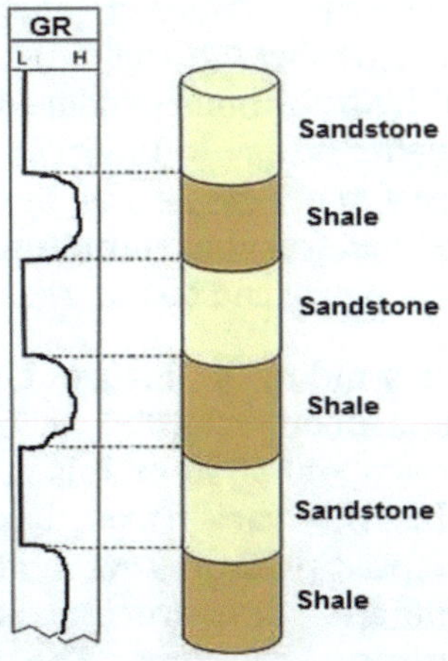

Figure 1.16: Schematic of a gamma ray (GR) log showing low responses (L) at sandstone intervals and high responses (H) at shale intervals

Reservoir – seal pair is an important factor for the migration and accumulation of hydrocarbon. The natural processes of high and low energy alternately transporting large and small grains at different times, both in terrestrial (land) and marine (ocean) environments, create the reservoirs and seals necessary to trap and preserve hydrocarbon. The geoscientist ascertains that these reservoirs are good and occur in the right geometry necessary for efficient trapping.

1.8 Timing of a Petroleum System

Among the factors considered for the successful trapping of hydrocarbons are the timing of generation relative to trap formation. For hydrocarbons to be successfully trapped, the trap must be formed before the hydrocarbons are expelled from the source rock. If structuration occurs after accumulation, the hydrocarbons will probably leak or are destroyed during deformation. It is believed that at the early stage of kerogen maturation, hydrocarbons are expulsed due to pressure and volume changes, migrate in an aqueous mixture (and possibly gaseous solution), and eventually as free oil molecules after segregation in the traps.

The distance of migration, the seal efficiency, and the aging of the system are also important. The shorter the hydrocarbons migrate, the quicker and easier they are trapped. The higher the efficiency of the seal, the greater the likelihood for hydrocarbons to remain trapped. Relatively older or mature sedimentary systems have higher possibility of generating hydrocarbons and get them to accumulate over time. The type of hydrocarbons produced will determine the quality after refining or the processes necessary to get the desired by-products. Crude oil is refined from heavier bitumen components to lighter petrol and cooking gas components.

The oil window in Figure 1.17 occurs at a minimum depth of 2000 m, extending up to depths of about 3000 m. It is important to note here that this varies from basin to basin. At these depths, pressures are moderate and aid in the generation of hydrocarbons under suitable temperature conditions. The minimum temperature required for hydrocarbon generation, as earlier stated, is about 65°C, which varies from basin to basin as well due to duration of *cooking*. Basins have different **geothermal gradients** (increase in temperature with depth).

At lower temperatures, biogenic gas is formed by biologic processes, as the kerogen is *undercooked* due to shallower burial depth and lower pressure. At temperatures between 65 and 150°C, and moderate pressures, kerogen is *well cooked*, allowing hydrocarbons to generate. When temperatures exceed 150°C and pressures become very high, the kerogen may become *overcooked* and thermogenic (or thermal) gas is formed.

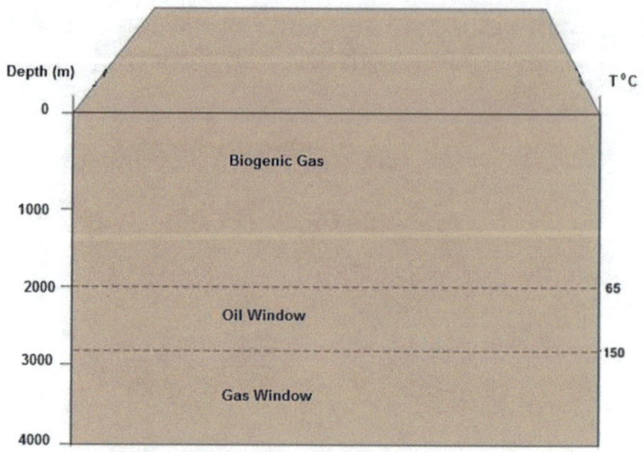

Figure 1.17: Oil Window showing the depth and temperature of maturation

Petroleum resources that are not formed within the oil window constitute part of **unconventional hydrocarbons**. Unconventional hydrocarbons may be sourced from an immature source rock that has not expelled the hydrocarbons. Technology today makes it possible for such hydrocarbons to be extracted using unconventional methods. Overcooked hydrocarbons are harder to make useful.

The heart of petroleum geology and the aim of petroleum geoscience is to understand petroleum generation, migration, accumulation, extraction and refining. As mentioned earlier, fine-grained sediments with high organic contents form kerogen (the initial material abundant in source rocks and necessary for the formation of hydrocarbons). A source rock, possible migration pathway and reservoirs with trapping potentials in a basin, constitute a **petroleum system**. This is discovered after an initial **exploration** (search for hydrocarbon), but sometimes there are already signs of oil leaks that could help to identify such a system prior to exploration.

EXERCISES

1. What is Petroleum? Describe the various forms in which it exists.

2. Describe the four groups of hydrocarbons. Use diagrams where necessary.

3. Explain the process of hydrocarbon formation, describing the role of temperature and depth of burial.

4. a) What are source rocks? Give two examples of source rock types.
   b) List the various kerogen types and describe each of them.

5. Describe the process of hydrocarbon migration.

6. a) What are reservoir rocks? Give two examples
   b) What are the two fundamental properties that make a rock a good reservoir? Explain them.

7. What are seals? What role(s) do they play in hydrocarbon accumulation?

8. a) Differentiate between traps and accumulations.
   b) List the various kinds of traps, giving examples of each.

9. Describe the importance of reservoir-seal pairs within a trapping system.

10. Why is it important to understand petroleum generation, migration, and accumulation?

REFERENCES & FURTHER READING

Aydin, A., 2000. Fractures, Faults, and Hydrocarbon Entrapment, Migration, and Flow. *Marine and Petroleum Geology*, v. 17, p. 797–814.

Dandekar, A. Y., 2006. *Petroleum Reservoir Rock and Fluid Properties.* Boca Raton, CRC Press, Florida, 460p.

Doligez, B., P. Ungerer, P. Y. Chenet, J. Burrus, F. Bessis, and G. Bessereau, 1987. Numerical Modelling of Sedimentation, Heat Transfer, Hydrocarbon Formation and Fluids Migration in the Viking Graben, North Sea, *in* J. Brooks and K Glennie (eds), *Petroleum geology of Northwest Europe*, Graham and Trotman, p. 1039–1048.

England, W. A., A. S. MacKenzie, D. M. Mann and T. M. Quigley, 1987. The Movement and Entrapment of Petroleum Fluids in the Subsurface. *Journal of the Geological Society*, v. 144, p. 327–347.

Evamy, D. D., J. Haremboure, P. Kamerling, W. A. Knaap, F. A. Molloy and P. H. Rowlands, 1978. Hydrocarbon Habitat of Tertiary Niger Delta. *AAPG Bulletin*, 62, p. 1–39.

Gluyas, J. and R. *Swarbrick*, 2003. *Petroleum Geoscience,* Blackwell Publishing, 359 p.

Hubbert, M. K., 1953. Entrapment of Petroleum under Hydrodynamic Conditions. *AAPG Bulletin, v.* 37, p. 1954–2026.

Hunt, J. M., 1990. Generation and Migration of Petroleum from Abnormally Pressured Fluid Compartments. *AAPG Bulletin*, v. 74, p. 1–12.

Hunt, J. M., 1996. *Petroleum Geochemistry and Geology*, Freeman and Company, New York, 743 p.

Moretti, I., 1998. The Role of Faults in Hydrocarbon Migration. *Petroleum Geoscience*, v. 4, p. 81–94.

Okui, A. and D. Waples, 1993. Relative Permeabilities and Hydrocarbon Expulsion from Source Rocks, *in* A. G. Doré, J. H. Augustson, C. Hermanrud, D. J. Stewart, and Ø. Sylta, eds., Basin Modelling: Advances and Applications. *Norwegian Petroleum Society Special Publication* 3, p. 293–301.

Schowalter, T. T., 1979. Mechanics of Secondary Hydrocarbon Migration and Entrapment. *AAPG Bulletin*, v. 63, p. 723–760.

Weber, K. J., G. Mandl, W. F. Pilaar, F. Lehner, and R. G. Precious, 1978. The Role of Faults in Hydrocarbon Migration and Trapping in Nigerian Growth Fault Structures: *10th Annual Offshore Technology Conference Proceedings, Society of Petroleum Engineers,* v. 4, p. 2643–2653

Whiteman, A. 1982. *Nigeria: its Petroleum Geology; Resources and Potentials.* London, Grahams and Trotter, vol. 1-2, 394 p.

CHAPTER TWO

SEDIMENTARY PROCESSES AND BASINS

2.0 Sedimentary Basins

Petroleum resources are found essentially in sedimentary basins. This means that most exploration and production works are carried out in sedimentary basins. This chapter looks at the process that forms basins, when sediments are transported by agents of erosion from high to low grounds. Low grounds are formed by structural processes that create accommodation space for arriving sediments in a basin. A study of sedimentary processes and basin formation should aim at an understanding of:

- Depositional environments
- Basin formation and types
- Extensive, compressive and relay processes
- Examples of world basins
- Petroleum basin analysis
- Roles of temperature and pressure

These are all discussed in this chapter, and their understanding is necessary for determining the source rock maturation, migration and trapping of hydrocarbons.

2.1 Sedimentation Processes

Sedimentation is the process whereby rock forming particles known as sediments are brought together to form a *soft rock,* which eventually become *hard rock* with time. Many factors affect the sedimentation process, but the two main factors, described below, are the **rate of sediment supply** and **accommodation space** for deposition.

1. *Rate of sediment supply:* The rate at which sediment is supplied to the lowland area affects the sedimentation process. The higher the rate of supply, the faster the sedimentation process. Pre-existing rocks are acted upon by *weathering* (which breaks them down) and *erosion* agents (which transports the broken down sediments), such as wind and water, carry them from a higher to a low ground known as accommodation *space.*

2. *Accommodation space:* The accommodation space is where the new sediments are deposited. A lot of processes lead to the formation of the accommodation space. Processes such as tectonics, faulting, subsidence, and other basin forming mechanisms create a tilt that allows sediments to be transported down slope. The steeper the tilt and basinal space, the higher the chances of sediments moving to it.

After sediments are transported from uplifted zones of orogeny (mountain building) to valleys where they are buried, they are compacted and cemented to form rocks (Reineck and Singh, 1980). Figure 2.1 explains the process, where weathering breaks down rocks (that were initially formed by other processes, such as igneous and metamorphic processes), and moving water brought about by rainfall transport the sediments down to the sea or ocean. They accumulate and are buried over time, allowing the sedimentation process to take place until the sediments are transformed to rocks.

During **burial**, sediments undergo a processes of normal **compaction** (compression of sediments leading to increase in density) and **lithification** (where the sediments compact under pressure, expelling or allowing the fluids in its pore spaces to react, and gradually bonding to becomes solid rock).

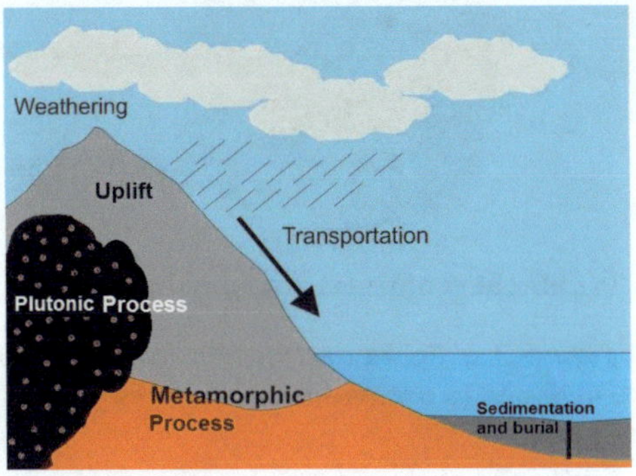

Figure 2.1: Sedimentation process showing agents of deposition (not to scale)

These processes form natural **cement**, which binds the sediment grains to each other, reducing pore spaces in the process as density increases. The pore spaces are filled with cement matrix over time. Clastic sedimentary rock reveals three parts when examined under a microscope – sediment grains, natural cement and pore spaces.

The illustrated example in Figure 2.2 displays a microscopic view of a rock sample, showing grains of sediments and pore spaces filled with natural cement. This is typical of a clastic sedimentary rock composed of broken pieces of older rocks formed by weathering and erosional processes.

No matter how packed together sediments such as sand grains are, there will be *pore spaces* between the grains. Once the grains have been buried in the subsurface, the pore spaces are filled with water that can be very salty at depth, and such water that conducts electricity is useful in oil and gas exploration, as oil and gas are non-conductors of electricity (they are resistive to electricity). The absence of *conductive water* often signifies presence of *resistive oil or gas*.

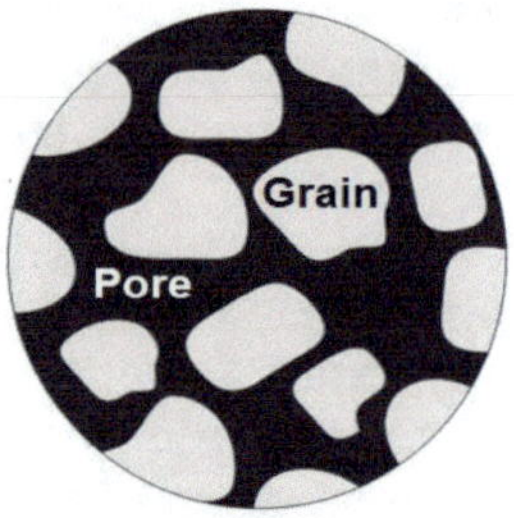

a. Rock sample with clean pores

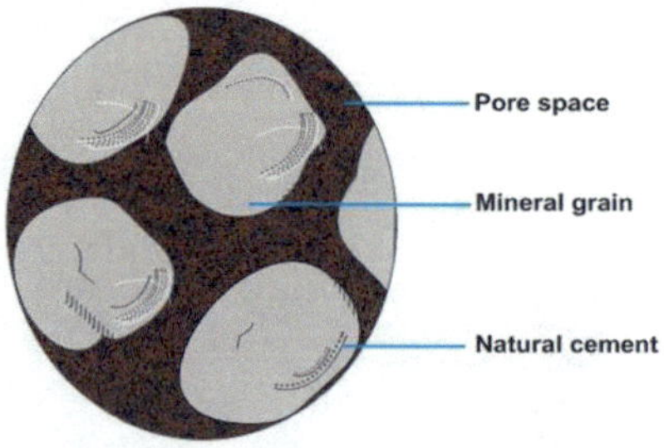

b. Cement contacts and lithified pore spaces

Figure 2.2: Rocks are formed by sediments that are combined as grains and pore spaces

Sedimentation occurs in a low-lying ground or in a basin, as the sediments would move from a place of higher domicile to that of lower, whether controlled by gravity, water current or wind. To have lowland or a basin, certain conditions would have to be in place. These may be erosion, *fractures* (cracks in the rocks, which when it result in the displacement of one side of the rock with respect to the other is known as *fault*) or folding (geological process that causes a rock layer to buckle or bend). Sedimentary geology is important in the petroleum industry because it is the sedimentation process that harbours the source materials that eventually mature to generate, transport and store hydrocarbons.

2.2 Depositional Types and Environment

The overall Earth structure is made up of hard rocks and soft rocks, and the oceans. The hard rocks often occur in mountains, or are alrburied. Between the mountains are regions with soft rocks, in which sediments from the highland are deposited. Broadly speaking, these are the basins. The lowlands under the sea are also basins.

Submarine canyons and fans are erosions made into the continental shelves and slopes, extending from shallow depths off the beach down to the bottom of the continental slope. Sands are deposited at the beginning of the slope canyons, along the channels to the deep offshore lobes (Figure 2.3). The *continental margins* transit from the *continental shelf* to the *continental slope* and *submarine fan*. This area marks the edge of the continent and the beginning of the ocean environment.

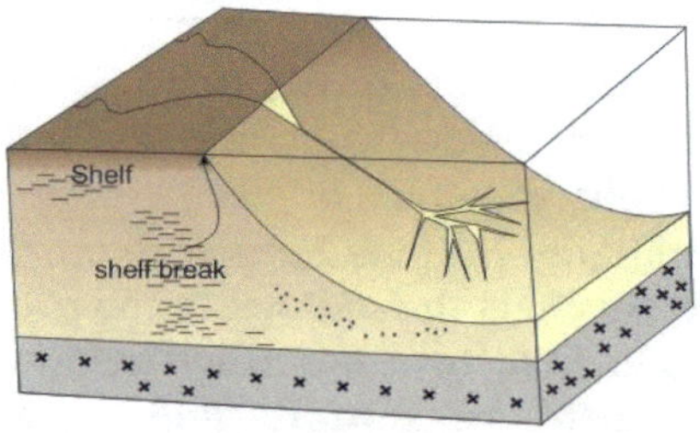

Figure 2.3: Continental shelf to offshore deposits

On the continental shelf and slope there are *v-shaped submarine canyons* formed by erosion, which extends from shallow depths off the beach down to the bottom of the continental slope. Sands are deposited at the submarine canyons on the slope where there is high water energy, which progressively leads deposition in the submarine fan, buried later with shaly or clay deposition when there is a drop in the water energy.

Sandstone reservoirs in *submarine canyons* and *fans* are main target of petroleum exploration. Sandstone *turbidity current* deposits are called

turbidites. Relatively thin sandstone separated by shales is characteristic of submarine fan reservoirs.

2.2.1 Depositional Systems

It is possible to identify depositional systems and the environment using several combination of geologic knowledge. Below are some of the systems that create the deposits you observe from onshore through the shelf to offshore.

Aeolian dunes are deposited in arid environments such as desert regions, where winds erode, transport and deposit fine-grained sands to arrange them in crescent shaped units that are easily recognizable in present day systems. Wind as an agent of erosion is effective in this case because of sparse vegetation and large amount of unconsolidated sediments. Wind depositions occur usually as *ripples* (sedimentary structures or features indicative of wind or water agitation), *sand sheets* (sand deposition formed where grain size are too large or velocity of wind is too low to transport sediments and form dunes; usually flat with minor undulations) and *dunes* (hilly loose sand deposit formed by relatively smaller grain sizes transported by relatively higher wind velocities).

Alluvial fans are systems put in place when precipitation and a mix of other natural wearing effects deposit cone-shaped sediments rich in mineral at the base of mountains, or in nearby valleys. Where they are not built up, usually by streams, but by debris flow (sediments with different shapes and sizes ranging from boulders to clay and filled with water rushing down a mountain), the term *colluvial fan* is used to describe the deposit so formed. Glaciations have been responsible for providing the energy that carries sediments from mountains down to today's lowlands. Whatever the system or means of transportation, deposition begins to occur as energy levels drop, usually downslope at less steep mountain base or nearby valleys.

Fluvial deposits are formed by rivers and other types of flowing water bodies, including meandering channels inland. Moving water erodes sediments, transport and deposits them on riverbeds; the coarser sediments will usually move closer to the bed (bedload), while the finer sediments will be carried in the water (suspended load). This process of sediment erosion and transportation will continue throughout the length of the river especially where the river flow is fast, until the river velocity decreases, where sediments begin to deposit. Fluvial deposits are among the most common in terrestrial domain; the rivers today are leaving sediments, rock fragments below, which will form reservoir deposits of the future.

Shelf deposits occur in shelf regions (the continental shelf, an underwater landmass extending from the continent covered by relatively shallow water) where different types of carbonates are formed by precipitation and reefal activities, thereby forming important reservoirs especially when their intergranular porosity is further enlarged by fracturing or dissolution. When mineral dissolutions occur in a limestone due to acid water breaking down the surface of the rock, creating large spaces within its lithology as the rock continues to degrade, the process is known as ***Karstification*** and the *karsts* or *spaces* produced form important post-depositional reservoirs in both limestone and dolomite.

Shallow marine and submarine deposits are, as their names imply, deposited seaward from land areas to the ocean. The v-shaped canyons act as the main route for sediment transport from the shallow marine to deep marine. During lowstand (low sea levels), flowing channels from shore induce turbidity currents caused by concentrated sediments in partial suspension below the sea and deposit high energy channel conduits in shallow marine (deposits formed as a result of turbidity currents are referred to as turbidites); the currents continue basinward to form the lobes (*fans* or fan-like lobular large sand bodies) in deep deposits, and provide massive reservoirs.

2.2.2 Sedimentary Features

Different *sedimentary features* may help to define the environment or nature of sediment deposit. Examples are **cross-stratification** and **graded bedding**.

Cross-Stratification

An example is *hummocky cross-stratification*. This is an undulating bedform emplaced by wave-generated oscillatory flows or combined flows resulting from passage of storms. It indicates storm environment, typical of areas between **fair-weather wave-base** and **storm wave-base**. Hummocky cross-stratification is characterized by a gently curved, low-angle cross-lamination arranged in a **hummock** (a convex-upward) and a **swale** (a concave-upward) pattern (Figure 2.4). Another example is the *herringbone cross-stratifications*, formed by current in opposing directions that are approximately equal in strength.

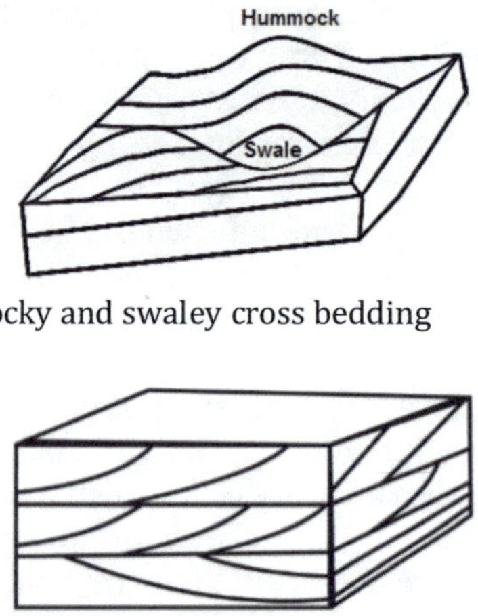

a. Hummocky and swaley cross bedding

b. Herringbone cross bedding

Figure 2.4: Different types of cross stratification

Cross stratification can help to see the direction in which the sediment was being transported. This can be counted and plotted in a *rose diagram* as shown in Figure 2.5; the graduation counts sediments trend in a given direction of flow. As the count is higher for the northern direction, it indicates that the dominant sediment transport that emplaced the deposit was northward, as shown in the inset figure to the right.

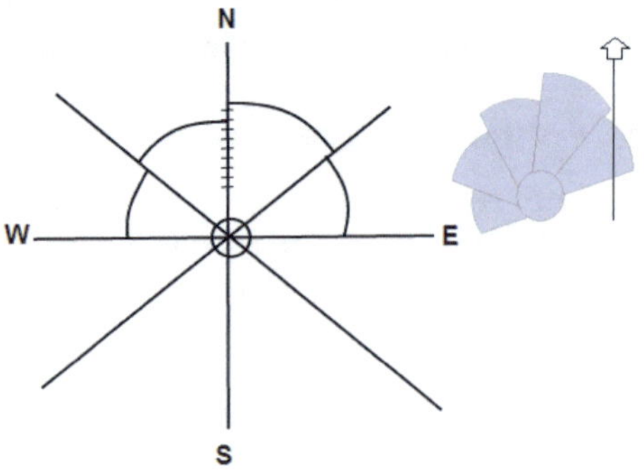

Figure 2.5: Rose diagram indicating direction of flow

Cross stratification or cross bedding is a feature that reflects the transport of sediments of gravel and sand fractions by (river or wind) currents that flow over the sediment surface, usually in river channels or coastal environments. This feature occurs at various scales, and is seen in *conglomerates* (coarse-grained clastic sedimentary rocks composed of rounded grain size fractions larger than 2 mm diameter) and sandstones.

The inclination of the cross beds indicate transport direction and current flow, while the size of the cross bedding can be used to estimate transport velocity of the sediments. Examples are *hummocky* (formed by combined flows or simultaneous occurrence of waves or currents) and *herringbone cross-stratifications* (formed by current in opposing directions that are approximately equal in strength).

Graded Bedding

Graded bedding is a layer in which the particle sizes change according to a systematic gradient in a vertical (or lateral) direction. In *normal graded bedding,* the particles become systematically *finer upward*. In *inverse graded bedding* the particles become systematically *coarser upward*. Decelerating sediment–laden current generates coarse to fine grading upwards, a common characteristic of turbidites deposited from waning density as energy reduces.

In Figure 2.6, sediments become progressively finer to the top. There is a vertical systematic change in the particle sizes deposited. Particle sizes are obviously larger at the bottom and those at the middle are larger than those above.

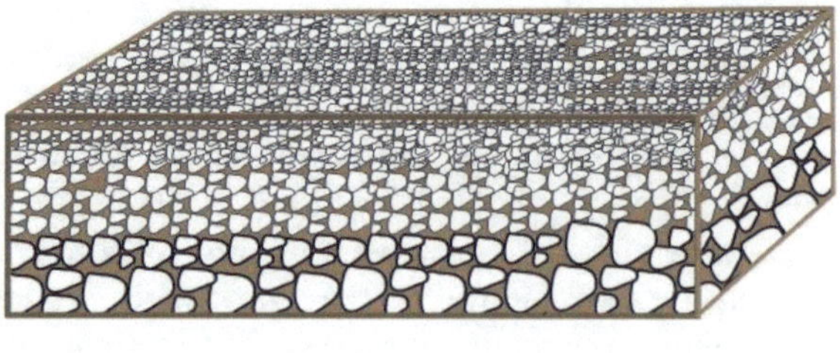

Figure 2.6: Normal graded bedding with coarsest sediments at the bottom and finer sediments progressively on top

Among the transporting agents of sediment before deposition are wind, water and gravity. The higher the velocity of transport, the less time the grains will have to be sorted (segregated into similar sizes) before deposition. As shown in Figure 2.7, the lower the velocity the better the sorting, and vice versa. Because of the low velocity, there is enough time for sediments to be sorted out into similar grain sizes before they are deposited.

Likewise, sediments transported over a long distance will have better roundness. *Angular sediment grains* suggest *proximal deposits* (situated nearest to the point of origin), while well-rounded *grains* suggest *distal deposits* (situated much farther from the point of origin). This is because as sediment grains travel during transport, their edges break off and the grains are gradually rounded.

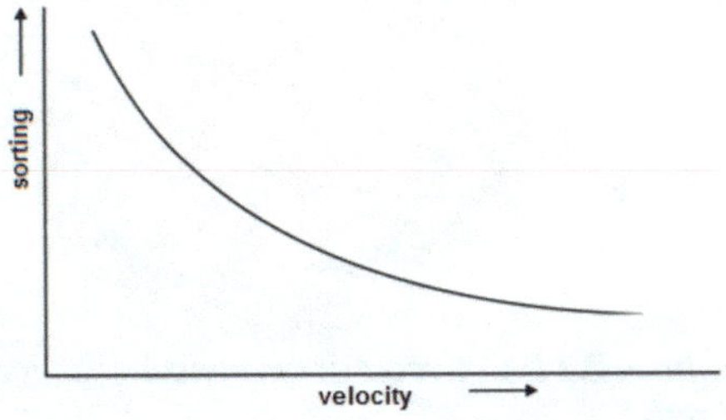

Figure 2.7 Relation between sediment velocity and sorting

The farther the transport, the more rounded distal sediments become. Sediments that are deposited near where they are weathered off remain unreworked and as such angular. Grains could be fully rounded, slightly rounded (subrounded) or even spherical. These descriptions are important in lithological description of oil field cuttings (fragments of rocks in the subsurface) that come to the surface from a well.

Figure 2.8 shows grain characteristics for proximal and distal sediments. Those for proximal sediments are usually angular while those for distal sediments are usually rounded (or subrounded) depending on the travel distance.

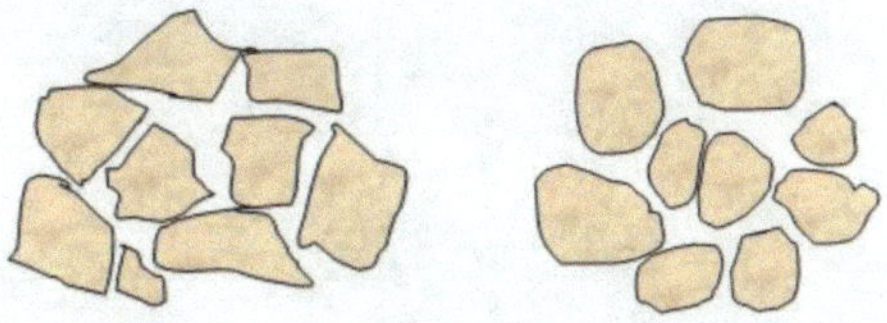

a. Angular grain b. (Sub) Rounded grain

Figure 2.8: Grain characteristics in proximal and distal sediments

The result of deposition of sediment over time is the association of *vertical and lateral facies* (vertical and lateral distinct rock unit characterized by its origin, composition and environment of deposition) which are matured over the years. This eventually leads to the formation of sediments from which the present day petroleum fields are built. These include the:

- Source Rocks
- Reservoir Rocks
- Seal Rocks

Most rivers flow through loops and bends called **meanders** (Figure 2.9). These are caused by friction of water flowing along the bottom of the channel. Water on the outside of the meander flows faster and erodes the outer bank. Water on the inside of the meander slows down, depositing the coarsest sediment that was suspended in the water. The crescent-shaped well-sorted sand bars deposited on the inside of the river meanders are called **point bars**. Buried point bar sandstones are often good oil and gas reservoirs

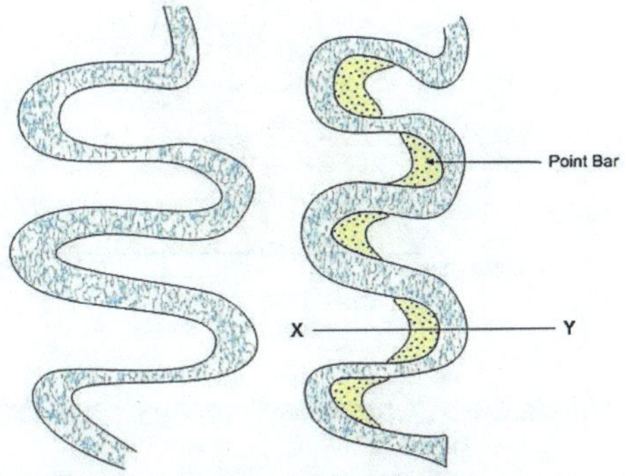

a. Meandering river and point bar sand

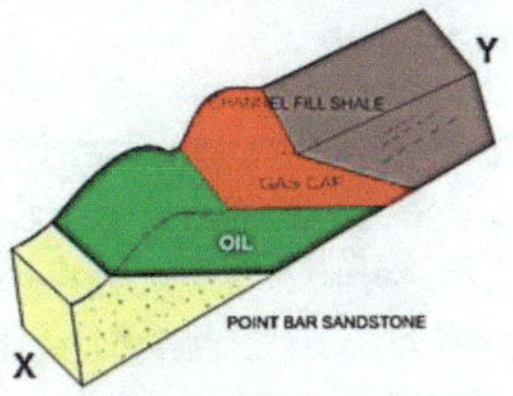

b. Cross section of point bar sandstone

Figure 2.9: Fluvial deposit of good reservoir (modified from Berg, 1968)

The stratigraphic layers produced during deposition of either source, reservoir or seal rocks makes them *correlatable* (establishing a mutual relationship between them) across the basin, in most cases. Despite the limitation of such correlation, sedimentary geologists are able to tie them and give specific thicknesses and lateral extension to petroleum fields and basins. They are also able to determine what type of rock it is, among the common rock types that form sedimentary basins.

Sedimentary Rock Distribution

Shale, sandstone and limestone make up close to 100% of the sedimentary rocks that constitute the Earth crust (Figure 2.10). Many sedimentary rocks are a combination of these three types. Their mixtures are described as sandy, shaly, and limey or calcareous. For example, in a sandstone-shale mixture, a sandy shale has shale as the dominant constituent whereas a shaly sandstone has shale as the dominant constituent. A rock with some limestone (calcium carbonate) constituent is said to be *calcareous.*

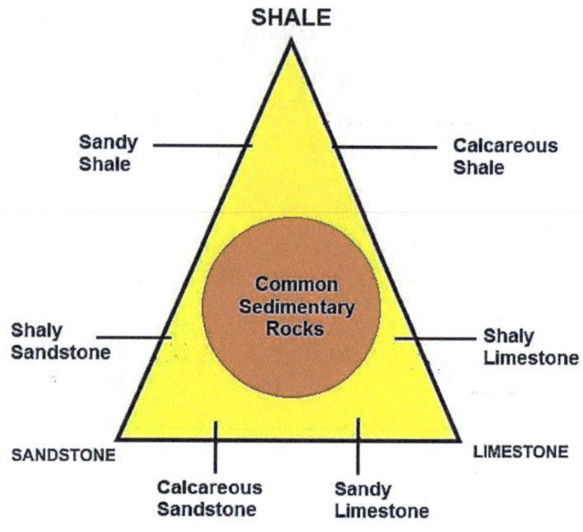

Figure 2.10: Common sedimentary rock distribution

2.3 How Basins are Formed

Basin formation

The Earth is made up of rigid plates composed of the *crust* and part of the *upper mantle*, termed *lithosphere*. These plates are in constant motion, and the nature of their movement leads to the formation of different types of basins. Movement of these plates is possible due to *convection currents* (causing the transfer of heat through fluid) in interior of the Earth, which causes the liquid mantle region known as the *asthenosphere* to allow for displacement of the upper units of the earth. When they move away from each other, they form an **extensional basin**. When they move towards each other, they form a **compressional basin**. When they move past each other they form what is known as a **strike-slip basin**.

The interior of the Earth (Figure 2.11) consists of the *core*, the *mantle* and the *crust*. A cell of flowing liquid caused by heating and cooling, causes the interior of the Earth to be in constant motion. Convection currents in the Earth's interior make it possible for the lithosphere to be in constant motion, giving rise to compressional, extensional and strike-slip basin formation. This occurs both on land and in the oceans.

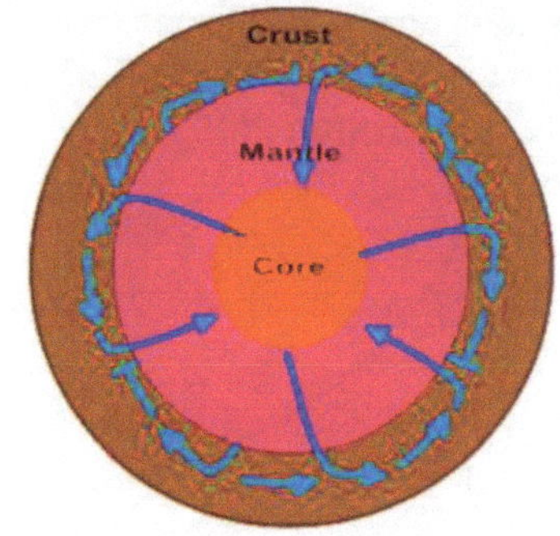

Figure: 2.11: Convection currents in the interior of the Earth

There is therefore *continental plate movement* (on land) and *oceanic plate movement* (in the oceans). World's longest mountain chain is located almost in the very centre of the Atlantic Ocean that can be traced over 60,000 km. The super continent, Pangea (Figure 2.12), previously had all the continents joined together before it broke up during the Mesozoic, about 150 Ma. That led to the formation of the present day continents, which include Africa, Asia, Europe, North America, South America, Australia and Antarctica.

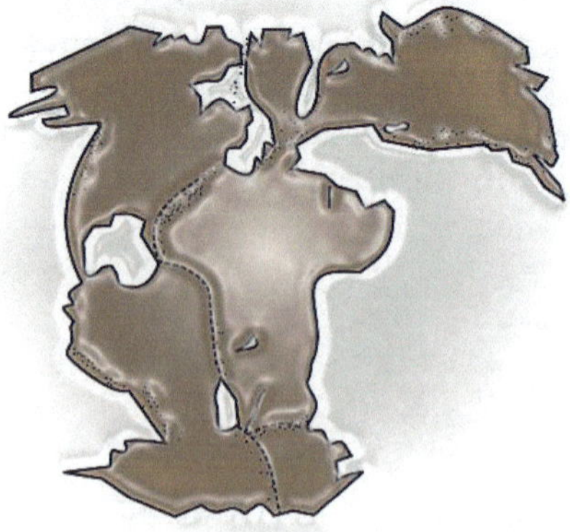

Figure 2.12: Pangaea - the super continent

The chain of mountain known as **mid oceanic ridges** (a long narrow natural elevation on the floor of the ocean) is formed when a rising hot current from the interior of the Earth cannot penetrate the crust of the Earth. It creates an arch in the crust to form the mid oceanic ridge (Figure 2.13). Where it is able to penetrate, it splits the crust at the crest of the ridge, and pulls the crust apart. The hot, molten current then divides and flows to either side of the mid oceanic ridge. It further splits the solid crust of the earth at the ridge crest and drags it to either side of the ridge. This leaves a trace where the initial continents, which were together, separated to form present day distinct continents. That super continent is known as **Pangaea**.

a. Hot current arching the crust

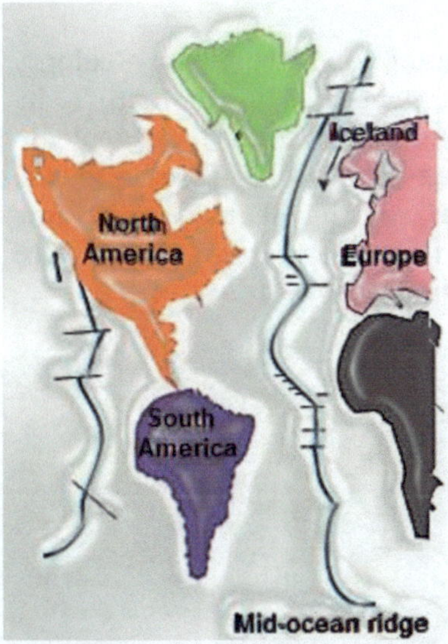

b. Continent showing the mid oceanic ridges

Figure 2.13: Formation of the mid oceanic ridge

Triple junction

The initial break up of a continent by plate tectonics can take the form of a *triple junction*. A triple junction has three rifts (arms) that join in the centre. Usually two of the arms unite and continue rifting to form an ocean (Figure 2.14). The third arm, which stopped spreading, is called a *failed arm*. The formation of the Benue Trough has been described as being associated to the triple junction that formed the Atlantic Ocean during the breakup of Pangaea (Benkhelil, 1986; Senghor, 1995).

Figure 2.14: Triple Junction, illustrating the two arms that form an ocean and the pointed end (to the left) that enters the continent

Figure 2.14 shows a triple junction, showing the point of breakage of a continent by plate tectonics. Two of the arms (at the distal and proximal ends of the image) continue to spread to form an ocean, while the third arm (on the left of the image) fails to spread.

Types of Basins

There are three major kinds of basins, which are described below.

Extensional Basins

This is formed by **divergent** (plate) movement. Examples include cratonic basins (intracratonic and epicratonic embayments), passive margin, rifts, and troughs. A failed rift, that is a broken rift, is known as aulacogen.

Major sedimentary basins in Nigeria (Figure 2.15) include:

- Chad Basin (also known as Borno Basin, which is the sub-basin in Nigeria)
- Iullemmeden Basin (also known as Sokoto Basin, which is the sub-basin in Nigeria)
- Bida Basin (also known as Mid-Niger Basin)
- Benue Trough (with has the Northern, Central and Southern, previously considered Upper, Middle and Lower Benue Trough)
- Anambra Basin (previously considered as part of, as it is situated west of, the Benue Trough)
- Niger Delta Basin.

These basins has been extensively discussed by many authors (Benkhelil, 1986; 1989; Benkhelil et al., 1988; Nwajide, 2013).

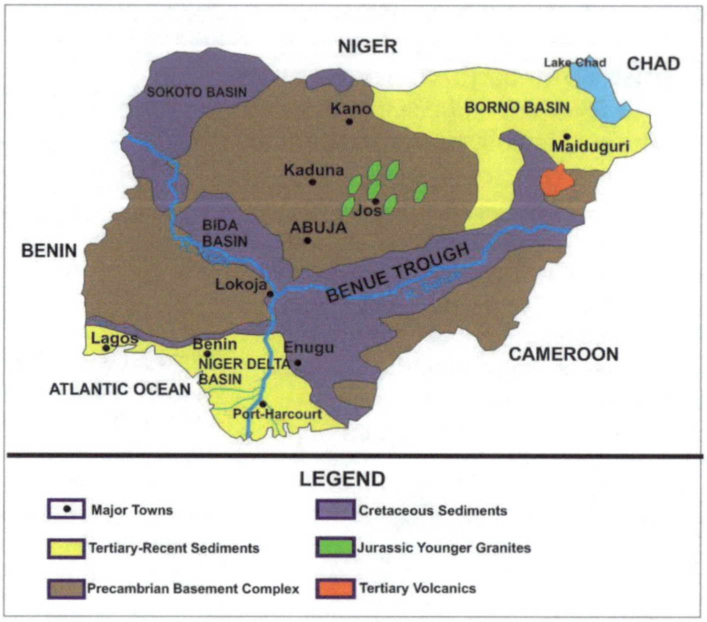

Figure: 2.15 Sedimentary Basins in Nigeria (Modified from Obaje, 2009)

There is also a stretch of the Dahomey Basin (Abeokuta Basin) to southwestern Nigeria.

Compressional Basins

These are formed by **convergent** (plate) movements. Examples include arc systems (fore- and back-arc), foreland basin and fore-land deep.

Strike-slip Basins

They are formed by **lateral sliding** along the strikes (of plates) or by "wrenching", where there is also a vertical component of displacement. Examples are *pull-apart* basins and *wrench fault* related basins.

Some basins are formed *inland* (**cratonic**) while others are formed in the *oceans* (**oceanic**). They are formed by Earth's upper units separating (rifting) or coming together (compression) as a result of natural crustal movement over the *mantle*, by processes known as *plate tectonics*.

African Basins

The basins in the southwestern part of Africa are related to the separation of the African continent from the South American continent. The Chain and Charcot fault zones, sometimes described as a triple junction discussed in the Chapter on Structural Geology, created the *lineaments* (a linear feature in a landscape expressing an underlying geological feature, the failed arm in this case) and weak zones that emplaced the Benue Trough and consequently the Niger Delta Basin. Figure 2.16 shows major sedimentary basins in Africa.

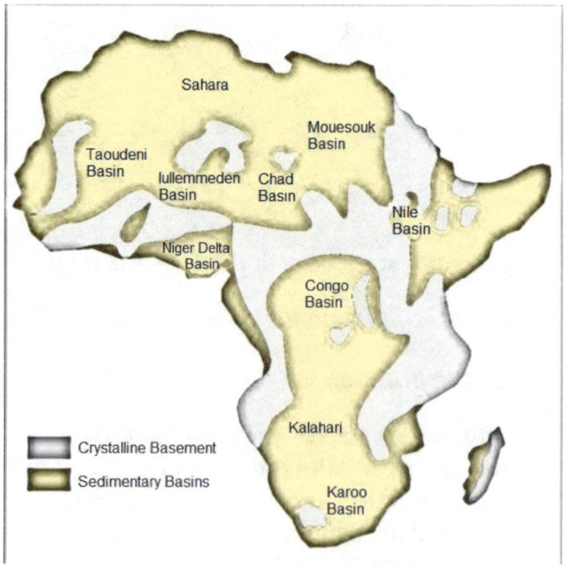

Figure 2.16: Sketch showing the approximate locations of sedimentary basins and basement in Africa

Examples include Taoudeni Basin, Iullemmeden Basin, Chad Basin, Congo Basin, Mouesouk

Basin and Karoo Basin (which have been described in several literatures. (Selby, 1985; Key, 1992; Reijers et al., 1998; Nwajide, 2013).

Niger Delta Basin

The Niger Delta Basin is a well-studied basin. It transitions from extensional to compressional domain, but is mainly an extensional basin at a larger scale. It therefore has local compression and compressional basinal features as well as the regional extension and extensional basinal features. Studies in gravity tectonics of a passive margin help to understand this as illustrated in Figure 2.17. The cross-section shows the activity from extension to compression in a gravity controlled basin deformation in the deep offshore.

Sometimes what is seen in the field or on *seismic* lines is a part of the basin, such as normal faults, troughs or depression from either *synclinal* or *anticlinal* flank.

A faulted basin is shown in Figure 2.18 in which sediments are transported from highland, depositing in the layers representing the active sedimentation, as younger deposits buries older units. The hanging wall (fault block resting on top of the fault line) moves down relative to the footwall (fault block lying below the fault line). As sediments are transported from the highland, they are subsequently deposited in the depression created by the normal fault, forming progressively younger to older sediments.

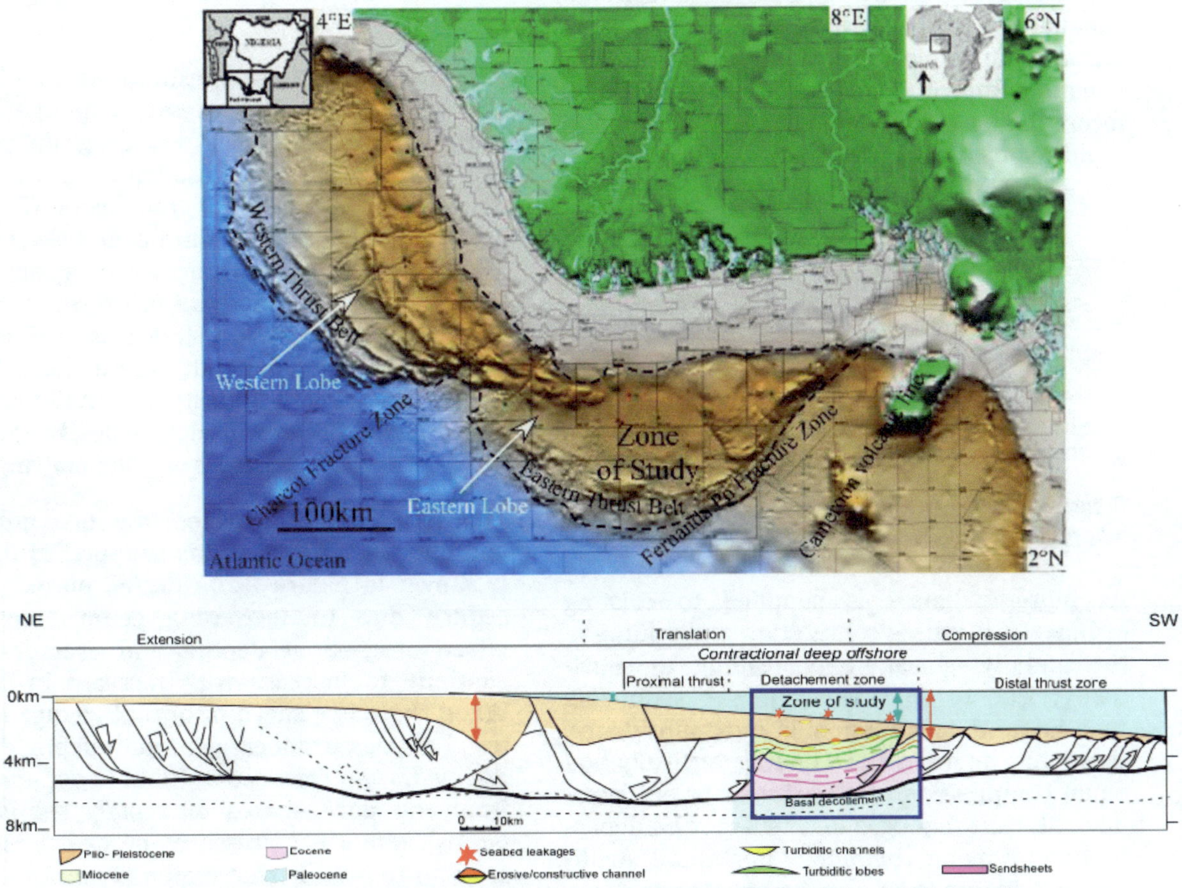

a. *Mini compressional basin in the extensional / passive margin offshore Niger Delta*
b. *Section showing a schematic illustration of activity from extension to compression*

Figure 2.17: Niger Delta showing gravity controlled basin deformation in the deep offshore (Nosike et al., 2008; Nosike, 2009)

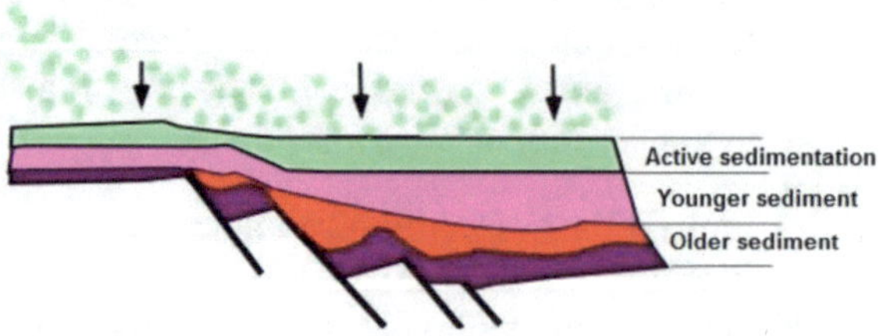

Figure 2.18: Basin created by normal faults, where the hanging wall is downthrown, creating a low

Subsidence

Basins are also formed when the weight of the overlying sediment concentrate in the centre of the pile, leading to that central part sagging or eventually folding as in a syncline. This process is known as **subsidence**. Subsidence leads to compaction (Figure 2.19) and compaction may lead to overpressure if the fluids are not adequately expulsed in the process, as is common in rapidly subsiding basins. This situation known as *disequilibrum compaction* forms the basis of pore pressure prediction in rapidly subsiding basin.

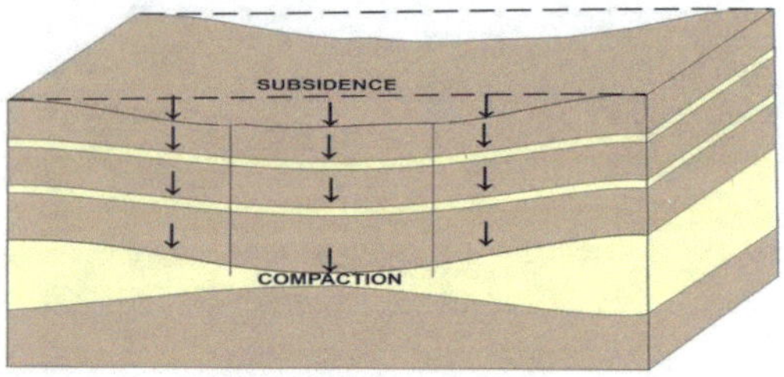

a. Subsidence leading to compaction

Figure 2.19: Cross section of a sedimentary basin whose centre sagged due to subsidence

As more sediment is supplied to existing sediments, it causes compaction and folding of the underlying sediments, leading to subsidence. Due to subsidence, newer sediments then fill up the depression. In pure subsidence, there may be increase in folding vertically and equal compression of all affected layers laterally. After the folding, a new bed may be deposited and remains completely horizontal. Artificial subsidence may also occur, that is when subsidence effect is man-made, like pumping out the subsurface water or petroleum that supports and prevent the layers from sagging. Removal of sediments and fluids during mining, or other similar processes, weaken the *sub-strata* and the overburden outweighs the support.

2.4 Petroleum Basin Analysis

A basin life starts as soon as sediment starts to fill and pile up in the basin. The basin will undergo such processes as *compaction*, *compression* and *re-folding – faulting* and other types of deformation. Different layers of sandy or low permeable lithology will be deposited, leading to what creates a suitable environment for petroleum formation and trapping. For a petroleum basin analysis, you ask:

- What is the structure of the basin?
- Is the basin mature?
- What is its burial and thermal history?
- What is its pressure profile?
- Any restructuration or secondary tectonics?
- Are there source and reservoir rocks, and seal?

The answers to these questions start by building profiles for the pressure, temperature, porosity and effective stress or grain matrix that supports the load (also known as overburden) within a structural framework (Doligez, 1987; Burrus, 1991; Osborne and Swarbrick, 1997). At a depth of overpressure generation, there is a sudden increase in pressure higher than the normal or hydrostatic pressure (Figure 2.20). This is the depth where there is no longer reduction in porosity, and fluid pressure bears part of the overburden, instead of normal increase in effective stress or grain matrix.

This relationships between pressure, porosity and effective stress (grains support) at depths is shown in Figure 2.20. Ideally, porosity will reduce due to increasing compaction and effective stress at depths, and pressure will continue to increase with increase in depth. When there is a sharp or abrupt change in the trend of pressure increment with depth, due to *disequilibrium compaction* (fluid is not expelled from the pore spaces of *rapidly subsidizing basins*), abnormal pressures or overpressures are said to exist in that region (*Transition Zone to Overpressure in Figure 2.20*). This concept is used to predict overpressured zones in the subsurface.

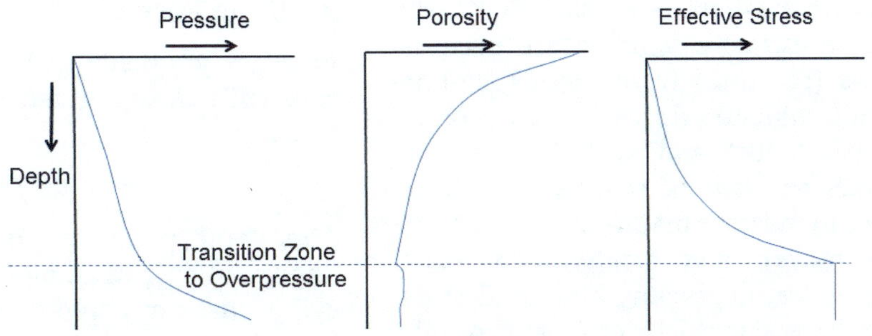

Figure 2.20: Impact of compaction on pressure, porosity and effective stress in a rapidly subsiding basin

Structural geologists analyse the structure of a basin and determine its tectonic settings and local structural features. Following an increase in heat with depth as more sediment is supplied, a basin is said to mature at a given temperature.

2.5 Basin Temperature

Subsurface temperature generally increases with depth, with the exception of where fluid has migrated from another unit to a confined segment. Subsurface temperature could be measured using oil well drilling equipment known as Bottom Hole Assembly (BHA). Example of tools coupled on BHA are:

- Well Formation Tester (WFT), which measures temperature alongside pressures.
- Logging While Drilling (LWD), which also has integrated temperature measuring tools

The temperature obtained from WFT is considered closest to the actual formation temperature, although the well has been disturbed and regions near the well, at the depth of the measurement probe, has been altered. LWD temperatures are also not correct because they are measuring the drilling fluid temperature in part. Wireline temperature recording provides reading close to the actual borehole temperature if well is left to begin to regain its temperature and pressure after circulation. A *Horner Correction* (explained with a plot in Chapter 13 on *Drilling and Geosteering*) is used to bring the measured temperature closest to the actual formation temperature.

However, each measurement (LWD or Wireline) will give a trend highlighting the geothermal gradient. **Geothermal gradient** form the bases for prediction of temperature and for understanding evolution of the petroleum system in a basin. Uses of downhole temperatures include:

- Establishing a regional geothermal gradient
- Basin modelling and assessment of petroleum system maturation
- Estimation of resistivity and water saturation

Since basin temperature typically increases with depth, the important thing is to know the rate of increase. Different basins have different rates at which temperature changes with depth, and this rate of change is usually constant during steady sedimentation for a given basin. The rate of change of temperature with depth is known as **temperature gradient** or **geothermal gradient**. Temperature gradient is important because it determines the maturity of a source rock (remember that temperatures of at least 65°C will generate oil) and whether or not a *petroleum prospect* (available source, reservoir, seal rocks together with traps) should be drilled.

Temperature gradient (TG) is the change in temperature (dT) over a given change in depth (dZ), given as:

$$TG = dT / dZ$$

An increase in temperature with increasing depth is illustrated in Figure 2.21. The rate of change of temperature with depth in this case is about 0.025 degree Celsius for every one metre (Temperature gradient = 0.025 °C/m). This can be higher in oil fields due to abnormal pressures resulting from rapid burial or tectonics, chemical reactions and changes in fluid property.

Temperatures in the subsurface can be measured using wireline (a tool that measures temperature) where there is a well. Other tests can be done to estimate temperature such as *illite reactions* (reactions from non-expanding clay crystalline minerals developed as a function of burial depth) and various chemical reactions, such as *vitrinite reflectance (%Ro)*, which is used to indicate maturity in hydrocarbon source rocks), and *biological markers* (measurable indicator of some biological state for assessing thermal maturity of source rocks).

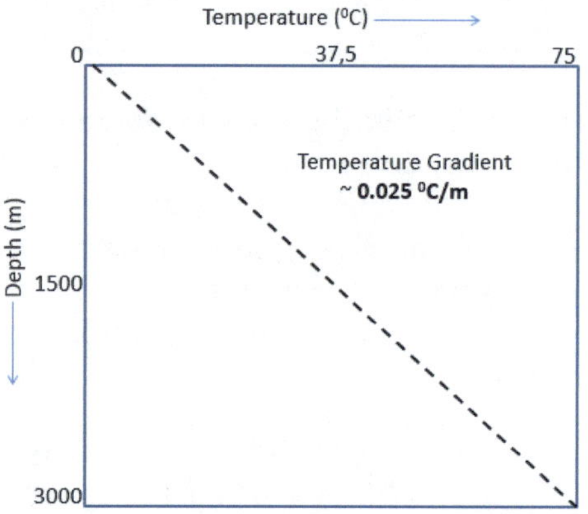

Figure 2.21: Temperature gradient, where temperature over a burial depth of 3 km is about 75 degree Celsius

A basins initial temperature may be modified by secondary tectonics. When *secondary tectonics* occurs, after the deposition and initial deformation of the sediments, the basin will be refolded, uplifted or sagged even more due to subsidence and reactivation of faults, which will lead to more compaction or decompaction (decompaction is a reversal of compaction during uplift). When the sediments are uplifted, the effect of compaction may not be fully reversed (only very little effect is reversed) and lost porosity is never fully recovered. Uplifted sediments are exposed and subjected to erosion because the sediments have moved upward above the level of other sediments, thereby exposing that area to fluid redistribution such as influx of fresh water, cooling and temperature changes. This makes prediction of temperature in reactivated basin more challenging.

Further basin modelling methods and software that use multiple data types, when available, help to project the evolution of basin temperature (as well as pressure) over time.

2.6 Basin Pressure

Pressure gradient (PG) is the change in pressure (dP) over a given change in depth (dZ), given as:

$$PG = dP / dZ$$

After depth, pressure is perhaps one of the most important data in an oil field. It is important at every phase of exploration and production. It helps in predicting hydrocarbon column and its prediction is important for drilling safely. Basin modelling attempts to predict pressure and its role at a larger scale.

Pressure can be measured directly using wireline *Well Formation Tester (WFT)* and a pressure gradient approximated for the basin, with corrections made to take into consideration the effect of the open hole (as an open hole or well dissipates the original pressure at that depth before drilling). *Repeat Formation Tester (RFT)* and *well tests* also give indication of pressure evolution within a basin. The **Terzargi equation** (*discussed in Chapter Twelve, on Pore Pressure Prediction and Interpretation*) gives a relationship between burial, overburden and the in-situ pressures in a pile of sediment. Eaton's equation also expounds the relationship for estimating these pressures, including the overpressure that results from disequilibrium compaction in rapidly subsiding basins.

Figure 2.22 illustrates pressure gradient, marked by increase in pressure with increasing depth. The rate of change of pressure with depth in this case is about 10 kilo Pascal per meter (Pressure gradient = 10 kPa/m). This is about 0.1 bar /m.

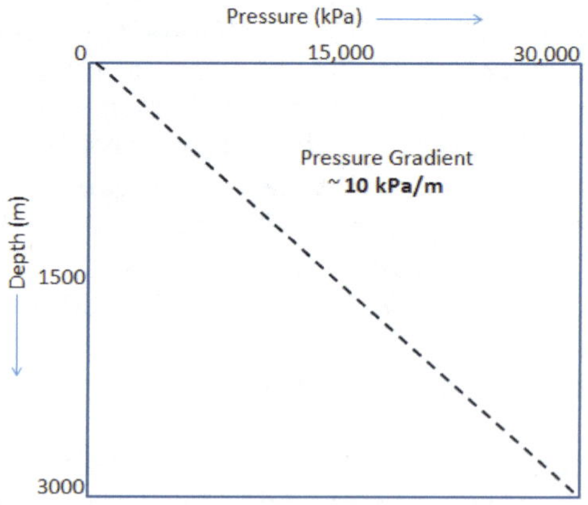

Figure 2.22: Pressure gradient, where change in pressure over a burial depth of 3000 m is about 30,000 kilo Pascal (or 300 bar)

Petroleum Kitchen and Source Rock Maturation

Basin temperature and pressure are directly related to source rock maturation, expulsion and eventually trapping. They create the equilibrium between oil field fluids and the surrounding sedimentary and structures. This is why basin history is important and modelling often takes pressure as well as temperature into consideration.

Temperature and pressure gradient are a function of depth and they control hydrocarbon maturation (Hunt, 1996). Figure 2.23 illustrates petroleum kitchen showing the role of depth in oil generation window. In this case, at depths of at least 2000 m, oil forms within the oil window. At temperatures of at least 60 °C and moderate pressure, kerogen reaches maturity (well cooked) and generates hydrocarbon within the oil kitchen. Gas is generated at more depth and higher temperature from 120 °C. However, this oil temperature window may vary slightly for different source rock and basins.

When produced, crude oil retains the ability to yield different products with changing temperature and pressure (Figure 2.24). When crude oil is heated until it turns to vapour and then allowed to condense at different temperatures, fractions of the crude oil begin to form, and can be collected at different outlets, with gas and bitumen having minimum and maximum condensation temperatures, respectively.

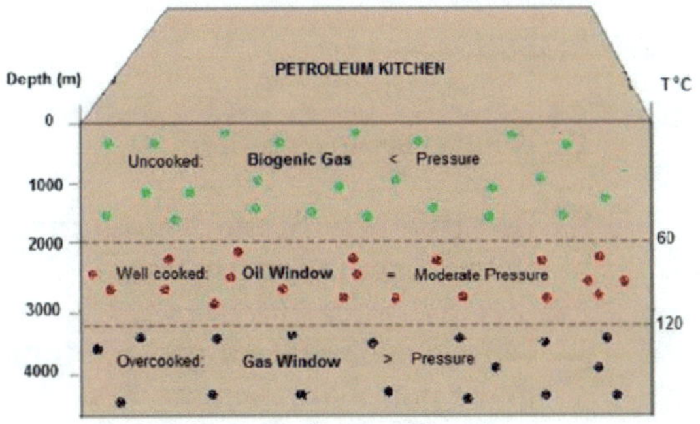

Figure 2.23: A petroleum kitchen showing the role of depth in oil generation window

Fractional distillation of crude oil yields different products, which as well depend on the kind of crude oil supplied to the refinery. Other fractions from the fractional distillation of crude oil include petrol, naphthalene, kerosene, diesel, fuel oil and lubricating oils and waxes, as well as left over bitumen. These may occur at temperatures within the range of 20 to 400 degrees Celsius, with the lighter and volatile petrol yielded before the heavier ones left over despite maximum heating.

This is similar to what happens in nature, where gradual heating and pressure conditions trap different types of crude at depth. *Wet* or *dry gas*, *condensate*, *liquid oil*, *black oil*, or other types of crudes, may be trapped at different depths in a natural basin. Many factors affect the generation of petroleum in a basin, going beyond temperature and pressure, and relating to both sedimentary and structural processes that take place before, during and after oil and gas trapping.

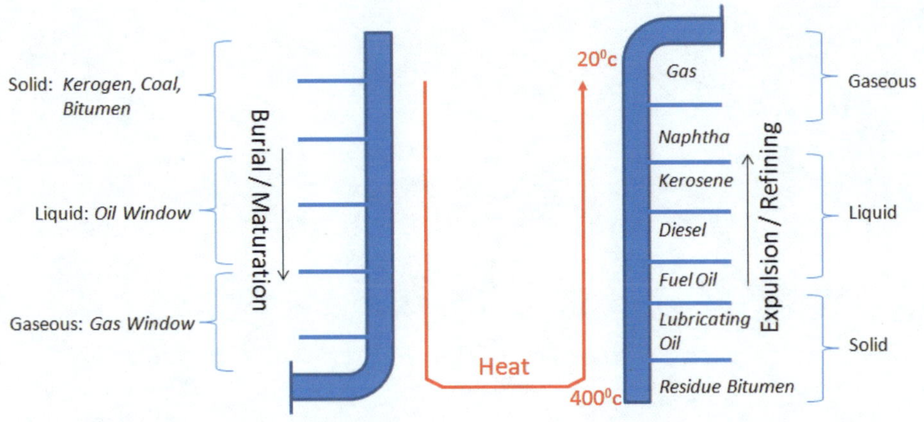

Figure 2.24: The role of heat in maturation of crude oil and expulsion of hydrocarbon, in a geological basin and in artificial fractional distillation

EXERCISES

1. What is sedimentation?
2. Describe briefly the factors that affect the sedimentation process.
3. Why is sedimentary geology important in the petroleum industry?
4. What are sedimentary basins? What roles do they play in hydrocarbon formation and accumulation?
5. List the three types of basins and explain how they are formed.
6. List the various rocks formed by sediments in basins from which the present day petroleum fields are built.
7. What role does subsidence play in basin formation?
8. Describe the role of basin temperature in the sedimentation process. How can this temperature be measured?
9. Describe the role of basin pressure in the sedimentation process. How can this pressure be measured?
10. What role does secondary tectonics play within a petroleum basin?

REFERENCES & FURTHER READING

Benkhelil, J., 1986. *Structure et evolution geodynamique du bassin intracontinental de la Benoue (Nigeria).* Unpubl. Thesis, Nice et Elf Nigeria SNEA (P), v. 2, 231p.

Benkhelil J., 1989. The Origin and Evolution of the Cretaceous Benue Trough (Nigeria). *Journal of African Earth Sciences*, 8, no. 2/3/4, p. 251–282.

Benkhelil, J., P. Dainelli, J. F. Ponsard, M. Popoff and L. Saugy, 1988. The Benue Trough: Wrench-fault Related Basin on the Border of the Equatorial Atlantic, in W. Manspeizer, ed., *Triassic – Jurassic Rifting: Continental Breakup and the Origin of the Atlantic Ocean and Passive Margins*, Part A. Elsevier, Amsterdam, p. 787–819.

Berg, R. R., 1968. Point-Bar Origin of Fall River Sandstone Reservoirs, North Eastern Wyoming, *AAPG Bulletin*, V. 52, p. 2116 – 2122.

Burrus, J., A. Kuhfuss, B. Doligez and P. Ungerer, 1991. Are Numerical Model Useful in Reconstructing the Migration of Hydrocarbons? A discussion based on the northern Vikin Graben. *In*: W. A. England and A. J. Fleet (eds.), *Petroleum Migration.* Geological Society of London Special Publication, 59, p. 89–109.

Damuth, J. E., 1994. Neogene Gravity Tectonics and Depositional Processes on the Deep Niger Delta Continental Margin. *Marine and Petroleum Geology*, v. 11, no. 3, p. 320–346.

Doligez, B., P. Ungerer, P. Y. Chenet, J. Burrus, F. Bessis, and G. Bessereau, 1987. Numerical Modelling of Sedimentation, Heat Transfer, Hydrocarbon Formation and Fluids Migration in the Viking Graben, North Sea, *in* J. Brooks and K Glennie (eds), *Petroleum geology of Northwest Europe*, Graham and Trotman, p. 1039–1048.

Doust, H. and E.M. Omatsola, 1990. Niger Delta, *in* J.D. Edwards and P.A. Santogrossi, eds., Divergent/Passive Margin Basins. *AAPG Memoir* 48, p. 201–238.

Elliot, T., Apps, G., Davies, H., Evans, M. J., Ghibaudo, G. and Graham, R. H., 1985. A Structural and Sedimentological Traverse through the Tertiary Foreland Basin of the External Alps of South-East France. Int. Sympos. on Foreland Basins, Fribourg, 2 – 4 Sept. 1985, *Excursion Guidebook*, P. Allen et al. ed., p. 39–73.

Hyne, N. J., 2001. *Nontechnical Guide to Petroleum Geology, Exploration, Drilling, and Production*, PennWell, 598 p.

Joseph, P. & S. A. Lomas (eds) 2004. *Deep-Water Sedimentation in the Alpine Basin of SE France." New Perspectives on the Grès d'Annot and Related Systems.* Geological Society, London, Special Publications, 221.

Key, R., 1992. An Introduction to the Crystalline Basement of Africa, pp. 29-57 in Wright, E. and Burgess, W. (eds) *Hydrogeology of Crystalline Basement Aquifers in Africa*, Geological Society Special Publication, No. 66, London.

Mulder, T. and P. Cochonat, 1996. Classification of Offshore Mass Movements. *Journal of Sedimentary Research*, v. 66. no. 1. p. 43–57.

Nosike, L., C. Wibberley, C. Jardiné, and Y. Pacalin, 2008. The Role of Tectonics in Geopressure Compartmentalisation and Hydrocarbon Leakage in the Deep Offshore Niger Delta: *YORSGET Extended Abstracts*, Trabajos de Geología, 6p.

Nosike, L., C., 2009. Relationship between Tectonics and Vertical Hydrocarbon Leakage: A Case Study of the Deep Offshore Niger Delta. *PhD Thesis*, University of Nice-Sophia Antipolis, France, 281 p.

Nwajide, C. S., 2013. *Geology of Nigeria's Sedimentary Basins,* CSS Bookshops Ltd., 565p.

Ojoh, K. A., 1992. The Southern Part of the Benue Trough (Nigeria) Cretaceous Stratigraphy, Basin Analysis, Paleo-oceanography, and Geodynamic Evolution in the Equatorial Domain of the South Atlantic. *NAPE Bull.* 7, p131-152.

Osborne, M. J. and R. E. Swarbrick, 1997. Mechanisms for Generating Overpressure in Sedimentary Basins: A Re-evaluation. *AAPG Bulletin*, 81, 1023–1041.

Reijers T. J. A., S. W. Petters and C. S. Nwajide, 1998. The Niger Delta. *in* R.C. Selley, ed., *Sedimentary Basins of the World 3*: Amsterdam, Elsevier, p.151–172.

Reineck, H. E. and I. B. Singh, 1980. Depositional Sedimentary Environments with Reference to Terrigenous Clastics, *Springer-Verlag*, Berlin, Heidelberg-New York, 549 p.

Selby, M.J. (1985). *Earth's Changing Surface.* Clarendon Press, Oxford, 607p.

CHAPTER THREE

STRATIGRAPHY

3.0 Layering of Rocks

Stratigraphy is one of the major building blocks of geology. Geology students become thrilled when they understand that rock or the ground below their feet is layered, and that geologists could establish the relationship of the layers among themselves. A geologist who specialises in stratigraphy is known as *stratigrapher.*

Over several millions of years, sediments deposited in basins have been compacted and some underwent chemical reactions to form present day rocks and minerals. Stratigraphy attempts to unravel their association spatially, vertically and laterally. To achieve this, the stratigrapher carries out a logging process where the properties of each rock body are recorded at different locations. Then similar rocks are tied, or linked together, across exposed and non-exposed regions of the earth to correlate the layers. Modern solutions, including seismic exploration techniques and software, have made the work of the stratigrapher much easier.

This chapter will look at:

- The concepts and principles of stratigraphy
- Different types of stratigraphy:
  - *Lithostratigraphy*
  - *Chronostratigraphy*
  - *Biostratigraphy*
  - *Seismic Stratigraphy*
  - *Sequence Stratigraphy*
- How stratigraphy relates to other geoscience disciplines
- Use of stratigraphy in correlation and petroleum prospecting
- Applications of stratigraphy to the petroleum industry

Stratigraphy helps to link rock layers to not only local, but also global vertical and spatial sedimentary deposits from the beginning of geologic time. It is the knowledge of stratigraphy that differentiates the geologist from others, in terms of how they interpret the physical rock masses and mountains around and below them.

3.1 The Concepts of Stratigraphy

Stratigraphy is the branch of geology that deals with the study of *rocks in layers*. This layering can be observed *vertically* (from bottom to top) or *spatially* (array of rocks having space between them) when studies are carried out on the rocks. These studies are made on rocks that have been formed since geologic time (Berggren, et al., 1995), from the *Precambrian* to *Holocene*, attributed to be a period of over *4.5 billion years*. Local basin studies however take place in rocks that are between a few million years to thousands of years.

Different scenarios could result in layering of rocks. Figure 3.1a shows different kinds of layering in sedimentary rocks, the form in which they are naturally deposited. Most stratigraphic studies are directed towards sedimentary rocks as these rocks occur in layers. However, rock layering can also be observed in igneous rocks as shown in Figure 3.1b. In this case, the layering of the igneous rock is within existing beds, where volcanic eruptions take place from time to time, causing lava (molten magma – liquid rock) to flow and solidify before and after deposition of sediments.

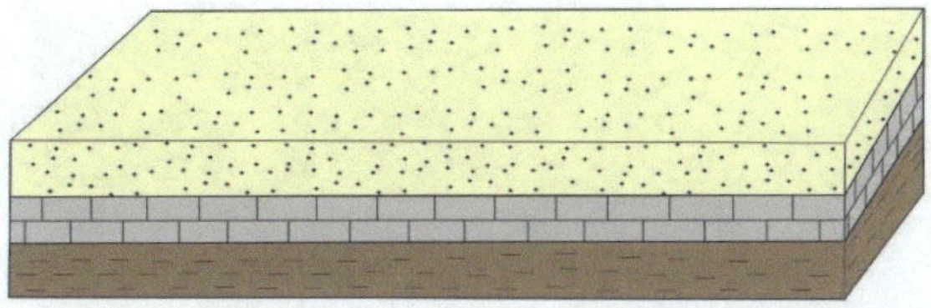

a. Normal layering of beds – undisturbed strata

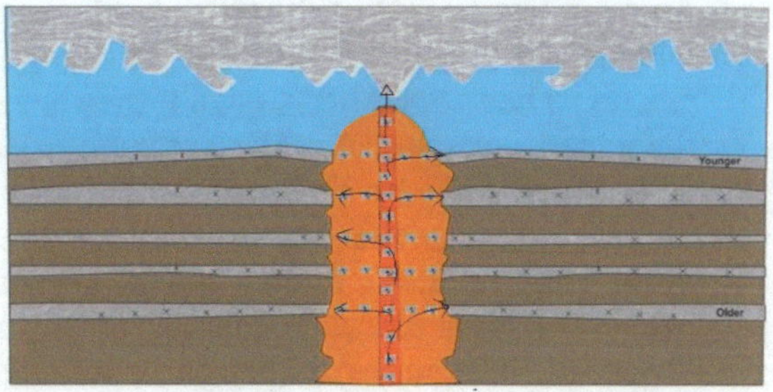

b. Layering igneous rock within existing beds

Figure 3.1: Strata as layered rocks

3.2 Laws of Stratigraphy

Several laws have been developed to show the relationships among strata. One of such laws is the **cross-cutting** relations. A geologic feature is younger than the bedding units it cuts. For example, faults are younger than the rocks they cut; accordingly, if a fault is found that penetrates some formations but not those on top of it, then the formations that were cut are older than the fault, and the ones that are not cut must be younger than the fault.

Likewise, intrusives are younger than the rocks they cut. An igneous intrusion (projection from the base in Figure 3.2) is an igneous rock formed within rocks already existing. These pre-existing rocks have cracks and magma from the subsurface may flow through these cracks and try to find their way to the surface. When it is impossible to get to the surface, they begin to solidify under suitable conditions, leading to the formation of intrusives. If an intrusive cuts across already existing rocks, it means that the rocks are older than the intrusive, or the intrusive is younger than the rocks it cuts, as the rocks were already there before the intrusion took place.

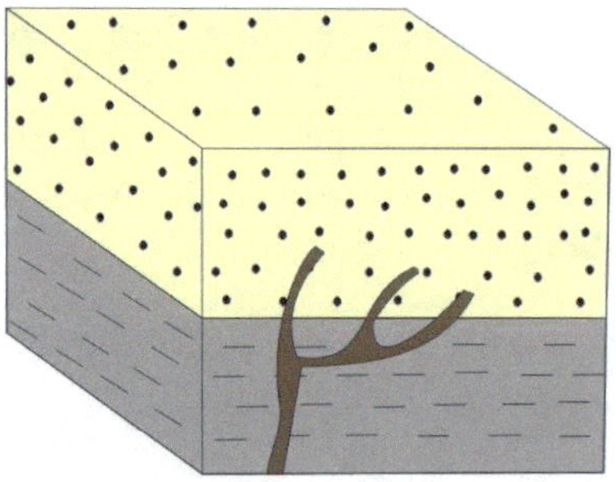

Figure 3.2: Cross-cutting relationship, showing that the intrusion is younger than the beds it cuts

Another law is that of **original horizontality**, which states that deposition of sediments occurs as horizontal beds, even if they are later deformed or folded. All sedimentary (or layered) rocks were horizontally due to effect of gravity on sediments as they settle in a basin (Figure 3.3a) before Earth movement processes take place. If folding or tilting is observed, it signifies a change in its original deposition (Figure 3.3b).

Figure 3.3a also describes the **law of superposition**, which states that in an *undisturbed sequence of strata*, younger beds lie on top of older beds. This means that the natural process of deposition is one in which a rock body deposited on top of another rock body must be younger than the rock body on which it is deposited, and the rock on which deposition took place must be older than the rock deposited on it. Remember that this only holds true in an "undisturbed sequence of strata" where the natural deposition has not been tilted. For a tilted or overturned sequence of strata, the orientation may change, and younger rocks can lie at the bottom of older rocks.

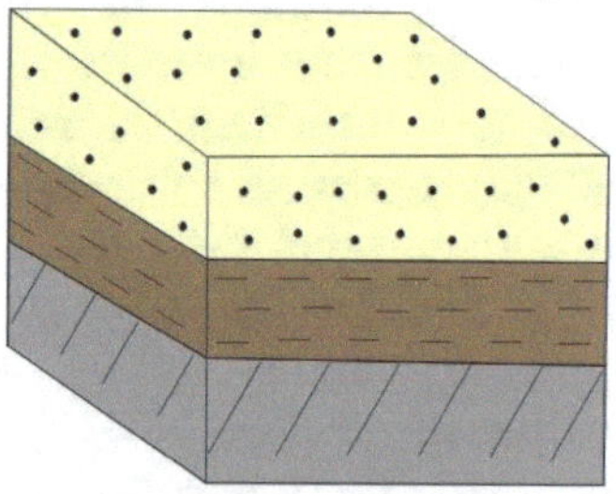

a. Horizontal strata after deposition

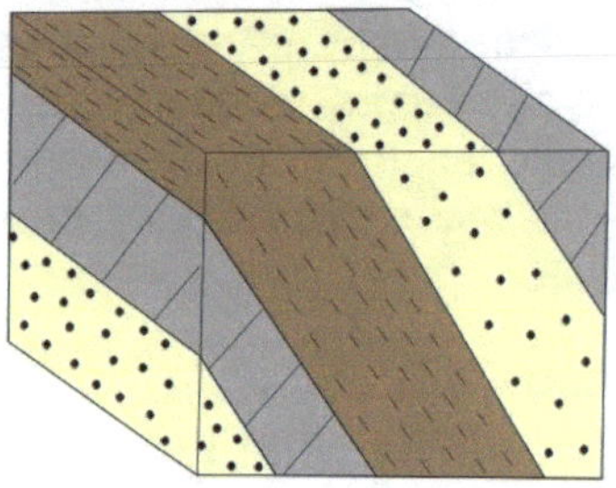

b. Tilted due to deformation

Figure 3.3: Strata originally deposited horizontally and later tilted during deformation

Sometimes a sedimentary unit is completely turned upside-down during folding, this is known as *upturned strata*. Yet stratigraphers use several indices to know which way was up. One of such indices is *mudcrack*, which leaves fissures filled on one side by sediments. When a unit is upturned, the filled side which was once facing upwards now faces downwards, as in the Figure 3.4.

Mudcracks are seen where the ground surface having mud solidifies and cracks (map view in Figure 3.4a). The cracks in the mud have some form of depressions referred to as fissures (as seen in the X cross-section of Figure 3.4b). These fissures are usually filled with sediments. When a sedimentary unit is completely turned upside down (upturned) during folding, the normal orientation of the mudcrack changes, and the orientation of the fissures filled with sediments is upside down (as seen in the Y cross-section of Figure 3.4b).

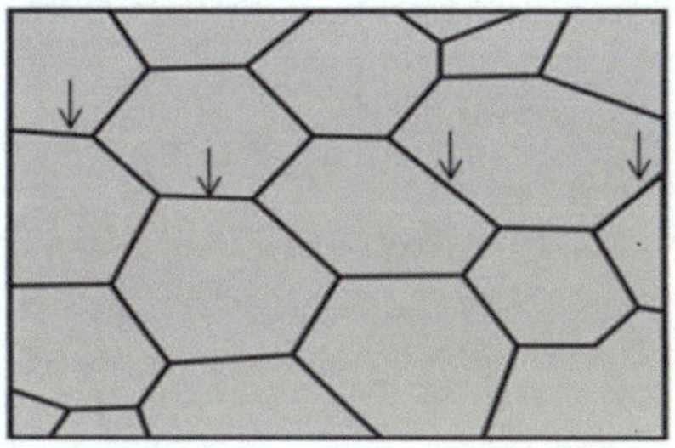

a. Mudcrack fissures

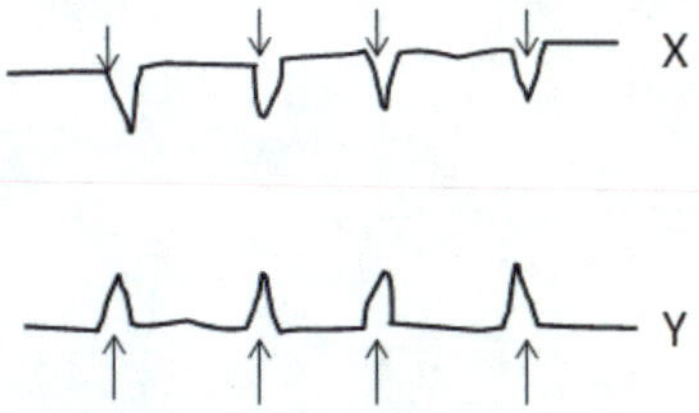

b. upside (X) and upturned bed (Y)

Figure 3.4: Mudcrack fill showing normal and upturned bed

When a rock unit is tectonically disturbed, the stratum is no longer horizontal, thereby indicating an interruption. Horizontal strata extend all over the space available for deposition. This introduces the law of **lateral continuity**, which holds that layers of sediment initially extend laterally in all directions; they are continuous except if terminated / truncated by structural event (example the flank of a basin or a fault as shown in Figure 3.5) or by erosional processes.

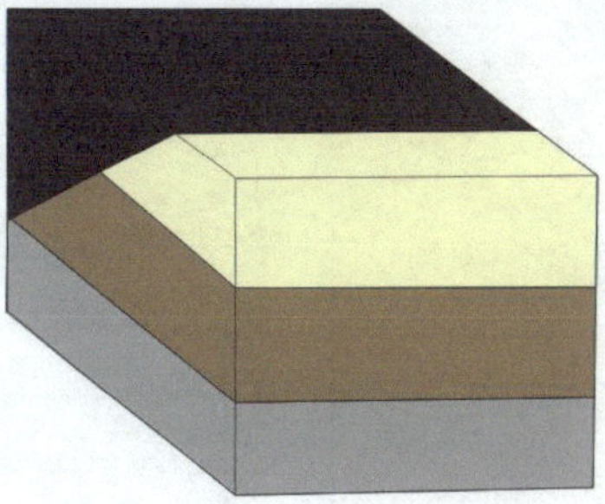

Figure 3.5: Lateral continuity of sedimentary layer until truncated by a fault (dark posterior region)

Another law, that of **faunal succession**, holds that *fossils* or *fossil groups* are succeeded by other younger groups through time. Faunal, or fauna, refers to animal life in a particular region or period. This means that since certain organisms existed at the same time or period throughout the world, their presence or (sometimes) absence may be used to provide a relative age of the formations in which they are found (Figure 3.6). If the age of the fossil is known, the age could be attributed to that of the rock, as the organism whose fossil is observed in the rock existed during the time of the deposition of sediments that formed that rock.

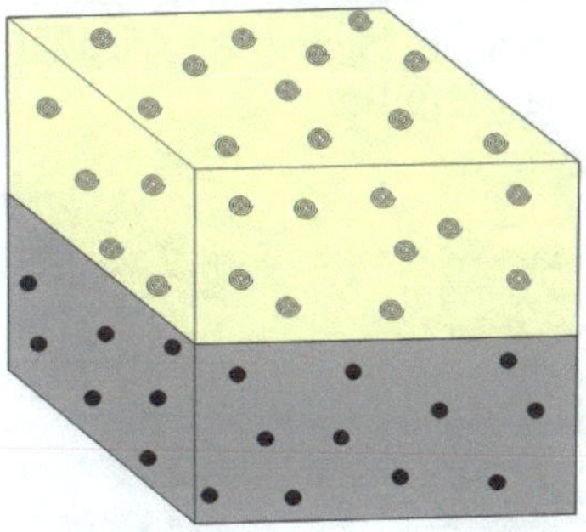

Figure 3.6: The fossil group in the upper unit is younger than that in the lower

Different types of stratigraphy use different methods to study the arrangement of sedimentary deposits in terms of time, lateral and vertical position, when and where they occur, and then identify missing bed. A bed may be missing due to erosion or it was just not deposited alongside others some distance away. A missing *stratum* is known as **hiatus**.

3.3 Types of Stratigraphy

There are many types of stratigraphy, including some more specific aspects such as *Pedostratigraphy* (soil), *Chemostratigraphy* (chemical) and *Magnetostratigraphy* (polarity). In fact, feature that could be tied to given strata, consistently, is a form of stratigraphy. The most popular stratigraphy types (which are often used in the oil industry) are discussed below.

Lithostratigraphy

Lithostratigraphy is based on lithology or rock type. It is probably the most basic of all stratigraphy. Rock units are considered in terms of their *lithological characteristics*, that is the physical contrast between the various rock layers with depth. *Logging* is used to carry out physical descriptions of rocks, where field studies are done from the bottom to the most recent deposit at the top in *unturned sediment.*

Reverse is the case in *upturned sediment.* Lithostratigraphy alone cannot give an indication of how these sediments were deposited over time; it needs to be associated with other forms of stratigraphy.

It is possible to link rocks deposited together at different places within a basin, based on their lithostratigraphic association. This process is known as correlation. Two lithostratigraphic logs obtained during field studies by the author is shown in Figure 3.7, drawn by observing lithological units.

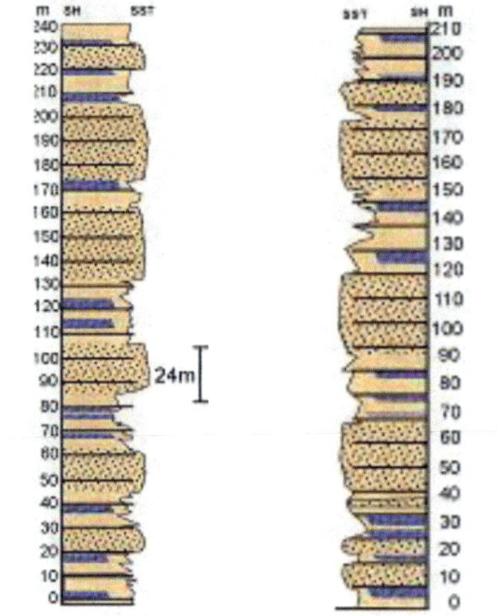

a. Left Wall b. Right Wall

Figure 3.7: Correlatable lithostratigraphic logs, drawn by observing lithological units.

These lithostratigraphic units are similar, and could therefore be *correlated* (establishing a relationship between them). *Attempt to correlate the strata in the figure from the left to the right walls.*

Chronostratigraphy

Chronostratigraphy is simply known as *time stratigraphy*, as it deals with the time at which the different rock layers were deposited. Rocks are therefore analysed in order to link them to a given time. It is often an end result of other types of stratigraphy; after physical description of the rocks with depth, and showing their vertical relationship to each other, it becomes necessary to know when (i.e. what time they were deposited).

The geochronology of the Earth (Figure 3.8) helps to date and determine the time sequence of events in the Earth's history. It is that part of geology that deals with time scale. It also shows long-term (black) and short-term (red) eustatic curves to the right.

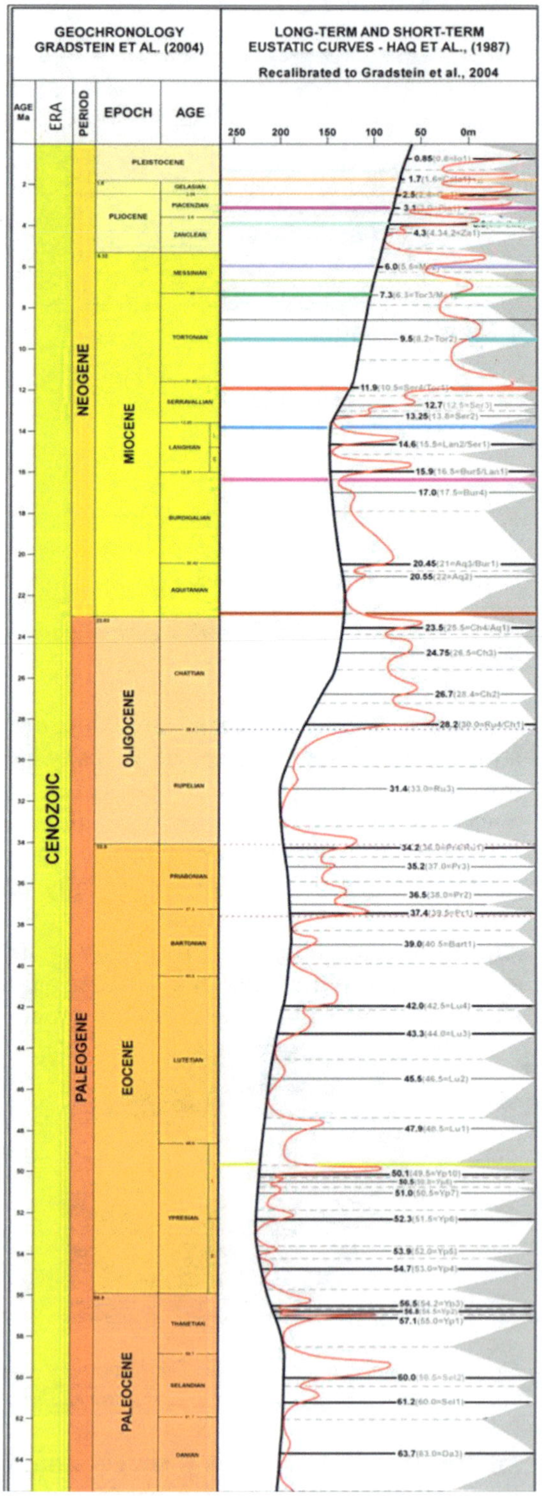

Figure 3.8: Example of chronostratigraphic chart of the Cenozoic Era, showing long and short-term

eustatic curves (Haq et al., 1987; Gradstein, et al., 2004)

These curves mark a uniform worldwide change in sea level, attributed to fluctuations in the amount of water due (especially) to melting of the polar ice caps, or change in the water retention capacity of the ocean basins known as **eustacy**. The varying retention capacity of the ocean, leading to varying water levels and sedimentation in the ocean, results in another type of stratigraphy. This type, using sea level rise and fall to determine sedimentary packages, is known as *Sequence Stratigraphy* (which is discussed below).

The Niger Delta Basin for example consists of *Cretaceous* to *Holocene* marine clastic sediments that overly the oceanic and fragments of continental crust (Burke, 1972; Corredor *et al*., 2005). The *Tertiary* section of the Niger Delta is divided into three formations, representing *prograding* depositional environments: the Akata, Agbada and Benin Formations.

The Akata Formation, at the base, is the oldest and is a predominantly shaly marine deposit, with a thickness of about 2000 m at the most distal part of the delta and up to 7000 m thick beneath the continental shelf (Avbovbo, 1978; Doust and Omatsola, 1990; Reijers et al., 1997). It is hot shale at depth, serving as a decollement layer for the major thrust faults, and may as well have been source rocks. It is about 5000 m thick in the deep-water fold and thrust belts.

The *Agbada Formation* whose deposition began around the Eocene is a fluviodeltaic deposit; it overlies the Akata and consists of a series of alternating sands and shales. The Agbada formation is more than 3500 m thick and is the most petroleum-bearing unit. It could serve as both source and reservoir rocks. The *Benin Formation* is the uppermost and youngest of these formations, and is a fully continental unit found onshore and in the coastal regions, including alluvial and upper coastal-plain deposits that are up to 2000 m thick (Avbovbo, 1978). It is neither source nor reservoir rock.

Chronostratigraphy of events determines the relative order and assigns relative ages to past events that took place. In Figure 3.9, a fault cuts across and displaces several layers of rocks (grey, dark brown and dull green from top to bottom), but does not cut across other layers (yellow and brown, deposited after). The fault is therefore older than the upper layers of rocks it does not cut across, but younger than the bottom layers it does cut across. The upper unit (yellow) is obviously above and younger than the fault, but the fault is younger than the brown sediment as it was already there before being offset by faulting. This is a **relative (**not **absolute) age** dating of the faulting event; it does not provide numeric dates for the rocks in this case.

Chronostratigraphy assigns a time to a geologic unit either *relatively*, that is saying which one is younger or older than the other, or it could be *specific*.

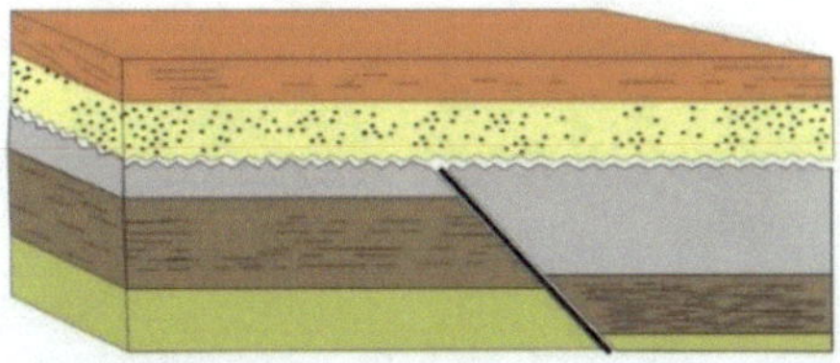

Figure 3.9: Relative age dating of fault based on chronostratigraphy of events

Specific timing of geologic unit, linking it to global depositional history in time, is known as **absolute timing** or **absolute dating**. Radioactive and isotope studies have been helpful in absolute dating, just as biostratigraphic remains have been helpful for relative dating in specific basins. Some fossils occur at specific time in history and are used for absolute dating.

Biostratigraphy

Bio- is a prefix pertaining to life; therefore, *biostratigraphy* refers to stratigraphic studies pertaining to life. It makes use of the principle that different organisms lived at different depositional times, and the evidence of their existence (**fossils**) occurs in different rock units from ancient to recent deposits. This evidence could be remains of the organisms that existed for a very short period of time (**index fossils**) or it could be an evidence of its existence such as footprints (**trace fossils**). Whichever is the case, the remains of the organism in the period that the organism thrived most is used to carry out biostratigraphic studies in a particular basin, to stratigraphically place the units in the order in which the organisms are found. A stratigraphic column is thereby established for the basin.

Shells of different gastropods (such as snails) are easily identifiable and big enough to be seen with the naked eye, but microfossils are seen only when magnified under a microscope (Figure 3.10).

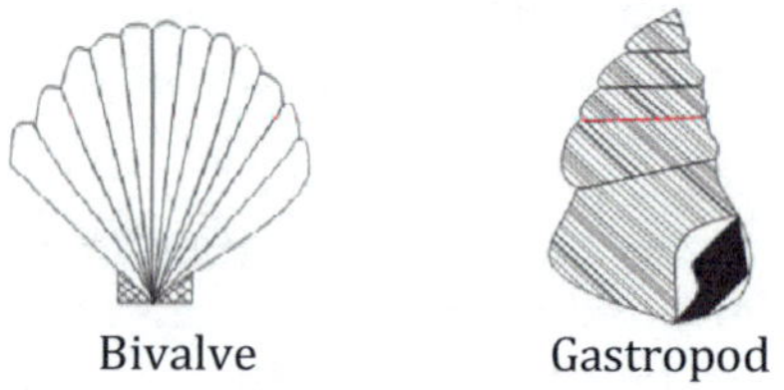

a. Examples of fossil

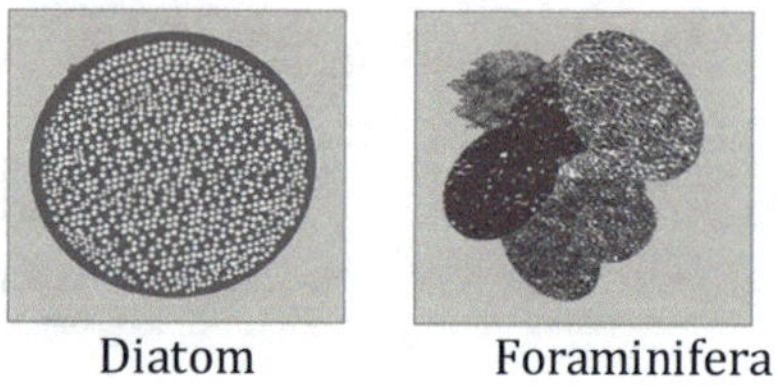

b. Examples of microfossil

Figure 3.10: Fossils

Biostratigraphy has been extended to not just establishing a *stratigraphic order*, but to the *evolution* of the organisms that transcended from their basic forms to a different form in the subsequent units (as time itself evolved). Therefore, for adherents, it has helped in developing the theory of evolution despite its several limitations and missing links. Because microorganisms live in a particular environment, biostratigraphy has been very useful is determining *depositional environment*, such as **terrestrial** versus **marine**, **water depth**, **saltiness** and **oxygen level** of **environment** and even **polarity** or **magnetic cycles.**

Through the concept of evolution of species with vertical layering, fossils can indicate which rock is older or younger, and relative dating can be used to establish the sequence of strata. As shown in Figure 3.11, if the fossil (Z) in the sandstone (yellow bottom unit) is known to occur in an older era (3), it means that the unit is older than the subsequent brown shale (2) and grey limestone unit (1), even if the vertical stacking could not be seen. The development of the fossils are expected to evolve from bottom to the upper younger units.

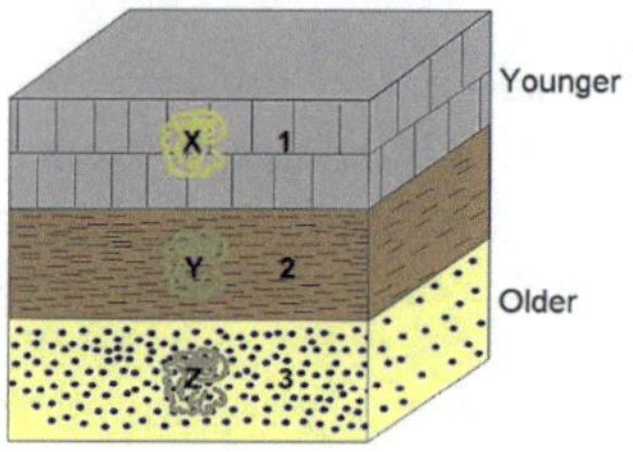

Figure 3.11: Relative age dating using fossils evolving from bottom to the upper geologic units

Organisms whose existence was *short,* coming in and becoming extinct in a short geological age, is better for biostratigraphy as they can be associated to a given period, as compared to those who lived across all the era. This means that the units in which they are found can easily be associated to a particular period (Blow, 1969). Ammonites, foraminifera, and other planktonic and benthic marine fossils are examples of fossils used for biostratigraphy in the petroleum industry (Table 3.1). These are however, not very available in *high-energy environments* such as fluviatile and lowstand systems tract, and desert or dunal deposits, where they cannot be deposited, or where they are easily destroyed.

Table 3.1: Some terrestrial and marine fossils used in biostratigraphy

Terrestrial - Fluvial - Lacustrine	Marine	
	Benthic	***Planktonic***
Spores Charophyta Gymnosperm pollen Angiosperm pollen	Ostracods Benthic forams Dasycladacean algae Phodophyta algae Codiacean algae	Conodonts Chitinozoans Acritarchs Radiolarians Dinoflagellates Nanofossils Planktonic forams Calpionellids Silicoflagellates

		Diatoms

Although petroleum wells provide notable fossil remains that have been helpful for correlations within petroleum basin, the major challenge in extracting fossil from petroleum well is that they are reworked before getting to the surface; the larger ones are easily destroyed during drilling. This is why *microfossils* (0.001-1 mm), *nanofossils* (0.005-0.010 mm), and *palynomorphs* (0.005-0.500 mm) are more widely used in the industry. Seismic stratigraphy is now helpful where biostratigraphy was used in the past, and the two combined can be a very valuable tool.

Seismic Stratigraphy

Seismic stratigraphy uses other features visible on the seismic data, apart from the sediments itself, to establish a stratigraphic sequence. These features include:

1. Seismic surfaces that are equivalent to time
2. Angular relationship of the seismic markers
3. Seismic signatures

If most *seismic reflectors* are time surfaces representing a deposit in time (Payton, 1977; Mutti, 1977, 1985), then *angular relationship* to each other can give their stratigraphic order. *Seismic signatures* (chaos, smooth, reflectors) give indication of what deposited them and in which kind of energy environment. Since *high energy* would normally precede a *low energy*, we could establish stratigraphic relationship in space and in time.

Seismic stratigraphy rely on seismic responses and uses images that are generated from waves that travel into the subsurface and are reflected (or refracted) back to the surface. The method depends on features that are visible on seismic cross-sections, traces that delineate layers, rather than the sediments itself, to establish a stratigraphic sequence. These features include:

1. **Seismic surfaces**: When a wave is sent into the Earth it will continue to travel (within a particular rock layer, the first layer) until it gets to another layer. Because the second layer is denser due to compaction or has a different property and therefore has a higher or different velocity from the first layer, it will cause some waves to be reflected back to the surface. These reflective surfaces are equivalent to time; it marks the period in time in which the first layer ceases to exist, and the period in which the second layer begins to exist. Some waves still go through and reflect at the third and subsequent layers below. The interface between two layers often results in reflection or refraction of waves, indicating layers and stratigraphy.

2. **Angular relationship of the seismic markers**: If most *seismic reflectors* are time surfaces representing a deposit in time, then their *angular relationship* to each other can give their stratigraphic order.

3. **Seismic signatures**: This can give an indication of what led to the deposition of the units, and in which kind of energy. These signatures could be rough or smooth, indicating *high* or *low energies* respectively. This often alternates, indicating different episodes of deposition and may help to establish a stratigraphic relationship. On seismic data, chaotic or high amplitude, continuous or discontinuous signatures, may be used to distinguish among facies. For example, chaotic or high amplitude may be indicative of high energy.

Consider Figure 3.12, the downlap marks the termination of younger overlying strata that are more steeply dipping against an older surface (or an older underlying stratum having lower dip). The toplap marks the termination of a younger surface (or overlying shallow dipping younger stratum) against more steeply dipping underlying older stratum.

a. Downlap b. Toplap

Figure 3.12: Seismic signatures related to stratigraphy

Seismic stratigraphy is often associated with *sequence stratigraphy*, as most of the *depositional packages* (or depositional systems) described in sequence stratigraphy are visible on seismic data. One aspect of sequence stratigraphy whose application is better achieved from seismic data is the **geotime map**, where each seismic marker on a seismic section (often representing the near shelf zone) is flattened and restored to show the time to which they correspond.

In the sequence near a shelf break, sediments may be deposited after an erosion or directly on sediments after they have slid and are now at an angular relationship with new bed (as in the relationship between younger beds 10 to 13 deposited above older beds 1 to 9) in Figure 3.13. Where no stratum is missing between 9 and 10, it is a *continuous concordance* and continuous in geotime. Going from 9 to 11 and to 12 and 13, on the right side of the figure for example, then a time is missing. A good geotime construction will leave an empty space to show or account for the missing time when that deposit was emplaced before erosion, or was never emplaced at that particular location.

Seismic Imagery-Aided Interpretation

The use of seismic has made stratigraphy a lot easier, as seismic imagery shows the layering of sedimentary rocks in a basin and any intersection by faults or hard rocks. Seismic data is useful for seismic stratigraphy and *near shoreline* (along the edge where a body of water shares boundary with land) sequence stratigraphy. Hydrocarbons are deposited in some sedimentary sequences more than in others, so understanding sequence stratigraphy and the various *sequence tracts* is useful in petroleum geoscience.

Sequence, Geotime, Hiatus

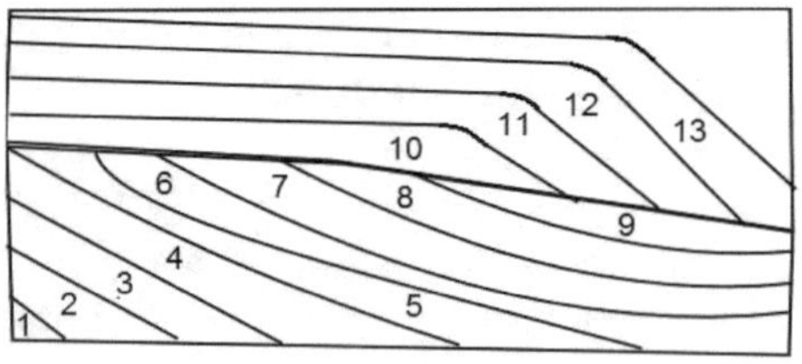

Figure 3.13: Seismic stratigraphic display of sequences around the shelf break (as seen in a near shore seismic image)

Sea level *rise* and *fall* control basin stratigraphy and their cycles are used to explain stratigraphic relationships and depositional environments. These are registered by seismic signals and their interpretation is aided by seismic imagery.

In Figure 3.13, sequences align in the same direction around the *shelf break*, of the continental margin (end of the continental shelf and beginning of the continental slope). Sediments 10 to 13 may have been deposited after erosion or directly on sediments 1 to 8; they downlap on sediment 9. Sediments 1 to 9 have slid or tilted due to a slope and are now at an angular relationship with new beds. This deposition is concordant on the left side, as it is continuous in geotime and no stratum is missing from 1 to 13. It is discordant on the right hand side, where 9 makes contact directly with 11, 12 and 13.

Sequence Stratigraphy

Sequence stratigraphy is the association of strata based on *eustatic sea-level changes* and *tectonic activity* that provided the *accommodation space* in which they were deposited. It subdivides and links sedimentary deposits into different stratigraphic units based on changes in relative *sea level*, the rate at which sediments are supplied and the rate of change in the *accommodation space*. The major elements of sequence stratigraphy, tools and its application as they are used in the petroleum industry, are:

1. The rising and falling sea level
2. Agents of deposition, energy and accommodation
3. Depositional signatures on seismic
4. Basin type and regional tectonics
5. Systems tracts and depositional packages

The Rising and Falling Sea Level

When the sea level rises, energy reduces and there is deposition of fine-grained sediments, such as clay (and silt). When they fall, energy is higher and deposition of coarse-grained sediments takes place, such as sand (and gravel).

Agents of Deposition, Energy and Accommodation

Sea level rise because there is possibly subsidence of the oceanic crust with respect to the inland, thereby creating accommodation space for deposition

Depositional Signatures on Seismic

As sediments go downdip at the transition between land and sea known as shelf margin, they form a signature on seismic known as downlap. They also form rough **chaotic** signature in high-energy sand deposit during low sea level and **calm** signature in low energy shaly or *hemipelagic deposit* (marine sediment that consists of fine-grained soil particles with constituents of plants and animals) during high sea level.

Basin Type and Regional Tectonics

Platforms basins (continental area covered by relatively flat or gently tilted sedimentary strata overlying earlier deformed basement rocks), **passive margin** (transition between oceanic and continental lithosphere that is not active) and **active margins** (active regions where the continental lithosphere deforms against the oceanic plate) will deposit different kinds of sediments. Plate movements, apart from its crushing or sliding effect of massive rock bodies, will ultimately create gravity tilt by relative tectonic uplift. Deposits take the shape of this tilt and produce *syndeformational* deposits.

Systems Tract and Depositional Packages

In the overall, sequence stratigraphy identifies the various deposits affected by the stratigraphic elements above. They are grouped from Lowstand Systems Tract (LST) to Maximum Flooding Surfaces (MFS). These packages will be seen if a section is taken along the depositional track, within incisions made by flowing turbidity current. Figure 3.14a shows how the incision is made when sea level is low and the current has high energy. They deposit the *lowstand systems tract.*

When sea level is high and the sediments are further to the deep offshore, shaly or hemipelagic sediments are deposited. These packages will be seen if a section is taken along the depositional track, within incisions made by flowing turbidity current (Stanley, 1975; Sultan et al., 2007). They will all fall on the lower part of the seafloor, such that you can only identify them by their lithology or types and time in the eustatic cycle. That is the aim of sequence stratigraphy.

Seismic stratigraphy is very useful in the interpretation of sequence stratigraphy. Figure 3.14b shows how it can indicate rising and falling sea level. Tracts 1 - 6 are different sediments deposited; while the level was falling, the shelf moves towards the sea (2 and 6) and while the sea level was rising, the shelf moving towards land (1, 3, 4 and 5) relatively. In analysing layers of deposits, stratigraphic analysis also identifies periods of *non-deposition*, known as *hiatus*, and periods where *erosions* have removed a roughened layer most likely tilting or being uplifted before subsequent deposition, known as *discordance* (a discordance is shown between the yellow and brown bed in Figure 3.17).

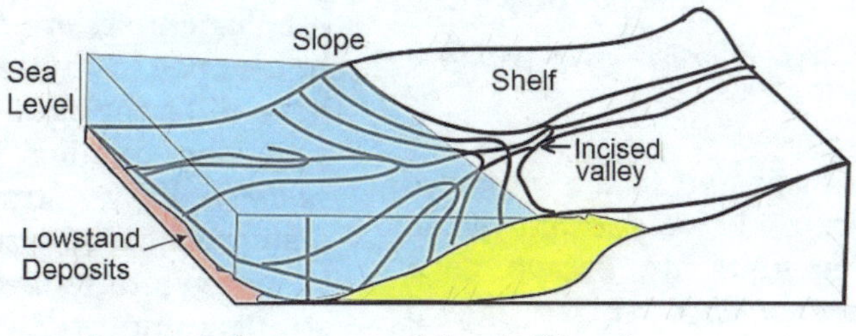

a. Different sediment packages around the shelf break

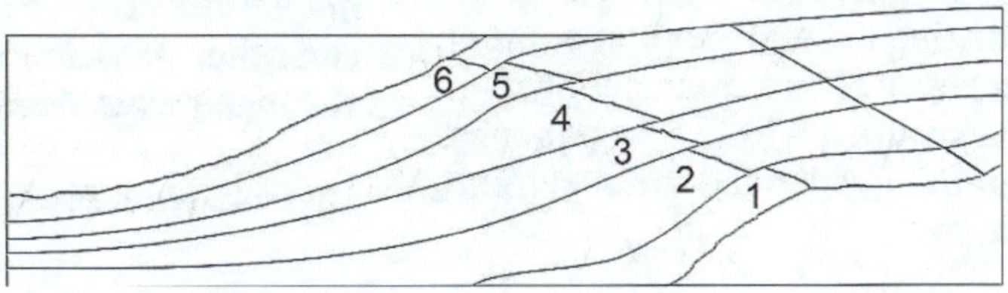

b. Sedimentary deposits during sea level changes.

Figure 3.14: Sea level rise and fall, stratigraphic cycle deposits and seismic signatures

The rise and fall of sea level has occurred in enormous cycles. A cycle represent a full period of sea rise and fall, as shown in Figure 3.15. However, there are smaller rise and fall within the main cycle, like a shorter up and down fluctuation during a major cycle. As such, there are shorter cycles on the large cycles.

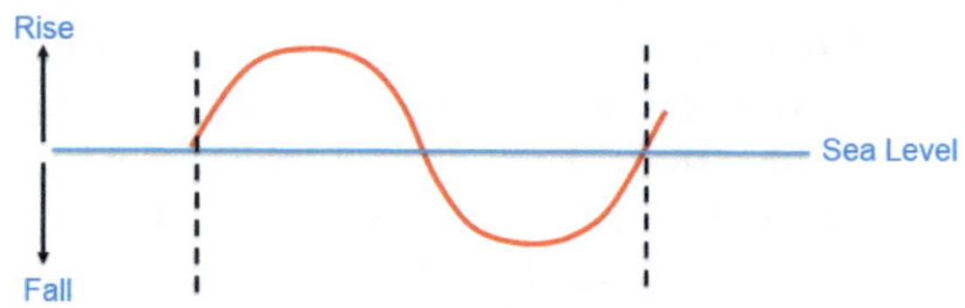

Figure 3.15: A cycle indicating a full period of sea level rise and fall

At least five orders of sea level cycles have occurred, with the shortest occurring every few tens of thousands of years. The longest is known as the first order cycle, occurring for over fifty million years; the largest cycle is known to occur every hundreds of millions of years. Second, third, fourth and fifth order cycles occur in about 3 to 5, 0.5 to 3, 0.2 to 0.5, and 0.1 to 0.2 million years respectively (Figure 3.16).

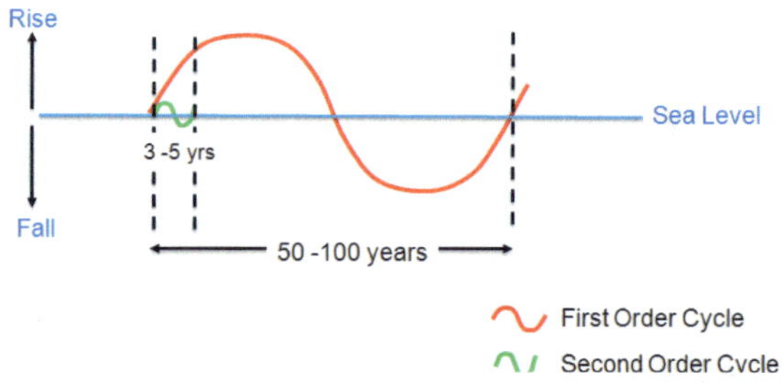

Figure 3.16: Smaller second order cycle within a major first order cycle

The *sequence* is the basic unit in the sequence stratigraphic system. Relatively conformable and related bedding units are known as a *parasequence*. A *parasequence set* describes a succession of genetically related strata that stacks to form a particular pattern. Each bedding unit is distinguished from one another by differences in physical and chemical properties, which can be grouped within the sequence. A sequence can be subdivided into *systems tracts*; these include:

- **lowstand** systems tract (which forms when the rate of sedimentation is higher than the rate of sea level rise, where seal level is low and rivers channelling sediments to the sea);
- **transgressive** systems tract (which forms when the sea level is rising and moving landward);
- **highstand** systems tract (which forms when the rate of sea level rise reaches a pick and only hemipelagites is deposited; it may last for a long period of time to deposit a condensed section made of thin clay or shaly sediment representing a long period of time);
- **regressive** systems tract (which forms in the distal marine part of the basin when the sea level begins to fall and moving seaward).

One major advantage of sequence stratigraphy is that it can help to predict depositional packages and to some extent petroleum results. Once a clear depositional sequence is established vertically and laterally, it highlights the cycle and what is next in the sequence could be known based on local sea level changes and global eustatic curves.

3.4 Applications in the Petroleum Industry

Relationship to other Geoscience Disciplines

It is impossible to discuss stratigraphy fully without relating it to the other geoscience disciplines. The basic layers that form stratigraphic patterns are *lithologic*, and within the lithologic units are *biological* remains that can be grouped in stratigraphic layers. These layers have different ages, leading to a *chronological* sequence relating to basin analyses as in *sedimentology*. Stratigraphic layering is often disturbed by *structural* evolution of a basin, requiring a tectono-sedimentary compliment in stratigraphic studies. This would also be coupled with *geochemistry* and eventually field *geophysics*. The relationship of stratigraphy to the other geoscience disciplines has influenced the major classification of stratigraphy.

Hydrocarbon Prospect

Hydrocarbon *reservoirs* are formed in particular types of sediment, essentially *clean sands* and *carbonates*. These sediments are deposited in given environments identifiable from sequence stratigraphy. The *lowstand systems tract* for example is characteristic of high-energy sands deposited by currents during sea level fall. The *maximum flooding surface* is known to form condensed organic matter rock unit, which could form source rock as well as regional vertical seals. This unit is also a good *marker bed* as it retains important microfossils for identification and correlation of strata.

In Figure 3.17, a distinct fossil is seen in each rock unit, and if such units are spread all across the world then they can be easily identified and compared as such. Such comparison, termed *correlation*, can be done using a marker bed (distinctive rock layer easily identified), where adjacent rocks above and below the marker beds can then be correlated. Microfossils are more accurate for correlation especially in regions complicated by faulting, facies changes, or unconformities.

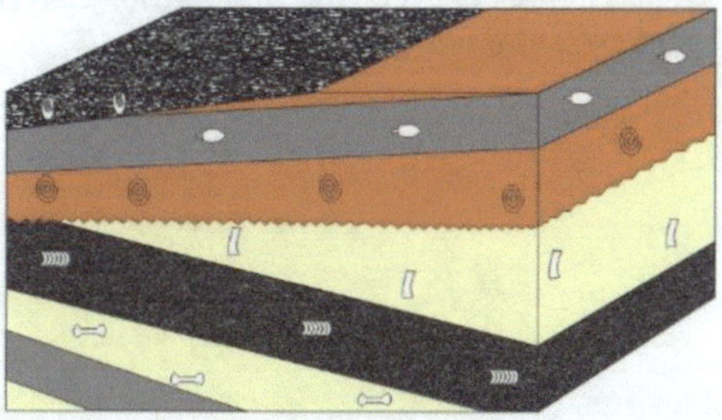

Figure 3.17: The use of microfossils for correlation between wells across a discordant unit

Other aspect of stratigraphy that is useful is the biostratigraphy, which helps to identify packages deposited at the same time when petroleum source rocks were prevalent. Depositional environment identified from biostratigraphy could tell where we are in a basin and help to limit uncertainties. The use of stratigraphy is so fundamental that it is part of every crucial aspect of petroleum basin evaluation and field production. While producing a field, *depletion* (reduction in hydrocarbon volume due to extraction), *injection* (production support by water flow techniques to improve recovery) and *reserve revisions* (improvement made on a declining field) are all tied to given stratigraphic units.

Correlation and Well Development

Almost every process followed during well development is based on important correlations (of reservoirs, markers, fluid continuity) and these are all tied to stratigraphy. In fact stratigraphy forms a major part of petroleum evaluation and chronostratigraphy often precedes any exploration study, and a *frontier exploration zone* (an area known or suspected to have petroleum systems but yet without a commercial discovery of hydrocarbons) for example, would have to be tied to petroleum generation ages. It is so important that sometimes wells are drilled just for stratigraphic confirmation reasons. Well may cut across marker beds that helps in correlation. Marker beds are identifiable even if some parts of the sediment have been eroded above or below (Figure 3.18).

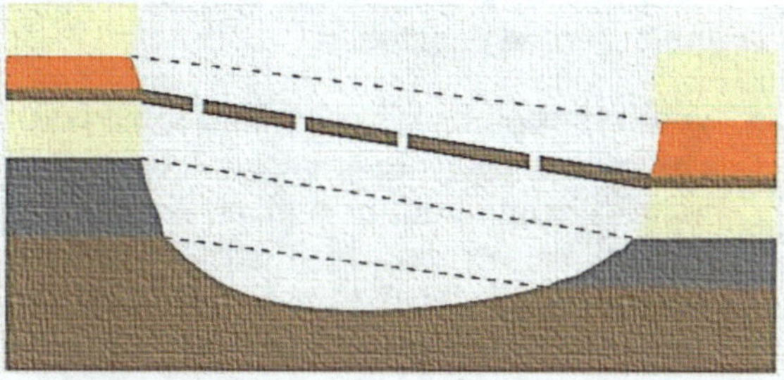

Figure 3.18: Use of a marker bed to start correlation

A key horizon, the top or bottom of a thick, distinctive rock layer, can be used to start a correlation. In areas complicated by faulting, facies changes, or unconformities, an accurate correlation is done by matching microfossils between wells. On the surface, correlation of sedimentary rocks is often started with a marker bed. Marker beds can also be used in subsurface correlation. A marker bed is a distinctive rock layer that is easy to identify, such as volcanic ash layers, thin beds of coal or shale, limestone or sandstone, or fossil zones, as long as they are distinct and easily identifiable. This identifiable characteristic makes correlation easier, even when some parts of the sediments have been eroded.

In Figure 3.18, it can be seen that the marker beds are not at the same depth and are distances apart, but their very nature makes them identifiable and useful for correlation purposes. After correlating the marker bed, the rock layers above and below the marker bed can then be correlated based on a physical similarity and their position in the sequence of layers. This is made by hanging the well logs from the same marker in each well and associating similar responses on the logs to delineate the rock units. Hence, correlation can be done on the (yellow) rocks below and (orange) rocks above the marker bed.

Remember the principle of *original horizontality*, which leads to the understanding that all stratigraphic units are deposited in horizontal layers. It means that for a deformed or folded strata, because the marker bed was originally deposited horizontally, the stratigraphic cross-section can be restored such that the maker bed is made to become horizontal. Where the marker bed is restored, it is said to be a **flattened horizon** (Figure 3.19). It shows the changes that have occurred to the units above and below, with respect to that maker in an undisturbed position.

A stratigraphic well or stratigraphic test well is drilled to test the model or to identify the stratigraphic units in a given basin. They are drilled in the central, thickest part or most stratified area of the basin where all the sediments will be represented. They cut across the strata and give information on the other elements necessary for the chronostratigraphy and geologic history, in addition to depositional environment. Further processes such as *coring* (process of collecting cylindrical samples of rocks from a formation) and *vertical seismic profiling* (seismic acquisition done down the well rather than on the surface) could be adopted during the drilling of a stratigraphic test well.

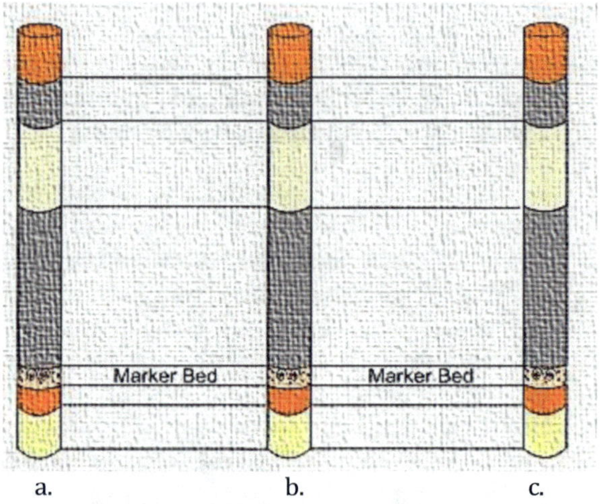

Figure 3.19: A stratigraphic cross-section with well a, b and c tied across a common marker bed restored to its horizontal position

EXERCISES

1. a) What is stratigraphy?
   b) State five laws of stratigraphy. Use diagram to describe each.

2. a) Discuss the concept of lithostratigraphy.
   b) How would you carry out a lithostratigraphic study of an outcrop?
   c) How are mudcracks used to identify an upturned stratum?

3. Discuss the concept of chronostratigraphy. How is it used in understanding the relative and absolute sequence of events in Earth's history?

4. a) Discuss the concept of biostratigraphy.
   b) What are fossils? Describe the various types.

5. Describe the three main formations of the Niger Delta Basin, stating the types of lithology, their thicknesses and hydrocarbon potentials in each case.

6. Give three examples each of terrestrial, benthic and planktonic marine fossils.

7. a) Discuss the concept of seismic stratigraphy.
   b. How are seismic surfaces and seismic signatures used to understand the stratigraphy of a region?

8. a) Discuss the concept of sequence stratigraphy.
   b. Differentiate between downlap and toplap.

9. a) What are the major elements and tools of sequence stratigraphy?
   b) List the various systems tracts found in a sequence.

10. What are the applications of the stratigraphy concepts in the petroleum industry?

REFERENCES & FURTHER READING

Avbovbo, A.A., 1978. Tertiary Lithostratigraphy of the Niger Delta. *AAPG Bulletin,* v. 62, p. 295–3000.

Berggren, W.A., D.V. Kent, C. C. Swisher and M.-P. Aubry, 1995. A Revised Cenozoic Geochronology and Chronostratigraphy. In: *Geochronology, Time Scales and Global Stratigraphic Correlation*, Society of Economic Paleontologists and Mineralogists Special Publication, 54, p.129–212.

Blow, W. H., 1969. Late Middle Eocene to Recent Planktonic Foraminiferal Biostratigraphy. *1st Internat. Conf. Planktonic Microfossils Proc.*, v. 1, p. 199–421.

Bolli, H. M., 1966. Zonation of Cretaceous to Pliocene Marine Sediments based on Planktonic Foraminifera. *Bol. Inf. Venez. Geol. Min y petrol.*, v. 9, n° 1. p. 3–32.

Bouma, A. H., 1962. Sedimentology of some Flysch Deposits: A Graphic Approach of Facies Interpretation. *Elsevier.* 162p.

Bouroullec, R. and C. Lansigu, 1999. High Resolution Sequence Stratigraphy of A Turbiditic System: Synsedimentary Deformation and Reservoir Implication, *AAPG Conference*, Birmingham.

Burke, K., 1972. Longshore Drift, Submarine Canyons, and Submarine Fans in Development of Niger Delta: *AAPG Bulletin*, v. 56, p. 1975-1983.

Claude D., C. Cazzola, J. Gerard, E. Guyotte, J.C. Navarre and F. Temple, 2000. Variation of Architecture, Facies and Depositional Models of Stratigraphically Controlled Turbiditic Deposit: Examples from the Lower Congo Tertiary Basin. *AAPG Poster*, April, New Orleans.

Emery*, D. and K. J. *Myers, 1996. Sequence Stratigraphy, Oxford, *Blackwell,* 297 p.

Gradstein, F. M., J. G. Ogg, A. G. Smith, F. P. Agterberg, W. Bleeker, R. A. Cooper, V. Davydov, P. Gibbard, L. Hinnov, M. R. House, L. Lourens, H-P. Luterbacher, J.Mc Arthur, M. J. Melchin, L. J. Robb, J. Shergold, M. Villeneuve, B. R. Wardlaw, J. Ali, H. Brinkhuis, F. J. Hilgen, J. Hooker, R. J. Howarth, A. H. Knoll, J. Laskar, S. Monechi, J. Powell, K. A. Plumb, I. Raffi, U. Röhl, A. Sanfilippo, B. Schmitz, N. J. Shackleton, G. A. Shields, H. Strauss, J. Van Dam, J. Veizer, Th. van Kolfschoten, and D.Wilson., 2004: A Geologic Time Scale. *Cambridge University Press.* ~ 500 p.

Haq, B.U., J. Hardenbol and P. R. Vail, 1987. Chronology of Fluctuating Sea Levels since the Triassic. *Science*, v. 235, p. 1156–1167.

Mutti, E., 1977. Distinctive Thin Bedded Turbidite Facies and Related Depositional Environments in the Eocene Hecho Group, South Central Pyrenees, Spain. *Sedimentology*, 24, p. 107–131.

Mutti, E, 1985. Turbidite Systems and their Relations to Depositional Sequences. Provenance of Arenites. *NATO ASI Series C.* ed. G. G. Zuffa, *Mathematical and Physical Sciences.* 148. p. 65–93.

Payton, C. E., 1977. Seismic Stratigraphy: Applications to Hydrocarbon Exploration, *AAPG Memoir 26,* 516 p.

Pickering, K. T., D. A. V. Stow, M. P. Watson and R. N. Hiscott, 1986. Deep-Water Facies, Processes and Models: A Review and Classification Scheme For Modern and Ancient Sediments. *Earth Science Review,* 22. p. 75–174.

Reijers, T.J.A., Petters, S.W., and Nwajide, C.S., 1997. The Niger Delta Basin, in Selley, R.C., ed., *African Basins – Sedimentary Basin of the World 3*: Amsterdam, Elsevier Science, p. 151-172.

Shaker, S., 2002. Sequence Stratigraphy: Key to Geopressure Profile Assessment. *CSEG Recorder,* p. 88–91.

Sprague, A. R., M. D. Sullivan, K. M. Campion, G. N. Jensen, F. J. Goulding, T. R. Garfield, D. K. Sickafoose, C. Rossen and D. C. Jennette, 2002. The Physical Stratigraphy of Deep-Water Strata: A Hierarchical Approach to the Analysis of Genetically-Related Stratigraphic Elements for Improved Reservoir Prediction. *National AAPG/SEPM Meeting Abstracts*, Houston, Texas, March 10-13, p.10–13.

Stanley, D. J., 1975. Submarine Canyon and Slope Sedimentation (Grès d'Annot) in the French Maritime Alps. *Proc. IX Int. cong. Sediment.*, Nice. 129p.

Sultan, N., M. Voisset, B. Marsset, T. Marsset, E. Cauquil and J-L. Colliat, 2007. Potential Role of Compressional Structures in Generating Submarine Slope Failures in the Niger Delta, *Marine Geology*, 237, p. 169–190.

CHAPTER FOUR

STRUCTURAL GEOLOGY

4.0 Structural Process

Structural process is any movement that makes a horizontal layered rock to bend, after the initial sedimentation and solidification. Initial sedimentation has been covered in chapter two. Layering of rocks, which are usually horizontal due to the role of gravity during final settling in sedimentary basins, has been discussed in the chapter three. This is then usually followed by structural movements, leading to a change in the existing rock layer's orientation. If a loosely consolidated sedimentary deposit bends before full solidification, usually due to a movement downslope, it is known as a *slump*.

Structural geology is the study of rock deformation; it is applied to the understanding of hydrocarbon generation and trapping. This chapter looks at structural processes and how to study them, including:

- Deformation and folding
- Faulting and fault types
- Plate movements / plate tectonics
- Formation of basinal structures
- Relationship between folding and shortening
- Structural restoration and timing of structural events
- Integrated basin modelling
- Relationship between burial, temperature and pressure

The interrelationship among geologic disciplines demands that petroleum generation, sedimentary processes and basin formation are discussed alongside structural processes. These are necessary to fully define the geologic formations, rock bodies and mechanisms that put petroleum systems in place. These therefore are the basics of petroleum geoscience, which have been covered at this stage, before going to the major exploration and production disciplines in the next part and subsequent chapters.

4.1 Deformation and Folding

Once sediments have been deposited, they are most likely to be affected by some form of deformation over time, either due to inherent *subsidence* with compaction or other tectonic activity. A strata or rock layer is deformed when it is no longer horizontal as it was after deposition (Rettger, 1935). The rock formation or strata is said to be *tilted* or *deformed*. Sediment supply, changes in temperature and pressure, lateral compression, fluid flow and other naturally occurring processes in the subsurface lead to deformation and formation of structures. Most petroleum pools worldwide are trapped in some form of structures.

Because of the thick piles of sedimentary rock, most petroleum basins have source rocks that have been buried deep enough in the geologic past to generate oil and gas. The deep part of the basin where the gas and oil forms is known as the *kitchen* or *oven*. After the gas and oil is generated, it flows upward in the overlying rocks where it is eventually trapped and concentrated by rocks that impede further movement. In Figure 4.1, the subsequently deposited, undeformed younger, beds lie horizontally on top of the older deformed strata. If the younger beds are impermeable, they will trap any fluid that eventually migrates into the (yellow) sediment in the centre.

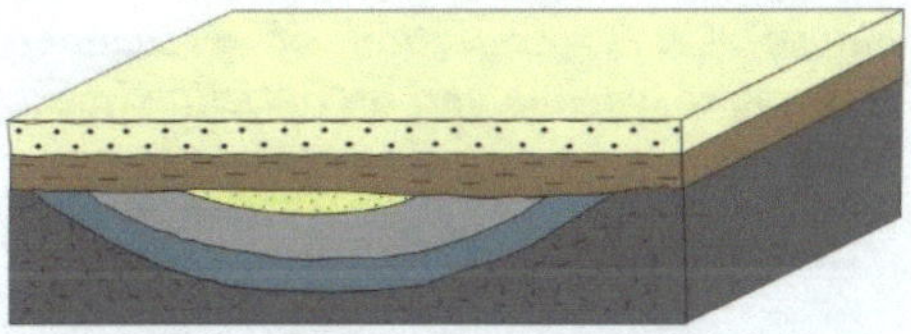

Figure 4.1: Sedimentary strata deformed to form a synclinal basin overlain by undeformed younger beds

The understanding of the deformation process, how subsidence affects basin maturation, has been employed in machine learning to automate basin modelling and the estimation of temperature and pressure gradients. Vertical and lateral variations of these parameters within an oil field have consequences on exploration and production of petroleum resources. There are many causes of deformation, some of which are:

Sediment Supply and Subsidence

Sediment supply creates load that can make a basin to sag in its central part, which is a form of subsidence. Subsidence can also result from tectonic movements, either extension or compression, that weaken the central part bearing the load and making it to sag further. As the load builds, it makes the previously deposited layers to sink due to pressure of new sediment supply (Steckler and Watts, 1978; Roberts et al., 1997).

The cross section of a sedimentary basin (Figure 4.2) shows subsidence due to *overburden* (surface soil that acts as overload) to form a syncline, and (possibly upturned) strata forming an anticlinal prospect. This is a prospect because it has a source rock, a substantial reservoir and a seal.

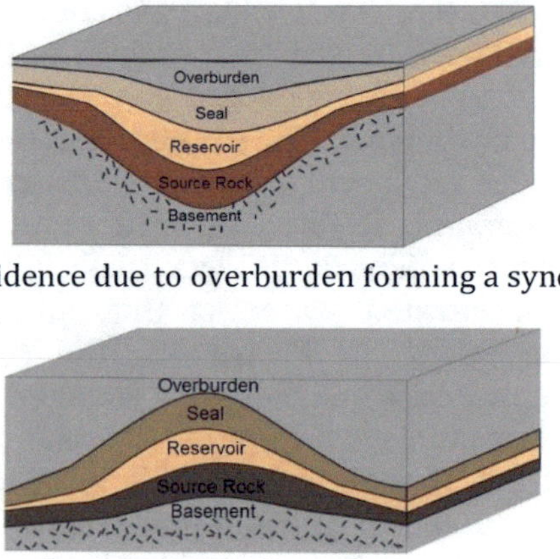

a. Subsidence due to overburden forming a syncline

b. An anticlinal prospect above a basement complex

Figure 4.2: Cross-section of a sedimentary basin showing a syncline and an anticline in a sedimentary basin

Compaction is a process, which may be related to subsidence, in which the weight acting on other sediments below leads to a reduction in pore spaces and in the vertical thickness of the sedimentary pile. A reduction in vertical thickness creates accommodation space and leads to more sediments arriving, until the central force is enough to cause a sag or a bump, creating synclinal and anticlinal structures.

Compression, Extension and Lateral Slip

The compression and extension of rock formation result in weakening and folding of the central part bearing the load. Compression implies pressures from either side of the basin, or from one side, making the sediment to push up in the central position and flex down at the edges to form anticlinal fold (Boyer, 1982). Extension will make the sediment to fracture or just weaken at the centre, creating sagging at the central position and synclinal fold Bilotti et al., 2005).

Extension (**x** and **a.**) happens when movement is in the opposite direction, away from each other, while compression (**y** and **b.**) happens when movement is towards each other (Figure 4.3). For lateral slip (**z** and **c.**), movement occurs side by side, slipping past each other.

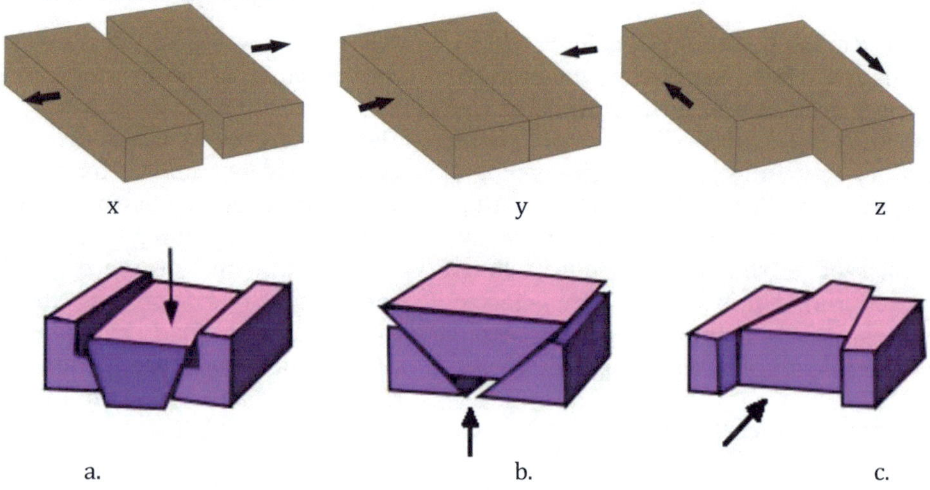

Figure 4.3: Extension, compression and lateral-slip movements (x, y and z show the concept while a, b and c show the process)

Folding

The tectonic activities leading to folding are several: *uplift, relaxation, faulting,* reactivation of old faults, and many aspects of plate movement. When uplift and relaxation takes place on deposited sediments, that is, a rise of land to a higher and lower elevation respectively by geologic forces, it will result in a folded structure. The sediments original position may never return to its original state after the uplift and relaxation processes cease. When old faults are reactivated and new faults are formed, sediments may be deformed by the displacement of the fault walls, leading to formation of structural features (Connors et al., 1998; Corredor et al., 2005).

Piggyback basins, rift basins, foreland basins, graben and other types of folds structures are all created by tectonic activity. Figure 4.4 is an example of ductile folding, which occur in incompetent beds that may not always fracture and create faults. Folding takes places essentially in rocks with ductile properties, as opposed to those of brittle that will fracture or break when forces are applied to them. When compressional forces acted on originally parallel sedimentary strata in Figure 4.4, they bent the strata into permanent folded structures of anticlines and synclines. Sometimes a rock may undergo faulting and some folding as well.

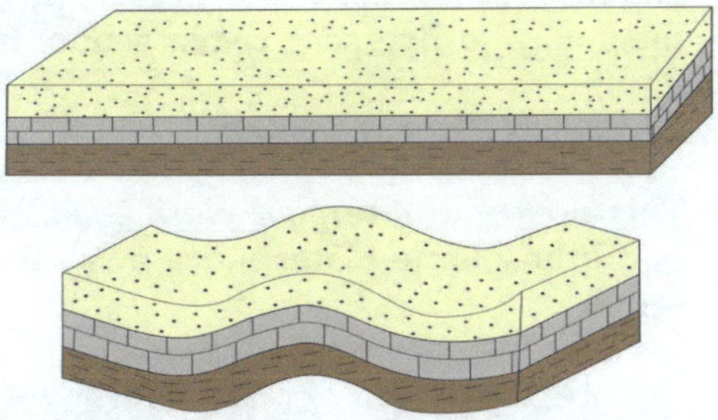

Figure: 4.4: Folding process shown as a deformation of geologic unit in an undulating manner (unfolded layer above / folded layer below)

Faulting and Fault Types

Where a competent rock unit is compressed or stretched, it will rather break instead of folding. If the strata are cut without the left or right side wall being displaced, it is said to be a ***fracture*** or a ***joint***. When there is displacement of one of the walls relative to the other, it is known as a ***fault***. Normal and reverse faults are the most common.

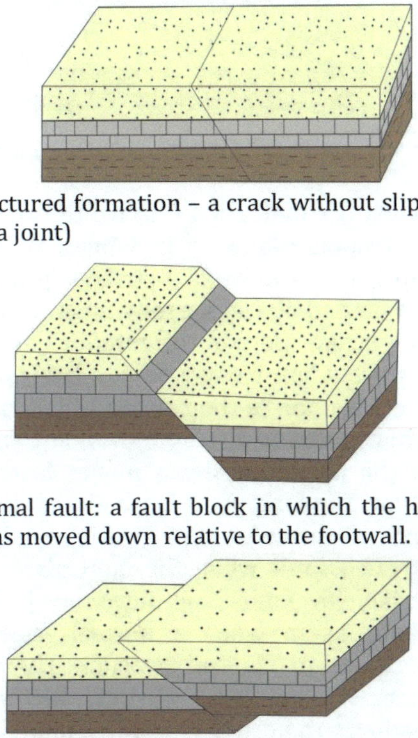

a. Fractured formation – a crack without slip of the walls (a joint)

b. Normal fault: a fault block in which the hanging wall has moved down relative to the footwall.

c. Reverse fault: a fault block in which the hanging wall has moved up relative to the footwall.

Figure 4.5: Fracture, normal and reverse faults

In Figure 4.5a, the formation has been *fractured*, without movement or displacement of the rock bodies on either side of the crack. When there is movement in an extension or compression tectonics context, it creates faulted blocks; the fault walls move up or down relative to each other as seen in Figure 4.5b and 4.5c. The side of the block on the fault dip, above the fault line, is known as **hanging wall**, while the side below the fault line, opposite the dip, is known as **footwall.**

When he hanging wall moves down, relative to the footwall, it is known as a **normal fault**, it is a **reverse fault** when the hanging wall moves up relative to the footwall. A **thrust (thrust fault)** is a low lying reverse fault in which relative movement of the fault block has occurred in such a way that the fault plane is inclined at an angle not more than 45°. Figure 4.6 shows a thrust fault where the hanging wall moves over the footwall. In this case, older beds will be seen lying on top of younger beds.

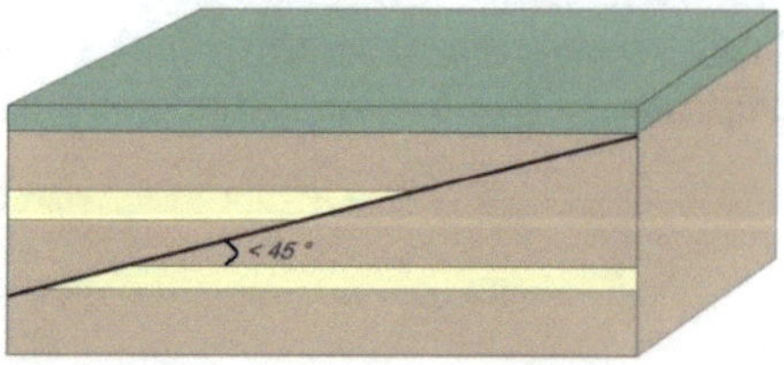

Figure 4.6: Thrust faults showing the low angle dip

When the fault slips relative to each other in a horizontal plane, it is known as **strike-slip fault**. It is known as a strike-slip fault because the faulted blocks are moving in the direction of the strike (lateral orientation of a geologic feature when viewed from a horizontal plane). It is common in the context of lateral tectonics, where there is a relay movement and one side of the displaced strata moves faster than the other. The fault plane is usually vertical, with no vertical throw or slip. This is also known as a **wrench fault** when the fault plane is vertical. Strike-slip fault may have an inclined fault plane, as in when a normal fault has been displaced in a lateral direction. This means that there was a local vertical tectonic regime leading to a normal or reverse fault, followed by a lateral regime leading to strike slip. Such example is shown in Figure 4.7, where the fault slip seems to occur at an angle (probably the vertical displacement has been eroded or is not seen on an outcrop).

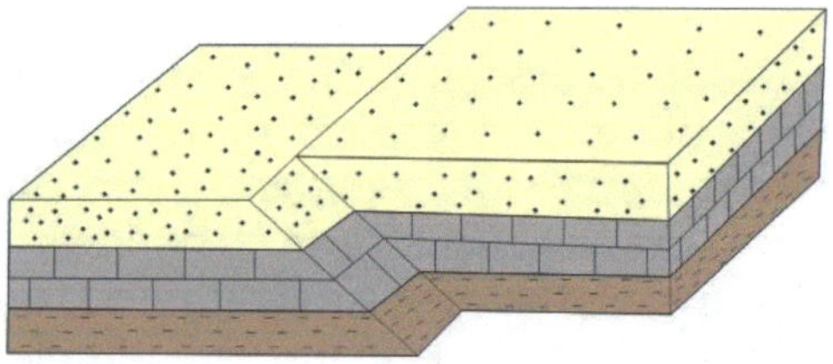

Figure 4.7: Strike-slip fault with an inclined fault plane

Standing on one side and observing the fault block on a strike-slip fault, if that block across appears to have moved to the right, the fault is known as **dextral** or *right-lateral*; if it appears to have moved to the left, the fault is known as **sinistral** or *left-lateral*. Looking at Figure 4.8, the fault in which the opposite block moves to the right, as in a., is dextral or *right-lateral* and the one that moves to the left, as in b., is sinistral or *left-lateral* strike-slip fault.

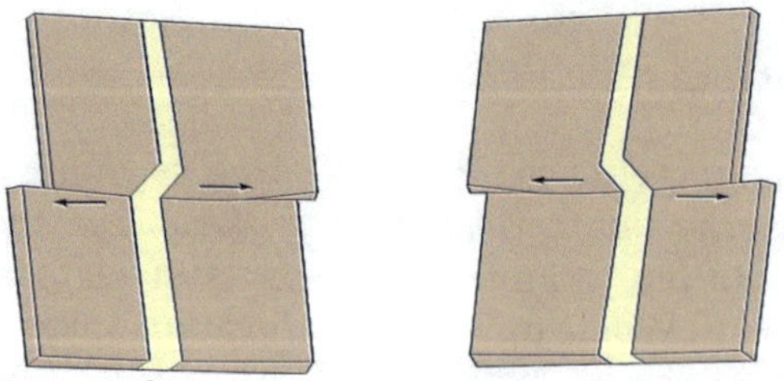

a. Dextral *b. Sinistral*

Figure 4.8: Strike slip fault showing dextral-right-lateral and sinistral-left-lateral configurations

4.2 Plate Movements

The Earth crust is made up of several plates, probably twenty or more. These include the continents, oceanic plates and any other part of the Earth crust that is "mobile" and not stationary (Hutton, 1975; Engelder, 1993). The upper mantle produces a convective current that seems to shift the plates in minute displacement per year and sometimes faster when there are major earthquakes. Plate movement is also known as **plate tectonics**.

During major tectonic activity, plates are crushed together to form mountainous push-ups, or slide below each other to form trenches. As these happen, smaller portions of the Earth crust such as local stratified rock layers may be faulted or folded. As the strata weaken, plutonic activities break through the upper layer and expulse magmas to form volcanic arc along the strikes of the layer. Plate tectonics are the main control of compressive, extensive and strike-slip movements in basins. As such, plate movement may be divergent, convergent or lateral.

Divergent Plate Movement

This occurs in *divergent plate boundary* when two lithospheric plates move away from each other.

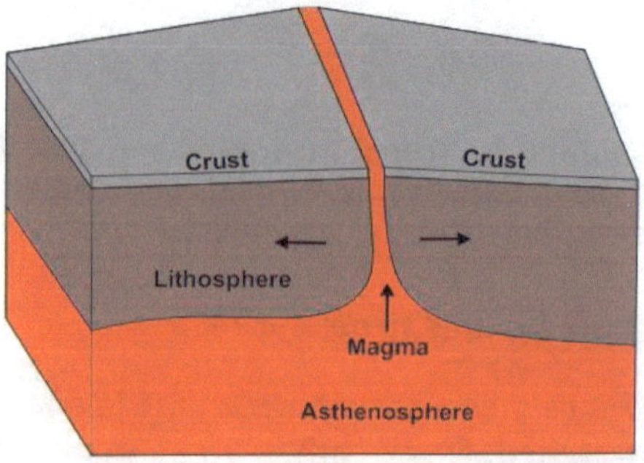

Figure 4.9: Magma upwelling and crustal separation in the zone of divergent plate movement

Figure 4.9 shows a divergent plate boundary; when this happens, magma from the subsurface finds its way to the surface, and solidifies to form a new crust. The asthenosphere is believed to be responsible for this plate movement.

Convergent Plate Movement

This occurs in *convergent plate boundary* when two lithospheric plates move towards each other. Figure 4.10 shows a case where a continental and oceanic lithospheres converge (convergence can also be between two continental plates or two oceanic plates). Because the oceanic lithosphere is denser, it subducts underneath the less dense continental counterpart, forming a trench in the process. As the oceanic plate continues to subduct, higher temperatures in the subsurface causes the rocks to melt, generating magma in the process. The molten rock finds its way to the surface, forming a chain of volcanic arcs.

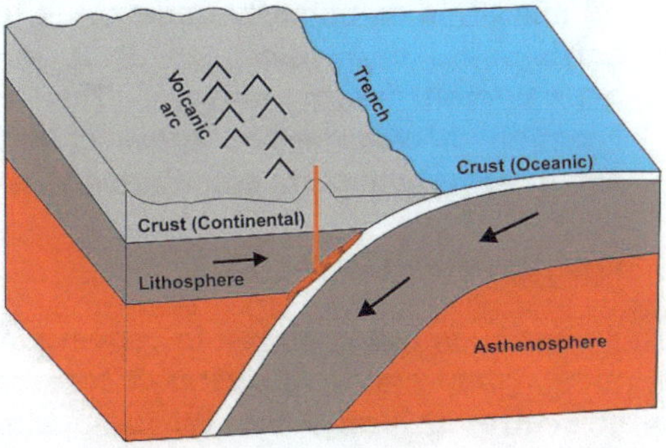

Figure 4.10: Subduction of an oceanic crust and formation of trenches and volcanic arc in the zone of convergent plate movement

Lateral Plate Movement

This occurs in *transform zones* when two lithospheric plates slides relative to each other. A *transform plate boundary* is a plate boundary characterized by plates slipping past each other, along strike lines.

These processes form different basin types discussed in the next section.

4.3 Formation of Basinal Structures

Basin Types

There are three major types of basins classified on the bases of their *geometry* and *formation process* during movements of crustal plates. The three types are **compressional**, **extensional** and **strike-slip basins**.

Extensional Basin

This is formed by divergent plate movement, generally grouped as:
1. Rift basins
2. Intracratonic basins

Rift basins are formed by *rifting* or *break up*, while plates or Earth layers separate. They are associated with *normal faults* and occur in *divergent* plate regions. Their activity is usually less aggressive compared to those at the convergent plate margins. Example is a graben basin. *Intracratonic basins* refer to basins formed on *land* or *inland*, as compared to those formed in the oceans.

Intracratonic basins and rifts may be covered by water when there is inflow from the sea due to *lows* formed by the basin. Erosion transports sediment from land through a network of delta to the basin. On the Atlantic side of the globe, this is known as *passive margin* or *continental margin*. Figure 4.11 illustrates a **horst** and a **graben** where the foot wall has been elevated on both sides (blue colour), leading to the formation of horst, while the hanging wall has been depressed in the centre (green colour), leading to the formation of a graben.

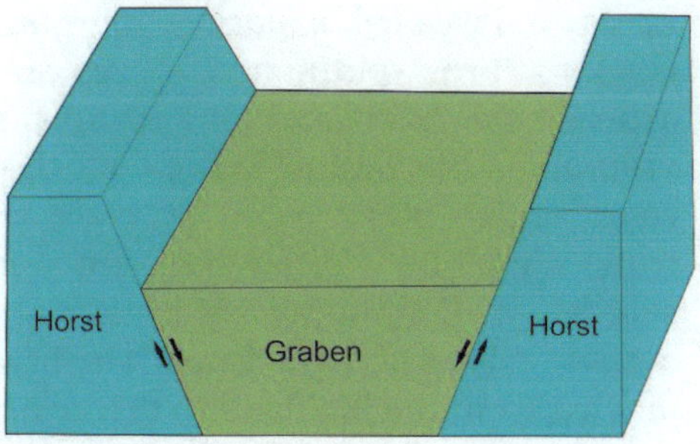

Figure 4.11: Horst and graben as normal fault blocks of the Earth crust that have been elevated and depressed respectively

There are situations where the system of horst and graben has been disturbed, is not fully developed. This is known as *half graben*, as illustrated in Figure 4.12; it is possible that either the feature is a full horst-graben system, but was not observable from an outcrop.

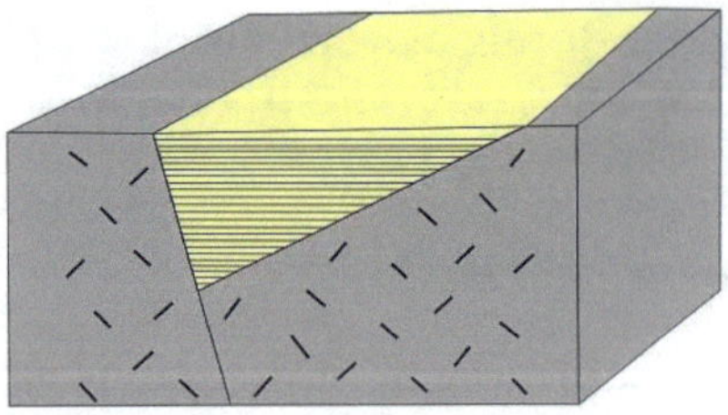

Figure 4.12: Half graben basin showing where one of the faults truncates the other

Continental Passive Margins

The continental margin consists of the continental shelf, shelf break, slope and basin floor (Figure 4.13). It is the edge of the continent and is still considered part of the continent even when it is submerged in shallow waters of the ocean.

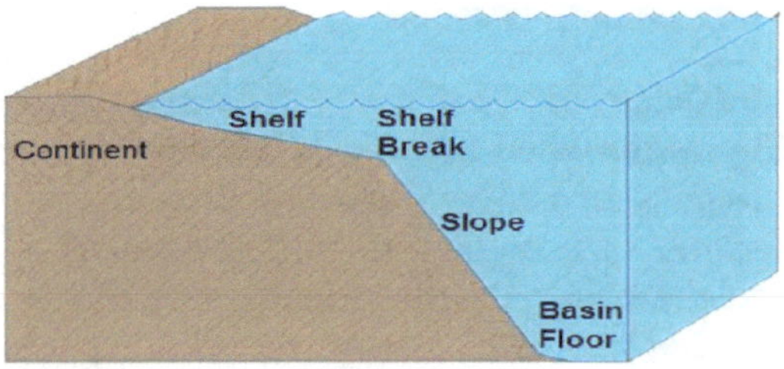

Figure 4.13: Cross section of a continental margin (not to scale)

Where *extension* has been so important that the basin created is low enough to be filled with seawater, like in the areas between two continents, it is called a *passive margin*. Passive margins are continents - sea boundaries experiencing extension on a regional scale. An example is the Gulf of Guinea, off the Nigerian coast, which was created by plate movement, triple junction formation and continental separation. Continuing separation at the mid-oceanic ridges, and gravity activity of sediments flowing from the highland to the low, has led to the expanse of the Atlantic Ocean and its sea flow deposits.

Deposition occurs due to transport of sediments from uplifted zones or *highs*, and near-shore soft rocks, to parts of the passive margin, including the platforms, shelf, rise, slope and deep offshore. Figure 4.14 illustrates a cross-section showing the path from high grounds to the coastal plain, where sediments are transported from mountainous regions to the continental margin areas. Sand deposition takes place first as the water energy is higher at the shelf and slope regions. Shales are formed offshore by fine-grained materials deposited at the basin floor, where water energy is lower. Sometimes the sands cut across shaly layers, as they are displaced further downslope, within incisions and channels created by turbiditic currents.

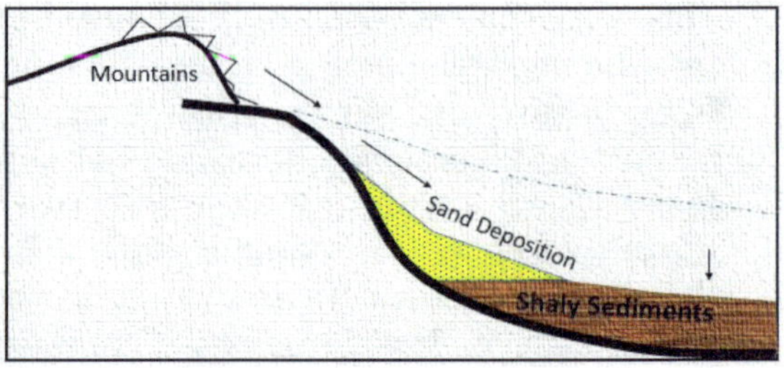

Figure 4.14: Coastal plain to deep offshore sediment infill

Sometimes further offshore, away from a passive margin, gravity tectonics may occur at a relatively low and local scale. When a continental margin is dominantly undergoing tectonics, earthquakes and volcanoes, it is known as **active continental margin**. This is more common in compressive edges of continents, along convergent plate boundaries.

Compressional Basin

As has been learnt above, in convergent movement, compression of rock unit form anticlinal structures at local scale. When anticlinal folds are formed, the low edges of the anticline may be sagging, thereby producing extension locally. Likewise, larger crustal compressive movement produces major *compressive uplift* as well as associated extension locally. Extensions could also occur during *relaxation*, after a subduction for example. This is exemplified in the **arc** system of deformation.

Examples of large-scale compressional basin types are:
1. Back-arc
2. Foreland

Back-arc basins are formed when one tectonic plate *subducts* under another, as in the Pacific Ocean where the convergent plates meet. The subduction creates a *trench* that subsides and weighs on the mantle, which is destroyed, thereby allowing magma to flow upwards to the surface. Back-arc region are major zones of *orogeny* (process of mountain formation) and mountain building.

Eventually mountains are built in subduction zones of compressional basins, but sometimes they are eroded or not high enough to be observed in present day. In Figure 4.15, as one plate subducts, it disturbs the sediments on the other, forming mountains in the process through the upward displacement of the overlying non-subducted rocks.

Foreland basins are formed close to a mountain when the load of the mountain cause the lithosphere to *bend* or *flexure*, thereby subducting below the mountain. The basins form at the base of the mountain and is usually filled by *fluvial* (river) and other *terrestrial* (land) sediments as well as *marine* (ocean) sediments where seawater flows in due to the low. *Lacustrine* (lake) and *deltaic* (funnel-like drainage to open sea) systems may also be present.

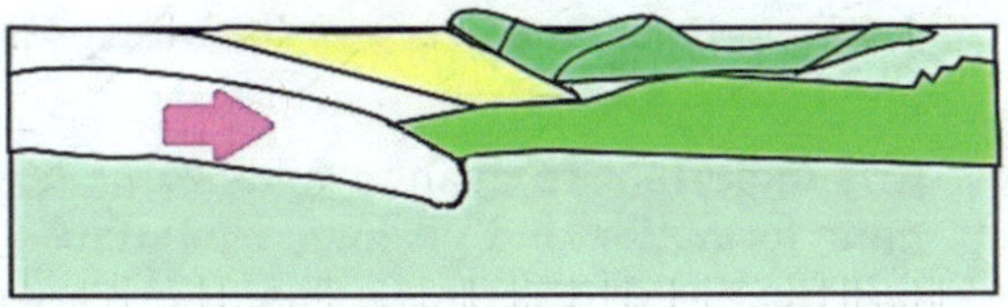

a. One plate subducts below the other, disturbing the sediments above

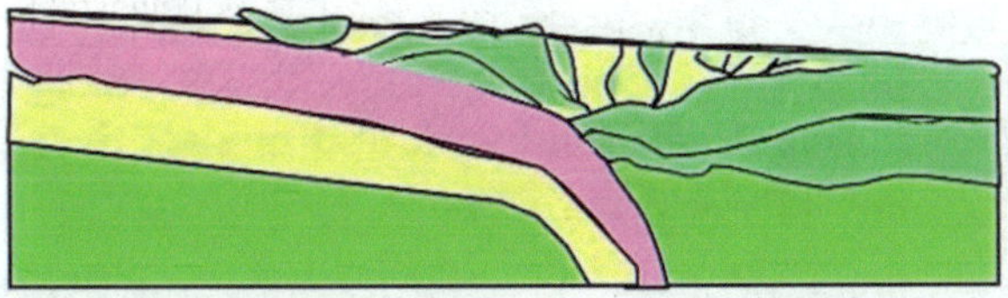

b. Continuous subduction results in trenches, local basins and wedges

Figure: 4.15: Formation of complex mountain chains in zones of subduction

After the formation of the mountain (topmost purple region in Figure 4.16) during lithospheric plate convergence, it can be eroded, and its sediments deposited beside it in low-lying areas known as **foredeep**. When the load of the basin increases, it will result to subsidence and further burial of the basin, with more complexities.

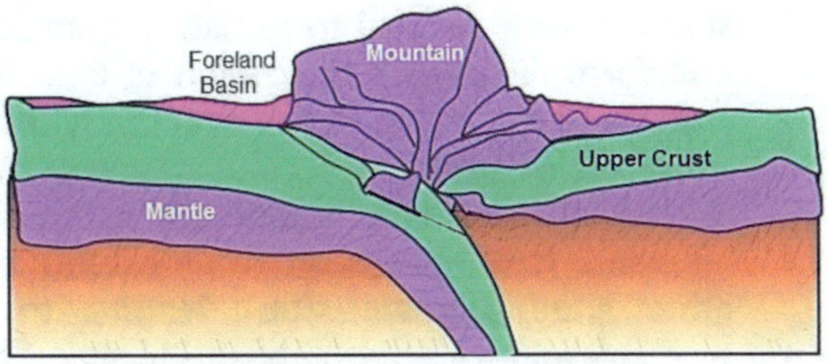

Figure 4.16: Mountain building in zones of subduction and convergent plate movement

The nature of the unit colliding with the other is important in determining the nature of fold, among other factors. Lithospheric convergence could either be *continent* against *continent*, *oceanic* against *oceanic* and *oceanic* against *continent*. Heavier oceanic crust tends to subduct, move below, lighter continental crusts, thereby forming major **arc/angle,** and these may happen relatively quick – leading to major earthquake compared to activities in passive margins.

Strike-Slip or Wrench Basins

These are formed by lateral slip along strike lines. The basinal structure is related to the strike-slip in which the movement of the fault block is in the strike direction, that is, slip occurs primarily in the horizontal direction.

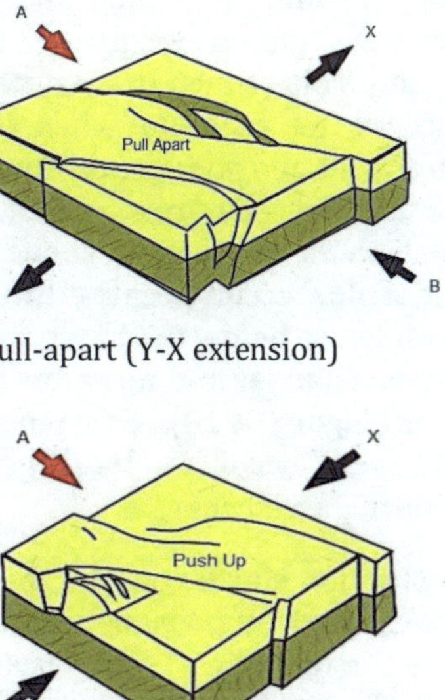

a. pull-apart (Y-X extension)

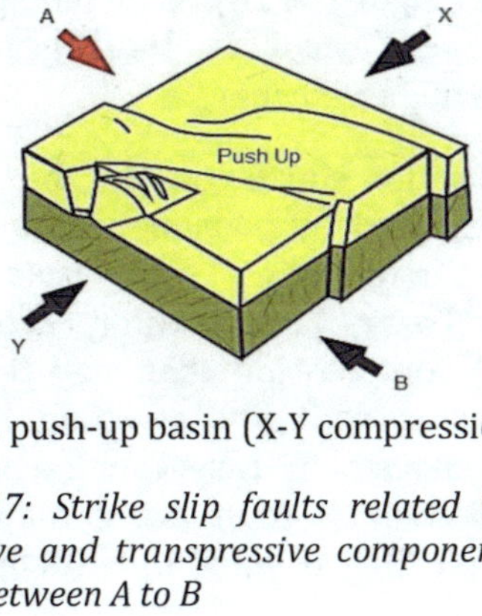

b. push-up basin (X-Y compression)

Figure 4.17: Strike slip faults related basins with transtensive and transpressive components, slipping restively between A to B

Wrench basin is used to explain the structures that form during the movement of fault blocks relative to the other, which eventually involves some rift or up-thrown compartment. As the units slide past each other, they compress on one end (**transpression**) and extend at the other edge (**transtension**), leading to *transpressive* and *transtensive* basins locally.

These are similar to extensive and compressive basins, but with a lateral slip component. *Pull-apart* and *push-up* basins are typical examples. Figure 4.17 shows pull-apart and push-up basins formed in a strike-slip situation. The pull-apart is formed by the orthogonal extension regime, while the push-up is formed by the compressive component as well perpendicular to the direction of movement.

Other forms of structures associated to strike slip basins are the *flower structure*, observable in seismic image. This occurs because the vertical fault zone is crushed at the upper part due to lateral resistance to movement, making the structure so formed to look like a flower. In general, the fault may look more vertically straight compared to those in the other basins.

Folding and Shortening

During a compressive and an extensive deformation forming anticlinal and synclinal structures, the lateral length of folded units is reduced. This reduction is known as *shortening.* It occurs, for example, when two lithospheric plates meet and subduction does not take place, such that they compress against each, or when a thrust sheet is moving above a decollement layer. Folds occur because the unit could not slip above or below the lower strata, acting as if there is a barrier and no *excess distance* in front (as in Figure 4.18). Shortening and folding could also occur in local gravity-controlled basins in passive margin.

Usually the shortening in the distal part is compensated by proximal folding or fractures above the (anticlinal) fold (Davis and Reynolds, 1996; Tozer, 2006; Butler, 2007). It creates lateral compression that may tighten or open fissures episodically, affecting fluid retention and leakages in reservoir respectively. Such crestal-collapse fault are a major concern in some oil filed trapped by anticlines. They may leak, trap or compartmentalize reservoir fluid pressure.

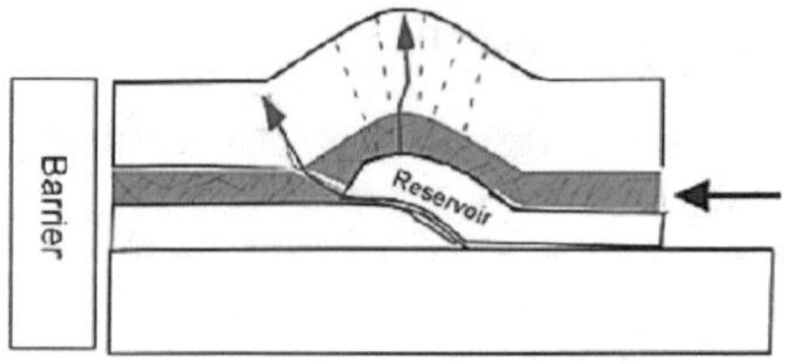

Figure 4.18: Accommodation of distal shortening by fold structures

In Figure 4.18, displaced strata fold due to the presence of a barrier (possibly another plate or lineament). The top of the anticlinal structure formed has been fractured and hydrocarbon where present in the reservoir can leak away or be compartmentalised. Shortening gives an idea of the rate of folding, as a given shortening distance over time is as well the rate of deformation over that period. Estimation of shortening is also important in structural **restoration**, **decompaction** and *basin modelling*.

4.4 Structural Restoration: Backstripping and Decompaction

It is important to analyse folds to understand their formation and geologic evolution. Before this can be done, certain assumptions are made to provide technical and representative approaches during analysis, to help understand the evolution of structures formed and how they can be used to develop a model for the basin. Principles employed are based on observations, fieldwork and models developed conceptually.

Some of them are used to carry out structural analysis, known as **restoration**, and also to undo the effects of sedimentation and subsidence over time, known as **decompaction**. *Flattening* of each unit, *timing* and *reversal* of fault where present are all processes of structural restoration. The recursive and computation involved in the process may be further aided by artificial intelligence and automation, since there are abundant data from structural restoration software and models to harness for the machine learning (training the computer systems to learn and apply the process independently).

Basin modelling involves a morphologic and systematic restaging of the structural and sedimentary processes that formed a basin. The structural aspect attempts to understand a deformational process by reversing or restoring the process, restitution of the original state before activity, inferring which process produced the final structures before possibly re-deforming to check their repeatability (Suppe 1983; Woodward *et al.*, 1989; Suppe and Medwedeff, 1990; Erslev, 1991; Woodward, 1995; Labrousse, 1998; Tozer *et al.*, 2006; Butler, 2007). In **backstripping** of detachment thrust and fold structures, the basic assumption is that the proximal translation is proportional to the accommodation by decollement, distal thrusting and/or thrust related fold.

Decollement is a process causing folding and deformation where some strata are partially detached from those underneath and sliding over them. The act of a group of strata sliding over others, due to a low-lying reverse fault, is known as *thrusting*. Figure 4.19 shows simplified steps in thrust fault deformation and folding, as outlined below.

Step 1: As geologic layers with alternation of competent and less competent or incompetent layers are being contracted, the competent layer fractures or is uplifted.

Step 2: Where the overlying layer is less competent, it folds. A fold without a major underlying thrust is known as simple decollement fold. This is common at the early stage of folding, in the absence of inter-bedded competent layer or at a slow rate deformation.

Step 3a: With increasing displacement of the upper compartment riding above the basal layer, the hanging wall of the central competent layer is faulted, forming a fault-propagation fold if the upper layer is not cut off.

Step 3b: Where the overall overlying strata are cut off, offset and displaced, a fault bend-fold is formed. Certain fault-propagation folds may never develop to a fault-bend fold, even with increasing displacement, depending on the rate of movement, rheology or ductile nature of the units.

There is a relationship between the rate of slip, the displacement, the resulting geometry and wall thickening. With *synsedimentation*, sediments overlying the *forelimb* or *backlimb*, lying parallel to the uppermost dotted line in Figure 4.20, are developed as **growth strata**. Such growth adjacent to fault flanks is exploited for timing deformation and knowing which unit moved relative to the other.

Inference on both positive and negative tectonic inversion can be made, since thinning and thickening in the same side of the fold may imply tectonic inversion in such condition (Woodward, 1995; Allmendinger, 1998). Assuming a steady deposition rate, rapid displacement and slow slip would result in increased folding and thinning of adjacent strata, while slow displacement would mean thickening of strata.

Sometimes shortening is considered as the reduction in length without necessarily folding, but that should be considered as *lateral compaction*. By dividing the units into different elements, providing the dimensions and trigonometry that define the different structures, the nomenclature guides in the architectures of the discretisation and digitalisation for automating the modelling processes.

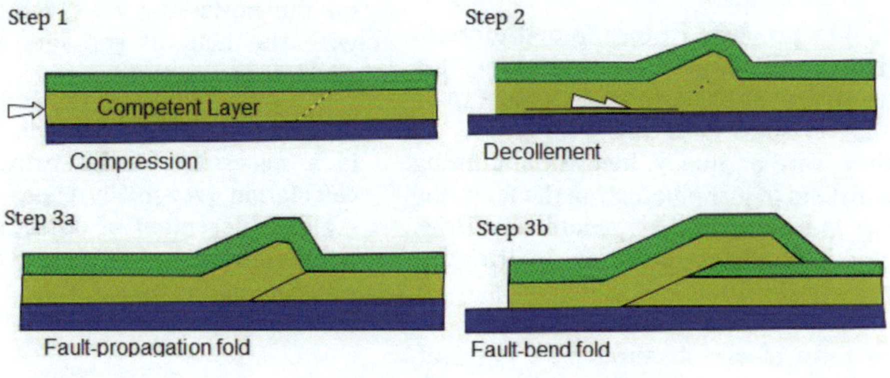

Figure 4.19: Accommodation of shortening by decollement, thrusting and folding

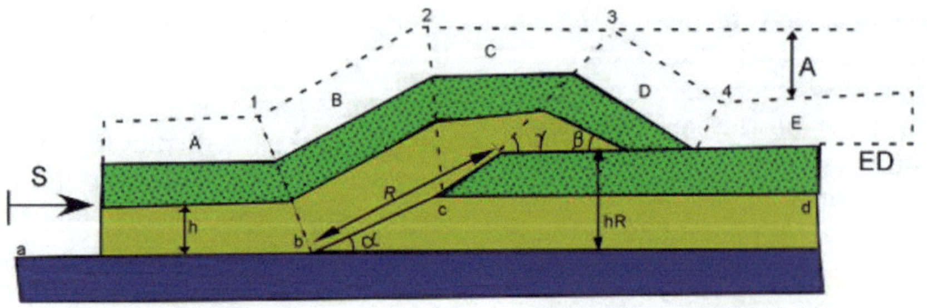

A: Backlimb
B: Inner flank
C: Upper compartment
D: Outer flank
E: Forelimb
ED: Excess Distance

1: Inner syncline
2: Inner anticline
3: Outer anticline
4: Outer syncline

Shortening (S) = original length – final length after folding

Figure 4.20: Elementary fold structure with consistent geometry and dimensions (sketch based on previous illustrations of fold modelling concepts discussed above)

Unfolding of Structural Units and Timing

If no major erosion has affected the folded layers, then it is possible to stretch the folded unit and recover their length before shortening. To unfold folded layer, there is need to *flatten* the horizons one after the other. This is because the units are not folded equally, so as the first layer above is flattened, it is possible to observe the level of fold remaining in the underlying layer. That way the amount of folding that occurred during the deposition or post deposition of each unit could be separated. By doing this, the timing of structuration is established. In most cases, bedding units thin out only during folding, implying that relative thinning and thickening indicate beginning of deformation.

Flattening and Fault Reversal

To be able to properly restore folded units to their original conditions before shortening, it is necessary to first shift back faults against their offset, that is slide back the faults walls to where they were originally, before the faulting process. Instead of just eliminating the fault, the fault is reversed to also eliminate any drag, displacement of adjacent wall or deformation associated with it.

This is a form of architectural fault reversal, which could be done on paper or with geomodelling software. It requires the separation of the faulted and the unfaulted units. The upper unfolded/unfaulted layer of may be removed to flatten and restore the lower faulted unit before modelling. Unfolding of structural unit is achieved by flattening the horizons or units one after the other. Some software allow the user to do this by just clicking on a marker and selecting "flatten" from the drop down menu. Once the faults have been reversed, it is possible to flatten the units and proceed to its decompaction.

To achieve a movement of small cells in modelling software, the vertical and horizontal axes of the cross-section of interest are divided into equal blocks. In Figure 4.21, smaller and larger blocks are exemplified in **2-3** on the vertical axes. Such smaller division is important to model features like the faulted area around **2** on the horizontal axes, which can be moved with the help of software tools to simulate geologic movements.

Gridding and referencing points in coordinates is a necessity in the principle of scientific calculation (especially those involving geometry and placement of object spatially), as each point represented in an x, y and z axes can be programmed to take different positions.

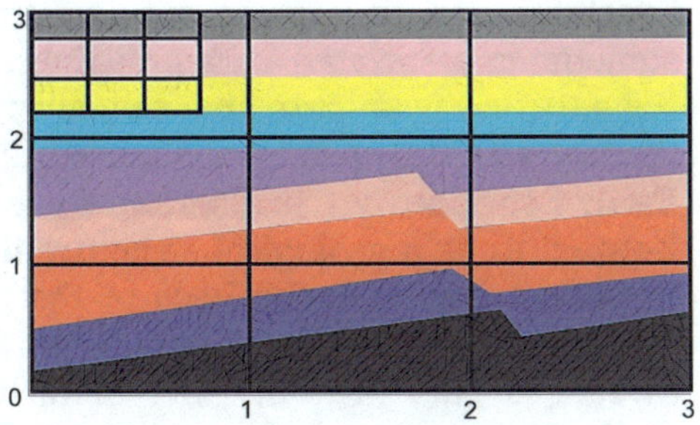

Figure 4.21: Graphic plane digitalization for modelling purposes

Vertical and Lateral Decompaction

Decompaction is a form of undoing the effect of compaction, sometimes including the associated subsidence and *isostatic effects* (a kind of gravitational balancing between the upper Earth layer and the less solid layer underneath it). It is like adding the lost porosities, thereby re-establishing the original thickness of the unit. A simple decompaction approach provided by geological software literally uses a *compaction log* to adjust and thereby predict a sediment's reduction in thickness vertically, due to compaction. A proposed method (Nosike, 2009) applies this laterally, where it is assumed that lateral shortening affects lateral compaction in a similar way to vertical compaction.

One way to apply the concept of decompaction is in gravity controlled tectonics, where the loading side creates burial and the unloading side creates an uplifted fold. In such scenario, fluid flow will be affected by the fault behaviour.

Figure 4.22 illustrates the effect of such movement on stress generation and release in normal and reverse faults, as has been demonstrated in laboratory simulations. Assuming a fulcrum migrating from point 1 to 2 as sediments are deposited in the basin, a rotation is implied from the *extensive domain*, which uplifts and relieves the *compressive domain* of its confining stress. Laboratory simulations (Nosike, 2009) shows that in most cases the reduction in porosity and permeability are not fully recovered by decompaction.

Particularly in *synsedimentary* context, the loading in the zone of relaxation increases the burial pressure (and could generate overpressure later transmitted to the zone of stress relief). The fault zone may also retain synsedimentary materials. This model is valid at both regional extensional and compressional basins, and in shallow synsedimentary collapse faults above displacing folds. Such deformation may retain both normal and contractional faults within the same zone or during different episodes of uplift and subsidence.

4.5 Integrated Basin Modelling

Structural restoration and decompaction are only precursors of a more robust basin modelling in most software, and for most explorationists. Basin formations are vertical stack of sediment deposited over millions of years, and whose history is unravelled by seismic and well studies where there are no exposures.

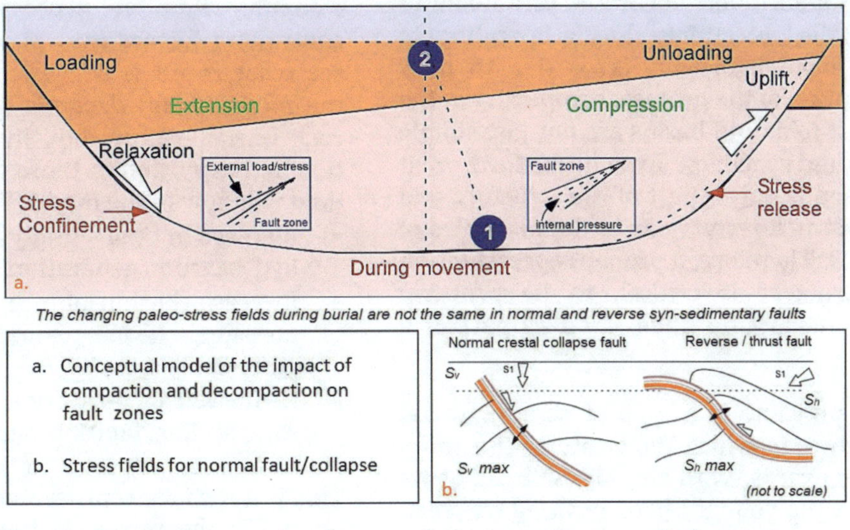

Figure 4.22: Different kinematics of normal and reverse faults in gravity controlled tectonics

Basin modelling attempts to understand how the basin has evolved since its formation, by the effect of tectonic stresses and other geological constraints. Older basins with sandstone reservoir rocks would them already altered, albeit there are higher chances that the source rock in them would be mature. A combination of mature source rock and non-altered or fractured reservoir is preferable.

The formation of traps would have to occur before the expulsion of such hydrocarbon to the reservoirs. This timing of trap formation is evaluated by a computation of isopachs (maps representing bedding units of same thickness with the same colour, to illustrate thickness variation), backstripping and decompaction, and other basin modelling methods. The interplay of several factors, few of which are listed below, is the subject of investigation in integrated basin modelling:

- *Maturity – basin formation and history*
- *Compaction and subsidence – sedimentation and petroleum system*
- *Heat flow – petroleum maturation*
- *Fluid flow – hydrocarbon migration*

Igneous intrusion and other crosscutting events make basin modelling more complex. In a diapiric or intrusive salt dome, the sediment associated with the structure might have been shifted and would not be easily represented in a model. Because every application of data from one point to the overall basin will require a proper spatial distribution, it is important to adequately understand the structure. In the industry, sometimes only a small portion of the basin is modelled and important care taken to represent the uncertainty during upscaling, in order to avoid enormous error due to poor representation of the geologic complexity of the basin. Most folds and basins are not just simple anticlines and synclines. Even in the field, what may be seen is only a part of the structure and anticlinal structures vary as in the examples of Figure 4.23. Therefore, a proper representation of the structure is critical; to be properly modelled, it must be understood as part of a basin.

A simple anticline is a type of compressional fold structure in which the strata of the rocks slope downwards, with the oldest beds at its centre and the younger beds forming its crest. Synclines are the lower side of the flanks, sometimes associated with anticlines. A plunging anticline is an anticlinal structure in which the hinge of the structure is not parallel to the Earth's surface, but inclined at some angle. A **homocline** is a geological structure that exists in such a way that the beds of the rocks dip uniformly (at the same angle) in the same direction. This may be part of an anticlinal flank. Knowing its relationship to the entire basin will be useful in properly modelling it.

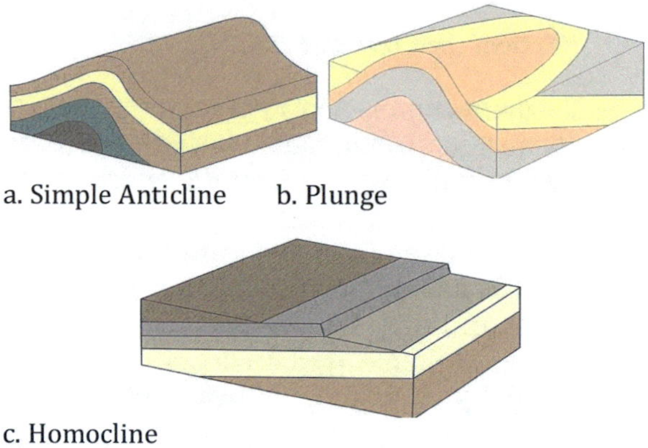

Figure 4.23: Anticlinal structures as part of a basin

Basin modelling adds the other factors in nature, such as effects of compaction on *porosity* and *permeability* and their impact on fluid flow; effects of burial on *temperature*, in order to understand the basin *maturity* and *hydrocarbon potential* among other reasons. Basin modelling may be *Iterative*, which is, executing the same set of instructions a given number of times until a consistent result is obtained. Modern integrated basin modelling uses computer iterative processes to put all the controlling factors into play and run them to see what result is suggested. It can integrate the *intrinsic* and *dynamic* factors that control rock burial, maturation, hydrocarbon generation and expulsion to those controlling how the fluid will flow in the media. For example:

- Increase in time – imply burial, maturation, hydrocarbon generation
- Increase in burial – imply reduction of porosity, heat generation, increase in stresses
- Reduction of porosity – imply fluid expulsion, folding, faulting, accumulation

This is not easily represented in a cross-section or in 2D; the power of iterative modelling is that it puts them all together, and to the closest possible means, the order in which they would occur geologically in nature. Older sediments are harder to model because the impact of changing temperature and pressure at given high temperature and depth is not fully understood. Once an initial model result is available, more data could be fed into the algorithm as time goes on and with further exploration, appraisal and availability of well results. The use of log, fluid properties, and specific depths and temperature pressure data enhance modelling results and the understanding of the basin. Integrated basin modelling result needs to be interpreted by taking the geological knowledge of the basin into consideration, to avoid misleading single-handed conclusion.

4.6 Relationship between Burial, Temperature and Pressure

This is probably the most important factor in geological basin modelling, as they control the other factors. The *interstitial water* (within pore spaces) during deposition of sediments is buried with it, such that if it were not expulsed in time it would come under pressure. Therefore, there is a direct relationship between depth and pressure, especially in *rapidly subsiding basin* and before generation of high temperature. Generally, temperature and pressure increase with increasing burial depth. Terzaghi's principle (*discussed in Chapter Twelve, on Pore Pressure Prediction and Interpretation*), gives a relationship between the **total stress**, the **pore pressure** and the **mean effective stress**. At a given depth, the sediment becomes heated up even more due to natural increase in geothermal gradient with depth and due to internal reactions within the sediment itself. That goes to say that such increase in temperature further hastens the reaction, which also increases the pressure alongside other causes of overpressure. At above 3 to 5 kilometres depth, sediments are under high temperature and pressure and burial alone is no longer suitable to accurately predict the expected linear or homogeneous compaction, pressures and temperatures for most basins, as such likewise the dynamism of the petroleum system.

There are cases where deeper structurers affect heat distribution at shallower depth, likewise pressure and fluid flow. Figure 4.24a shows a dome-like or **diapiric structure** formed as a result of ductile materials intruding into brittle materials above them, thus pushing and breaking such materials in the process and creating adjacent trapping structures. In such cases, not just burial, but nature of underlying plutonic activity or basement results in quicker maturation and expulsion as well as trapping of hydrocarbons. **Granite wash** is a potential reservoir formed by the *weathering of granite*. After the granite and granite wash have been buried in the subsurface, oil and gas can form in source rocks at lower elevation and migrate up into the granite wash (Figure 4.24b). These are special cases.

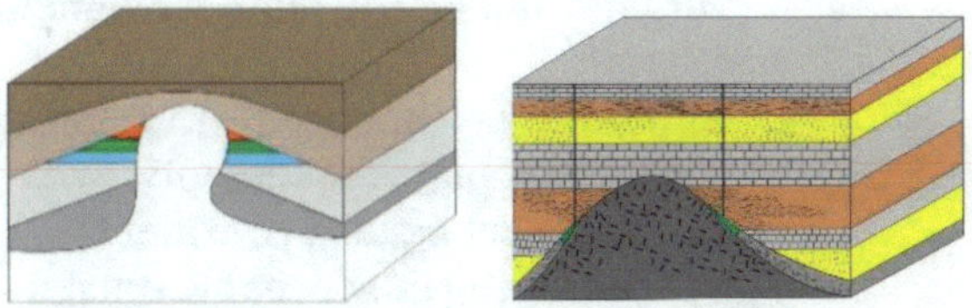

a. salt dome b. granite wash

Figure: 4.24: Migration of oil and gas into traps affected by deep structures

Anomalies and special cases can however be calibrated from basin to basin. Basin modelling approaches are used to test the various hypotheses of what comes to play at different basin types, petroleum system, trap structure, depths, temperature and pressure zones and fluid behaviours and maturation varying over each interval and time. It is a stacked, grouped, interpretative approach based on the modelling specification for each interval, which is later harmonized. Such approach will have to take into consideration all the incoming parameters along the line; these include *enhanced permeability* (not present at the onset), *fluid expansion* by oil cracking for example, and *fluid type transformation* alongside normal *chemical reactions* and *tectonic activity*.

Once a well is drilled, direct pressure measurement and well post-mortem could help to further understand the history of the basin and interplay of pressure versus burial and temperature. Fluid inclusions, evaluation of Pressure – Volume –Temperature (PVT), Gas Oil Ratio (GOR) and direct interpretation of petroleum result also give useful information that can be re-fed to the model for improved results.

EXERCISES

1. What is rock deformation (folding and faulting)?

2. Explain how the following cause deformation
   a) Sediment supply and subsidence
   b) Compression, extension and lateral movement

3. What is plate tectonics?

4. Discuss the processes of divergent, convergent, and lateral plate movements.

5. Differentiate between the following convergent plate movements:
   a) continent against continent
   b) oceanic against oceanic crust
   c) oceanic crust against continent

6. What is the role of structural restoration and decompaction in basin modelling

7. Explain the following processes of structural restoration

8. Three layers of rocks A, B, C have their lengths measured at 22, 27 and 34 m respectively. However, careful observation during structural restoration shows that these layers have undergone shortening of 3, 5 and 8 m respectively.
   a) Using a diagram, illustrate how to derive a formula to determine their original lengths
   b) Using the formula, what are the original lengths?

9. What is the importance of basin modelling?

10. What is the relationship between burial, temperature and pressure?

REFERENCES & FURTHER READING

Allmendinger, R. W., 1998. Inverse and Forward Numerical Modeling of Trishear Fault-Propagation Folds. *Tectonics* 17, p. 640–656.

Bilotti, F. and J. H. Shaw (2005). Deep-Water Niger Delta Fold and Thrust Belt Modelled as a Critical-Taper Wedge: the Influence of Elevated Basal Fluid Pressure on Structural Styles. *AAPG Bulletin*, v.89, no. 11, November 2005, p. 1475–1491.

Bilotti, F. D., J. H. Shaw, R. M. Cupich and R. M. Lakings, 2005. Detachment Fold, Niger Delta, in J. H. Shaw, C. Connors and J. Suppe, eds., Seismic Interpretation of Contractional Fault Related Folds. *AAPG Studies in Geology* 53, p. 103–104.

Boyer, S. E. and D. Elliott, 1982. Thrust Systems. *AAPG Bulletin,* v. 66, p. 1196–1230.

Butler, R. 2007. Tectonics and Mountain Building. *Geoscientist* 17 (10), p. 14–15.

Childs, C., T. Manzocchi, J. J. Walsh and C. G. Bonson, 2008. A Geometric Model of Fault Zone and Fault Rock Thickness Variations. *Journal of Structural Geology,* while in press., 11 p.

Connors, C. D., D. B. Denson, G. Kristiansen and D. M. Angstadt, 1998. Compressive Anticlines of the Mid-Outer Slope, Central Niger Delta (abs.). *AAPG Bulletin,* v. 82, p. 1903.

Corredor F., J. H. Shaw, and F. Bilotti, 2005. Structural Styles in Deep-Water Fold and Thrust Belts of the Niger Delta, *AAPG Bulletin,* v. 89, no. 6, June 2005, p. 753–780.

Davis, G. H. and S. J. Reynolds, 1996. *Structural Geology,* Hoboken, New Jersey, John Wiley & Sons, Inc. p. 400.

Engelder, T., 1993. *Stress Regimes in the Lithosphere*. Princeton University Press, Princeton, New Jersey. 457 p.

Erslev, E.A., 1991. Trishear Fault-Propagation Folding. *Geology* 19, p. 617–620.

Heinö, P. and J. R. Davies, 2006. Degradation of Compressional Fold Belts: Deep-Water Niger Delta. *AAPG Bulletin*, v. 90, no. 5, May 2006, p. 753–770.

Hutton, J., 1795. Theory of the Earth, with Proofs and Illustrations, *in* Reijers, T.J.A., 1996. *Selected Chapters on Geology*. S.P.D.C. Warri, p. 9.

Knipe, R.J. 1989. Deformation Mechanisms – Recognition from Natural Tectonites. *Journal of Structural Geology,* v. 11, p.127–146.

Kostenko, O. V., S. J. Naruk, W. Hack, M. Poupon, H-J Meyer, M. Mora-Glukstad, C. Anowai and M. Mordi, 2008. Structural Evaluation of Column-Height Controls at a Toe-Thrust Discovery, Deep-Water Niger Delta. *AAPG Bulletin,* v. 92, no. 12, p. 1615-1638.

Kristensen, M. B., C. J. Child and J. A. Korstgard, 2008. The 3D Geometry of Small-Scale Relay Zones between Normal Faults in Soft Sediments. *Journal of Structural Geology*, 30, p. 257–272.

Labrousse, 1998. Modélisation Numérique des Plis de Rupture ; Applications à des Exemples Naturels, *Thèse de doctorat*, Université de Pau, *N° : 98 PAU 3012*, p. 364.

Lehner, P. and P. A. C. De Ruiter, 1977. Structural History of Atlantic Margin of Africa. *AAPG bulletin, v. 61, no. 7*, p. 961–981.

Nosike, L., C., 2009. Relationship between Tectonics and Vertical Hydrocarbon Leakage: A Case Study of the Deep Offshore Niger Delta. *PhD Thesis*, University of Nice-Sophia Antipolis, France, 281 p.

Obaje, N.G., 2009. *Geology and Mineral Resources of Nigeria*. Springer, Berlin, Germany, Pages: 221.

Onuoha, K. M., 1999. Structural Features of Nigeria's Coastal Margin: An Assessment Base of Age Data from Wells: *Journal of African Earth Sciences*, v. 29, no. 3, p. 485–499.

Ramsay, J. G. and M. I. Huber, 1983. The Technique of Modern Structural Geology, *Volume 1: Strain Analysis*. Academic Press, London, 307p.

Rettger, R. E. 1935. Experiments on Soft-rock Deformation. *AAPG Bulletin*, v.19, p. 271–292

Roberts, A. M., E. R. Lundin and N. J. Kusznir, 1997. Subsidence of the Vøring Basin and the Influence of the Atlantic Continental Margin. *Journal of the Geological Society, London*, 154, p. 551–557.

Sibson, R. H., 1995. Selective Fault Reactivation during Basin Inversion: Potential for Fluid Redistribution through Fault-Valve Action, *In* J. G. Buchanan and P.G. Buchanan,

Eds., Basin Inversion. *Geological Society (London) Special Publication* 88, p. 3–19.

Steckler, M. S. & Watts, a. B. 1978. Subsidence of the Atlantic-Type Continental Margin off New York. *Earth and Planetary Science Letters*, 41, p. 1–13.

Suppe, J., 1983. Geometry and Kinematics of Fault-Bend Folding, *American Journal of Science* 283, p. 684–721.

Suppe, J. and D. A. Medwedeff, 1990. Geometry and Kinematics of Fault-Propagation Folding, *Eclogae Geologicae Helvetiae* 83, p. 409–454

Tozer, R. S. J., R. W. H. Butler, M. Chiappini, S. Corrado, S. Mazzoli and F. Speranza, 2006. Testing Thrust Tectonic Models at Mountain Fronts: Where Has the Displacement Gone? *J. Geol. Soc.* London, 163, p. 1–14.

Woodward, N. B., 1995, Thrust Systems in the Tamworth Zone, Southern New England Orogen, New South Wales, *Australian Journal of Earth Sciences*, v. 42, p. 107–117.

Woodward, N. B., S. E. Boyer and J. Suppe, 1989. Balanced Geological Cross-Sections: An Essential Technique in Geological Research and Exploration, *Short Course in Geology* v. 6, American Geophysical Union, Washington, DC 132 p.

Wu, S. and A. W. Bally, 2000. Slope Tectonics-Comparisons and Contrasts of Structural Styles of Salt and Shale Tectonics of the Northern Gulf of Mexico with Shale Tectonics of Offshore Nigeria in Gulf of Guinea, *in* W. Mohriak and M. Talwani, eds., Atlantic Rifts and Continental Margin. Washington, D. C., *American Geophysical Union,* p. 151–172.

PART TWO

EXPLORATION AND PRODUCTION DISCIPLINES

This part deals with the various exploration and production disciplines. It starts with the basic process of field acquisition to the use of aerial photos and remote sensing in petroleum prospecting. Oil companies acquire blocks and go through the process of identifying and following a lead to find a *prospect*, by the identification and study of *plays*. When a discovery is made, an *appraisal* process is used to further refine the *reserve* volume as well as provide field and reservoir targets for commercial development.

To achieve these, knowledge from the various exploration and production disciplines must be used. This includes the different aspects of exploration geology and geophysics and several methods for the identification of petroleum traps during the exploration phase. Oil field and block evaluation ascertains the vertical and lateral *compartmentalization* to some levels of certainty. Calculation of *hydrocarbon in place* is an important step before committing to the development of the field.

This part looks at the above processes and concludes by detailing the process of prospect evaluation, risk and uncertainty analysis, as well as seal assessment. Sealing processes are explained with the concept of sealing efficiency and integrity, indicating seal retention or breaching by various processes. Top seals control vertical boundaries; just as faults play major roles in lateral boundaries, and there are methods for assessing their impact on trap filling. Meanwhile, unconventional hydrocarbons have put a new dimension in mining of petroleum resources. Understanding all this is important before engaging in well operations and field development.

CHAPTER FIVE

PETROLEUM PROSPECTING AND BLOCK ACQUISITION

5.0 Processes from Acquisition to Field Development

When embarking on a petroleum prospecting process, the prospecting company would need to have a field development strategy in place, either by themselves or by an operator company in case of exploration success. The initial evaluation of block or acreage may take few years, likewise the exploration to development stage. Production and field management usually take longer, up to 10 to 20 years in most cases. Decommissioning and restoring the field at the end of production may as well take few years.

The processes involved in the value chain of petroleum prospecting (discussed in this chapter) are:

- Block Acquisition
- Exploration
- Discovery
- Appraisal

The other processes that follow (discussed in details in the later chapters), which may be foreseen and planned during petroleum prospecting, are:

- Field Design
- Development
- Operations
- Production
- Field Management
- Abandonment

Oil block acquisition requires bidding and contractual agreements. *Field Investment Decision (FID)* follows discovery, usually confirmed by a well test. Before and after the beginning of production, the operating company continues to acquire and analyse data to determine the available reserve or resources, produced volumes and future producible hydrocarbons. The studies carried out on the original condition of the field before production started is known as *static evaluations* while those carried out after production has started is known as *dynamic evaluations*.

Some of the data, with examples, to be acquired or estimated are:

a. Remote sensing and aerial photography (satellites and radar images, and aerial photos)
b. Geophysical data (seismic data, magnetic and gravity data)
c. Geological data (maps, lithology, mud or cutting samples)
d. Petrophysical (logs, porosity, net-to-gross, fluid saturations)
e. Pressure Volume and Temperature (PVT – compressibility, expansion factor, recovery factor)
f. Well testing (production index, skin, field permeability)
g. Production (pressure changes, aquifer strength, rate of production/injection)

For these data sets to be acquired, the block or field would have to be a good prospect. Prospecting is preceded by large-scale data acquisition, starting from surface maps to subsurface information.

Map information and similar useful images are obtained from fieldwork and by acquisitions from distance known as *Remote Sensing*. Coupled with seismic information, explorationists are able to class some fields as having petroleum plays because they have potential for oil and gas resource. Only good petroleum plays will have measurable pressures, porosity, permeability, oil saturation, and it will only be worthwhile if the field has positive economic petroleum results. The possibility of profitable exploitation is assessed during block appraisal. The appraisal process covers several aspects of oil exploration and production, focusing on answering the question of quantifiable results.

5.1 Remote Sensing

Geoscience exploration starts by knowing where we are, the world, the space, and the Earth. Capturing the globe, its form and images, from a distance is known as *remote sensing*. Two major types of remote sensing is aerial photo and satellite imagery. Aerial photo refers to a picture taken by a camera mounted on a plane or any flying object; satellite images are taken by satellites at the orbit and transmitted digitally to Earth. They are important for the initial identification of structures in petroleum basins and for locating wells all through the life of a field. In the remote sensing image shown in Figure 5.1a, major geologic features, such as the anticlines and the synclines, can be identified. *Compare with the geologic map in Figure 5.1b and notice how the remote sensing image highlights topography and structure better.*

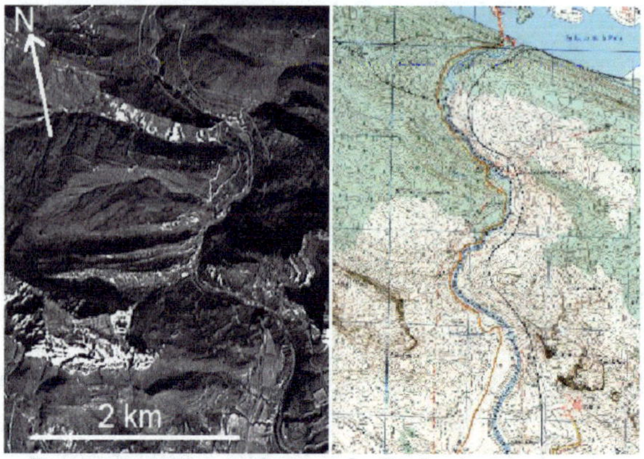

a. Remote sensing image b. Geologic map

Figure: 5.1 Remote sensing image versus and geologic map of the same area (Spot Image, 2006; Carte topographique IGN, 2004)

Geologic maps illustrate structures such as anticlines and synclines with symbols (not visible in this example of Figure 5.1), but remote sensing image may show this naturally. The acquisition of *aerial photos* and *satellite images* is based on the same principle as ordinary photo images, as if taking a picture from an aeroplane. However, in the case of satellite imagery, it is done from a *satellite* and different types of radiation are used. Successful acquisition of satellite imagery demands that atmospheric components and interaction are taken into consideration.

Remote Sensing and Electromagnetic Waves

Remote sensing is one of the many ways of acquiring information about the Earth; the only difference from other methods being that information is obtained without physical contact with the Earth. It makes use of *satellites* that have sensors, which receive *propagated signals* (*electromagnetic waves*) reflected by the object under investigation, to detect features (such as rocks, rivers, valleys, mountains) on the Earth's surface. Detection can be through signals emitted by the satellite (*active remote sensing*), or reflection from sunlight (*passive remote sensing*). Some form of electromagnetic wave is reflected on the film or photo equipment, which records the image from a distance.

Remote sensing images could be both coloured and black and white, and can be viewed in 2D or 3D, with the 3D sometimes requiring the use of a **stereoscope** for viewing. A *stereoscope* is a device that makes it possible to see two photographs of the same object, taken at slightly different angles, together, thereby depicting three-dimensional view. The shades in Figure 5.2 help to see the remote sensing image in 3D; this can be noticed in both colour and black and white images. They help in describing the topography and landforms of the area.

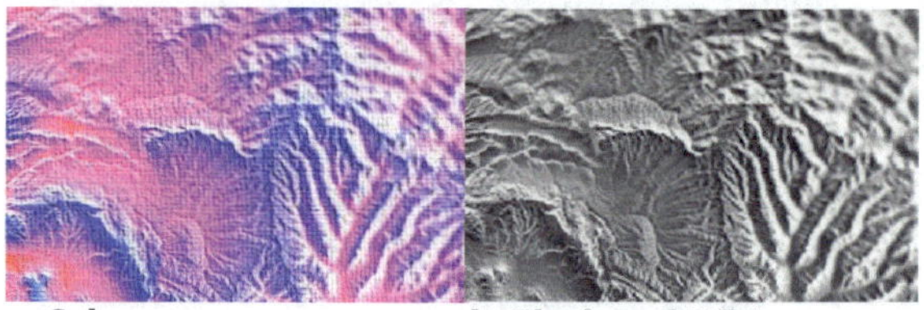

a. Colour b. Black and White

Figure 5.2: Example of Remote Sensing (RS) image, highlighting high areas and low areas (Spot Image, 2006)

The two main types of satellite imagery used in remote sensing are based on **optical** and **radar** principles. *Optical images* uses the visible wavebands and provide images that are similar to how the human eye sees pictures. Radar images are produced by penetrating radio signals, which are not affected by weather, but they are less like real picture.

Principles of Radar Imagery

Radar systems work by transmission and reception of signals or pulses to provide information on size of object, polarisation or change in phase of the signal on return, and the *two-way time* taken for the signal to travel to-and-from the object be detected. This is similar to radar systems used on land, such as *speed detectors* for traffic control. There are two types of sensors, active (providing their own illumination by a transmitter) and passive (only receiving radiation reflected by the object being detected). Example of active radar systems are scatterometers and altimeters, while examples of passive systems are microwave radiometers. The two common types of imaging radars are *Real Aperture Radar* (RAR, which is also known as *Side Looking Airborne Radar* - SLAR) and Synthetic Aperture Radar (SAR – which synthesises recorded signals in such a way as to augment coverage, as if elongating the aperture virtually).

There is the issue of resolution. ***Azimuthal*** resolution is a degree of detection *radially*, away from the linear position, to which a camera could capture an image. For example, SAR radar have an azimuthal resolution of half the real antenna size. With good processing, resolution may be as high as 1 meter for airborne radars and 5 meters for space-borne radars. *Relief* and *imaging errors* are often related to flight and must be corrected, as well as *interferences* with targeted object, which may be *spiky* or *warped* to picture being taken by a captor in motion.

There are different *types*, *mission* and *geometry* of satellite images as well as *characteristics* and *techniques*. An example is the emulsion type image capture where *Colour InfraRed* (CIR) or *camouflage-detection* film allows its emulsion layers sensitive to green, red and near-infrared radiations, to detect features that would not be otherwise caught by conventional colour films.

5.2 Geographic Information System (GIS)

The *Geographic Information System* (GIS) is a group of computer systems that allows the user to obtain geospatial information (geographic location of features of the Earth) in the form of digital maps or models. With a GIS software, it is possible to tie these digital versions to topographic or manually constructed maps from fieldwork, while making corrections for mismatches and highlighting important objects at different locations. GIS is to the geographer or geospatial geologist what the telescope is to the astronaut.

GIS may also be used to georeferenced, store and interpret aerial photos and remote sensing image. Once stored as spatial or geospatial data, they can be easily visualised, managed, manipulated and analysed, making them relatable to one another using *location* as its key variable index, and *coordinates* such as representing longitude, latitude and elevation. The principle behind GIS is to stock several layers of information known as "themes" into a single profile, so that information on rock types, location, habitation, communications networks and other useful information can each be separated as a layer making up a map database.

GIS software packages such as 3DEM and TNTlite allow the user to use standard *shapes files* and to manipulate *Digital Elevation Model (DEM)*. Illustration software with GIS facilities such as Canvas (PC World, 1998) may be used to represent images in geographic coordinates (3DEM, version 20.7). A current major GIS software used for georeferencing in the petroleum industry and for working with map is the **ArcGIS** (ArcGis, version 10.7.1)

An important aspect of maps, adapted for analysis with GIS software, is to know the exact location on the Earth's surface, and this is where the *Global Positioning System* (GPS) is most useful. GPS is a global navigation satellite system that provides information on time and geographic location anywhere on Earth. It serves as the basic referencing system on which map and related graphics are recorded and retrievable at all times. This is important in spatial monitoring and petroleum exploration, where precise information on position of objects of interest (including oil well position) in map is required. In Figure 5.3, the blue cube located in the central position of a spherical shape or globe can be represented in a spatially identifiable X and Y coordinates.

Any object at a position on any part of the Earth can be located and identified on the X and Y (and possibly Z, denoting elevation) coordinate system, as conceptualised in Figure 5.3. The location represented is not only a 2D or surface view of the Earth, but also 3D with additional information on altitude or elevation.

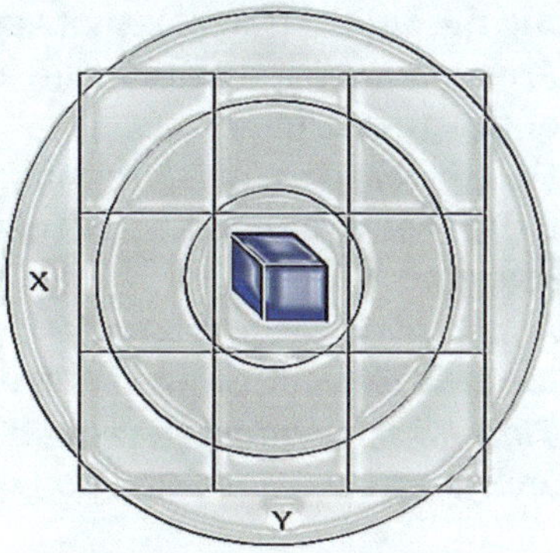

Figure 5.3: Location concept in GIS software

A digitized map showing elevations is known as Digital Elevation Model (DEM). With a software

such as 3DEM, this can be manipulated alongside "themes" to merge, analyse, and study displacement of Earth material, and solid or fluid materials overlying it, over time. This is one of the ways in which satellite images and digitised aerial photos are studied.
The group of studies that involves the collection, storing, analysing and applying geospatial information is known as *Geomatics*, and it involves the principles of mapping (including surveying and cartography), GIS, GPS-DEM, and Remote Sensing in general.

5.3 Processing and Application of Remote Sensing (RS) Data

There are several tools for processing remote sensing images. One of the simple RS image processing software, with a free student version, is TeraVue (TeraVue, 2006). With TeraVue, it is easy to manage, analyse, and digitalise satellite images. It also simplifies time-lapse studies by supervision or "classification" of environmental changes in layers. Like other similar software, it provides enhancement including filtering, 3-D plotting and *colour composition* of images to represent changes over time.

The techniques and the principles behind the image processing are:

- Constitution and visualisation – the basics of imaging, representing geospatial features on a plate or software screen
- Image improvement and filtering – removing noise, errors and highlighting most important features
- Colour composition – separating different properties with different colours

Among the application of remote sensing are:

- *Prospect finding*: location and identification of subsurface prospects
- *Sustainable development*: conservation of natural environment at risk of energy sourcing
- *Prediction of disasters*: societal and oil field hazards prevention and control
- *Time lapse monitoring*: recording and analysing environmental pollution

The use of GIS applications in processing remote sensing data has shown importance in geodesy, involving the measurement of the Earth's geometries and shapes, as well as its changing orientation in space. Geodesy is useful in the study of oil acreages and prospects, when it is applied to the restoration of structural events and large geobodies within a petroleum basin.

5.4 Oil Acreages and Location of Prospects

Geospatial coordinates of petroleum basins, oil fields and blocks are delineated by satellite images. The coordinate types and referencing are achieved by standard location system and recorded during survey of well. Every well, especially offshore, has a *Unique Well Identifier* (UWI), a 16-character code which gives the hole location and significant drilling information of the well. Regulating authorities, like the Department of Petroleum Resources (DPR) in Nigeria, usually keep these locations and references properly to associate every well position to the petroleum field (Nosike, 2019).
Each block division usually has alphabets and numbers used to identify it. Figure 5.4 shows an illustration of Oil Prospecting License (OPL) and Oil Mining Lease (OML) block divisions known as block acreages. The contour on the northeastern part of the image shows the highland areas while areas at the southern part mark out offshore regions with thick lines.

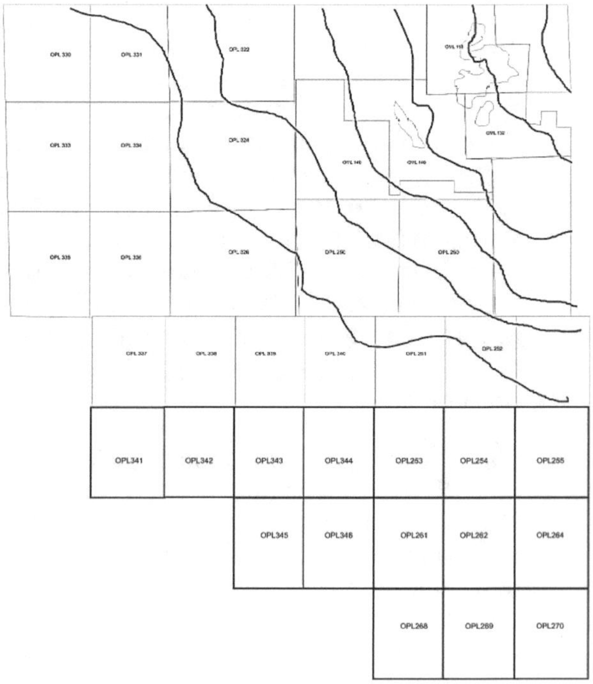

Figure 5.4: Illustration of OPL and OML block divisions known as block acreages

5.5 Mining Block Acquisition

Whether it is to search for or recover oil, the first step is getting a *license* to explore (prospect for oil) a given block, or produce (mining) oil from the block if already discovered. This can be on land (onshore) or below the ocean (offshore). That will remain the case unless oil is found elsewhere other than below the Earth surface. There are different kinds of oil licenses or oil exploration and production contracts, some of which are (Nosike, 2017):

- *Concession:* Land with prospective resources, or blocks, is given to the oil company on certain purchase terms. After bidding, a successful company obtains land in which it explores and owns every finding. The country or authorities that conceded the oil field benefit by way of taxes or percentage royalties for profits.

- *Joint Venture (JV/JOA):* Joint venture involves an agreement to work together between the host country and the operator or oil company on an oil asset. The voting superiority in terms of business decisions rests with the party owning the highest equity. The framework of the JV is often known as a *Joint Operation Agreement (JOA)*. Sometimes the term *Joint Operation* is used to emphasize that parties in the contract start the operation jointly. The parties are entities who come together and agree on an arrangement that involves not only sharing in the asset, but also participating in the control of the activities. It carries with it rights, obligations and liabilities shared by the parties involved.

- *Production Sharing Contract/Agreement (PSC/PSA):* The authorities of a country with a prospective oil field specify terms and give the Oil E&P Company the right to explore, produce and use the proceeds from successful discoveries and production to offset their expenditures. Only the profit remaining after that will be shared with or taxed by the authorities. *Production Sharing Agreement (PSA)* is similar to PSC, but PSA is based on negotiated terms specific to the contract. In some countries, the PSA may imply better tax benefits and royalty terms; a country may enter into a PSA agreement with the indigenous companies, with advantageous terms, giving them preference as part owners of the oil blocks.

- *Risk Sharing Contract (RSC):* This is a variant of the PSC, where the risks are shared rather than left for the oil company alone. For example, it may apply to a marginal field where there has been known discovery. It employs a tighter follow-up of the oil company to ensure that it produces, recovers oil and further maintain profit for all the parties.

- *Service Contract (SC):* In a service contract, the contracting authorities engage and pay an oil firm as a party to render its services in the exploration and production process.

- *Marginal Field Concession (MFC):* This is a type of less demanding agreement, which the government reaches with smaller, often indigenous companies in Nigeria, to operate mainly in the onshore.

Parties decide which oil licenses or oil exploration and production contracts best suit them, based on the terms and agreements of such licenses and contracts.

Exploration areas are acquired both in *known petroleum fields* and in *frontier exploration zones*. A frontier exploration zone is an area known or suspected to have petroleum systems but without a commercial discovery of hydrocarbons. Blocks in frontier exploration zones could be easier to acquire, but once petroleum has been discovered in commercial quantity, they become more competitive. Block acreage can be obtained by two main processes: **bidding** and **farm-in** & **farm-outs**.

Bidding

In bidding, conditions for obtaining an oil block include financial pricing and technical competence. Other conditions are social responsibilities and provision of certain infrastructure in the country leasing the license. Usually the highest or best bidder obtains the license. Most of the exploration lands are owned by the government and managed by the ministry of energy resources, who organises a licensing round known as ***bidding***. Individual landowners are however common in USA, onshore. The land is divided into blocks or areas isolated by geographical features, and they become a *petroleum concession* – a right or privilege that has been granted – dedicated for that exploration.

Sometimes a geographical block is carved out by dividing an area with virtual grids (a pattern of regularly spaced horizontal and vertical lines). Offshore, these are the lines that bound the blocks and run along the latitudinal and longitudinal gridding system. Each block has an alphabet and a number to name it. Although it is not possible to show details and clarity of such a large landmass in this small map, Figure 5.5 gives an idea of the intricate exploration (OPL) and production (OML) areas in the southern part of Nigeria, particularly in the Niger Delta regions. The names are not clearly visible in this figure and in practice, they may change with changing exploration results, and renaming, each year or every few years. The aim of this example is to show how the blocks are carved out and split within a geographical and geological landmass, illustrated by the traces on the map.

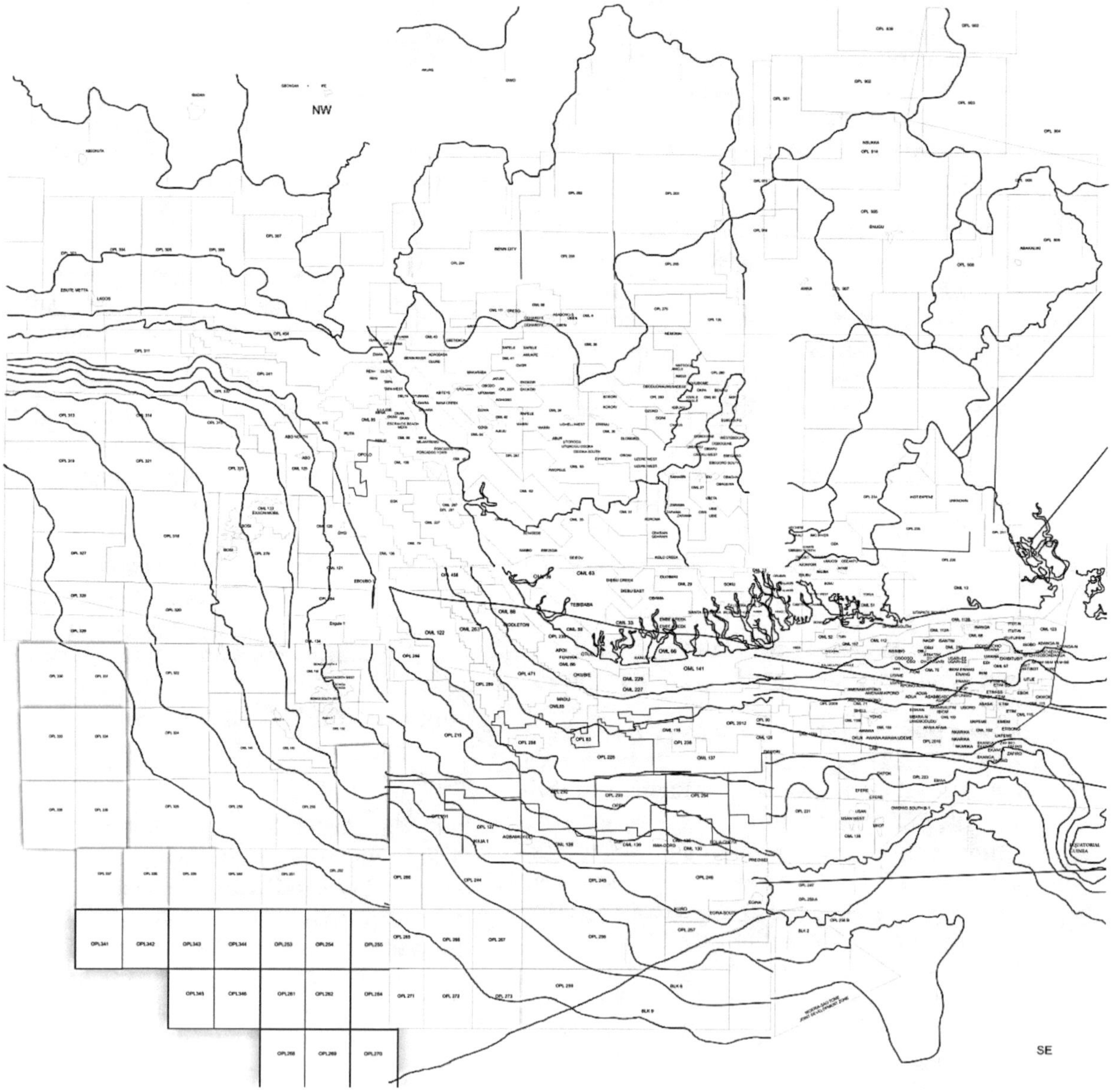

Figure: 5.5: Southern Nigerian example of Exploration and Production areas (OPL and OML)

Licenses exist for a certain period of time after which a well is either drilled or the right to explore the block is returned to the original owner. Fees during the exploration phase are targeted towards encouraging the company holding the lease to invest in exploration quickly, or receive some penalties. For example, a well may have to be drilled within four years of exploration or the block is reverted. Upon successful discovery and production, blocks are maintained by paying certain revenues to the government or original block owner. Some agreements allow the operator to recover some expenditure made during exploration, once the production phase has started. Subsequent bidders in the same region where oil has been discovered would have to pay higher fees to buy in.

Farm-In and Farm-Outs

Farm-in refers to the case where a company invites other companies to buy into their acreage. In that case, the new company evaluates the potentials for the block and then decides whether to buy in. The idea is that the second company may see what the first did not or that the first sees the potential but is unable to process it to fruitfulness. The company giving out the equity is said to be **farming-out** while the one obtaining it is said to be **farming-in**. Sometimes the original company may only invite a different company to be a technical adviser or to promote and be the full operator, to handle the technical and infrastructural aspect of the development for some *shares*. In rare occasions, a farm-in of one acreage could follow a farm-out of another by the parties, or vice versa, where it is virtually a trade-by-barter.

An oil company may farm out some of its acreages in order to concentrate effort in other projects. Some companies farm out because the block is considered marginal and no longer worth their effort at a higher developmental scale. *Marginal fields* are discoveries that have not been exploited or harnessed in full either due to its small size of reserve, or lack of infrastructure and technological constraints, or even political instability and government levies on them. They become commercial fields if economic and technical conditions become favourable.

Smaller oil companies may farm-in to a marginal field to save the initial cost of exploration and to optimise their budgets. Some of such marginal fields have turned out to, in some cases, produce longer than expected, probably because of a reduction in the risk and high cost of investments involving collaboration, co-ventures and other forms of partnership, which are a continual process in the oil and gas industry. In all cases, it is a risk and until the end, no one can be sure who will gain or lose.

5.6 Lead and Prospect

The aim of acquiring a mining block is to find hydrocarbon resources. This objective may be achieved by searching out *leads* and *prospects*, sometimes in a known *play* with potentials for the *discovery* of a substantial *reserve*.

Lead

When some basic component of a petroleum pool is identified in a geological unit, it is known as a *lead*. Once there is a lead, there is need to refine it to ensure it is ripe. There are leads that explorationists can hardly ignore. Leads come in different forms depending on the data available. On seismic this could be a *channel fairway* or a *conduit*. Examples of leads are:

- An anticline in a known petroleum field
- High amplitude under a 4-way closure
- "Flat spot" overlain by a local "bright spot"
- Local amplitude anomaly overlain by a fluid chimney
- A known reservoir unit offset by fault

After geological and geophysical investigations in the **exploration** zone, areas indicating possible hydrocarbon potentials are marked out as a lead. Further investigations on the **lead** could show more components that allow petroleum to be trapped if generated under suitable conditions, and the combination is marked out as **prospect**. The relationship between *exploration*, *lead* and *prospect*, where one process results in another, is shown in Figure 5.6, where the latter is a subset of the former respectively. Once a prospect is identified, the approval team will ask for more studies. Top management and regulatory authorities will ask for further evaluation.

Figures 5.6: A common illustration of the relationship from exploration through a lead to a prospect

Prospect

For each of the above examples of leads, it has not yet been ascertained if they have the four factors that make up a *prospect*:

- *Source rock*: where hydrocarbon is generated
- *Reservoir rock*: where hydrocarbon is stored
- *Trap structure*: where hydrocarbon is accumulated
- *Seal lithology*: which prevents hydrocarbon from further movement

Reservoirs and seals could be identified in a seismic from the nature of their reflection. A seal, shaly seal for example, usually has poor reflection while a sandy fluid-filled reservoir would tend to show a strong anomaly. Regional studies could indicate the presence of seal and provide information on charging history. Traps are seen as structural or sedimentary closures.

An anticline forms a geometry whose area can be estimated in a cross-section, as well as its volume when extrapolated along the strike line. In Figure 5.7, an anticlinal trap is seen, where the reservoir rock (in yellow) is covered by a seal (in grey). The seal envelopes the reservoir, which may extend to the bottom across a *Water Oil Contact (WOC)* and a *Free Water Level (FWL)*. Figure 5.7b shows a circle and the reservoir is part of it, forming a segment of the circle. It becomes possible to estimate the area of the segment, which can be multiplied by the length of the anticline along the strike direction to have an idea of the volume.

Where the reservoir rock is saturated with oil from the crest of the structure to its base, extending to the bottom across a WOC and FWL, the waste zone has to be taken into consideration. The area within the WOC and the FWL is a waste zone, which has dispersed oil that cannot be easily produced.

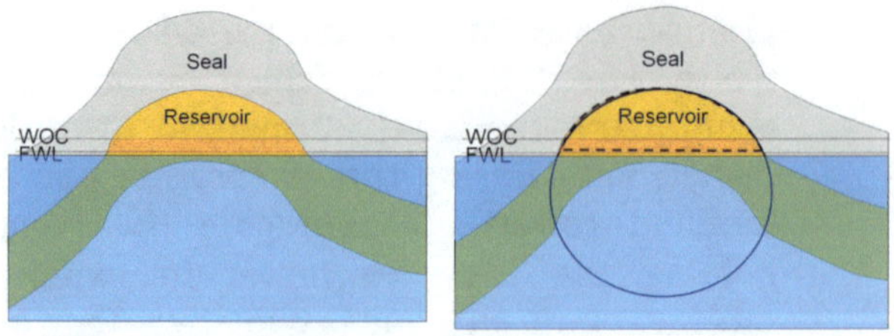

a. Reservoir rock under a seal

b. Reservoir rock as part of a circle

Figure 5.7: Oil-filled reservoir rock and ideal seal combination or cover

The waste zones and the dimensions of the segment illustrated in Figure 5.7 are taken into consideration when assessing the trap. The difference between FWL and WOC might not be overly emphasized during exploration; simply WOC is used often to show the transition.

5.7 Play and Fairway

Play

A play is a region where a petroleum system is thought to produce several prospects or oil fields. In such oil fields, suitable *structures* and *reservoir lithology combination* showing the same trend for hydrocarbon accumulation would have been identified. The word *play* may be used by an oil company to vaunt their reserves or to farm out a block, by implying that where there has been a proven petroleum field nearby, there is a tendency that the other plays would be positive.

Fairway

The seal considered in a play may be large, but the reservoir underneath it may be smaller. In a deep offshore field for example, regional shale could seal the overall units below, but not all these units below are sandy reservoirs. The channel, lobes or other sand bodies that form the reservoirs are localized, running in linear pattern or forming lobular feeds. The area where a play is likely to form a prospect, where the lithology is promising, carved out in (usually linear) geometry is known as a fairway.

The contents of the different lithology are questions asked during block evaluation. Figure 5.8a shows a play highlighting a fairway running across a larger rock unit. The play marks a combination of structures and reservoir lithology, while the fairway marks the channel, lobes or other sand bodies that form the localized reservoirs, which run in linear or sinusoidal pattern or form lobular feeds. Figure 5.8b shows a lateral fluid contact with a fairway which itself will have a contact with a larger rock unit. Lateral fluid contacts can be observed where oil and water are present across a media or structure.

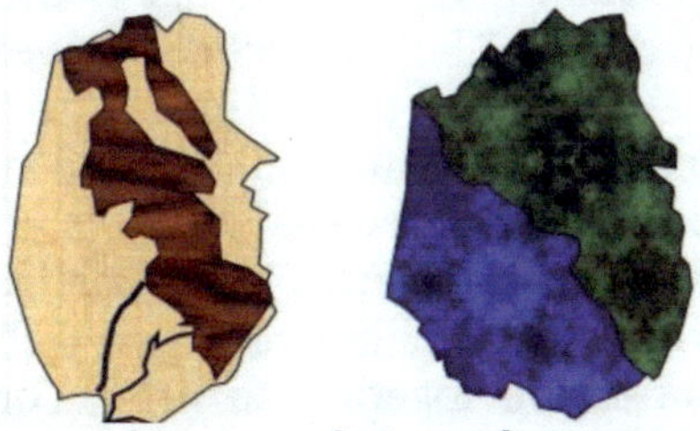

a. Fairway within a rock unit   b. Lateral contact (water blue / oil green)

Figure 5.8: Play highlighting a fairway (channel, Sand drape or Shale plug) in a larger rock unit

5.8 Discovery and Reserve

Discovery

After the initial exploration phase, if a well is drilled and hydrocarbon found after exploration, a *discovery* is said to be made. A discovery refers to some oil or gas pool in a field without specification to the exact position and levels to be produced. The thickness of the column where oil has been found, and the quality of the reservoirs, could give an indication of whether it occurs in a commercial quantity or not. This is especially if other studies have been able to extrapolate the lateral extension of the unit drilled.

Reserve

When hydrocarbon in place has been determined, and multiplied by the factors of recoverability, the estimated amount of substantial oil and gas that can be recovered is known as *reserve*. This can be summed up for various wells to estimate the reserve of a basin. Once one or few wells are drilled, the oil company may use the information obtained from the wells to estimate the total oil that could be in that field. Such information includes the *permeability* or the capacity of fluid to flow in the rock *pore spaces* and the *formation factor*, a factor limiting actual volume size relative to that found in a specific structure type at depth. The total oil in place, in the subsurface, may not all be extracted eventually and brought to the surface and another factor, the *recovery factor*, indicates how much could be produced. Recovery factor depends on technology, so it is relative and variable while the reserve is the standard used to judge a company's petroleum equity.

A well path may be planned (shown with blue lines) to strategically target different depths (shown with dotted orange lines) and lateral extensions (dotted green line) to allow maximum recovery of oil from a reserve (Figure 5.9). The vertical depth is the True Vertical Depth (TVD) of the Well, while the actual distance of the deviated well is the Measured Depth (MD). Several wells may be drilled from a surface *manifold* or one well is drilled and *sidetracked* to target multiple areas. The latter is known as multilateral wells, some of which may have a slip sleeve as part of their completion. Slip sleeves are sliding screens that may be opened and closed to produce different intervals.

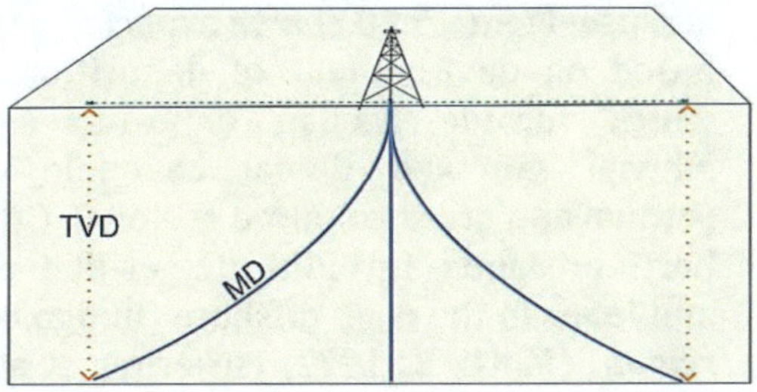

Figure 5.9: A well plan targeting different depths and lateral extension as a recovery strategy for an oil reserve

Reserves are a major consideration in the decisions relating to exploration and production and the nature of regional and local seal in an area, and how they are properly evaluated play a major role in the estimation (once the initial wells have been drilled to get tangible data). They also help to define and calculate the degree of uncertainty.

5.9 Reservoir Expectation in View of Field Development

Oil Field

An oil field is the land area or region where a petroleum system has been established, where oil is being explored, or where oil has been found. A *petroleum system* is a body of sediments (usually within a basin), where there is a source rock that has matured and migrated to charge adjacent rock bodies. Usually, oil seeps, indicators on seismic data and a nearby proven oil field development, are used to determine a petroleum system.

Expected Reservoirs

The unit bearing the oil, usually sand or limestone is known as the reservoir, and they could occur as several layers in the subsurface. Before drilling a well, and for subsequent wells after a discovery, hydrocarbons are associated with specific reservoirs in specific formations in a petroleum field that are expected to be encountered at a given depth. Such specific oil and gas objectives are known as *targets*. When acquiring a block the geoscientists are thinking of what geologic scenarios may have such reservoirs in the subsurface.

Some environments of deposition are indicative of reservoirs, which can only be deduced by geoscientists working on a field or blocks acreage. Figure 5.10 shows expected reservoirs based on environment of deposition, among which include aeolian deposits, estuarine, alluvial fan and fluvial channels onshore (brown and green coloured regions). Others are barrier islands, turbiditic finger-like channels and lobes in the deep offshore (indigo coloured region) (Mutti, E., 1977; Pickering, et al., 1989; Mulder and Cochonat, 1996). These are all targets for onshore and offshore oil field development.

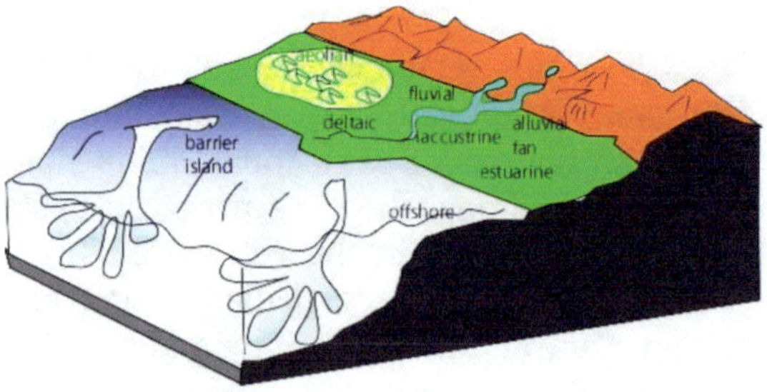

Figure 5.10: Environment of deposition suggestive of reservoir lithology and trap formation

5.10 Block Appraisal

When an exploration well is drilled in an oil block, either a discovery is made (find reasonable amount of hydrocarbon) or it is a *dry hole* (absence of reasonable hydrocarbon). Even where it is a dry hole, the results are often useful to the understanding of the field and refining approaches to further exploration. Where the result is positive, the discovery will be *appraised* in order to prepare the development of the well – that is the various drilling of producing and injection wells and the eventually production of the crude.

The process whereby discovered petroleum is assessed for its commercial value is known as **appraisal**. The difference between appraisal and prospect evaluation is that appraisal happens only after oil has been discovered, whereas prospect evaluation is on a yet-to-be discovered project. Appraisal actually may involve the drilling of more wells, tying the information together to verify connectivity or compartmentalization, and estimation of actual potential of production vis-a-vis market prices and delivery costs. For an appraisal to be effective, new well data, follow-up seismic, improved processing and interpretation might be necessary.

In the initial stage, the nature of the fluid, oil or gas, and their quality is determined. This is achieved by evaluation for a global picture of the oil pool based on:

Thickness of the Reservoirs

The thickness of a reservoir can be measured geometrically from a well or seismic data. In oil appraisal, it is most commonly represented on a map that is interpreted alongside 1D depth provided by and tied to the well. They are mapped out from a seismic block, whose time values (results from seismic data acquisition) would have been converted to depth in an important process known as time-to-depth conversion. With the right depth and seismic block, the fairways in the field are carved out and the oil pool envelop established.

Several wells can help to understand the trend of the reservoir laterally, their varying thickness and quality. The quality of reservoir is obtained by extracting properties based on amplitude anomalies and facies from seismic data. Interpretational software allows the user to extract and observe several of these properties, including *coherency* (continuity or discontinuity of response, leading to orderly and consistent alignment of areas of the reservoir), *chaotic signature* and *averaging of amplitude* to highlight anomalies from outliers (that indicate sand bodies which could serve as reservoirs). Others include *Amplitude Variation with Offset / Amplitude Variation with Angle* of incidence (*AVO/AVA*), which are indication of differences in lithology and fluid content.

Non-Sand Reservoir Oil Containers

Petroleum prospecting today has gone from *conventional* to *unconventional* hydrocarbon. Oil can be produced from almost any geologic structure, where it is generated or to where it has migrated. This includes *basement complex* or other hard but fractured rocks. A situation where there are a combination of reserve in conventional reservoirs, unconventional seal and basement rock is illustrated in the Figure 5.11.

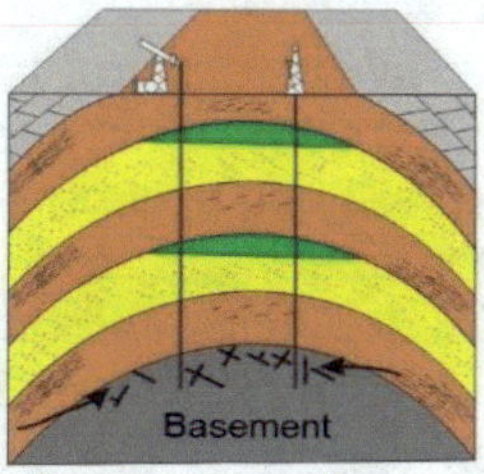

a. Reservoirs and fractured basement

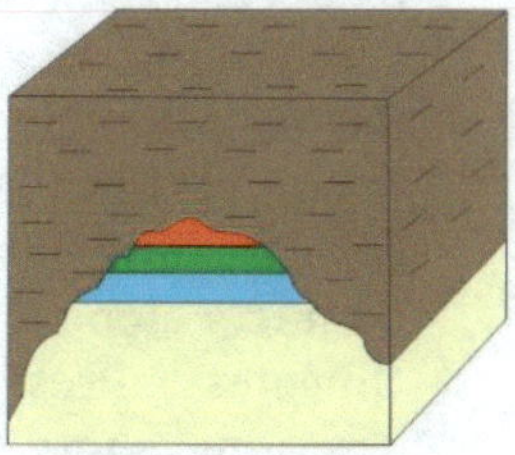

b. Reef structure

Figure 5.11: Accumulation of gas and oil in fractured basement rock on an anticline and on a reef structure

Hydrocarbon will flow to areas of least resistance; they can be exploited if accumulation is economical. Any fractured rock, even fractured basement, can be a reservoir rock. When an anticline or dome is originally folded, the brittle basement rock is often fractured along the crest of the fold. Oil and gas form in source rock along the flanks of the anticline at a lower elevation than the crest and migrate up into the fractured basement rock. A reef is a bar of rock, sand or coral lying beneath the surface of water. Examples include *barrier reefs*, *atolls* and *pinnacles*.

Delineating of Vertical and Lateral Boundaries

The amount of oil in a pool is not only determined by the vertical extension, but also by the lateral extension as no pool of oil runs indefinitely. These vertical and lateral boundaries define the limits or boundary, and the extents of the reservoirs. Vertical boundaries are usually in the form of regional seals, while lateral boundaries could be in the form of faults. The faults cutting through the reservoirs can be mapped across the seismic data and used to proffer the limits where they serve as *sealing faults*, and as well used to establish connectivity where they serve as *conduits* (leaking faults). Fault analysis to establish sealing and leaking is handled by experienced geoscientists or structuralists in the team.

It is important to establish these features in their right places in both the surface and subsurface maps at each level. Most of the interpreting software also provides contours automatically, making it easier to not only qualify these features, but also quantify and place them relative to their position in geological *time* and *space*. A software may make it easy to relate the role of a fault to the thinning and thickening of a unit during tectonics. Software also provide ***isochore*** (line or map showing areas of equal *vertical thickness* of a layer) and ***isopach*** (line or map indicating areas of equal *stratigraphic thickness* of a layer). It is the *true stratigraphic thickness (TST)* that is extrapolated along a dipping or folded reservoir layer, to get the gross rock volume, between the upper and lower lithological boundaries. The difference between isochore and isopach is shown in Figure 5.12a. Another example is the representation of events in stages, to show evolution of bedding units, by *geological restorations* of structural and sedimentary features.

Vertical and lateral reservoir boundaries define the reservoir extent where hydrocarbon can be trapped. Faults acting as lateral boundaries can dissect a continuous reservoir lithology, making it to be compartmentalized, more expensive and difficult to produce, as shown in Figure 5.12. In a., oil (green) trapped below a regional lithological seal, which serves as vertical limits or boundary of the reservoir. In b., illustration of lateral compartmentalization leading to additional cost of field development as more wells are required to drain the reservoirs. A geologic knowledge of the environment of deposition helps to locate where to expect reservoir limits and to assess the value of the field, block or acreage.

The point bars (green) and channels (light blue) shown in Figure 5.13 are usually sandy and trap the reservoir against shaly non-reservoir walls.

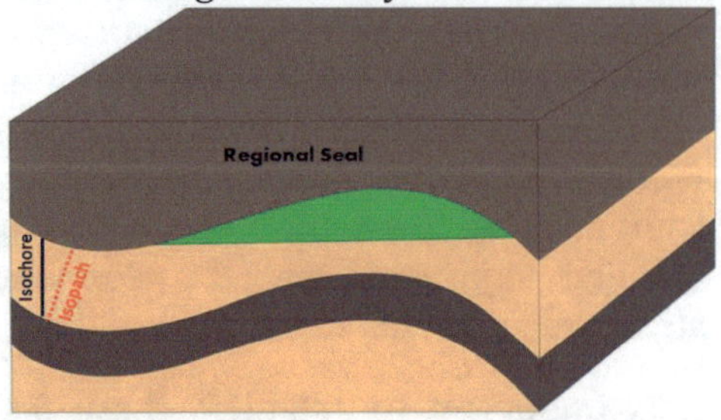

a. Vertical lithological seal or limit (stratigraphic boundaries)

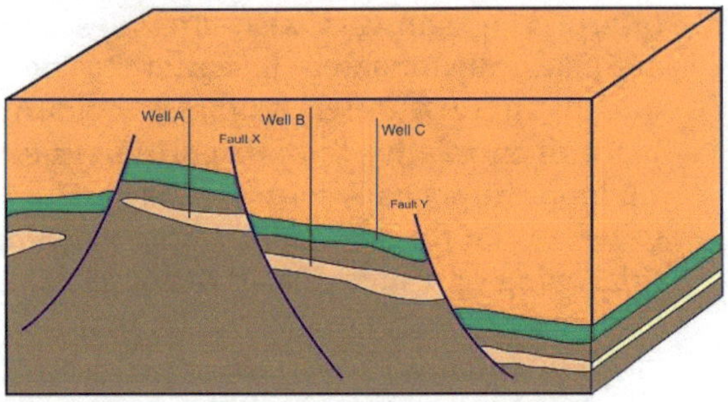

b. Lateral structural limits or seal (fault boundaries)

Figure 5.12: Vertical and lateral reservoir boundaries – a major consideration for estimating the value of an oil block

The *fluvial deposit* is formed from a *braided river channel* and could define the lateral and vertical limits of fairways (when it cuts through) within a non-reservoir unit that buries it. For example, a sand boundary at the *levees* may result from braided river flowing within a shaly unit.

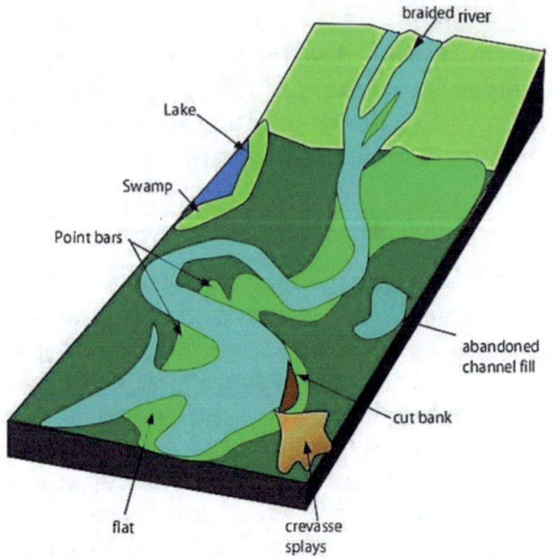

Figure 5.13: Fluvial deposit related reservoirs showing lateral and vertical limits of fairways

Establishing the *water-oil-contact* and possibly *gas-oil-contact* is important in properly appraising an oil field. With a well and pressure points from logging, gradients could indicate the fluid contacts along a well path, hence defining fluid thickness. Pressure measurements and fluid sampling show which region of the well has what pressure or fluid, but it is necessary to identify at least three points in a pressure-depth plot to obtain its gradient.

Establishing Contacts and the Hydrocarbon Heights

If the seal is *efficient* (no hydrocarbon leakage through the seal) and hydrocarbon is filling it to spill, then the *spill point* is another very important factor to identify during appraisal. In which case the spill is used to estimate the available volume trapped. The spill point, unlike the contacts, is not always a plane, but a location somewhere along the circumference of the pool. Therefore, it is important to assess it on a map; for a three-way closure, or a dome only open at one side below, it is where the contour did not close at that depth (Figure 5.14a).

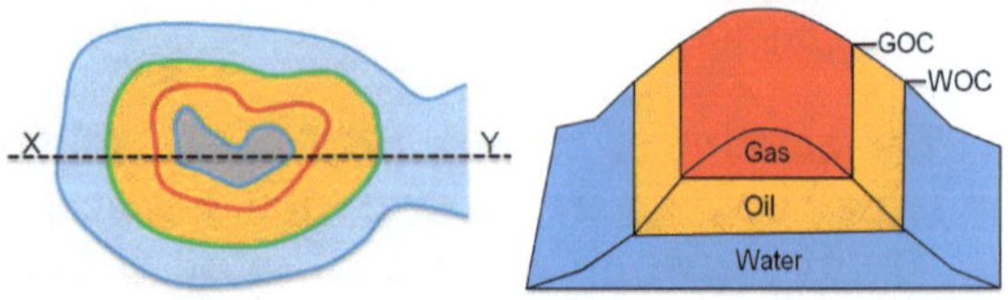

a. b.

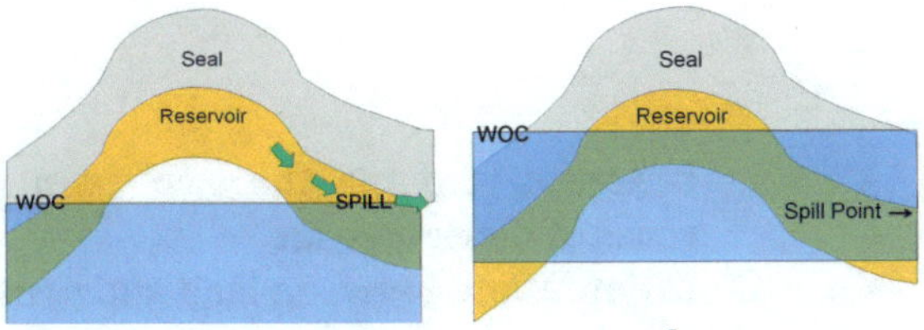

a. Map view with contour showing oil spilling towards 'Y'
b. Block view showing possible estimation of volumes
c. Trap filled to spill with oil (from top to spill point)
d. Trap not filled to spill with oil (spill point within water)

Figure 5.14: Map with contour (in a.) and contacts (Gas Oil Contact and Water Oil Contact in b.) and spill / spill point (in c and d.) for an anticlinal trap

Figure 5.14 c. and d. shows scenarios where a trap is filled to spill and the other where a trap is not filled to spill with oil. The spill point is located at the base of the structure, which marks the possible deepest part of oil accumulation. Knowing the GOC and WOC can help to determine the thickness and volume of the oil to be produced. In block form, gas, oil and water are identified (gas is red, oil is orange and water is light blue, Figure 5.14b) and one begins to see that the volume can be estimated with basic mathematical formula.

5.11 Commercial Evaluation of Oil Field

At the end of an appraisal process, with all the factors put together, risked and uncertainty determined, the major question of whether the developmental project will be economical is asked. Several factors will be considered along with their costs, often in millions of dollars. These will include, among others, the cost of rig site installations, actual drilling, HSE, tax and revenues paid to the government, welfare and transport. These costs would need to be offset by produced oil and gas, and as such will depend on the rate of production and oil price over the lifetime of the field.

A barrel of oil usually goes for a range of prices from high to low. It is important to go by the lower oil price during the calculation, as there will be no loss if oil price eventually rises, rather than lose money because prices went down. The initial calculation of volume of oil in place, which depends on the area and extent of the traps, is important, although actual recovery over time will vary.

An idea of the gross volume of oil in trap can be estimated by multiplying the area of the trap (in Figure 5.15, it is a value close to 30 x 100 m) by the lateral extent if known. This gross or bulk volume includes the rock mass, and must be multiplied by the porosity, and by the oil saturation of the Gross Rock Volume (GRV), to get the actual volume of oil trapped.

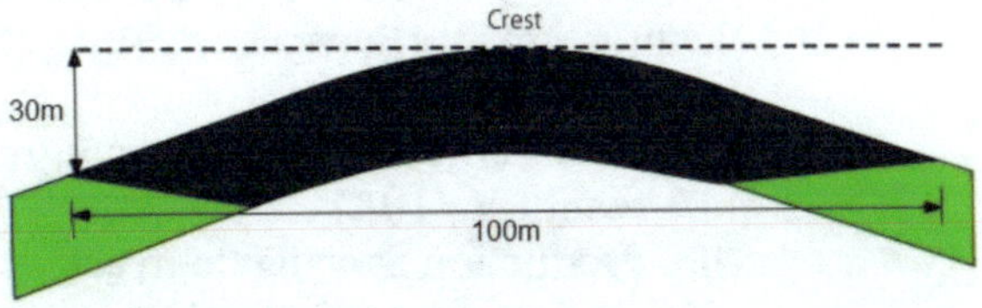

Figure 5.15: Structural envelop and the dimensions necessary to estimate the area and eventually volume of crude oil (in black) in an oil trap

Once the available oil resources are known, the times it will take to reach *plateau* production and *decline* are evaluated and compared with the trend of the markets. This aspect is usually managed by the corporate business department in conjunction with the technical geosciences and reservoir department. The actual cost of drilling the wells is seen as fixed expenditure limited in time and known as *capital expenditure* (CAPEX) while the subsequent cost of running and producing the well is known as the *operating expenditure* (OPEX). The geosciences and drilling departments will have to provide the technical backing to check this, in collaboration with the corporate business department in an oil company.

If it is not economically favourable under the current technical and economic conditions, where the sum of CAPEX and OPEX is greater than the income from producing the field, an oil company may not proceed with the production of the oil field. It is possible it becomes commercially profitable to produce in the future when technological advances and oil prices make it favourable to produce.

EXERCISES

1. What are the processes in field aquisition and development?

2. Explain the principles of remote sensing and GIS.
   i. What is stereoscopy?
   ii. Mention different types of satellites and satellite images.
   iii. Give basic aerial photo techniques and software.

3. Explain the different methods of block acquisition and licences:
   i. Concession
   ii. Joint Venture / Operations Agreement (JV / JOA)
   iii. Production Sharing Contract /Agreement (PSC/PSA)
   iv. Risk Sharing Contract (RSC)
   v. Service Contracts (SC)
   vi. Marginal Field Concession (MFC)

4. Differentiate between petroleum fields and frontier exploration zones.

5. Describe in detail the two main processes of block acreage
   a) Differentiate between lead and prospect
   b) What are the four factors that make up a prospect?

6. Differentiate between play and fairway.

7. Differentiate between a discovery and a reserve.

8. Differentiate between a field and a reservoir target.

9. How is the commercial value of an oil field evaluated?

10. a) What is the difference between appraisal and prospect evaluation?
    b) Describe the importance of the following in block appraisal:
    i. The thickness of the reservoirs
    ii. Delineating of vertical and lateral boundaries
    iii. Establishing the contacts and hydrocarbon heights

REFERENCES & FURTHER READING

ArcGIS, Environmental Systems Research Institute (ESRI). Version 10.7.1, June, 2019, http://www.esri.com/software/arcgis

Carte topographique IGN, 2004. Carte topographique IGN à 1 : 25,000. Institut géographique national, ed. 2. 107 rue la Boétie, Paris.

3DEM, Digital Elevation Model, version 20.7, *Free Geography Tools,* https://freegeographytools.com

England, W. A.., A. S. MacKenzie, D. M. Mann and T. M. Quigley, 1987. The movement and entrapment of petroleum fluids in the subsurface. *Journal of the Geological Society*, v. 144, p. 327–347.

Evamy, D. D., J. Haremboure, P. Kamerling, W. A. Knaap, F. A. Molloy and P. H. Rowlands, 1978. Hydrocarbon habitat of Tertiary Niger Delta. *AAPG Bulletin*, 62, p. 1–39.

Mulder, T. and P. Cochonat, 1996. Classification of offshore mass movements. *Journal of Sedimentary Research*, v. 66. no. 1. p. 43–57.

Mutti, E., 1977. Distinctive thin bedded turbidite facies and related depositional environments in the Eocene Hecho Group, South Central Pyrenees, Spain. *Sedimentology*, 24, p. 107–131.

Nosike, L., 2017. Overview of the Petroleum Industry: *Delizon Publishers,* 191p.

Nosike, L., 2019. *A Career in the Petroleum Industry: In a Time of Energy Transition*. Delizon publishers. 253 p.

PC World, 1998. Transparent Publishing with Deneba's Canvas 6. December 31. Archived from the original, 19 October 2012. www.pcworld.com

Pickering, K. T., R. N. Hiscott and F. Hein, 1989. Deep marine environments: Clastic sediments and tectonics. Chapman and Hall, London, 416p.

SPOT, Satellite Pour l'Observation de la Terre. *Centre national d'études spatiales (CNES)*. Retrieved June 2006.

TeraVue, 2006. Le site TeraVue. *Éditions de la Boyère*. http://www.teravue.com

CHAPTER SIX

EXPLORATION GEOLOGY AND GEOPHYSICS

6.0: Generalities of Exploration

Much literature exists on exploration geology and geophysics. So, the aim of this chapter is not to go into details of geological and geophysical techniques, but to summarise the principles behind them. The chapter looks at the aim of exploration, a phase that identifies important factors in a petroleum system, in the search for oil and gas; it reviews direct hydrocarbon indicators with examples. To be learned in this chapter are:

- Aims of exploration
- Exploration Phase
- Hydrocarbon Indicators
- Exploration methods
  - *Gravimetry,*
  - *Magnetography,*
  - *Downhole logging*
- Seismic method
- Seismic interpretation
- Correlation with offset wells

Structural and sedimentary mapping approaches are discussed, with example of field work and oil field exploration case studies. All this plays a role in the understanding and correlation of subsurface data, from onshore through the shelf to the marine environment, to consolidate on exploration results.

6.1: Aims of Exploration Geology and Geophysics

Exploration geology and geophysics have several applications worldwide. In the oil and gas industry, geologic and geophysical principles are used to find new petroleum reserves that can be exploited. This is done by searching for new sedimentary basins where there is an oil pool, and sometimes existing petroleum fields for not yet discovered levels of hydrocarbons and hydrocarbon bearing structures. Explorationists try to identify the process of the generation of hydrocarbon from the source rocks and their trapping within a stratigraphic or structural unit.

The word *exploration* means a careful systematic search. *Exploration geology and geophysics* imply the careful systematic search using basic principles of geology and geophysics. In the oil and gas industry, these methods are applied in the search for hydrocarbon within suitable structures in the subsurface. The aim here is to know where those structures are, and if such structures have accumulations that were formed and transported within suitable rocks. Usually, exploration is carried out in *sedimentary basins*; this is an environment or location where broken down particles of rocks known as sediments have accumulated over time and formed layered rocks called sedimentary rocks (Figure 6.1).

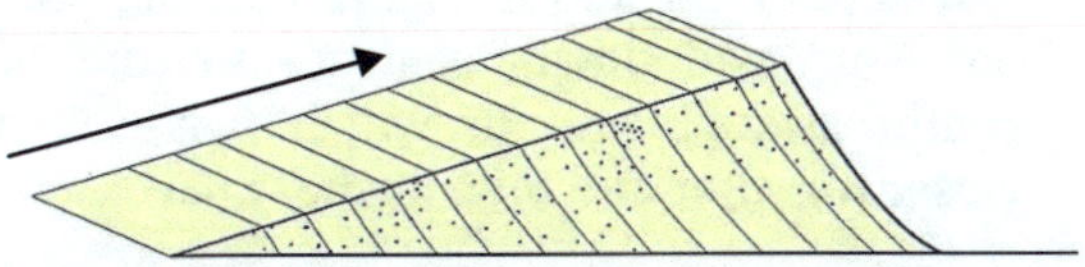

Figure 6.1: Ideal near-vertical geologic layers where every unit is exposed on the surface

One of the duties of the exploration geoscientist is to make field observations by studying exposed rocks. These exposures are known as *outcrops*, which constitute part of a much larger feature below the Earth surface. Information obtained from this can provide an idea of what structures and lithology could occur at depths. As the exploration geologist moves in the direction of the arrow in (Figure 6.1), he knows that there is a relationship between what he sees on the surface and what is at depth, if the rock were to be cut.

Some of the principles of exploration are:

1. Physical observation as a means to predict the unseen rock types at depth
2. Vibration targeting materials at some distance or depth
3. Event recording of measurements and feedback

Physical Observation

Physical observation and measurement as means of predicting the unseen at depth is the most ancient form of geological exploration. It covers the observation of *attitudes of beds* (strike and dip), *oil seeps* (natural leakage of hydrocarbons across geological layers to the surface) and *field mapping* (field activities aimed at collecting structural and geologic information that might be useful in identifying a petroleum system).

Vibration and Wave Travel

When an object is made to vibrate without the object itself being deformed, then a wave has moved through the object. Wave travel in seismic exploration can be *compressional* (particles of the medium oscillate in the longitudinal direction of wave propagation) or *shear waves* (particles of the medium oscillate at right angle to the direction of wave propagation), which are the two main modes of *acoustic energy propagation* in solid Earth. For shear waves (as indicated in Figure 6.2) the particles move up and down, contrary to that of the compressional case in which particles move closer together and then farther apart.

Before the wave travel, at time T_1, the material remains undeformed. As waves begin to propagate through the material, at time T_2, the material begins to vibrate without necessarily being deformed. As soon as the wave propagation stops, at timeT_3, the object returns to its original shape.

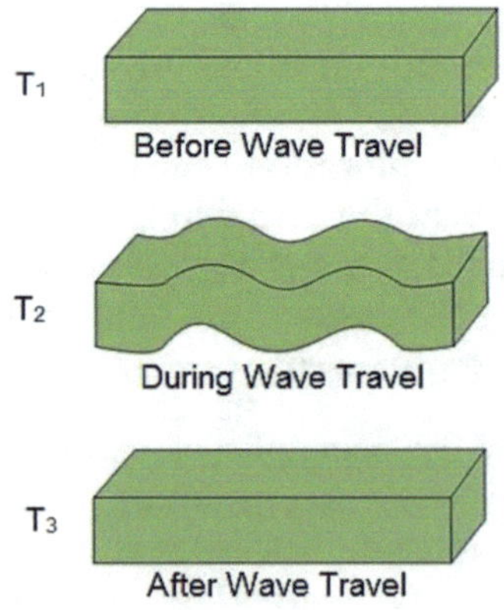

Figure 6.2: Time T_1 to T_3 showing wave travel in a material without deforming it

Event Recording

The principle of exploration requires that natural events *observed or induced* and measured are recorded on the surface laterally and at depth, if the depth is inferred. This recording of depth or predicted depth based on surface observation or measured induced events forms the bases of modern exploration tools. With seismic acquisition, the seismic data are displayed to represent the vertical cross section of the Earth. The original vertical scale is in seconds but this is eventually converted to the equivalent depth. Across the top of the record are the *shot point* locations. On the side of the record will be a header displaying information such as the seismic line number and how the data was acquired, processed and displayed.

During exploration, it is required that natural events observed, which could also be induced, are measured and recorded on the surface laterally and at depth vertically. The records of information at the surface and at depths based on what is naturally observed or artificially induced forms the bases of modern exploration tools, with the seismic acquisition tool being the most widely used for hydrocarbon exploration, albeit expensive. With seismic acquisition, the seismic data can be displayed to represent the vertical cross-section of different layers of the Earth (Bouvier et al., 1989; Brown, 1996).

The original vertical scale for seismic records is usually in seconds, or milliseconds, but these are eventually converted to the equivalent depth. Figure 6.3 shows a seismic event recording format on the surface and at depths. Across the top of the record are the *shot point* locations at different distances on the surface, where energy is transmitted so that wave can propagate into the earth and received at some distance away from the shot point. Adding a *header* at the top of the record could provide o other information such as the seismic line number and how the data was acquired, processed and displayed.

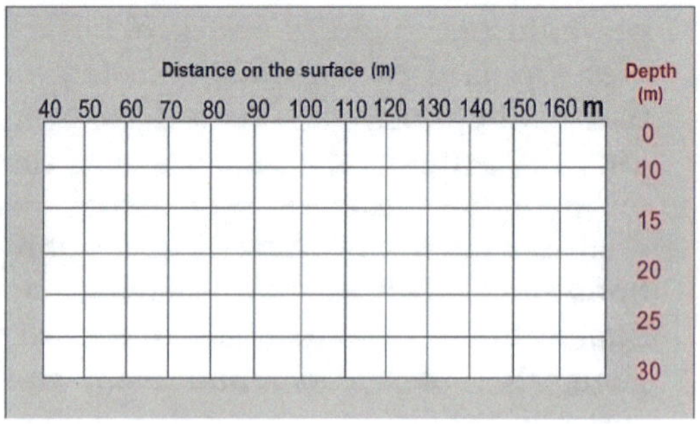

Figure 6.3: Event recording format showing distance on surface and the depths

The aim of petroleum exploration is to find suitable hydrocarbon traps. Once a structure that may contain hydrocarbon found, the exploration geoscientist asks questions on whether the structure has good charging of either oil or gas, or both, within a reservoir that is well sealed. The age of the basin, and source rock, and the various tectonic processes, burial and subsidence it has undergone is important. Surface and environmental location in present day is key to the success of exploration and accessibility for subsequent development in case of success.

It is the duty of the petroleum geoscientist to investigate the sedimentary rocks within the basin, to know if there are suitable structures present that will allow for hydrocarbon to accumulate, and if oil and gas so formed are prevented from leaking away through rocks which serve to prevent such from happening. Sedimentary basins differ based on certain conditions within each basin, which may make it favourable for hydrocarbon generation and accumulation compared to other basins. Some of these conditions include the age of the basin (older basins are "more mature" and have higher tendencies to generate and accumulate hydrocarbons), source rock maturity (which determines whether hydrocarbon has been formed under suitable conditions), tectonics, burial and subsidence (which determines the environment for deposition of sediments at various depths). Surface and environmental locations are considered, as they can contribute to the success or failure of exploration.

6.2: The Exploration Phase

Search for a Prospect

Four elements are important for hydrocarbon to form and accumulate, and their combination has been defined as a *prospect*. The objective of the exploration phase is to be able to identify these elements. These include:

1. A petroleum source rock
2. A reservoir lithology
3. A petroleum trap
4. A suitable seal

For hydrocarbon to generate there must be a petroleum **source rock**. An example of a source rock is shale. For a rock to be termed a source rock, it must have *organic matter* buried under suitable *temperature* and *pressure* conditions and *depths* in the subsurface. Under these conditions, the organic matter within the source rock undergoes various processes before it is transformed to hydrocarbon. Once hydrocarbon is formed, it is expelled from the source rock into adjacent carrier strata and eventually settle in a **reservoir rocks**, if they get to a **structure** where they begin to accumulate. A suitable structure or combination of features for petroleum accumulation is known as a **trap**.

In a successful trap, the reservoir rock must be overlain by a *cap rock* or ***seal***, which is an impermeable rock layer that does not allow fluids to flow through it. Hydrocarbon is prevented from escaping away from the trap by **seals**. Without a cap rock or seal, the gas and oil would leak up to the surface. For a seal to act as a barrier to flow, despite having pore spaces as well, such pores need to be very tight. Rocks with smaller grain sizes usually tend to have smaller pore spaces. Shales serve as seals. Hydrates and coals are other common examples.

Figure 6.4 illustrates an oil pool and a gas cap showing the profile of impermeable cap rock (brown folded layer of rock) and trapped fluids (gas, oil and water). The gas oil contact (GOC) and water oil contact (WOC) are also shown. The trap so formed is an *anticlinal trap*; its duty is to allow hydrocarbon accumulate within the reservoir rock. Without an impermeable cap rock, migrating hydrocarbon would leak away to the surface, forming what is known as oil seepage, and there may be no accumulation at all.

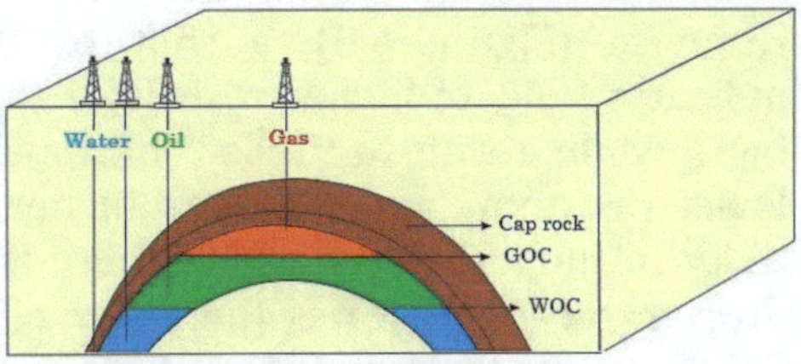

Figure 6.4: Oil pool and gas cap in a petroleum trap

Use of Well Logs to Identify Hydrocarbon Intervals

One way in which geologists tell if there is a good reservoir, with oil and gas in a trap while drilling, is to use *logs* (a detailed record of geologic formations penetrated at depths in a well), a combination of *electrical* or *radioactive* logs, for example. Two common examples are the *Gamma Ray* (GR) and the *Resistivity* (Rt) logs, which are radioactive and electrical logs, respectively. Figure 6.5 shows GR and Rt log responses.

GR log response can differentiate reservoir lithology from non-reservoir lithology. This is because reservoir lithology such as sandstone will have lower radioactive materials and therefore lower radioactive responses, compared to non-reservoir lithology such as shales that have higher GR responses due to higher presence of radioactive materials in them. For the GR log response (Figure 6.5), a shift to the right indicates higher radioactivity (shale), while a shift to the left indicates lower radioactivity (sandstone).

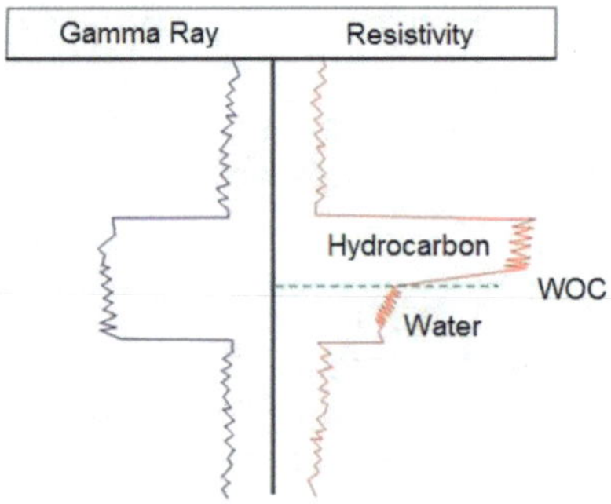

Figure 6.5: GR and Rt logs showing low gamma (good sand) and high Rt (presence of hydrocarbons)

One of the things Rt log response can do is to differentiate hydrocarbon fluid from formation water (with high salinity) known as brine, as hydrocarbon is non-conductive compared to conductive brine. Therefore, in the Rt log response (Figure 6.5), a shift to the right indicates fluid of higher resistivity (hydrocarbon), while a shift to the left indicates fluid of lower resistivity and therefore higher conductivity (brine). The area where there is a sharp drop in the Rt indicates the water oil contact (WOC) within the reservoir.

Hydrocarbon bearing intervals may be correlated across different prospects, where the well already drilled is used as an "offset well" with its log guiding in the target of a new well.

Mapping Traps

In order to carry out investigations of a trap, there is need to provide a representation of the objects in map and section view (Woodward et al., 1989; Tearpock and Bischke, 1991; Mansfield and Cartwright, 1996; Wood and Ruppel, 2000). The map view shows the extent of investigation on the surface of the Earth, while the section view shows a representation of the environment at depths. Exploration geophysicist provides the seismic data by carrying out data acquisition campaigns, processing them to create downhole imageries.

An example of a trap is an anticline, which is a common geologic structure associated with hydrocarbon accumulations in the oil and gas industry. To identify and characterise an anticline, the geoscientist has to provide a visual aid representing the system on a surface and section. Figure 6.6 shows an oil rim on a map and on a section. The free gas cap is shown with the producible oil rim below it. The map view is the *lateral* view across the surface of the oil accumulation, while the cross section view is the *vertical* view that tends to slice the oil accumulation so that it can be seen as a section at depth.

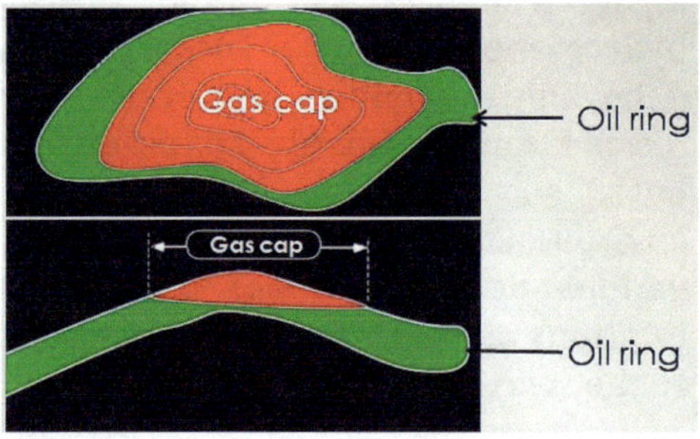

Figure 6.6: Map and Cross Section view of an oil accumulation

The image in Figure 6.6 shows the areal extent of a gas cap (red) and an oil rim (green) as will be observed on the surface. The oil rim lies below the gas cap; such separation of both fluids is due to their different densities (water will be below the oil rim where present as it is denser than oil). Below the map is the cross-section view, where the separation of the free gas cap and oil rim is much easier to identify or describe; it shows the different fluid levels, both located within an anticlinal trap.

Sometimes there is need to divide formation into layers for a more indebt investigation, often provided by software. Figure 6.7 is an example of a layer-based investigation, involving the division of the area of interest (a sedimentary channel / sandstone reservoir) into different layers (1 - 4), both vertically and horizontally. This is done so that minute or smaller scale *heterogeneity* (diversity or dissimilarity) and *anomaly* (deviation from the expected) could be identified. Such investigation may be aided by use of software.

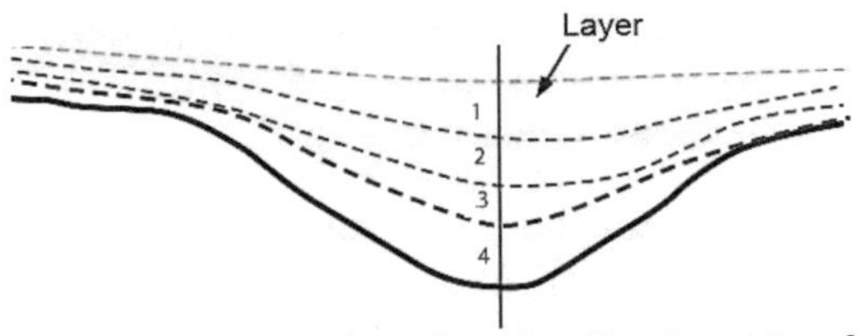

Figure 6.7: Layer-based investigation involves dividing the area of interest in layers both vertically and horizontally

The exploration geoscientist applies different geologic and geophysical techniques to carry out studies and generate maps, which are interpreted and worked upon together with drillers and reservoir engineers to drill wells at specified locations.

6.3 Hydrocarbon Indicators

How are oil and gas traps identified? In the early days of exploration, oil seeps on the surface were a major indication of presence of hydrocarbon. There was not much advancement in technology that could identify petroleum systems in the subsurface. Early explorationists used indications of oil seep, where natural flow of hydrocarbon through various geologic layers occurs at the surface. Few explorationists were lucky, as the oil seep was close to the accumulation in the subsurface, and as such exploited it by drilling wells a few metres deep; others were not as lucky, as the oil seep had travelled far away from the accumulation zone (Thrasher et al., 1996; Heggland, 1998). The need to identify accumulations in the far subsurface became a necessity, and so advancement in technology led to developing different methods used to explore and identify hydrocarbon accumulations at depths (Le Chevalier, 2002; Posewang and Mienert, 1999).

Among indicators of oil trapping are:

- Oil Seeps
- Statistics
- DHI (Bright Spot, Dim Spot, Flat Spot and Polarity flip)

Today the use of seismic data with identifiable oil signatures in the amplitude serves as *Direct Hydrocarbon Indicators* (DHI). DHI could indicate to a reasonable degree, the lithology, structures and possible fluid types in the subsurface. **DHI** include ***Bright Spot*** which is a strong amplitude contrast in a porous lithology suggesting more fluid content such as oil or gas (or even water). Where there is a horizontal line cutting across lithologies and structure/fold, it is said to be a ***flat spot*** indicating fluid contact, essentially water oil contact.

A weaker amplitude than its surrounding it is said to be a ***dim spot***, suggesting changing response which in some cases is due to changing fluid or introduction of hydrocarbon. Strong signal on seismic cross section is known as *peak*, while weak signal is known as *trough*, and a change in their continuity is known as *polarity flip*. Presence of hydrocarbon may also lead to a ***change in polarity*** of seismic response, where a dark trend becomes bright or vice versa.

Other solution includes *statistics* showing that oil has been found often in such circumstances. For example, an anticlinal four-way closure (domal structure) with a thick regional seal, above a known reservoir unit in an established petroleum system, will have a very high chances of trapping oil and most oil companies will drill it in their oil block. There is always some level of uncertainty until the first well is drilled. Despite detailed studies using modern-day methods and technologies, there still always remains some level of uncertainty as to whether or not hydrocarbon is present until the well is drilled.

Figure 6.8 shows active oil seeps in Ugwueme, Enugu, Nigeria, sampled by the author during a fieldwork (iESog, 2018). The petroleum potential of the area has been described by various authors (Reyment, 1965; Ekweozor, 1990). This seep is made possible by the presence of active spring, perhaps *perched* or *artesian aquifer*, which washes out the crude oil, indicating not only a petroleum system but suitable traps and possibly slightly compromised seal rock above an overpressure reservoir with natural drive. Detailed analysis is however required to *mature* (that is to confirm) this prospect.

Figure 6.8: Oil seeps in Ugwueme, Enugu, Nigeria

The DHI in Figure 6.9 indicating the boundary and presence of oil and water, above which there should ideally be local high responses (dark shades known as *bright spots*) that can indicate the presence of hydrocarbons. The well path intended to produce the oil is marked by the red line, across the image, from top to bottom. The DHIs in this case are water oil contacts (WOC).

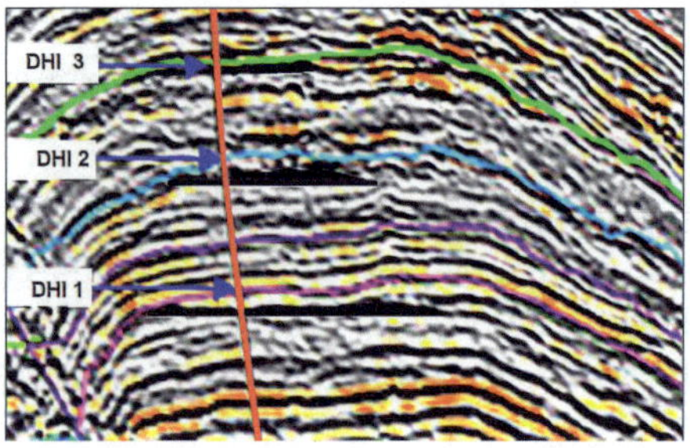

Figure 6.9: Water oil contacts on seismic shown as DHIs (below the markers indicated by the blue arrows)

Despite the exploration and exploitation of several basins in the world for petroleum, some regions of the world and marine environment still have untapped hydrocarbons. Environmental issues are key factors limiting exploration in the Arctic, as the oil industry is becoming keen in only finding oil in areas where it will be less deteriorating for the environment. Exploration and production methods are therefore getting more stringent and requiring experienced geoscientists and engineers.

6.4 Exploration Methods

Several methods are used to acquire data that indicate subsurface elements that serve as accumulation structures or barriers to hydrocarbon flow. These may serve as a guide to whether or not oil is trapped or just provide the initial geological information. Some of the methods are discussed below.

Field Mapping

Geological field mapping involves going to the field to establish areas where rock bodies are exposed on the surface as *outcrops*. Outcrops could have been exposed due to processes of weathering and erosion, uplift or other surface and subsurface Earth processes. The relationships between the rocks are identified and tied together to create a map that describes the surface environment and indicate possible structures beneath the Earth.

Sometimes these outcrops may be covered with grasses or soil, making them difficult to identify; where they are not covered, they become exposures and actual logs and rock descriptions can be carried out on them. Field work however has its limitations, as it cannot easily identify abnormal features hidden below the Earth surface. For example, if the limbs of an asymmetrical anticline (Figure 6.10) are only partially exposed on the surface, results of field mapping may suggest a tilted bed rather than an anticlinal structure. In this asymmetric anticline, the limbs are not regular in shape or orientation. Relying on only fieldwork would result in missing the accumulation of resources (green) trapped within the structure.

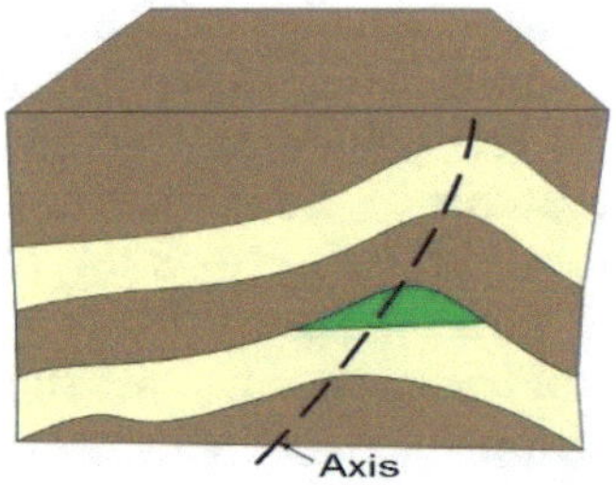

Figure 6.10: Asymmetric anticline showing limbs that are not symmetrical

In order to have more valuable information about the area of investigation, field work should be integrated with other data such as logs, wells or remote sensing, so that a clearer structure may be produced. Parameters sampled along a given plane will have to be represented in 2D and 3D, presenting a visual representation for proper understanding of the parameter under investigation. Take a map for example, it gives surfaces that correspond to the same depth (depth map), or surfaces on which different degrees of parameters can be qualified (such as temperature, pressure, etc.). A prominent feature on maps is a *contour line*, which is an imaginary line that is made to represent, and follows the same feature evaluated, especially the depth values. Generally, contours are used to indicate areas of highs and lows, which could be above sea level or below sea level.

The contour map of an anticlinal structure is shown Figure 6.11, representing depths below sea level. The 500 contour line indicates 500 m, while the 1000 contour line indicates 1000 m. The lines between the 500 and the 1000 are four, each representing 100 m intervals from 600 to 900 m. As these contours represent depths below sea level, their values should rightly be represented with a negative sign to indicate that it is at depth, which are -500 to -1000.

Figure 6.11: Map using contour to designate high and low grounds on an anticlinal structure

Some of the characteristics of a contour line include:

- A contour line cannot cross itself or another contour line.
- It must also form a closed surface (although this may be beyond the bounds of the map).
- Contours appear to merge on vertical surfaces and appear to cross on overhanging surfaces, but in 3D space the contour lines lie one above another.

Not all contour lines are created equal. A simple guide to contouring, to enable easy construction and subsequently easy understanding by the reader suggest that you always mention the depth value of a given contour, or one of the contours, represented in your map. Such would be a reference for reading the other contours or their separation.

The **index contour**, every 5th line for example, should be thicker and in bold to make it easy to read. The lighter contour lines, which are usually four in number, falling between the index contours are referred to as **intermediate contour** lines; they do not have their depth (or elevation, temperature or whatever parameter measured) values indicated. Sometimes **supplementary contour** lines, usually dashed lines, are used to indicate expansively flat terrain or where there is little change in elevation; they are usually half of the elevation between the contour lines surrounding them.

Structural and Sedimentary Mapping

Mapping for oil prospecting are used to establish the presence of structural trapping geometry and the presence (or absence) of good reservoirs and seals – a combination that forms a prospect. It further goes on to evaluate the structural settings of the area under investigation. Information sought for include fractures in rocks of a given lithology (sandstone and limestone for example), and if these fractures are cutting across the seal. A *prospecting map* will be generated, which should be good enough to highlight the geologic history, the possible origins of petroliferous shales and source rock maturity. Visible oil leakages have been very useful, especially onshore, before the advent of seismic maps, hence the importance of carrying out field mapping.

Figure 6.12: (a) Panoramic view showing present day observation of landscape and (b) a sketch in an attempt to reconstruct its geologic history based on field observations

A panoramic view is a wide-angle view from an altitude or distance of present day observation of a landscape. A good attempt, as shown in Figure 6.12, when integrated with regional geology, may help in reconstructing its geologic history based on field observations. These surface features could indicate related lithological and structural features at depth.

With the use of seismic images, both surface and cross-sectional depth images are provided and tied to their stratigraphic layering and structural orientations, thereby highlighting suitable traps and fluid properties. Figure 6.13 highlights the scale of seismic events with respect to observation during field studies. As seen in the figure, structural prediction from surface as seen in field highlights what is obtainable at depth when looking at seismic image. The scales suggest that only the shallower normal faults (red lines) are seen in this field example; the deeper thrust fault (black line) occur at a much larger scale.

Field work has been used to identify oil seeps; to identify likely petroleum basin by field study of exposures, their organic content analyses and the evaluation of structures that indicate possible fluid migration pathways in the subsurface. It requires the structural mapping and establishment of stratigraphic units, measuring the dip and strike of outcropping beds (the attitude of the beds) in order to predict where they will form suitable petroleum structures at depth.

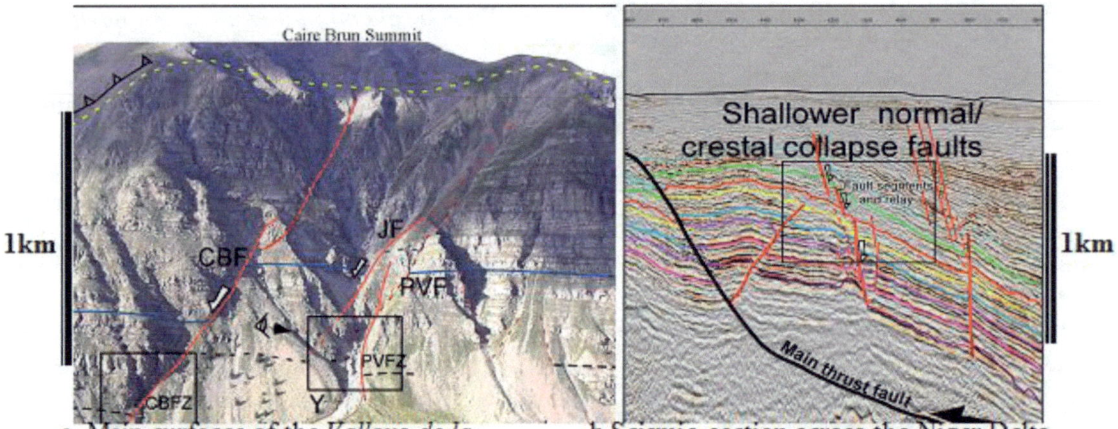

a. Main surfaces of the *Vallons de la Moutière* faults

b Seismic section across the Niger Delta study block showing main fault surfaces

Figure 6.13: Fault planes represented from a distant panoramic view and on seismic section

Aerial Photos, Remote Sensing and Seismic Depth Profiling

Use of pictures taken from aircraft (aerial photos) and those formed by the detection of the Earth surface from satellite (remote sensing) are very useful, especially in the initial reconnaissance (preliminary investigation) of a basin. Satellite images provide even more comprehensive *trends* (measurement of the orientation of a linear geologic feature) and *lineation* (linear structural features within rocks) on the surface which could indicate suitable structures at depth (sample figures of *satellite image* have been shown in the previous chapter). Similar lineation, uplifts and topography buried at depth are shown by *seismic maps*, extracted from a seismic block, as in Figure 6.14.

When points of equal time, age or their depth equivalent is joined by a line, it is known as an **isochron** or **isochrone**. In seismic interpretation, the term refers to a map or contours showing seismic events of same time (or depth equivalents). By showing same event or time with the same colour, isochrones show variation of colours that represent differences in two-way travel times and consequently depths between two or more seismic events (Figure 6.14).

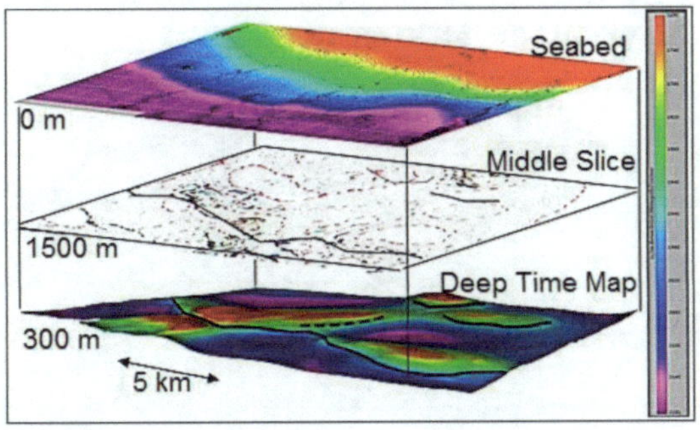

Figure 6.14: Seabed, slice across an intermediate level, and deep isochron map (Nosike, 2009)

The seabed, as the name implies, marks the bottom of the sea, covered with water, at 0 m burial depth from the sea floor (Figure 6.14). From the seismic block, a slice is taken at an intermediate level to highlight visible structural features, at 1500 m depth. The deep isochron displays more features and variation in topography, at about 3000 m depth in this case. Presence of lineation, uplifts and topography changes buried at depth, is highlighted with colour differences.

Gravimetry and Magnetography

Gravity and magnetic methods serve as *reconnaissance tools* for hydrocarbon exploration, before or preceding the use of seismic as the main exploration tool. Data obtained from them help to delineate regions with possible favourable conditions for hydrocarbon generation and accumulation. Gravimetry works by detecting *variations in densities* of rocks, while magnetography measures variation in *magnetic susceptibility*, which is the ability for some Earth materials to be magnetised.

Basement rocks have higher magnetic susceptibility than sedimentary rocks, and so the magnetic method measures magnetic susceptibility from the surface to the basement depth or to magnetic intrusions in the sedimentary rocks. Both methods are more effective when combined during interpretation; they could show the subsurface shape and structure within the basin, as well as indicate areas within the basin having thick sedimentary pile, a requirement for suitable conditions for hydrocarbon generation. Studies carried out using these methods are followed up with seismic exploration, where there are positive indicators, in other to refine the observation and be more precise in the interpretation. Gravimetric and magnetic data acquisition over a fault is shown in Figure 6.15.

In the figure, the magnetic and gravity responses (blue and green lines respectively) would have been straight in the absence of a fault, as there would be no displacement of any of the fault blocks. However, due to displacement, readings that would have been straight are seen to change at a particular distance along the fault zone, but become continuous to the end of the region. The gravity and magnetic equipment (gravimeter and magnetometer) are able to detect this, delineating structural features (faulted regions) as well as various sedimentary thicknesses across the region.

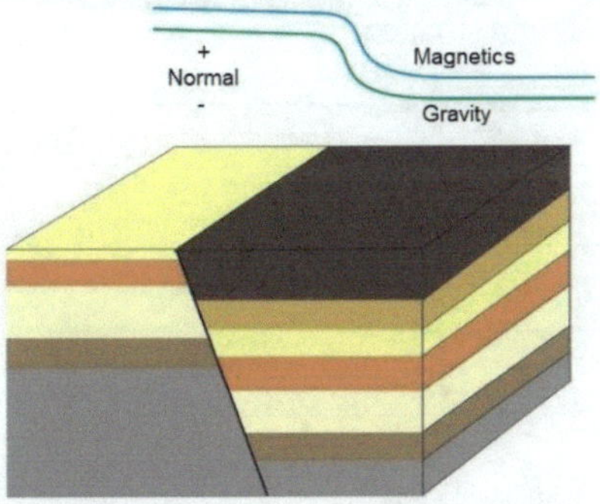

Figure 6.15: Gravimetric and magnetic data acquisition over a fault wall

6.5 Seismic method

This is by far the most widely used and reliable of all the methods (Bouvier, et al., 1989; Brown, 1986). Seismic data is acquired by sending **seismic waves** from a source to the surface and subsurface. They are acquired on both land and water; they provide the most reliable information used to explore the offshore and deep offshore regions. The waves either reflect or refract and then get to a sensor known as **geophone**. The signals are processed and summed together. A next step known as *imagery* is used to provide surface and subsurface maps to indicate geologic structures that can generate, transport and trap hydrocarbons. One advantage of seismic image is that it provides information at all stages of prospecting, highlighting basin scale structures and local reservoirs.

Waves are generated from a source, with *vibroseis* and *dynamite* as few examples, after which the waves travel down until they get to a reflector (a different rock type), from where they are reflected back to the surface and received by geophones (Figure 6.16a). Sometimes the waves travel to multiple reflective surfaces, resulting in different wave types reaching the geophones. These waves include direct, reflected, refracted and transmitted waves (Figure 6.16b).

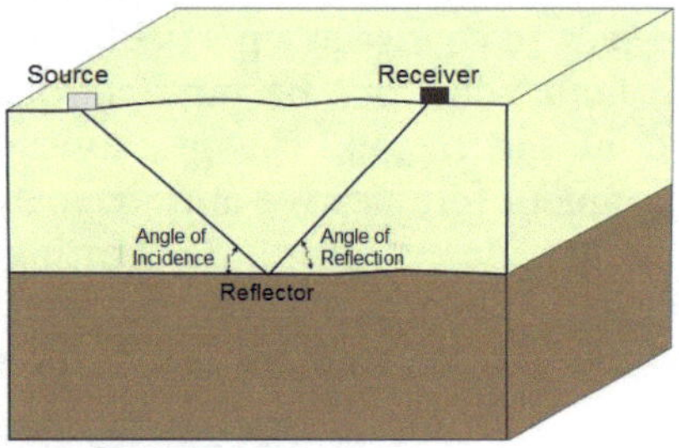

a. Single surface

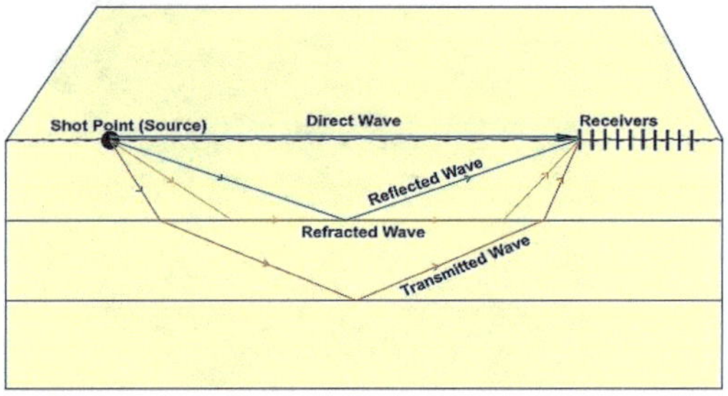

b. Multiple surface

Figure 6.16: Seismic methods showing source and geophones and reflected and refracted waves

At the exploration stages, geoscientists may proceed to initially acquire 2D seismic showing just a cross section of the block in study. Then they go ahead to acquire a 3D if the area is of interest, applying improved but usually more expensive processing methods. Other information rather than maps or images could be obtained from seismic, and may include pressure information by analysing the speed of the waves.

Seismic Method at Sea

The source is towed in the water behind a boat that streams at about 5 nautical miles per hour. The seismic energy is powerful enough for much of the sound to penetrate the ocean bottom. The returning reflections are recorded on vibration detectors called **hydrophones** contained in a long plastic tube (streamer) that is towed behind the boat. Figure 6.17 shows seismic acquisition at sea.

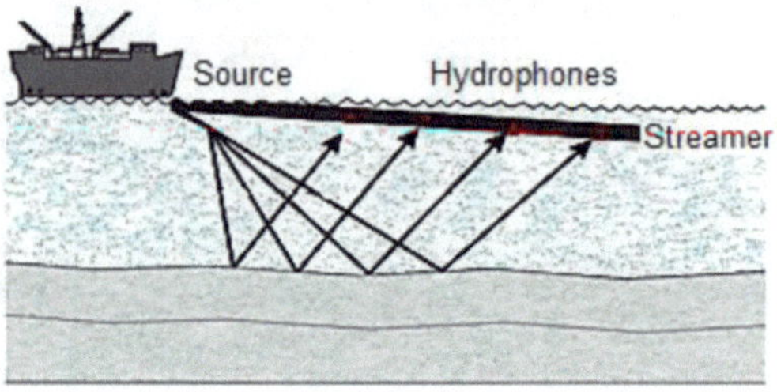

Figure 6.17: *Seismic acquisition at sea*

Some waves will move past the first surface (seabed) and reflect on the next. The wave front is a surface, but only the points captured on the hydrophone are recorded. A specially-equipped vessel is utilised; as it moves, it tows one or more long plastic tubes known as *streamers* at a distance away from the vessel just beneath the surface of the water, containing a series of hydrophones at constant intervals, positioned to receive travelled waves. The source is also towed in the water behind the vessel, just before the series of hydrophones. Seismic waves that are powerful enough for much of the sound to penetrate the ocean bottom is generated from the source, through the use of an airgun (seismic energy source having highly compressed air, used to acquire marine seismic data), and received by the hydrophones as they travel back to the surface.

Other seismic acquisition methods are those employed while already drilling. **Seismic while drilling** has the advantage of capturing a better image. This method, in which the source or geophone is placed within the well, rather than on the surface as previously described, is known as **Vertical Seismic Profiling (VSP)**. Some of the configurations are shown in Figure 6.18.

The profiles in Figure 6.18 are used for correlation with surface seismic data. *VSP without offset*, also known as zero-offset, has sources close to the well; *VSP with offset* has sources some distance away from the receivers in the well. *Walkabove VSP* is used in a deviated well where each receiver is in a different lateral position with the source directly above it. *Walkaway VSP* has a source that is moved progressively farther away while the receiver remains in a fixed position.

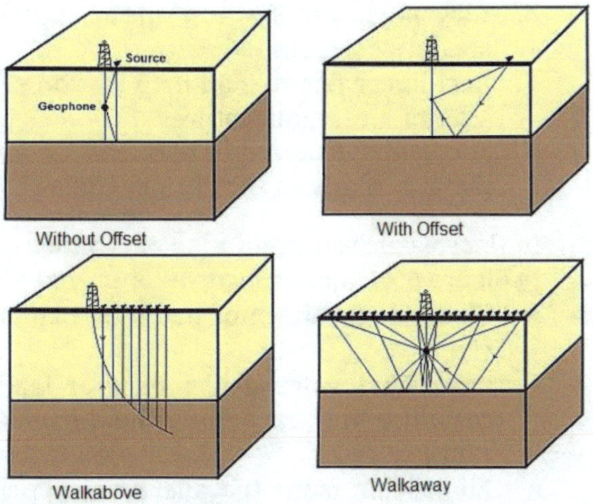

Figure 6.18: Conventional downhole seismic acquisition profiles

Seismic Processing

The result from seismic data acquisition is a record of the time taken for the wave to travel into the Earth and reflect back to the surface. For this data to be used, it has to undergo several seismic processing steps. The first step involves transformation of signals from *time domain* to usable digital forms in the *frequency domain*, using the *Fourier transformation*. One of the reasons for this is that while time domain shows point measurements, the whole signal shape can be represented in frequency domain. Filtering and other mathematical operations are better done in the frequency domain.

In Figure 6.19 the *amplitude* indicates the maximum vertical distance of points on a wave from the centre line to its *crest* (highest amplitude) or *trough* (lowest amplitude). The *frequency* is the number of waves passing a point at a certain time. The higher the energy of the wave, the higher its amplitude will be, and vice versa.

Separation of the initial signal and the effect of the Earth media it has passed through is carried out in a process known as *convolution* (or its inverse known as *deconvolution*). There is a noise *filtering* and *frequency optimisation* to improve the data and ensure that undesired interferences are not integrated with actual data.

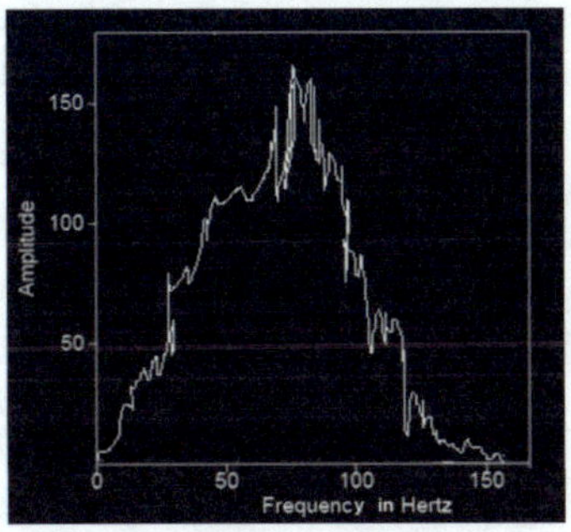

Figure 6.19: Signal recording showing amplitude and frequency

This is necessary in order to obtain the ideal signal that ought to have been recorded. Sources of noise include air wave, waves that typically travel on the surface to the receiver without going into the Earth, *multiple reflection* resulting from more than one reflection of a wave, and noises from moving vehicles and ships, all of which are different forms of waves that can be received by the geophone or hydrophone. Data points are then summed up so that they represent data taken from the same point or levels. A final restoration of the signals to position in space is done such that they provide an image in 2 or 3D.

Seismic processing is a long sometime complex process, involving:

1. Transformation of signals to usable digital forms in the frequency domain, using the **Fourier transformation**, in which the signal is changed from the time to the frequency domain.
2. A separation of the initial signal and the effect of the Earth media it has passed through, in a process known as convolution (or its inverse known as deconvolution) processes
3. Noise filtering and frequency optimisation, which improve the data and ensure that undesired interferences are not integrated with actual data.
4. Summation of data points so that they represent data taken from the same point or levels
5. A final restoration of the signals to position in space such that they provide an image in 2 or 3D

6.6 Seismic Interpretation

The aim of seismic interpretation is to be able to describe the geologic history of the area under investigation using seismic maps generated from seismic processing. In terms of the geologic profile, three key features are to be identified during interpretation - **lithology**, **structures** and **stratigraphy**. While lithology will provide information on potential reservoir rocks, structures and stratigraphy will provide information on traps and areas most favourable for hydrocarbon accumulation. The experience of the interpreter is crucial to the identification of these features.

Despite the usefulness of seismic, it is the interpretation that highlights these features in seismic, and it involves:

1. Eye observation and identification of geological features on the seismic from experience and deduction based on geological principles
2. Tracing out the reflectors on a 2D or 3D seismic image or block (on paper or in the computer) and then tying them together to form a map
3. Analysing the empirical values from seismic data, for example, speed or amplitude values and applying them to causal factors.

Possible geologic features (fault? flank of an anticline?) can be recognised from the simplistic seismic image in Figure 6.20. That is a process known as seismic interpretation. There are many techniques for *visual* and *empirical* seismic interpretation. While empirical is software-aided, visual observation highlights terminations and shapes that give a clue to the geologic events.

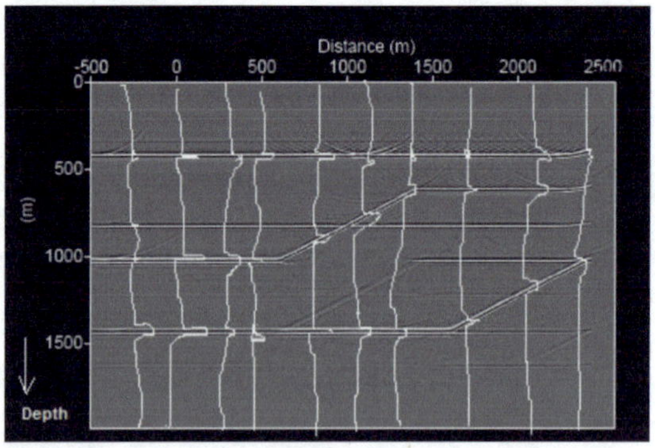

Figure 6.20: Representative of signals in position and space such that they provide seismic images

Among others, to interpret seismic, there is need to identify:

1. Vertical lineation – mainly fractures
2. Horizontal lineation – seabed, fluid contacts such as WOC and GOC
3. Diagonal lineation – top laps and downlaps, shelfal signatures
4. Rectilinear features on map and on section - faults, structural features
5. Sinusoidal features – horizons or ancient Earth surfaces on section, channels on maps

In the seismic representation of seafloor shown in Figure 6.21, among features observed are:

- Channels (outlines of paths or bends of a river)
- Paleo-mud volcano (remnant of landform resulting from eruption of mud from a volcano)
- En echelon faults (parallel or near parallel, closely spaced overlapping faults)
- Glide track (evidence of track formed by an avalanche, possibly by moving or gliding sediments)
- Fractured fold/ridge

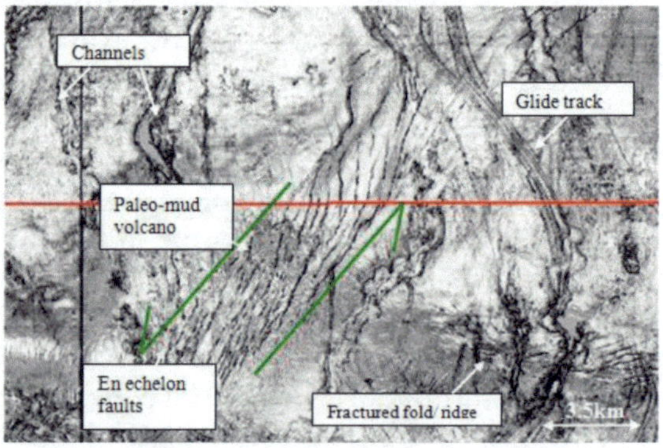

Figure 6.21: Seafloor observations of sinusoidal and rectilinear features (Nosike, 2009)

Sequence stratigraphic related seismic signatures are common in shelf margins. Their use in understanding transgression and regression surfaces/deposits is known as seismic stratigraphy. Figure 6.22 shows seismic signatures indicating deposits related to different geologic events.

The *toplap* marks the termination of overlying shallow dipping younger stratum against more steeply dipping underlying older stratum. The *onlap* marks a sequence in which the dip of the underlying horizon is greater, while in the *downlap* the dip of the underlying horizon is less than that of the more steeply dipping younger overlying strata. The *concordance/sheet* marks the termination of overlying younger beds against underlying older beds lying parallel to each other.

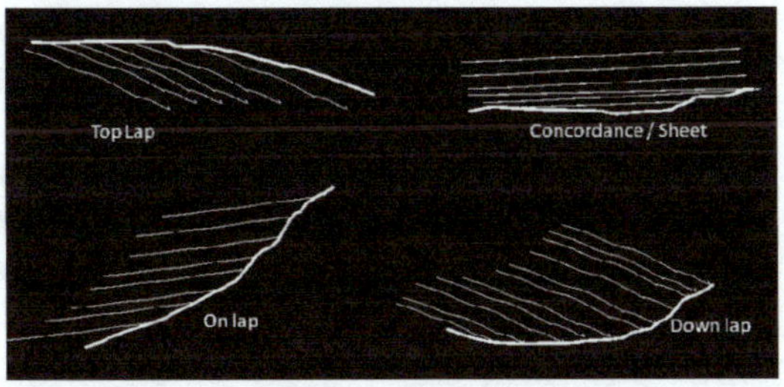

Figure 6.22: Seismic terminations and shape indicating known geologic events

Since the aim is to find reservoirs, in the deep offshore with turbiditic fills, channels and channel complexes are a major target for exploration. Channel fills vary and describing them in the less disturbed near surface deposits is very helpful for understanding their occurrence at depth. How this may be observed on seismic is shown in Figure 6.23. A lens for example may be formed by sand surrounded by shaly envelop, in which case an erosive channel and constructive filling may be seen in a cross section.

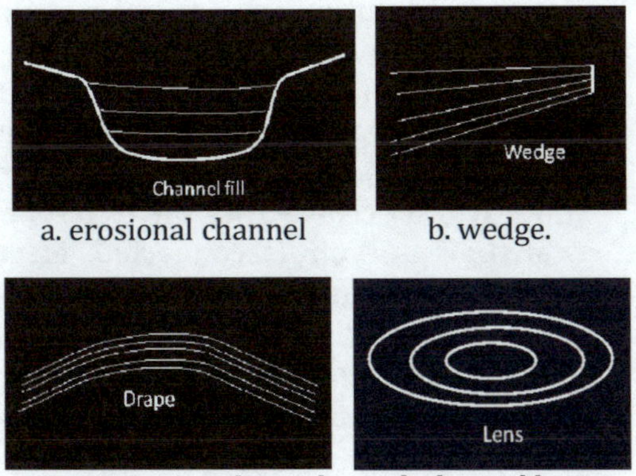

a. erosional channel b. wedge.

c. constructive channel d. channel lens

Figure 6.23: Channel fills and wedge as may be observed in cross section on a seismic image

In (a), erosion digs a channel in which case the bottom is rough. Smooth base may be created by slower initial shaly fill. In (b), a wedge is identified, in which a geologic layer or layers tend to reduce in thickness at one end. In (c), after a channel fill there may be a constructive phase draping the channel, where a number of rock layers having the appearance of a fold is formed. In (d), a draped looks like a lens in section view; formed by sand in a shaly envelop.

When looked at in details, several fills may be identified for a single channel. It is more obvious in the shallow deposits, before compaction and before the limits of seismic resolutions harder to optimize at depth. Figure 6.24 shows an example of a recent channel complex with different episodes of fills dissipating fluids along fault tips and at levees (natural embankments formed when a river floods). During the formation of a channel complex, there is an episode of deposition of fills, which may lead to compaction and dissipation of fluids.

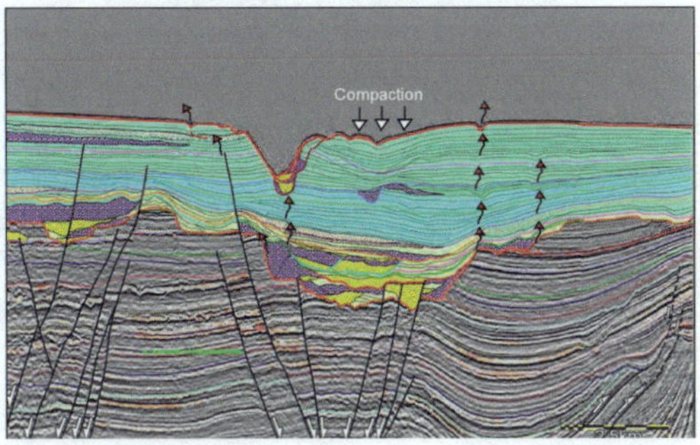

Figure 6.24: Example of recent channel complex with different episodes of fills dissipating fluids along fault tips and at levees (Nosike, 2009)

One basic step in seismic interpretation is the elimination of *artefacts*, observations that are not real but introduced by errors, usually man-made errors. Among other things to be isolated before interpretation, or at least identified, are *multiple reflections* and *tuning effects* resulting from interferences of signals. In general, seismic images are interpreted by geophysicist, geologist and engineers – it requires a collaborative work to make the most of it. It is possible to extract amplitudes, cut seismic blocks to vertical or horizontal slices, and play them as movies using specific software. Eventually, volumetrics of hydrocarbon, migration pathways, amount of reserves, and information of a petroleum system is derived during seismic interpretation.

There are other exploration methods but the use of seismic and statistics, comparisons with previous and neighbouring wells, have become dominant in the industry, even in frontier zones.

6.7 Correlation with Offset Wells

Where there are available data from nearby areas close to the zone of investigation, such data can be used to carry out correlation, to compare or establish if there are similar subsurface conditions, especially if such areas have a known petroleum system. In this case, seal and reservoir conditions, reservoir types and hydrocarbon source can also be compared. Such data could be from any of the other methods, be it seismic, log data, fluid sampling, core data and pressure or temperature values.

Information obtained from a nearby field is to modify correlatable data of a particular field of interest, suggesting reasons of similarities or discrepancies, and how such information can help to predict the subsurface. Figure 6.25 shows correlation of mapped horizons in the shelf margin region to the channel complex forming down the slope.

Maximum Flooding Surface (**MFS**, black line), Highstand Systems Tract (**HST**, purple line) and Lowstand Systems Tract (**LST**, green line) are sequence stratigraphic elements observed between sequence boundaries (red lines). Note that while shelfal deposition timing may be correlated with deep-water events, the deposits may not be organically linked further offshore.

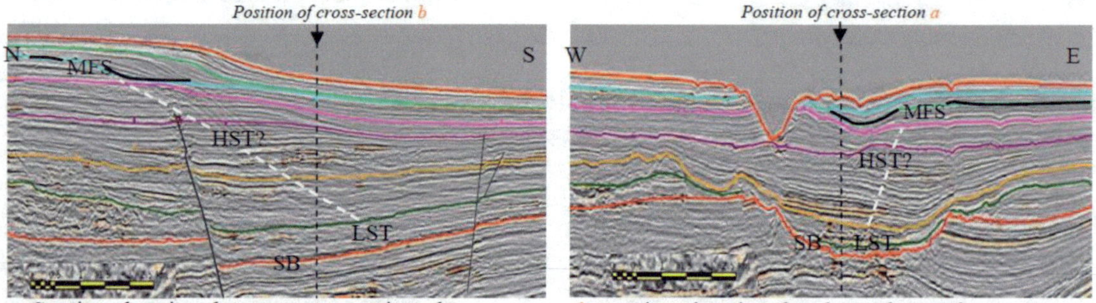

a. Section showing the sequence stratigraphy
SB = Sequence Boundary
LST = Lowstand System Tract

b. section showing the channel complex
HST = High Stand System Tract
MFS = Maximum Flooding Surface

Figure 6.25: Correlation of mapped horizons in the shelf margin region to the channel complex forming down the slope (Nosike, 2009)

Understanding shallow events and depositional process in present day helps in understanding similar processes that deposited sediments at deeper levels in the past.

One way to correlate units from different wells is to look at their pressure or overpressure. In Figure 6.26, four wells show shift in pressure at different depths, highlighting three units (1 to 3) that are correlatable in all the wells.

Pressure evolution with increasing depth was observed in wells 1 - 4, with units 1, 2 and 3 (green, blue and purple colours, respectively) in each well showing similar pressure values. This therefore indicates that the same reservoir units 1 to 3 occur across the different wells or that the reservoir is connected. Meanwhile a regional seal must be separating them vertically, such that there is vertical compartmentalisation but lateral (across the wells) communication

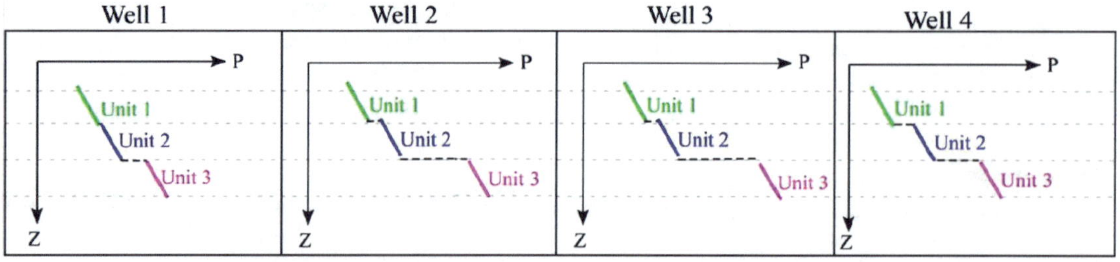

Figure 6.26: Overpressure breaks estimated from reservoir pressures along wells (indicating same reservoir units 1 to 3 occurring in the different wells)

EXERCISES

1. Why is exploration in the oil and gas industry carry out? What are the different methods of exploration?

2. List the subsurface features of interest in exploration.

3. How does the geoscientist know oil or gas is in there? List the different types of hydrocarbon indicators

4. What is the full meaning of DHI and what purpose does it serve? Give examples with illustrations

5. Describe the use of the following exploration methods:
   i. Field mapping
   ii. Geological structural and sedimentary mapping for oil prospecting
   iii. Aerial photos and remote sensing
   iv. Gravimetry and Magnetography

6. What is the difference between aerial photography and remote sensing in terms of data capture?

7. a) What is a contour line?
   b) What are the rules in contouring during field mapping?

8. Describe the acquisition and processing of seismic data. What is fourier transformation?

9. What information can be obtained from a seismic profile?

10. Why is there need to correlate data from a nearby or offset well to an area of interest?

REFERENCES & FURTHER READING

Bouvier, J. D., C.H. Kaars-Sijpesteijn, D. F. Kluesner, C.C. Onyejekwe and R.C. Van der Pal, 1989. Three-Dimensional Seismic Interpretation and Fault Sealing Investigations, Nun River Field, Nigeria. *AAPG Bulletin*, v. 73, no. 11 (November 1989). p. 1397–1414

Brown, A. R., 1986. Interpretation of Three-Dimensional Seismic Data. *AAPG Memoir* 42, 194 p.

Ekweozor C. M. and G. I. Unomah, 1990. First Discovery of Oil Shale in the Benue Trough, Nigeria. *Fuel*, Vol. 69, pp 502 – 508.

Fannin, N.G.T., 1979. The Use of Regional Geological Surveys in the North Sea and Adjacent Areas in the Recognition of Offshore Hazards. In: *Offshore Site Investigations, Proceedings of the Society of Underwater Technology Symposium, London,* (ed. D. A. Ardus), Graham and Trotman, London, p. 5–21.

Heggland, R., 1998. Gas Seepage as an Indicator of Deeper Prospective Reservoirs. A Study Based on Exploration 3D Seismic Data. *Marine and Petroleum Geology 15*, p. 1–9.

Le Chevalier, V., 2002. Application of Seismic Attribute for Deep Offshore Seabed Features Characterization: Preliminary Results for the NERIS Project, *Internal Report, ENSPM.*

Mansfield, C. S. and J. A. Cartwright, 1996. High Resolution Displacement Mapping from 3-D Seismic Data: *Journal of Structural Geology*, v. 18, p. 249–263.

Posewang J. and J. Mienert, 1999. High-Resolution Seismic Studies of Gas Hydrates West of Svalbard. *Geo-Marine Letters*, 19, p. 150–156.

Reyment, R. A. 1965. Aspects of the Geology of Nigeria. *University of Ibadan Press* P.48-60.

Stierman, D. J., 1984. Geophysical and Geological Evidence for Fracturing, Water Circulation and Chemical Alteration in Granite Rocks Adjacent to Major Strike-Slip Faults. *Journal of Geophysical Research* 89, p. 5849–5857.

Susan, E. N., N. L. Haskell, C. T. Steiner and K. L. Coterill, 1999. Debris Flow Outrunner Blocks, Glide Tracks, and Pressure Ridges Identified on the Nigerian Continental Slope Using 3-D Seismic Coherency. *The Leading Edge*, p. 595–600.

Tearpock, D. J. and R. E. Bischke, 2002. *Applied Subsurface Geological Mapping* with Structural Methods (2nd Edition). New York, Prentice-Hall, 800 p.

Thrasher, J., A. J. Fleet, S. Hay, M. Hovland and S. Düppenbecker, 1996. Understanding Geology as the Key to Using Seepage in Exploration: The Spectrum of Seepage Styles. In: *Hydrocarbon Migration and its Near-Surface Expression, AAPG Memoir 66* (Schumacher, D., M. A. Abrams eds), p. 223–241.

Wood, W. T. and C. Ruppel, 2000. Seismic Investigations of the Blake Ridge Gas Hydrate Area: A Synthesis. Proceedings of the Ocean Drilling Program (C. K. Paull, R. Matsumoto, P. J. Wallace and W. P. Dillon eds.). *Scientific Results, 164*, p. 253–264.

Woodward, N. B., S. E. Boyer and J. Suppe, 1989. Balanced Geological Cross-Sections: An Essential Technique in Geological Research and Exploration, *Short Course in Geology* v. 6, American Geophysical Union, Washington, DC, 132 p.

CHAPTER SEVEN

OIL AND GAS FIELD ASSESSMENT

7.0 Processes of Field Assessment

This chapter explores the tools and processes in oil block assessment leading to individual prospect evaluation. This is similar to petroleum prospecting processes in a block, but in this case, it is at field level. In a bid to present factors that are reviewed during oil and gas field assessment, this chapter will touch on some of the factors, like source rock, maturity and migration, and traps, discussed earlier in Petroleum Geology chapter. Reservoir quality characterisation and seal assessment are further explained. The processing of the assessment would include:

- Initial ranking
- Exploration and appraisal
- Assessment of compartmentalisation
- Analyses of fluid properties
- Assessment of filling and trap sealing
- Geological field modelling
- Estimation of recoverable oil
- Reservoir evaluation and targets

This part of this book focuses on the first two stages in the E&P value chain: *exploration* and *appraisal* of oil fields. After this, the aim of the later part of the book is to look at the last two in the chain, and make meaning to the long and demanding process of exploration to the culmination of the efforts through *development* and *production*.

7.1 Block Ranking

Oil field assessment is done on blocks acquired for exploration or for appraisal (Saugy, 2003). An initial ranking to choose blocks for indebt prospect analysis is necessary. Before a block is chosen for exploration in the oil industry, or bought during bidding, that block needs to be ranked in order of its importance. This ranking will determine whether it is chosen with respect to others. Ranking is not done arbitrarily; many factors are put into consideration by interested oil companies and investors. They are interested in information such as:

- The geologic age and chronostratigraphy of the block with respect to the basin
- The geological context – environment of deposition
- Physical location of block: onshore, offshore or deep/ ultra-deep offshore
- Previous prospectivity and nature of the traps (seals and reservoirs)
- The stratigraphy and nature of tectonic versus sedimentary activities
- Available seismic (2-D, 3D, 4D, VSP) and their processing
- The structures, sequence stratigraphy, biostratigraphy, lithostratigraphy and geologic modelling
- The temperature and pressure of the prospects (or offset wells)
- Source rock type, maturity, charging and unswept potential vs hydrocarbon type
- Play types, Probabilities of Success (POS), statistics, partnership, investments
- Surface facilities and Environmental Impact Assessment (EIA)

When all these factors are considered, they determine the importance of the block, helping investors to decide whether to bid for it, and how to bid – whether to put in higher or lower offers. Sometimes, the opportunity presents itself for internal ranking to be done by the oil companies. The field of geoscience and its general understanding is very important in this respect, as it is not easy to know all the answers regarding the viability of a particular block from the onset, even after exploration and production have taken place.

Oil blocks are further divided based on seismic data into zones, to carry out detailed study that makes more information available to help in decision making. A seismic block allows the user, using the right software, to view seismic cross-sections and horizons (Nestvold, 1987). The horizons are maps, which can be seen as lines on specific cross-sections. When a seismic block or cross-section is cut across several horizons, it is known as a **time slice** (as was done from 4 to 5 in Figure 7.1 to cut or slice across several geological time); when it is cut along the horizon, it shows a normal display of a **geologic surface**, representing a time in geologic history.

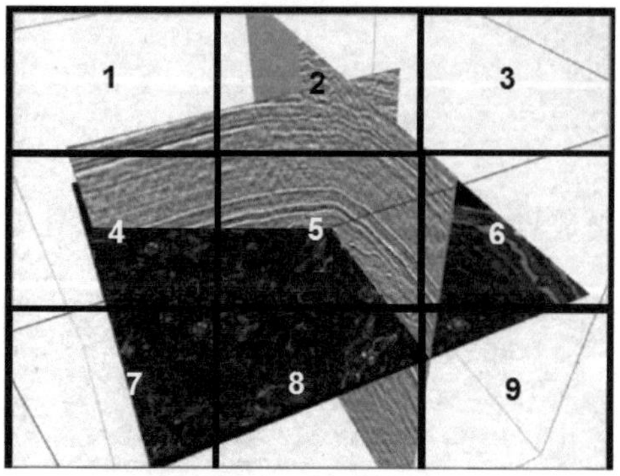

Figure 7.1: Division of oil field into zones based on seismic data for in-depth study

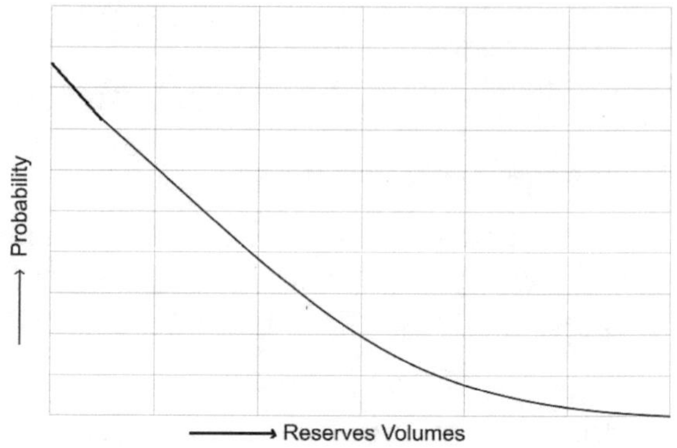

Figure 7.2: Probability of discovery of substantial volumes, showing a case where the relationship is inversely proportional

Once ranking is over and field or block acquired, the next step is to evaluate it for potential prospects, which will be recommended for drilling. The oil field is further divided into zones and each critically studied based on seismic data and as oil wells are drilled.

7.2 Oil Field and the E&P Value Chain

After the acquisition of an oil block for the purpose of prospecting and mining, the geoscientists are required to employ principles of geology and geophysics in assessing the oil field. Field structural mapping techniques, geophysical and petrophysical methods are employed in estimating the viability of the field. They determine the estimated total amount of oil in the reservoir, known as the *oil in place (OIP)*.

This may not all be extracted; the *recovery factor*, the percentage amount of possible recovery, gives how much of the OIP will eventually be extracted. Application of the geological and geophysical principles will also suggest potential reserves for drilling. Of most concern at this stage is whether or not there is crude oil, and if it has accumulated in commercial quantity; the higher the probability of discovery, the better the chances of developing the well.

Figure 7.2 shows the probability of discovering substantial oil volumes. The relationship is *inversely proportional*, highlighting a common trend in known basins where oil is explored. Overtime the chances of major discovery is low, such that wherever there is a higher chance of discovery, the lower the chance of it being economically viable volumes.

The language change used at the different levels of the E & P value chain varies, so people working together should learn such differences. Same word is used to mean different things. For example, the word "**hydrostatic**" to the driller is talking of a *column of drilling fluid* in the well while the exploration geologist is talking about the *pressure exerted by water* (or brine) at a given depth. Further, **hydraulic fracture gradient** for the driller refers to *resistance of formation wall* in a well, which guides in the safe drilling of the well with a given mud weight; for the exploration geologist it has to do with the *integrity of a seal*, showing oil retention limit in the subsurface before drilling.

The value chain in the oil and gas exploration and production business is:

Exploration → *Appraisal* → *Development* → *Production*

During oil field assessment, the geoscientist needs to have an overall picture of where his work fits into the business. Studying the viability of the overall process is known as **appraisal**.

Exploration Phase

The objective, and consequently the mind-set at exploration is that of optimism. There is need to keep working on the expectation without losing hope. Exploration could be protracted (relatively long and tedious in duration) at the initial stage as some "guess work" is permitted. Most of what has been covered in exploration geology and geophysics fall within this stage. When onshore, and for field-based *prospectivity*, a field trip may be required at the exploration stage for geological mapping. A guide for an effective and detailed mapping project (iESog, 2018) is outlined below:

1. Take a panoramic view of the outcrops, make quick observations, move round the outcrop if where possible, and locate best exposures where the most features of the outcrop are exposed.
2. On the field notebook and maps, the following necessary preliminaries should be noted:

a. **Date:** record the date and year of the field mapping and add other headers identifying the project.

b. **Location:** using a GPS system, locate and describe the place the outcrop was found. Those days, in the absence of the GPS, this was done by estimating the distance of the outcrop location from landmarks (recognisable natural or artificial features; sometimes trees, rivers and electric poles may serve). Other natural features (such as a reference mountain for example) which are not easily removed or changed over a long period were used, to aid future identification of same location.

c. **Location number:** for easy indexing, there should be a serial numbering of each outcrop location visited. This is to prevent any mix up and always know from which place or location a particular outcrop was seen (and possibly the direction from which outcrop observation was made).

d. **Plot in the location coordinates:** tie the coordinates given by the GPS to the physical maps used for the field mapping. It is a kind of georeferencing, to be able to locate the place where the outcrop was found on the map, and then plot it on the map. For future map exercise, where the location description is vague, coordinates can help pinpoint outcrop locations.

3. Establish lithological logs, geometries and structures.

a. **Bed attitude:** measure the strike and dip and establish structures, such as anticlines, synclines, faults and other trap types.

b. **Lithology:** major lithologies, such as sandstone, shale, coal should be identified. This is done after identifying the terrain, for example whether it is a sedimentary or an igneous terrain.

c. **Geometry and tectonic structure:** record the dimensions of the lithological units, which include vertical thickness and lateral extent in different directions. Structures such as anticlines, faults and joints or fractures may also be identified at this time.

d. **Colour:** true colour of the rocks is obtained from fresh samples. Use a hammer to dig into non-weathered samples.

e. **Texture:** record the grain size of the rocks, whether fine, medium or coarse grained. Also check the impact of compaction, whether they are friable (easily broken into smaller fragments) or consolidated (into a solid mass).

f. **Sorting** and **roundness**. Describe the grain sizes of the rocks. Are the grains relatively equal (well sorted) or not (poorly sorted). Are they grains rounded or angular? Spherical or planar?

g. **Sedimentary structures**: this should include sedimentary structures such as different stratifications / cross-beds, laminated layers and sediments affected by bioturbation (different from tectonic structures: folds, faults, joints or fractures)

h. **Fossils:** This includes body and trace fossils, presence of moulds and casts.

4. Take representative samples from the outcrops and then label them accordingly, with respect to the aforementioned location number. These will be analysed and tested in the laboratory eventually.
5. Take a snap shot of the outcrop, with an acceptable scale object in place. For close up pictures on important features, small scales such as pens can be used.

The above short field guide should help in addition to more standard methods of field assessment discussed with examples in the subsequent sections.

In oil field exploration work, physical topography and geomorphological features are assessed for easy well placement, which are all factors of cost and oil field viability. Initial feasibly only provide enough field information needed to build a base for further assessments. Further assessment may use seismic or other invasive exploration methods. When oil is found, the geoscientists move on to the appraisal stage.

Appraisal Phase

At the early stage of exploration, not much data are available. Yet several factors needs to be assessed to see how promising the oil field may be. Table 7.1 shows the presence, quality and assessment/weight given to factors used to determine the probability of success of an oil field. The Weight given to each item depends on the assessment for a particular field. The probability of success will depend on the individual probabilities of each factor. The higher the assessment/weight for each factor, the more the probability of success will be.

After an exploratory well is drilled, if the petroleum result is positive, the next step is to appraise it for its commercial quantity. These factors are further assessed with more available data from the well, as part of the appraisal process. During appraisal, more wells are drilled in order to assess the potential of the reservoirs before the producing the oil or gas field. This is an important step in the process of asset development in oil companies.

In addition to the estimation of quantity and assessment of the quality, other characteristics of the oil field, such as safety and environmental challenges, are considered in determining whether the oil field will be developed. Well tests provide the major information regarding the economic quantity of an oil reserve. When proceeding with a developmental project, a document known as the *Field Development Plan* (FDP) lays out the terms, the number of wells to be drilled, well types, and specific reservoir targets.

Table 7.1: Factors of oil field assessment showing that the Probability of Success depends on the individual probabilities of each of the factors (Nosike, 2008)

Presence	Presence and Quality	Assessment / Weight(1 to 10)
Source Rock	Type, amount, proximity to reservoir	
Maturity	Age, burial, temperature, tests.	
Migration	Pathway, drainage area	
Trap	Structural or stratigraphic, extension	
Timing	Structuration relative to charging, storage duration	
Reservoir	Quality, diagenesis, fracturation	
Geometry	Seismic quality, volumetric, accessibility	
Seal	Lateral and vertical, retention capacity, integrity/ efficiency	

Operations for drilling, putting facilities in place for drilling and production of petroleum are targeted during development. The cost of all these will determine whether or not a field will be economical. During oil field evaluation, *cost* and *safety* are the most important factors considered; whether the field can be developed and produced safely, and if the result would be commensurate to efforts targeted at development and production.

Development Phase

At the development stage, available technology is used to plan, design well architecture and successfully drill a well in the shortest possible time. Pipes that will be used to convey or transmit the oil to the point where it will be sold or to storage tanks from where it will be refined are laid at this stage. The development geoscientist cannot do this alone, and will have to collaborate with the drillers and other field staff (Nosike, 2019).

The operator handles all that is related to planning the wells, organizing the logistics of rig type to be used, and actually drilling the well. The well plan will depend on how segmented the reservoir is, as well as its architecture. The well geometry (the shape of the well), and the various well types (injection, observation, production wells) are planned at this stage of development. In addition, information that may not have been available during the appraisal stage could be obtained at this stage, allowing for better interpretation of results.

Oil assessment further re-evaluates these factors (listed in Table 7.1 for the appraisal) in the development stage, with more data and sometimes development wells drilled to confirm prognosis. This includes basic observations made at the beginning of exploration. For example, changes in seafloor fluid signatures. It has been mentioned that the presence of an oil seep is an indication of a petroleum system, and gives an indication on the behaviour of several of the factors of oil assessment in a petroleum field (Jennings, 1987). It indicates the presence of source rock, maturity, carrier strata and even seal at some areas. Whether this results in substantial accumulation may be confirmed during the development stage.

In Figure 7.3, an oil seep indicates the presence of a mature source rock, carrier strata or reservoir bed, possibly a trap and a breached seal. Breached seal may be local and does not eliminate substantial accumulation. During oil field assessment, it is the rate of hydrodynamic charging, or the size of the accumulation, relative to leakage, that is important. A minor leakage through a fault, or a seepage driven by spring water, may only indicate an intact accumulation some distance away. Yet the development process may activate faults and further leakages, and lead to major environmental challenges if not properly handled.

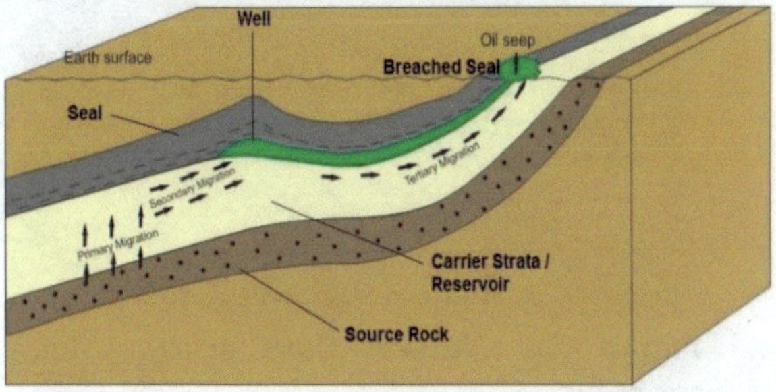

Figure 7.3: Oil seep indicating factors of a petroleum system for re-evaluation during the development phase

Therefore, the planning for the development stage may be started during exploration and appraisal. Observations at the early stage may become relevant later. Among the necessary factors for a successful well development are continual analyses, post-mortem and positive reserve evolution.

Continual Analyses

Geological, geophysical and reservoir geology analyses (sometimes known as 3G analyses) should be a continuous process. Subsurface data obtained during development or production should not be taken for granted, but should be archived, hierarchized and analysed. Sometimes major information comes up that will help to improve that reserve and others.

Post-mortem

As in a medical post-mortem, it is important to analyse the success or failure of a well. Like for a plane crash, each failure also provides invaluable information that may help to prevent such in the future. Post-mortem will therefore examine and reveal whether a well is successful or failed because of factors already foreseen. If not, then it becomes a new challenge to understand, and could lead to an innovation. If it fails or succeeds in the direction of prediction, better risking and application of the result becomes the only usefulness.

In success and failure analyses, it is not only an issue of whether a well had oil or not, it is also a question of whether things happened the way they were expected (like if the reservoir was found but empty, instead of a missing reservoir). Another aspect of failure also taken into consideration in the statistics is *logistical and environmental concerns.* A well may as well prove a successful discovery but with a shallow gas or other environmental hazard.

Positive Reserve Evolution

Reserve evolution refers to any actual changes in the estimated reserve, due to either depletion, losses or miscalculation discovered during development. As the expected recovery and actual recovery changes, the reserve is said to evolve. Any reduction in the estimated reserve will affect production and the overall profitability of the field. The aim of **positive reserve evolution** is therefore to make the reserve increase with time. This can be achieved by adding non-considered reservoirs previously or changing recovery possibility and methods.

This is known as *revision* or *addition of reservoirs*. The pool size of a field has often been found to be larger than expected, seemingly increasing after re-evaluation. The inverse could be the case, many reserves have evolved positively and their lifespan seen to extend even longer in both early and later developmental phases of a basin.

Production Phase

Once a well is completed and pipes are laid, channelled to convey hydrocarbon, it marks the start of the production phase. Most of the injection wells planned during the initial development stage to enhance hydrocarbon recovery would be put in place; more could be added during production (known as *infill wells*) if the reservoir behaviour evolves. During production, the actual dynamic behaviour of the well is understood (which is not possible if the well is *static*, not flowing). Once production starts, the way the well flows, its flow history, is plotted continually to be able to assess its performance.

One of the major factors that determine cost at this stage and viability of a field is the *drive mechanism* that will be used for production. As seen in Table 7.2, the *water drive* mechanism has the highest possibility of recoverability compared to those of *gas cap injection* (average recoverability), *natural depletion* (average recoverability), and *gravity and natural buoyancy* having the lowest possibility of recoverability. Natural depletion could lead to more substantial recovery in overpressured reservoirs. Water drive is preferred in deep offshore turbidite reservoirs where the channels serving as a conduit maintains the water pressure from the aquifer to the oil pool. Gas cap injection may eventually result in gas recycling when the oil is depleted. That is, same gas injected is recovered.

Table 7.2: Drive mechanism as a factor of consideration in oil field assessment

	Drive Mechanism	Recoverability
1	Water Drive	High
2	Gas cap Injection	Average
3	Natural Depletion	Average
4	Gravity and natural buoyancy	Low

7.3 Major Steps in Oil Field Assessment

Oil Field Compartmentalization

A petroleum pool does not necessarily refer to a pool of oil and gas filling the whole basin or rock interval. Think of it rather as small droplets within rocks that need to be forced out to form a pool. There is also a variation of hydrocarbon fluids within the liquids contained in the basin or trap structure, as well as different *vertical* and *lateral* barriers created by geological features. As such, the fluid compositions could vary vertically and laterally. It is therefore necessary to establish these distributions and nature of fluid variations in order to determine the expected quantity and methods that will be used to extract them.

Hydrocarbon Fluid Variation in a Field

Fluid within a reservoir would be separated into water, oil and gas vertically due to buoyancy, where the three components are present (as in Figure 7.4). Petroleum composition could vary vertically because of segregation following aging since the migration into a reservoir; there could be gravitational, thermal and biodegradational segregation. Natural separation of fluid types, due to differences in density, will make oil occur below gas and water below oil, in the conventional manner as shown in Figure 7.4. Water is most dense and is at the bottom, while gas is least dense and is at the top.

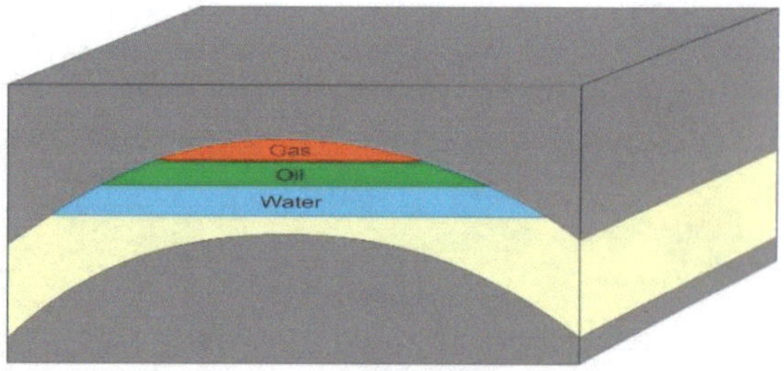

Figure 7.4: Natural fluid overlays within a reservoir

Other reasons for variation could be the source of hydrocarbon and their migration history. To understand the variation of fluid types and quality in a field, there is need to test properties and processes such as:

- Fluid density and viscosity in different wells
- The gas composition (C1 to C5)
- The Gas to Oil Ratio (GOR)
- Condensate to Gas Ratio (CGR)

- Presence of other gases, such as N_2, CO_2 and H_2S
- Crude separation and test (PVT)

When investigation shows that there is variation in salt composition in different aquifers, it could signify that the oil pools are different, or that the oil pools belong to different systems. Sometimes, these variations may not be so obvious, as the hydrocarbon fluid mix up and homogenize over time where there is communication between the barriers that were supposed to separate them.

PVT Related Fluid Variation

The depth of a reservoir, invariably controlling the temperature and pressure, also determines the nature of the fluids downhole. Figure 7.5 shows the relationship between temperature and pressure for downhole fluids. The **bubble point** curve shows the limit at which liquid, at a given reservoir temperature, begins to exsolve (vaporise to) gas, with reducing pressure. The **dew point** curve shows the limit at which gas, at a given reservoir pressure, begins to condense to liquid with reducing temperature. The point at which the fluid exists as both liquid and vapour is the **critical point,** which corresponds to the critical temperature (Tc**)** and critical pressure (Pc); the slightest change in temperature or pressure can cause the fluid to tend to either direction. Once the fluid gets to the surface, it changes composition to stable phase, separating to give the gas to oil ratio (GOR).

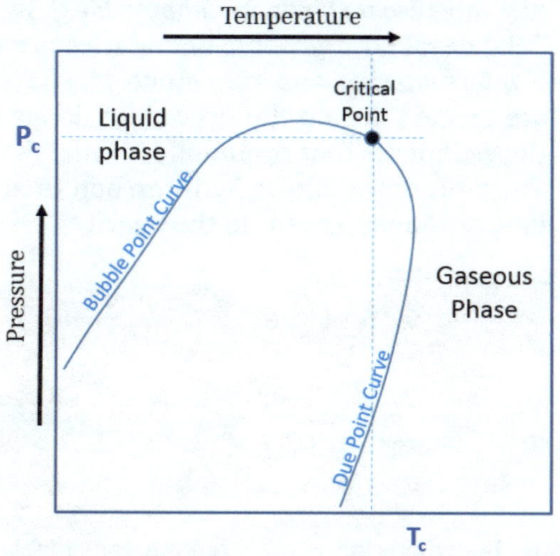

Figure 7.5: PVT phase diagram showing the bubble point curve, the dew point curve and the critical point

Reservoir fluid that was liquid under pressure may become gaseous on getting to the surface, leading to an increase in volume. This expansion is taken into consideration by a **formation volume factor**. Oil or gas formation volume factor (b_o or b_g) relates the volume of oil in the reservoir to what it becomes at standard surface conditions.

7.4 Oil Field Lateral and Vertical Limits

Since oil pools occur within geologic bodies with vertical and lateral boundaries, there is an impact of their distribution across the boundaries on recovering the hydrocarbons. The understanding of this distribution is a major part of oil field evaluation. That fields are divided into separate compartments is the reason several wells are needed to drill them. Minor and major barriers, *structural, stratigraphic* or *sedimentary* limits, make oil pools to be connected or disconnected in different places within the same field.

A simple reduction in porosity and permeability (possibly due to differential compaction) locally can make it difficult to drain a reservoir across its extension spatially; while one side of the reservoir may have significant porosity and permeability, the other side may have very low porosity and permeability. It is therefore important to identify some factors, discussed below, during evaluation of oil block.

Hydrocarbon Fluid Flow Direction

Past and present migration pathways should be considered, not just the initial primary or secondary migration from source rock to the reservoir, which might no longer be ongoing. At this stage, focus is on *tertiary migration* (hydrocarbon flow out of reservoir rock to another accumulation), *leakages* to the surface (outflow or discharge of fluid from the trap, usually through the seal) and *dynamic transfer* (fluid transfer from one pool or depth to another due to overpressure differences) across the boundaries. These would essentially be upward migration in the presence of vertical faults or inefficient seal, but would also be spatially across leaking faults and stratigraphic limits (Perkins, 1961). This may determine the hydrocarbon height, filled level of trap and possible spill. Figure 7.6 shows the trap-filling concept, where anticlinal traps are underfill, filled to brim and overfilled to spill.

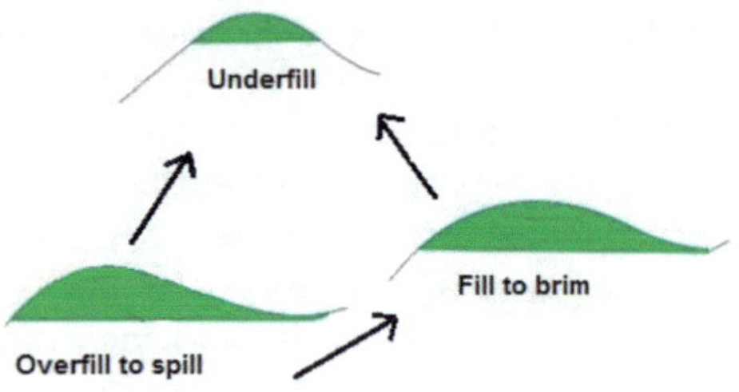

Figure 7.6: Anticlinal trap filling in an oil field showing overfill, fill to brim and underfill

Traps may be underfilled if the source material does not charge the trap sufficiently, or if there are leakages through the seal. If filled to the brim and the trap is still charged with hydrocarbon, it will overfill to spill, hence tertiary migration. Knowledge of this can help to determine the hydrocarbon height, and eventually its volume. For example, if a seal is efficient in an active petroleum system, it could suggest that a trap may be fill to brim; likewise a prospect found to be filled to brim below a regional seal will suggest that the other traps in the field have a good seal.

From pressure variation, showing the stage of homogenisation, it is possible to predict if pressurization is recent or not. This is important for not only understanding the pool, but also in understanding how they might behave and react to stresses during production.

Flow Boundaries or Pressure Cell Limits

A major control on flow is faulting, whose offsets cut across reservoirs. *Differential porosity and permeability*, and secondary reactions such as *cementation* could also limit flow. Presence of inflow of shale in *mobile shale* environment could form barriers to flow. Major factors controlling flow either allow or limit fluid pressure and are therefore known as **pressure boundaries,** which divide fields in pressure unit or cells.

Another way to know the dynamic behaviour of a well during well testing is to make diagnoses for flow boundaries, which involves identifying the nature or cause of flow boundaries during well testing. Well test can indicate the presence of fault or other limitation to flow. In well tests, wells are allowed to flow, or fluid is injected into it before allowing it to return to its original state; the rate at which pressure changes reveals flow behaviour and barriers to flow.

Lateral Structural Boundaries

These are mainly structural limits, major faults and thrusts. They may also be stratigraphic or sedimentary pinchouts. Faults will seal when they have high clay content in their gouges. This is why studies target testing out how much clay is within a fault (e.g. Clay Smear Factor, CSF and Shale Gouge Ratio, SGR). *Cataclasis,* which is a frictional deformation during faulting, leads to grinding of gouge materials in fault zones. Nature of fault displacement, fault segmentation and diagenesis all play a role in fault sealing and control lateral fluid limitation.

In Figure 7.7, a lateral structural boundary caused by a fault (thicker black line) is shown to play a role in sealing, thereby allowing hydrocarbon to accumulate.

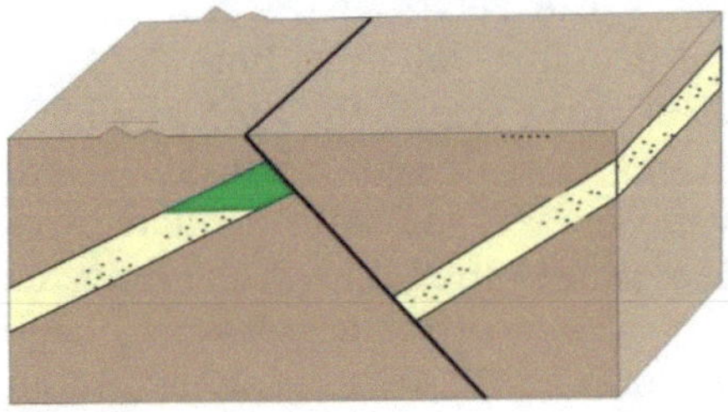

Figure 7.7: Lateral structural boundary due to faulting

Lateral Stratigraphic and Sedimentological Boundary

Stratigraphic and sedimentological boundaries occur where there is a thinning or an abrupt stop in continuity of a stratigraphic unit *(examples have been shown in Chapter Five).* In Figure 7.8, lateral stratigraphic boundaries are shown – *interfingering* and sandstone *pinchout.* They are formed during the process of deposition of the sediments that eventually formed the rocks. Their presence allows hydrocarbon to accumulate, as shown (green) in the figure.

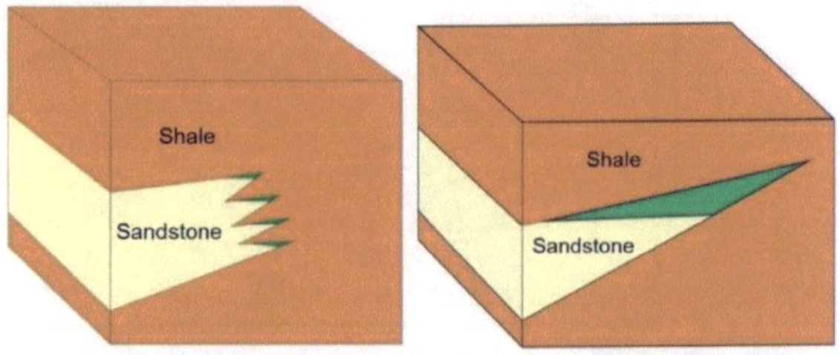

a. Interfingering b. Sandstone pinchout

Figure 7.8: Lateral stratigraphic boundaries serving as fluid flow boundary

Vertical Lithological Boundaries

These are essentially *regional seals* (as is different from any seal lithology or local intercalation). Regional seals are usually thick and are affected mainly by major faults (e.g. thrust faults) at the extreme lateral boundaries. Example of regional seals which are vertical flow boundaries are thick layers of lignite, coal, salt, with the most common being shale or claystone.

In Figure 7.9, a vertical lithological boundary is shown (similar to that shown in Chapter Five: Figure 5.12). It consists of a thick regional seal spread across an oil field. In this case, the figure shows how the region between the boundaries extend in the dip and strike directions. *Note that the size of the envelope between the boundaries does not correspond to fill or accumulation volume.* The oil (in green) only filled a portion of the available accumulation space. Boundaries of this type serve as good covers that allow sufficient hydrocarbon to accumulate by limiting vertical flow, in the absence of faults cutting across them.

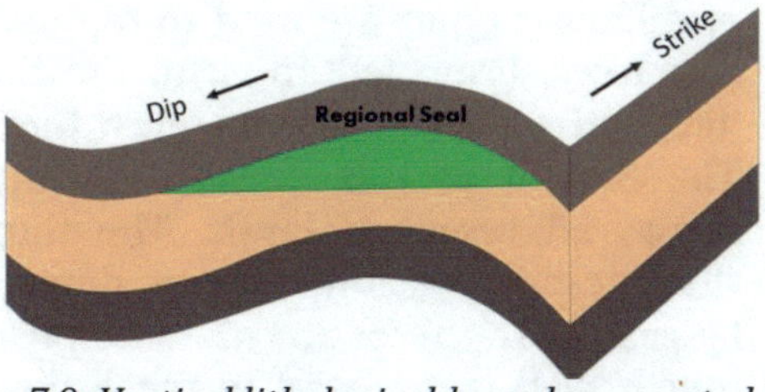

Figure 7.9: Vertical lithological boundary created by a regional seal across an oil field

Other examples are cemented and tight sections of reservoirs and siltstone, which could form partial barriers to vertical flow only limiting flow in rate over time; flow may take place but the magnitude of flow is limited in time. They are drilled across by wells and are often well known and correlated between wells. These are unlike faults that are difficult to drill across and as such often assessed based on inferred properties, geometrical evaluation from seismic and from field analogues.

Sometimes a tilted reservoir would have its near vertical crest limited by some near horizontal fault or thrust, as if the flanks of the anticline or folded reservoir is limited by a lithological unit. So vertical and lateral flow boundaries evaluation is not just a straight process, it is a combined check on various factors and the conclusion is a geological synthesis rather than a clear-cut single representation.

Reservoir Compartmentalization

The overall aim of understanding compartmentalization of an oil field is to know where the reservoir begins and ends vertically and laterally. The basin may be large, but the reservoirs are often smaller sand units. Lateral stratigraphic and sedimentological boundaries, as well as vertical lithological boundaries, are responsible for cutting them into compartments. In the deep offshore environment, reservoir channels are limited laterally by thrust faults, and above by regional fine-grained marine shales. Differential charging or leakage from different compartments could make them to have different overpressures, thereby dividing them into units known as pressure cells.

In the example in Figure 7.10, pressure cell units A, B and C are shown as lateral compartments. Across these are vertical pressure cell , which are reservoir units 1, 2, 3 and 4, from top to bottom, separated by different seal lithologies (green, blue, purple and black colours respectively) overlain by the seabed (red colour). The impact of lateral structural boundaries and top lithological seals result in lateral and vertical compartmentalization of the field. Vertical fluid flow through faults and fractures on the distal crest of the structure, and lateral fluid flows across the units, constitute lateral communication and vertical leakages across the pressure cells.

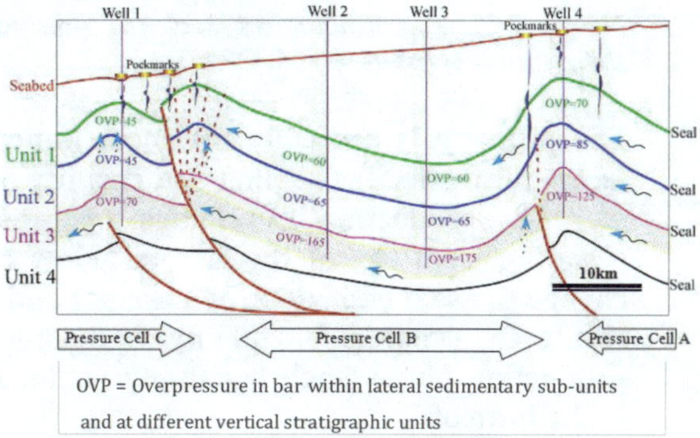

Figure 7.10: Field compartmentalisation showing vertical and lateral leakages from pressure cells (Nosike, 2009)

In deep offshore marine environment, evidence of leakages may be seen as *pockmarks* (in the form of a crater or mounds caused by fluid escape through sediments) on the seabed. Wells drilled across the units as shown in Figure 7.10 will tap available hydrocarbon accumulations and show overpressure shifts to indicate the compartmentalisation.

Understanding reservoir compartmentalization could help to understand how oil retention occurs and how they leak from the reservoirs at each cell boundaries and vertically across vertical lithological seals. Fluid property variation across reservoir compartments may indicate charging process in a dynamic petroleum system. If a reservoir is too compartmentalized, it becomes costlier to develop as there will be need to drill many wells or several horizontal extended reach well to target all the reservoir compartments, which could undermine the commercial potential of the field.

Pressure and Seal Evolution

Before the drilling and production, there is need to evaluate the type of seal, lithological or structural seal. Pressure seal and other types of seal are discussed in subsequent chapters. In Figure 7.11, two seal types are shown - vertical lithological seal and lateral structural seal. They prevent fluid flow vertically and laterally respectively.

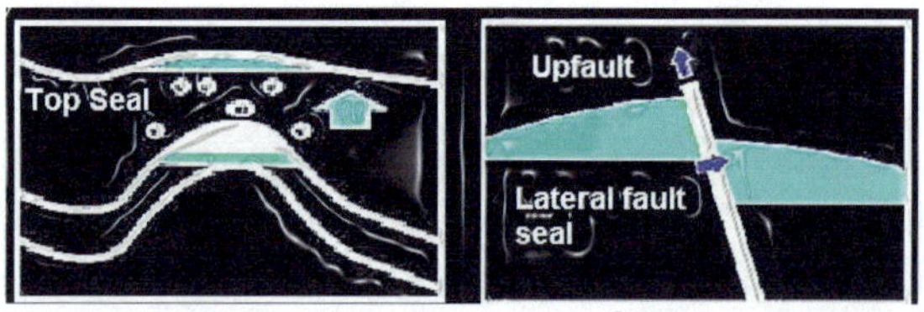

a. b

Figure 7.11 Top lithological seal (a) and lateral structural seal (b)

The *pressure* is probably the most important factor (also affecting sealing and stability of the well) in production. A change in pressure resulting from any of the factors below could change the seal behaviour of a system and may result in eventual instability. During the production phase, such instability could occur in the form of:

- *Dynamic transfer*: general movement of fluid from one compartment to another due to pressure changes.
- *Drilling kicks and losses*: sudden gushing in of formation fluid or fluid leakage into the formation respectively while drilling.
- *Induced failures*: introduction of water or gas by injection during production could increase pressure and lead to seal breaching.
- *Depletion*: loss of fluid from reservoirs by the well penetrating and producing from it, leading to a reduction in pressure.

While dynamic transfer refers to the general movement of fluid from one compartment to another due to pressure changes, such pressure changes are induced by the well during the production of hydrocarbon. It is possible for fluid drilled nearshore to be dynamically transferred from layers deep offshore. In the figure below, deviated (non-vertical) wells are used to target fluid whose reservoir extends to deeper zones, with overpressure units most likely transmitting fluid to shallower intervals. Complex wells, inaccessible site and sidetracking lead to higher cost of developing a field, which is taken into consideration during oil field assessment In Figure 7.12, deviated wells of different types are used to tap hydrocarbon resources. From left to right, the first well is deviated across a fault to reach the reservoir. The second well is deviated so that it cuts across all reservoir levels. The third well is sidetracked, possibly due to downhole challenges that necessitated sidetracking. The fourth well is deviated to reach the reservoir and avoid a non-reservoir facies; the fifth well did not get to the reservoir as it probably missed the target. The sixth well is a cluster, having several well extensions deviated from one wellhead area or clustered manifold offshore (the wells take off at different positions on the surface, all within nearby location to form a cluster).

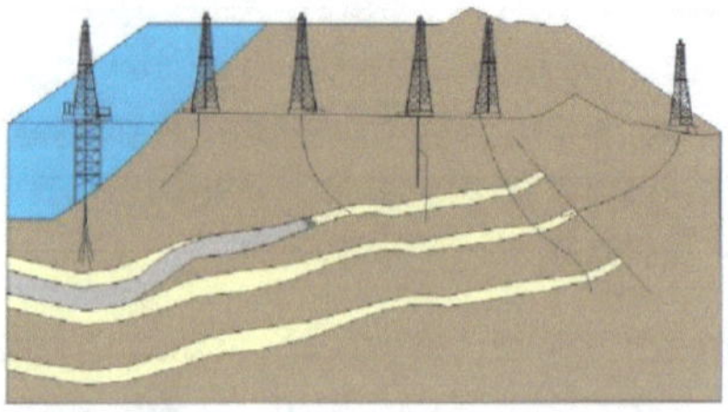

Figure 7.12: Deviated wells and possible dynamic transfer of fluid from deeper to shallower intervals along the same reservoirs (modified from Beebe, 1961)

Introduction of water or gas by injection during production could increase pressure and lead to seal breaching, lateral communication or capillary leakages, a form of **induced failure**. **Drilling kicks** (sudden gushing in of formation fluid while drilling) and **losses** (sudden leakage into the formation while drilling) are other kinds of seal breaching that interests the drillers most. Reservoirs become **depleted** by the loss of fluid to other reservoirs directly by the well penetrating it, or indirectly by feeding other wells due to pressure difference.

7.5 Geological Field Model

Geological field modelling is used to represent all evaluated geological parameters, usually on a smaller scale, in order to provide a visual relationship that is functional and representative of most if not all the parameters measured. The model will predict how the system would work before it is put in place. Results from the well tests will later be used to refine the model, so it can be used or applied in similar situations. The geological modelling referred to here is that done at local or field level, and not that at basin level. At basin level, focus is more on basin history, maturation, initial generation and charging of hydrocarbon and then sedimentation and subsequent tectonic events (Ungerer, 1990; Okui et al., 1998).

At a local level, focus is more on reserve volumes, compartmentalization and reservoir behaviours, how they will behave during production. The local model is targeted to drilling, establishing the prognosis through the integration of well log data, seismic data and known fluid gradients and properties in the field. When the term static and dynamic model is used, it is focused at production for the local scale. Models should be 3-dimensional and should highlight structural and sedimentary relationships, showing how they relate to the petrophysical and fluid properties.

Following oil field assessment, a reservoir model may show vertical and lateral compartmentalization, and the need for sidetrack (ST) wells. In Figure 7.13, a reservoir model is shown graphically during oil field assessment. From a well C, sidetrack wells A and B are generated to tap vertical and possibly lateral reservoir compartments. When the result is not a direct product of data input and output, it is known as a *conceptual model*, often a depiction of concepts that may or may not be validated by data.

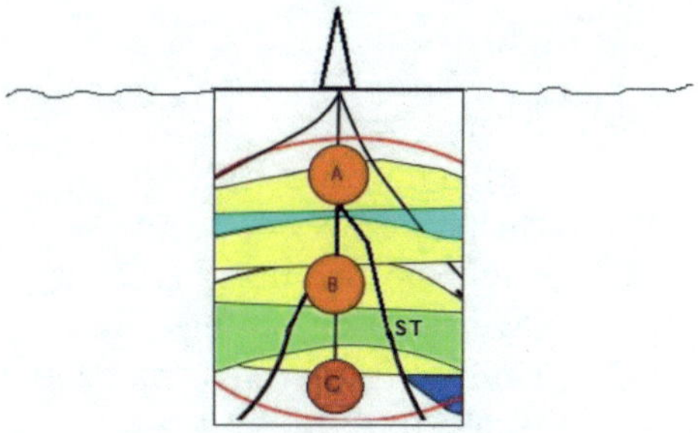

Figure 7.13: Reservoir layering model showing compartmentalization and need for sidetrack (ST) wells

A block can be modelled to explain the organization and the controlling factors of the petroleum system. The local model is often targeted to drilling, establishing the prognosis. They integrate well log data, seismic data and known fluid gradients and properties in the field. They should be 3-dimensional and should highlight the structural and sedimentary relationship, and show how they relate to the petrophysical and fluid properties.

Modelling at local field scale hardly represent such factors as leakages in the field. They may already be assessed at the initial seismic interpretation stage. Most of the leakage types observed on seismic section are indicated by fluid *chimneys* (seismic anomaly interpreted to be vertical fluid flow structure) or amplitude anomalies. However, not all chimney features or vertical anomalies are actually due to fluid leakage. Fluid leakage is confirmed by the presence of:

1. Direct leakage indicators: for example, high amplitude along the chimney
2. Pull up/push down due to change of seismic velocity because of the presence of fluid
3. Presence of pockmarks or other seabed expression of leakage overlying the anomaly
4. Geochemical analyses of samples above the conduit

Table 7.3 shows structural and sedimentary leakage formation mechanisms, as well as non-formation mechanisms. They denote causal factors that determine movement of fluid in a field or basin.

Table 7.3: Showing casual factors for fluid movement, including seabed pockmarks observation and channel conduits forming reservoirs at depth (Nosike, 2009)

	Formation mechanisms		Non-formation mechanisms
	Structural	*Sedimentary*	
1	Presence of fault/fault opening or displacement	Paleo and Recent channel/ levee conduit	No fault opening, tight or smeared fault zone
2	Anticlinal crest/ridge or thrust tip	Debris flow, slightly buried local erosive depression	Local shallow slides and glides
3	Fault rooting in reservoir or layer with migrating fluid	Biogenic activities, concretion and carbonate mass	Shallow fault /rooting away from reservoir
4	Fault or fracture intersection weak zone/dilatation	Erosion of conduit fault tip near shallow sediment cover	Pressure transmission away from zone
5	Stress rotation, lateral-frontal ramp offset junction, tears	Buried glides, tracks and grooves	Absence of underlying reservoir
6	Anticlinal plunge, crestal collapse faults.	Buried regional slides fed or generating biogenic fluid	Sealing of fault below (not reaching) surface
	Mixed		
7	Gas hydrate dissolution		
8	Deep overpressure build-up		

7.6 Oil In Place and Recovery

The geometric factor will depend on the shape of the trapping structure, and the net pay thickness (portion of reservoir saturated with hydrocarbon) will depend on the hydrocarbon fill level of the reservoir, since the fluid contacts will determine the actual hydrocarbon portion of the fill. Certain reservoir characteristics would have to be evaluated. Below are some of the factors to take into consideration in the estimation of oil in place.

Mapping of the Structural and Sedimentary Features

To identify the area available for trapping hydrocarbon, there is need to map the structural and sedimentary features. This is usually achieved by the interpretation of seismic data, where an anticlinal structure, for example, would be highlighted by contours and colour grading. It is from this that the geometry of the reservoir will be carved out. Figure 7.14 shows different structural and sedimentary features that form different traps for hydrocarbon accumulation. They include faults, unconformity and anticlinal fold, whose geometries have implication in well development. These reservoir bodies can be represent in a map, and important step for estimating trapped volumes of hydrocarbons.

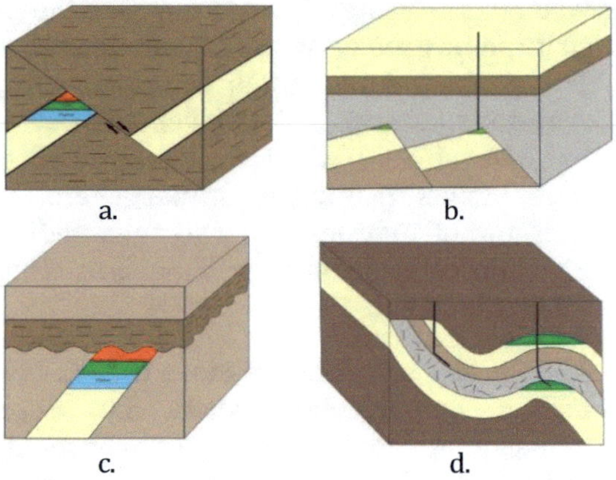

(a.) Trap due to offset of fault block
(b) Trap structures at tip of fault blocks
(c.) Unconformity related trap
(d) Crooked well path in folded structure

Figure 7.14: Structural and stratigraphic traps showing local oil pool on walls of a fault

Mapping of the Fluid Features

Fluid features on a seismic block may be highlighted by amplitude anomalies. How these fluids could show a *bright spot* or *flat spot* on seismic section has been mentioned above in the section on *lead.* The vertical fluid distribution and contacts are identified on the cross-section while the reflective differences could help show the spatial size (areal extent) of the pool on the maps. When shown on a contoured map, the clear trace of the pool is visualized.

On seismic data, various fluid signatures such as "pipe", chimneys, pull up and pull down, can be identified to show recent fluid features and migration pathways. Thicker fluid, muddy flows or mud volcanoes leave visible cones on the surface and in the subsurface. This has been mapped and used for the timing of over-pressure generation and active fluid movement in oil fields. The time of overpressure generation in excess of the sealing capacity is closely related to the age of the layer on which the expulsion is expressed.

In Figure 7.15, different episodes of mud volcanoes/cones (purple colour), mudflows (pink), and conduit for flows (yellow), marking expulsion of fluid from deeper compartments are shown on seismic. Expulsion at the western mud volcano is relatively higher (with more burial rate), compared to that of the eastern mud volcano that is fewer (with less burial rate).

Establishing the Fluid Types and their Contacts

If there are appraisal wells drilled closer to the crest of an anticline or shallowest part of a tilted reservoir, samples would be collected to show the fluid types and contacts: **Gas-Oil-Contact (GOC), Gas-Water-Contact (GWC)** and **Water-Oil-Contact (WOC)**. That is in addition to indication from *log* responses that could also provide information about fluid types. Where the well is drilled at the flank, it may not encounter the shallowest part of the fluid. Likewise, if the total depth (TD) did not penetrate the deepest part of the reservoir, the base of the fluid is not known. The fluid shallowest depth will be recorded as **Up-To** and the deepest depth as **Down-to** levels. Examples are Oil-Up-To (OUT) and Gas-Down-To (GDT), which means the top of the oil column and the base of the gas column were not encountered in the wells, respectively. These are important factors to consider in estimating the **net pay**.

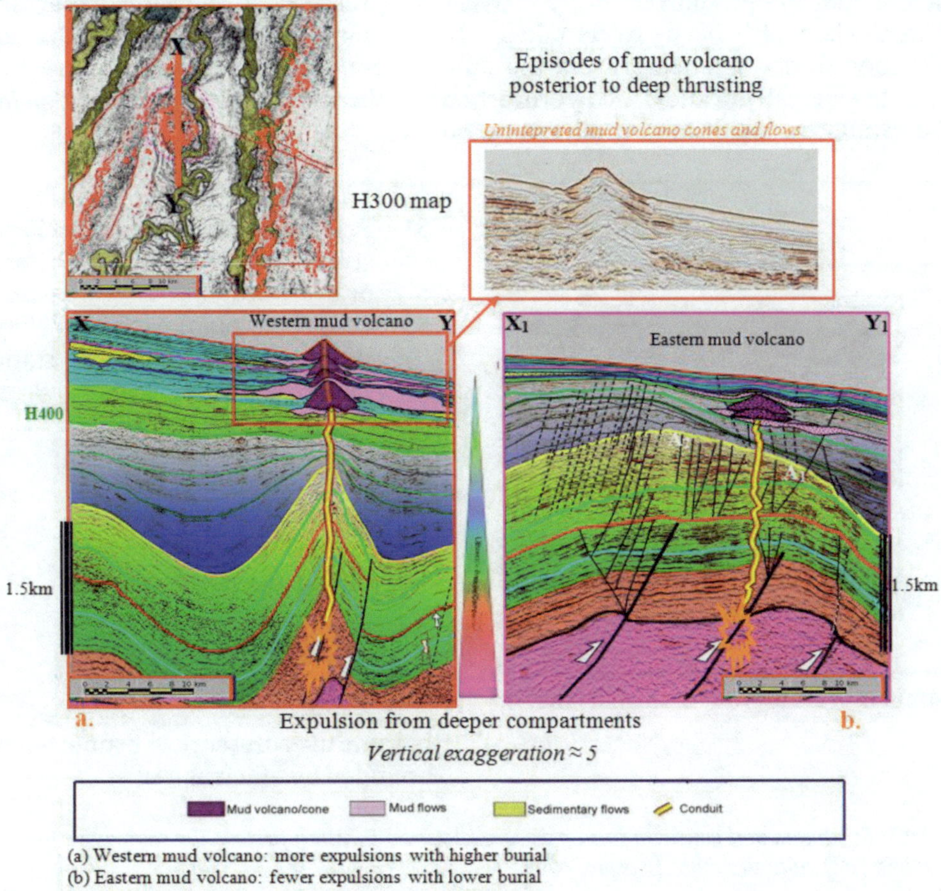

Figure 7.15: Fluid signatures on seismic showing the combined role of fault and anticlinal top in the expulsion of mud volcanoes (Nosike, 2010)

Another factor to be considered in estimating the net pay is a **waste zone**. This is a transition zone from the reservoir to the non-reservoir lithology or seal, with poor porosity and permeability, but sufficient for little droplets of hydrocarbon to penetrate or diffuse through. Such unit if found is neither fully a reservoir nor non-reservoir lithology, and has to be taken into consideration in determining the overall volume of hydrocarbon present.

Pressure data measured with **Well Formation Tester** (WFT) tools can be used to estimate the contacts in a field. This is because oil and water gradients vary. If the same reservoir, where the oil and water gradients cross, or where the oil and gas gradients cross, the gradients will give their most likely contacts. The oil-water-contact given by pressure points may not clearly show the actual level of water and oil where there is a gradational transition or waste zone. There is a *free water level* below the producible and non-producible oil by conventional methods, which needs to be taken into consideration during appraisal. From the fluid contacts, the actual thickness of hydrocarbon may be estimated. With known oil saturation (percentage amount of oil in the pore spaces), the actual oil that is in place can be estimated.

Pressure-Stress Relationship

In order to know what amount of stress the formation can be subjected to during drilling before it fractures, a test is done to know the *minimum stress* (S_3) that would fracture the formation, denoted by the red line in Figure 7.16a. Once this is established, drilling would have to be done so that the formation is not stressed enough to allow it fracture, and so a *drilling window*, above the green line and below the red line is used to safely drill through the formation. The *hydrocarbon height* can be deduced from the vertical distance between the top-seal retention limit and the fluid contact. It is below that the filling occurs, and leakage at the top-seal makes the contacts shallower. Figure 7.16b shows how these flows are controlled by differential burial, where higher pressures in the synclinal areas lead to up-dip migration of fluids to the anticlines, often emplaced by and above thrust faults (*note that there is a vertical exaggeration in the figure*).

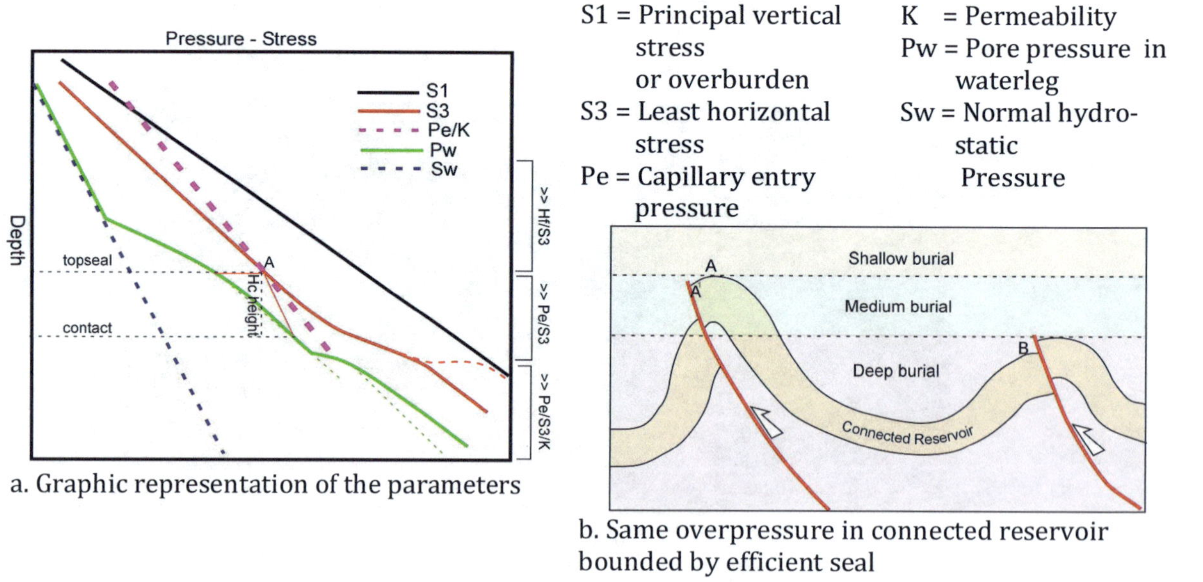

a. Graphic representation of the parameters

b. Same overpressure in connected reservoir bounded by efficient seal

Figure 7.16: Gradients and contacts from pressure plot and limiting factors for seal retention capacity as well as hydrocarbon (Hc) column height (Nosike, 2009)

When the retention capacity is estimated for an oil field at different depths of prospect locations, it helps in knowing the overall viability of the field. As shown in the pressure-stress relationship with depth, in Figure 7.16, the blue dashed line is the *normal (hydrostatic) pressure* (S_w) trend with increasing depth. Any change in this pressure could be because of an overpressure from burial of excess water trapped in the formation; such a change is seen, marked by the green line (P_w) that initially conformed to the normal pressure but changed due to *abnormal pressure*. The black line is known as *overburden* (S_1), and is due to the weight of all the lithologies acting on the reservoir and pressured zones of interests with depth. The purple dashed line is a limiting factor to fluid flow, due to the matrix and grains of the rock (capillary entry and permeability, P_e/K).

Qualifying and Quantifying the Petrophysical Properties

Qualifying and quantifying the properties of the reservoir are the essential aims of carrying out hydrocarbon reservoir evaluation. Petrophysical properties to be considered include porosity, permeability, shaliness or clayiness of the reservoir. They are used for modelling, and together with available logs provide cut-offs that describe the reservoir capacity to retain oil; *pay zones* are marked out using pay zone flags for each reservoir unit. Reservoir evaluation uses geometry, types and intrinsic properties of reservoir to determine hydrocarbon in place and its production. The distribution of porosity and permeability affect the initial oil in place.

Recovery Factors

Recovery is enhanced by artificial methods, such as *acidification* or *water injection*. Water injection is usually used to optimize oil production, though over-injection could lead to sudden water flooding, thereby losing the well.

When the natural pressure that drives the hydrocarbon to the surface depletes, water is passed through the injection wells (usually at the WOC/OWC) in order to drive the oil upwards through the production well where they can be recovered. Figure 7.17 illustrates the water injection technique used to enhance hydrocarbon recovery. Two injectors are placed at the flanks of a producer.

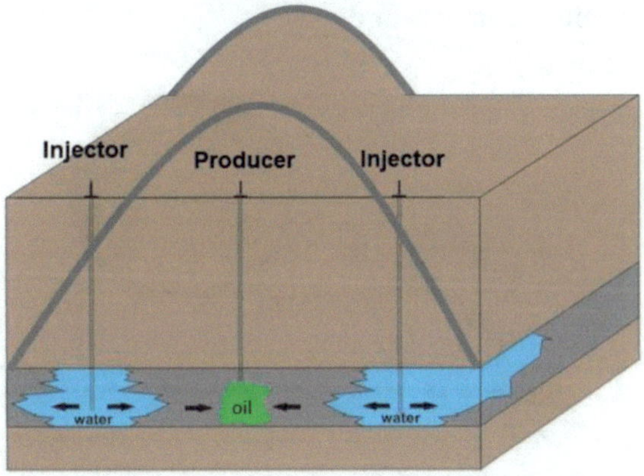

Figure 7.17: Two injection wells used to support a production well in the centre

Other factors of recovery are the *intrinsic property* of the reservoir and that of the fluid. Among those related to depth (temperature and pressure) are:

a. The recovery factor: estimated percentage of hydrocarbon recoverable from a reserve (with modern technology this can be up to 60% for oil and 90% for gas).

b. Oil formation volume factor (Bo): change of volume of oil when moved from higher pressure and temperature at depth to atmospheric pressure or in the storage tank.

c. *Gas formation volume or expansion factor (Bg):* change of volume of gas when moved from higher pressure and temperature at depth to atmospheric pressure or in the storage tank.

d. The initial (BO_i) and *Final (BO_f) oil* compressibility.

e. Bubble point (P_b), also known as *saturation pressure.*

The actual volume of fluid recovered is the **Stock Tank Volume**, or **Stock Tank Barrels (STB),** which is volume measured when fluid is in the stock tank on the surface. Due to expansion, this may be larger than the in-hole volume. The expansion is given by Gas formation volume factor (Bg).

Figure 7.18 shows an illustration of what happens when fluid is moved from subsurface temperatures and pressures to near surface or to surface atmospheric temperatures and pressures. When moved to surface it expands, but contracts at depth.

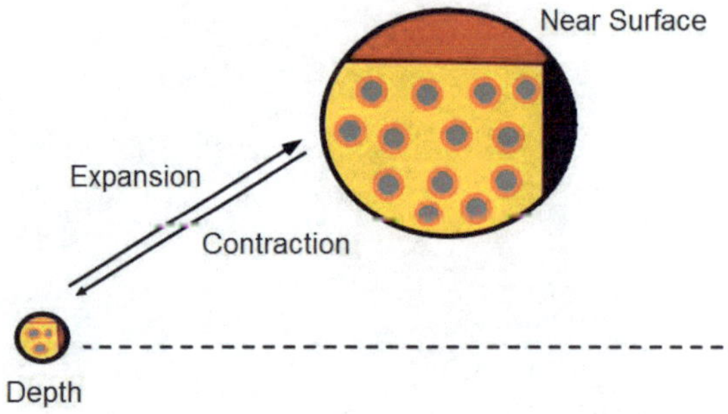

Figure 7.18: Illustration of fluid expansion when moved to surface and contraction at depth respectively

Calculating Hydrocarbon Initially In Place

Hydrocarbon Initially In Place (HIIP), either the *Original Oil In Place (OOIP)* or *Original Gas In Place (OGIP*), is dependent on the amount of space saturated by the oil or gas, and the net fraction of the reservoir section it occupies in the overall or gross rock volume. The first step in determining the HIIP is therefore to evaluate which part or fraction of a rock volume represents the reservoir facies. For example:

Step 1: Presence of reservoir lithology (visible lithology of reservoir facies)

Available Reservoir Presence (ARP)
= Gross Rock Volume (GRV) * Fraction of lithology of reservoir facies

This gives the *Net reservoir fraction volume* and can be expressed in percentage with respect to the GRV. It will be 100% if the overall rock volume is a reservoir lithology. Fraction of lithology of reservoir facies is the *Net-To-Gross (NTG).*

Step 2: Quality of the reservoir lithology (porosity and hydrocarbon saturation)

Hydrocarbon Pore Volume (HCPV)
= Net reservoir fraction volume * Net pore volume * net hydrocarbon saturation
= V * Phi * hc

This is then multiplied by the inverse of the formation volume factor – FVF (*bo* or *Bg*), to estimate what the volume will be in the stock tank. This standard volume is known as *Stock Tank Oil Initially In Place (STOIIP)* or *Gas Initially in Place (GIIP),* a synonym for OOIP or OGIP, respectively.

Therefore, Hydrocarbon Initially In Place (HIIP) is given as,

HIIP = V * Phi * Hc * 1 / FVF

Affirming the Reserve

When hydrocarbon in place has been determined, and multiplied by the factors of recoverability, the estimated amount of substantial oil and gas that can be recovered is known as **reserve**. Several variables determine the recovery factor (**RF**), including fluid type, mobility of fluid within the reservoir and drive mechanism.

Recoverable oil or gas = HIIP * RF

This can be summed up for various wells and filed to estimate the *reserve* of a basin. Reserves are a major consideration in the decisions relating to exploration and production and the nature of regional and local seal in an area; how they are properly evaluated plays a major role in the estimation (once the initial wells have been drilled to get tangible data). They also help to define and calculate the degree of uncertainty.

7.7 Reservoir Evaluation

Reservoir evaluation aims at describing the characteristics of a reservoir such as porosity and permeability, their size, shape and general geometry. Different types of reservoir show different characteristics. Some of these characteristics are intrinsic, meaning that they are inherent in the reservoir. Others relate to how they will behave with respect to flow or how they relate to each other, which are described below.

- *The main reservoir properties:* this will include properties inherent in the reservoir, such as porosity.
- *Reservoir lithology:* lithological description will highlight the various lithological facies.
- *Size and structure:* this will include such important factors as the size and geometry

of the reservoir bodies vertically and laterally.

- *Reservoir quality:* the reservoir quality groups several factors, including fracturation for example, degree of cleanliness and to some degree its flow properties such as permeability.
- *Clean and dirty reservoir:* this is the evaluation of the degree of shaliness of a reservoir (presence of clay means dirty while absence means clean) and the presence of cements and other signs of lithification.

Useful tools and methods of reservoir evaluation include:

- Field Reservoir Analogue
- Offset wells
- Correlation with and across fields
- Seismic *neural classification* (analyses of facies, provided by some seismic interpretation software, that sees them as "neurons")
- Reservoir modelling (using engineering software)

At the end of an overall oil field assessment, there should be a synthesis showing the reservoirs, potential and cost effectiveness of developing and producing the field. Figure 7.19 is a graphical illustration of a 3D synthesis showing the relationship between tectonics and vertical hydrocarbon leakages. It also shows how they affect reservoir potentials and cost effectiveness of field development and production. Examples of 1 – 6 in the figure show potential leakage indicators and weakness for reservoir fluid retention.

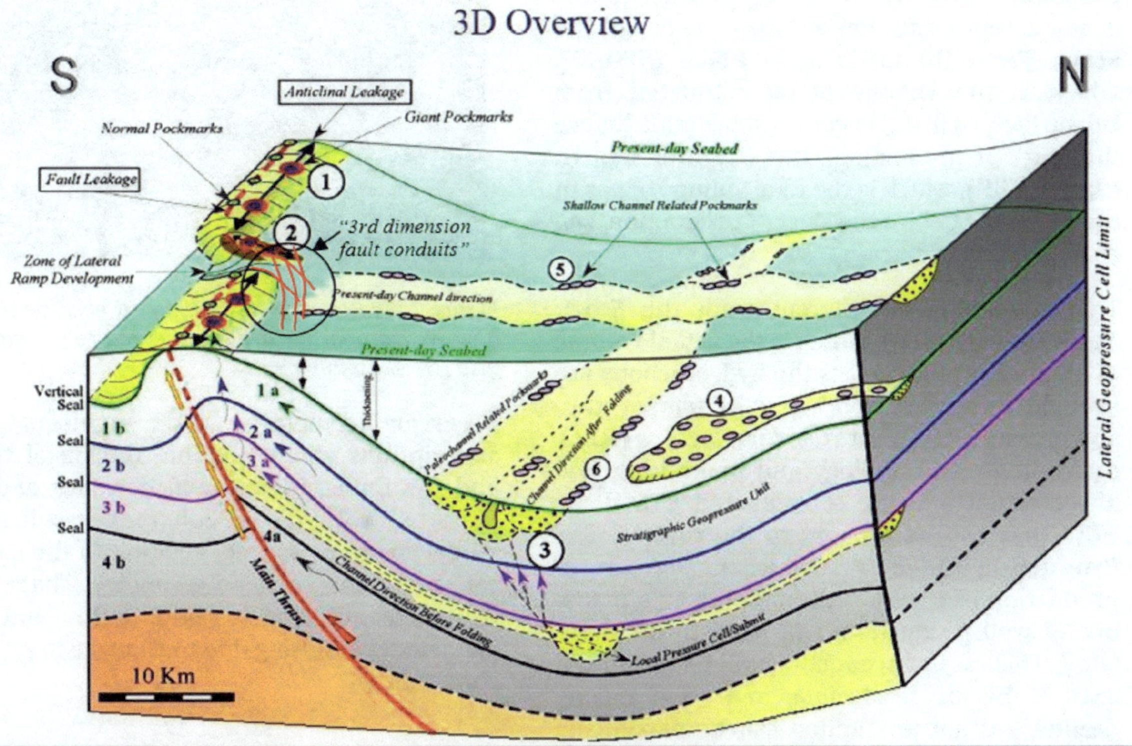

1. Pockmarks on top of anticlines 2. Increased fluid escape from en echelon faults in the zone of lateral ramp development	3. Leakages into channel from fractures 4. Pockmarks above debris flow 5. Pockmarks trailing levees of recent channels	6. Pockmarks trailing paleochannels

Figure 7.19: 3D synthesis of the relationship between tectonics and vertical hydrocarbons leakage in the deep offshore (Nosike, 2009)

Reservoir Modelling at Oil Field Block Level

The purpose of reservoir modelling at block level is to put in a representative and functional scale (often visual), the relationship between most if not all of the reservoir parameters evaluated. To carry out reservoir simulation, divide the field into small cells with properties relatively easier to qualify and quantify; it is over these small discrete units that parameters such as permeability, pressure, mobility are modelled, and these is made to run across the cells to simulate how they will behave during production.

To carry out reservoir modelling, the geoscientist divides a field into grids. Existing offset or development wells and their petrophysical properties are imported to propagate across the entire grid while respecting geologic principles (*further discussed in subsequent chapters*). This will help in predicting the geologic representation of future wells and the **Stock Tank Oil Initially In Place** (STOIIP), which is the volume of oil estimated from subsurface, as it will occupy a stock tank, before the start of production. The **Original Gas In Place** (OGIP), which is the total volume of gas in the reservoir before the start of production, can also be estimated.

This estimation is different from the **Stock Tank Barrels** (STB), which is the actual volume of oil after production, as the hydrocarbons are brought to surface conditions of temperature and pressure. Different **recovery factors (RFs)**, depending on technology and operations, will affect actual amount of estimated STOIIP or OGIP that will be brought to the surface. The **Production Index** (PI), which is the rate of production of a well with pumping or a measure of well production, will need to be simulated. This rate of production will be optimised until it begins to decline, and even during decline, so that production can be maximized over the lifespan of the well.

The use of reservoir top and base to create a layer grid is shown in Figure 7.20. Wells are positioned such that the properties of the well are propagated to populate the grid. This is a visual representative and functional scale model of the block, from which parameters such as permeability, pressure, and mobility are evaluated, and made to run across the cells to simulate how they will behave during production.

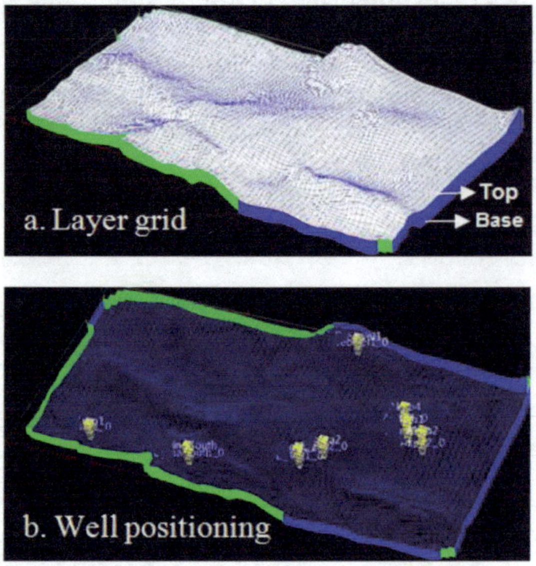

Figure 7.20: Reservoir grid and well position showing how well properties are propagated to populate the grid (Nosike, 2009)

Reservoir geologists and engineers make assumptions about possible mixing of the oil and gas fluids, testing which values and flow patterns will give a coherent result. Static model concerns aspects relating to the geological conditions such as geometry, shape, size, while dynamic models relate to flow and other behaviours resulting due to changes in pressure condition.

EXERCISES

1. What is the E&P value chain in oil and gas exploration? Describe the exploration and appraisal phases. Describe the development and the production phases.

2. What is oil field compartmentalization? Why is it important to establish the distribution and nature of fluid both vertically and horizontally?

3. What are the natural fluids within a reservoir? How and why are they separated in the subsurface?

4. Describe the following factors needed to successfully develop a well:
   a) Continual geological, geophysical and reservoir analyses
   b) Post-mortem
   c) Positive reserve evolution

5. Why is it important to identify hydrocarbon flow direction (present and past), flow boundaries or pressure cell limits, and reservoir compartmentalization during evaluation of oil block? What are the factors affecting sealing and instability of a well during production?

6. Why is there need to develop a geological field model during oil block evaluation? Why is there need for proper management of a producing field? Differentiate between CAPEX and OPEX. What role do they play in determining whether an oil field should be produced?

7. What is oil in place? How does it relate to oil recovery? Describe the following factors taking into consideration in the estimation of initial oil in place:
   i. Mapping of the structural and sedimentary features
   ii. Mapping of the fluid features
   iii. Establishing fluid types and their contacts
   iv. Qualifying and quantifying the petrophysical properties of the reservoir

8. Differentiate between the following controlling factors:
   a) The recovery factor (RF)
   b) Formation volume factor (Bo)
   c) Expansion factor (Bg)
   d) The initial (BO_i) and Final (BO_f) Oil compressibility
   e) Bubble point (P_b)

9. Determine the Available Reservoir Presence (ARP) given that the Gross rock volume (GRV) is 30 m^3 and Fraction of lithology of reservoir facies is 17 m^3

10. Determine the HIIP given that the net reservoir fraction volume is 3000 m^3, net pore volume is 25% and net hydrocarbon saturation is 60%

REFERENCES & FURTHER READING

Beebe, B. W., 1961. Drilling the Exploratory Well, in G. B. Moody, ed., *Petroleum Exploration Handbook*, McGraw-Hill Book Co., New York, p. 17-36.

iESog, 2018. Annual Geoscience Training and Field Trip. January 13 - 19. *www.*iesog.com

Jennings, J. B., 1987. Capillary Pressure Techniques: Application to Exploration and Development Geology. *AAPG Bulletin*, v. 71, p. 1196–1209.

Nestvold, E. O., 1987. The Use of 3-D Seismic In Exploration, Appraisal and Field Development. *12th World Petroleum Congress Proceedings*, v. 2, p. 125–132.

Nosike, L., C. Wibberley, J.-L. Montenat, C. Jardiné and Pacalin, Y. 2008. Relationship between Thrust Fault Architecture, Pressure Cell Stacking and Hydrocarbon Leakage: Deep Offshore Niger Delta, *AAPG International Conference and Exhibition*, 26-29 October, Cape Town, South Africa.

Nosike, L., 2009. Relationship between Tectonics and Vertical Hydrocarbon Leakage: A Case Study of the Deep Offshore Niger Delta. *PhD Thesis*, University of Nice-Sophia Antipolis, France, 281 p.

Nosike, L., C. Wibberley, and Jardiné, C. 2010. Relationship between Tectonics, Pressure Compartmentalization and Vertical Hydrocarbon Leakage in the Deep Offshore Niger Delta, *AAPG Africa Region Annual Conference*, Abuja, Nigeria.

Nosike, L., C. Wibberley and Delpeint, A. 2016. Application of Regional Tectonics to the Understanding of Local Reservoir Compartmentalization in the Deep Offshore Niger Delta: Implication for Well Design and Optimization of Field Development, *NAPE Extended Abstract*, 11p.

Nosike, L., 2019. *A Career in the Petroleum Industry: In a Time of Energy Transition.* Delizon publishers. 253 p.

Okui, A., R. M. Siebert, and H. Matsubayashi, 1998. Simulation of Oil Expulsion by 1-D and 2-D Basin Modelling - Saturation Threshold and Relative Permeabilities of Source Rocks, *in* S. J. Duppenbecker and J. E. Iliffe, eds., Basin Modelling: Practice and Progress. *Geological Society of London Special Publication* 141, p. 45–72.

Perkins, H., 1961. Fault-Closure Type Fields, Southeast Louisiana. *Trans. Gulf Coast Assoc. Geol. Soc.* 11, p. 177–196.

Saugy, L., and Eyer, J.A. 2003. Fifty Years of Exploration in the Niger Delta (West Africa), *in* M. T. Halbouty, ed., Giant Oil and Gas Fields of the Decade 1990-1999: *Amer. Assoc. Petrol. Geol. Mem.*, 78: 211 - 226.

Swarbrick, R. E. and Osborne, M. J. 1996. The Nature and Diversity of Pressure Transition Zones. *Petroleum Geoscience*, 2, p. 111–116.

Ungerer, P., J. Burrus, B. Doligez, P.-Y. Chenet and F. Bessis, 1990. Basin Evaluation by Integrated Two-Dimensional Modelling of Heat Transfer, Fluid Flow, Hydrocarbon Generation and Migration. *American Association of Petroleum Geologists Bulletin*, v. 74, p. 309–335.

Warbrick, R. E. and Osborne, M. J. 1998. Mechanisms that Generate Abnormal Pressures: An Overview, in B. E. Law, G. F. Ulmishek and V. I. Slavin eds., Abnormal Pressures in Hydrocarbon Environments. *AAPG Memoir 70*, p. 13–34.

CHAPTER EIGHT

PROSPECT EVALUATION

8.0 Prospect Evaluation Process

Much of *prospect evaluation* is about asking questions, and there will be several question marks in this chapter. The factors of recovery, target reservoirs, static modelling and dynamic simulation will be discussed. This chapter will build on what has been learnt in the previous chapters to answer the questions. Some of the questions that need to be answered, before going into the evaluation of individual features, are:

- What is a prospect?
- How do geoscientists evaluate it?
- What is the difference between risk and uncertainty?
- How are individual reservoirs evaluated?
- How is the recoverable hydrocarbons estimated?
- What are the main aspects of reservoir modelling?
- What is the relationship between well prognosis and post-mortem?

At the later stage of prospect evaluation, the geoscientist delivers a *well prognosis* to the engineering team for drilling. After drilling, *post-mortem* analyses are used to check results of prospect evaluation. Updates are then made to *static* and *dynamic models* built during the prospect evaluation, and new *history matching* and *forecasts* are made for management of existing and future wells.

8.1 Petroleum Prospect

A combination of some features are necessary to have a petroleum accumulation. Once these features are combined, the only missing piece to make up the "five fingers of the palm" that holds the black gold is the petroleum itself. Again, these four features are:

1. The Source
2. The Reservoir
3. The Seal
4. The Trap

These features have been defined in the first chapter, on Petroleum Geology. The current chapter looks at the presence or not of each of them. The combination of these features, even before ascertaining if the resource actually migrated from the source rock to the reservoir and is sealed to form a trap, is already a prospect. Figure 8.1 shows a source rock, reservoir, anticlinal trap and a seal on top of the trap, a combination that forms a petroleum prospect. Ascertaining the accumulation is another process, known as prospect evaluation. In view of this, it can be evaluated if the top lithological seal in the figure traps the resource while the barrier on the right traps the fluid laterally.

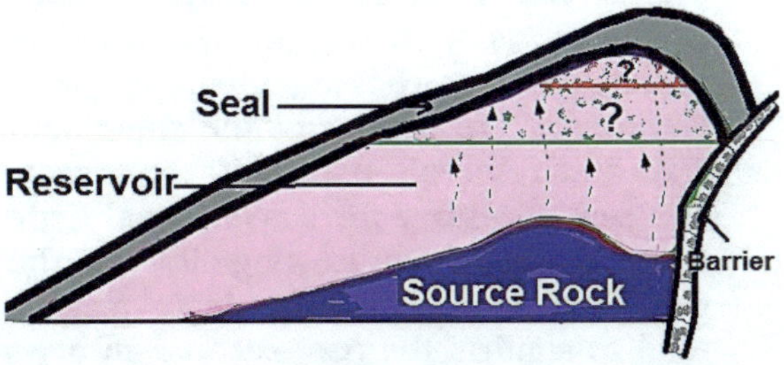

Figure 8.1: Petroleum prospect (with oil pool and gas cap?)

The lateral seal barrier will therefore prevent lateral fluid movement from the trap while the top lithological seal prevents vertical movement. In order to fully determine the viability of this prospect, its source rock volume and drainage area (part of the source rock that can get to the reservoir) will have to be evaluated in 3D; this is necessary to ascertain that it has the potential to sufficiently feed the reservoir.

Identification of the features of a petroleum prospect is possible using seismic data and a cross section. It is easy to identify and carve out a trap based on its geometry, curvature and shape. Reflectivity of the seal and reservoir rock can be used to differentiate them; seals have finer grains, forming the rock matrix in low energy environments and show *calm low energy reflectivity.* Reservoirs, on the other hand, are rather rough with coarse grains forming the rock matrix in high-energy environments, and will therefore show *high-energy chaotic reflectivity,* especially where fluid is present within the rock matrix.

The source rock and seal could be similar, only that the source is expected to be more massive and often underlying or adjacent to the reservoir to ensure proper arrival of expulsed hydrocarbon fluids. The identification of source rocks requires a broader understanding of the basin, the petroleum system and the possible migration pathways. This overall task of ensuring that each of the elements of a prospect is present in a prospective petroleum field is the requirement in prospect evaluation.

8.2 Prospect Generation

When oil field is assessed after block acquisition and potential structures are suspected for drilling, the geoscientist is zooming in on individual bodies that may have oil pools. A source rock and evidence of mature source rock is known as a petroleum system. The lead geoscientists searches and identifies each of the features of prospects and the combination of trap and reservoir with suitable source rock and seal necessary for conventional hydrocarbon resources. In an existing oil field, the lead geoscientist may identify single factors that tend to confirm the rest, such as an overpressure signal in a 4D Seismic monitoring data, any evidence of by-passed oil or a fault barrier that cuts off a target within a confirmed oil pool. This search is known as *prospect generation.*

As indicated before, a viable petroleum prospect was generated by the right source materials of petroleum that migrated to suitable rocks, where they accumulated over *geologic time.* Geologic time scales are in millions and billions of years. It probably took millions of years to deposit, mature basins, generate and expulse hydrocarbons, then keep them in traps. It therefore means that hydrocarbon deposits exploited today were formed millions of years ago. That is why petroleum is not a renewable source of energy.

Some active events help to visualise the processes taking place in an active petroleum system and suggest a prospect. For example, a natural petroleum leakage can help picture migration (tertiary migration in the form of an oil seep on the Earth surface), indicating presence of source rock and probably traps. *Prospect generation* is carried out to identify these features, trap, seal, reservoir and source rocks, while prospect evaluation confirms their combination and quality.

Figure 8.2 shows factors that could affect a prospect. Irrespective of the petroleum accumulation result, factors such as economics, cost, statistics and technology will need to be considered, as they contribute to determining the viability of a prospect. It is a surprise to the geoscientist that despite their arduous task in evaluating the petroleum prospect reserves, other factors such as politics and national policies, economic sanctions and challenges with engineering sometimes reduce their prospect to lowest viability.

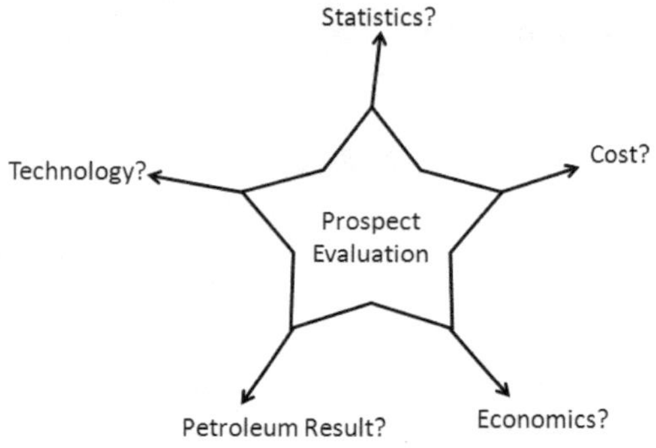

Figure 8.2: Questions asked during prospect evaluation

8.3 Prospect Evaluation

After acquiring a block and assessing the field for presence of prospects that could be drilled, *prospect evaluation* is then carried out. Looking at the main factors mentioned in the definition of a prospect, it is hoped that the source rock in question has charged the reservoir, and that the charging occurred after the emplacement of the structure, with a good seal over it. Prospect evaluation involves establishing the relationships between the source rocks, reservoirs, traps and seal lithology.

The initial stage of the prospect evaluation asks and tries to resolve certain questions already highlighted for oil field assessment. This is now done at a local scale, focussed on the prospect, with the aim of answering questions such as:

1. Is the source rock mature?
2. What is the evidence that generated oil has migrated to the reservoir?
3. How good is the reservoir?
4. Is the trap structure fractured?
5. Was the trap available before the charging or has the oil been migrated away before the formation of the structure?

6. Any recent tectonics that might have destroyed the prospect?
7. What is the seal efficiency and integrity?
8. Has the oil or gas leaked?
9. What are the main risks and uncertainties?
10. What volumes are trapped?
11. What is the economics, costs versus sales figures?

It might be wise to first to estimate the hydrocarbon volume in place and the economics before proceeding with further studies. This is because without enough volumes, the economics will be poor, and there will be no need to waste time and resources on the prospect. Where the volume is promising, further evaluation is carried out and a more in-depth Stock Tank Oil Initially In Place (STOIIP) or Gas Initially In Place (GIIP) estimation per panel or per compartment in the reservoirs will be used to sanction (approve) the project.

Initial Estimation of Volume

Roughly quantifying the initial volume of hydrocarbon present can be done using the gross thickness and shape or geometry of the trap. This estimation is done in 3D, as the sectional area and lateral extension will have to be considered. The volume estimated at this stage is an abstract volume, to provide an idea of the potential prospect to be drilled. Other factors will have to be considered when estimating the volume of original oil and gas in place. A gross thickness was estimated in a previous example (*in Chapter Five: Figure 5.15*), by multiplying the area of the trap covered by a structural envelope. This is shown again in Figure 8.3, this time highlighting how different shapes with mathematical formula is used to estimate the lateral extent (Figure 8.3b).

As shown in Figure 8.3, gross volume estimation of an oil pool in an anticlinal structure can be done by first obtaining information on the sectional area, by multiplying the length of the oil pool and its thickness. After that, volume estimation is done by multiplying the sectional area with available information on the lateral extent of the pool covered by the structural envelope. Standard shapes are used to estimate the lateral extensions of a prospect in order to estimate its volume. These shapes could be circular and oblong (full prospects), or semi-circular and half-oblong (half prospects - prospect possibly truncated by faults). Simple mathematical formulas are used in estimating these shapes.

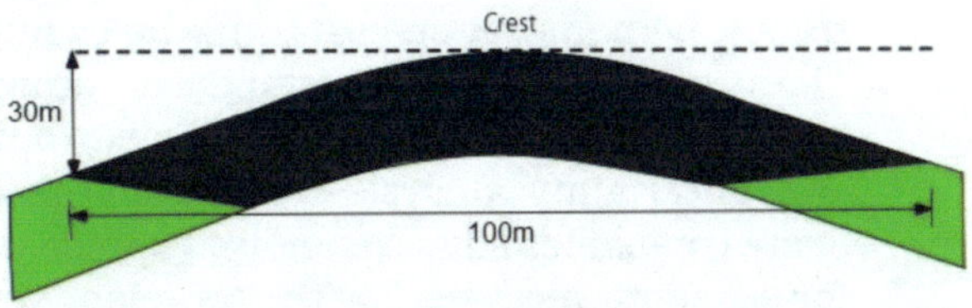

a. Structural envelope showing sectional area

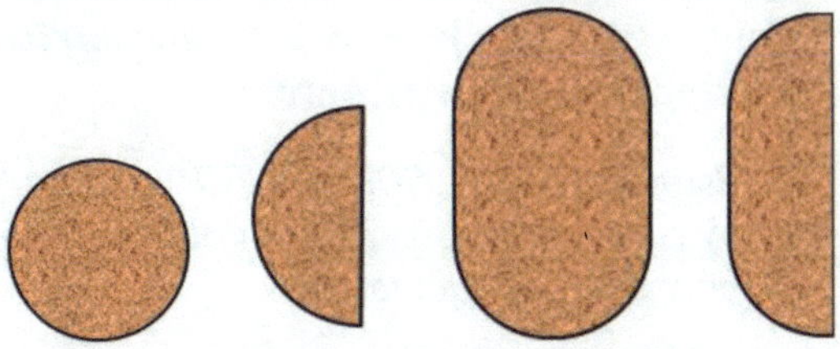

Circle Semi-Circle Oblong Half Oblong

b. Standard shapes showing different lateral extents

Figure 8.3: Vertical and lateral extensions of a prospect for volume estimation

The envelopes in Figure 8.3 exemplify circular shape / domal structure and oblong / elongated anticlinal structure. As has been shown in the chapter on Structural Geology, there are many more shapes and structural traps. Four-way closure or three-way closure are ways of explaining how the closures of the structure may affect the volume of a prospect; other factors to consider include whether it is a domal or elongated anticlinal structure (Cobbold et al., 2009; Kostenko et al., 2008).

In order to carry out a complete prospect evaluation, a broad experience and a general understanding of the concepts are necessary. Only through such experience and understanding will one know that, for example, oil may trap without a visible structure, as in the case of hydrodynamic traps. It has been mentioned that apart from geology-related concerns, other issues such as logistic, political and environmental concerns can prevent the drilling of a prospect. Oil company explorationists are therefore always busy identifying new prospects, having long-term focus despite targeting individual results at present.

Petroleum Economics

While the geoscientists focus on good accumulation of resources in a petroleum prospect, the petroleum economist focuses on factors such as project planning, procurement of equipment and contracting. National policies, including national content or workforce, oil-marketing trends, with impact on costs, may be evaluated. These evaluations use standard economic principles, based on probability distributions and uncertainty analysis for decision-making. Some companies have customised spreadsheets for economic analysis. For the geoscientists, the critical task in his role is the volumes and their value, which are known as ***Volumetrics.*** The two steps in volumetrics are:

- *Estimate HCIIP (OOIP or OGIP) and multiply by current (or forecasted) dollar price of oil barrel or gas equivalent.*
- *Find out if this value is more or less than the financial investment needed to drill the well or develop the field*

The geoscientist proposes scenarios for possible results and producible hydrocarbon volumes, based on technical studies. He delivers these to the economic management teams of the various parties in the joint venture or partners in the oil field. The operator of the field has adequate structure to handle the infrastructure and taxation processes, to manage the CAPEX and OPEX and eventual revenues as well as expenditures.

Among the economic indicators that are used to assess the viability of a prospect development project is **Value On Investment (VOI)** and the **Payback Period** (whether it is **production-based** or **fixed** in years). A key consideration is whether the project is single **operation-based** or requires **multiple activities** (follow up activities, such as *walkover* or other *intervention* after drilling a well, can lead to unexpected cost). Another tool used to assess return on equity is the **Before Tax Cash Flow Analysis**, which favours the investor when there is a net income before tax payment (as is the case is some *Production Sharing Agreements - PSA*). Just as the geoscientist models the reservoir, the petroleum economist performs a full *financial modelling*, covering cash flows and fiscal requirements.

Among the information from the geoscientists and engineers that will serve as input to such modelling are the **STOOIP**, **Production Rat**e, **Decline Curve, History Matching** and **Forecast**. The economist therefore uses parameters such as **Net Present Value (NPV)** and the **Internal Rate of Return (IRR)** to carry out **Investment Project Appraisal**. The NPV is the difference between the cash outflow and inflow value over the given time due to an interest, whereas the IRR is the rate at which the earnings intended by the invested capital is achieved. Where multiple wells will be drilled with additional capital investment, the returns which depend on additional unit of capital asset is known as **Marginal Efficiency of Capital (MEC)**. At the end, a successful prospect will give a positive **Expected Monetary Value (EMV)** (Nederlof, 2019).

The **Present Value Net Cash Surplus (PVNC)** gives the result of the cumulative cash flow and focusses on successful exploration wells. The cost of the wells drilled to evaluate a structure before it is abandoned should as well be taken into consideration, and this is known as **Abortive Exploration Costs (AEC)**. A formula for an economic evaluation of a project is given below:

$$EMV = POS * E[\,PVNCS\,] - (1\text{-}POS) * E[AEC]$$

Where,

EMV	=	Expected Monetary Value
POS	=	Probability of Success
PVNCS	=	Present Value Net Cash Surplus
AEC	=	Abortive Exploration Costs
E	=	Expectation (...)

Evidently, the prospect development that requires less cost, but will produce the same EMV as another prospect with more cost, is preferred. These is a major consideration before investment in a prospect, even if it has been rated positive based on geoscientific evaluations. Prior to this time, geoscientists were less involved in petroleum economic evaluation; however, the time now demands that they participate fully.

Risk and Uncertainty

The aim of this stage of evaluation is to refine every prediction made in the first stage. The language is "risk" and "uncertainty". Usually a + and – range is given for every value so that it is within the limits of certainty, or at least nearly so. The focus is on how to ensure that the well is drilled properly, reaching the right targets and used to conclude the understanding of the reserve.

Risking a Prospect

Petroleum exploration and production is a risky business, not only in terms of safety, but also in terms of the nagging possibility of failure and losing big money. Despite enormous studies that show the likelihood of a petroleum prospect having sufficient hydrocarbon volume, the chances of success are still not certain; there are even greater human, financial and environmental risks, and a chance that what was once considered a success could turn out to be a failure and complicate the situation.

The exploration geologist tries to limit such risks and factor in all stages of the business, and all aspect of the evaluation, by defining the uncertainty at each stage and for each feature, to become more precise on the chances of success. Figure 8.4 shows the Probability Of Success (POS) and the factors that determine the success of a prospect.

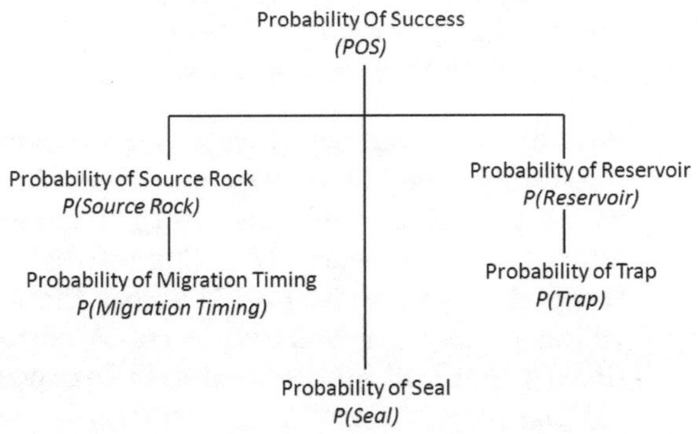

Figure 8.4: Probability of success as a factor of prospect features and migration

The probability of migration timing is one of such factors, which is dependent on the probability of source rock generating hydrocarbon. Another factor is the probability of trap, which is dependent on the probability of reservoir allowing fluid to flow and be stored in its pore spaces. The probability of seal determines what is necessary to prevent hydrocarbon flow away from the trap. Individual probabilities are multiplied to get the possibility of overall success.

Risk is defined in percentage: *percentage success* or *failure / probability of success* or *failure*. A well that has a 60% chance of success obviously has a 40% chance of failure. In real cases, this 60% chance of success does not depend on one factor, but on the four or five already mentioned factors of a prospect, (sometimes migration is added as a process that as well will nullify the prospect if not well assessed). Therefore, the question to ask is, what are the percentage chances that there is a good source, reservoir, trap and seal.

Each of these factors could be independent of the other, so the need to multiply their individual percentages and then divide by the total to get an estimated average risk. If the risk of something is 100% that is zero chance of success, then it automatically annuls every other value and the project will be too risky to undertake. This means that if the chance of success of one of the factors is 0%, even though all other factors have a success rate of 100%, the POS will be zero (i.e. 0 x 100 x 100 x 100 x 100 = 0). As such, it will be highly risky to drill such a prospect. Remember that all factors have to be in place before hydrocarbon generation and accumulation become possible.

Therefore,

$$POS = P(SR) * P(R) * P(MT) * P(T) * P(S)$$

Where,

POS	=	*Probability Of Success*
SR	=	*Source Rock*
R	=	*Reservoir*
MT	=	*Migration Timing*
T	=	*Trap*
S	=	*Seal*
P ( )	=	*Probability of ( )*

Once the POS is estimated with values for a given prospect, the highest possible and lowest possible values are plotted as shown in the figure below. How these are analysed and used varies for different companies, but the examples below are simplified to give the concept. Figure 8.5 shows a relationship between probability of success and distribution (reserve volume in this case). P-10 is high case (tending to maximum), P-90 is low case (tending to maximum), and P-50 is most likely case for volume to occur.

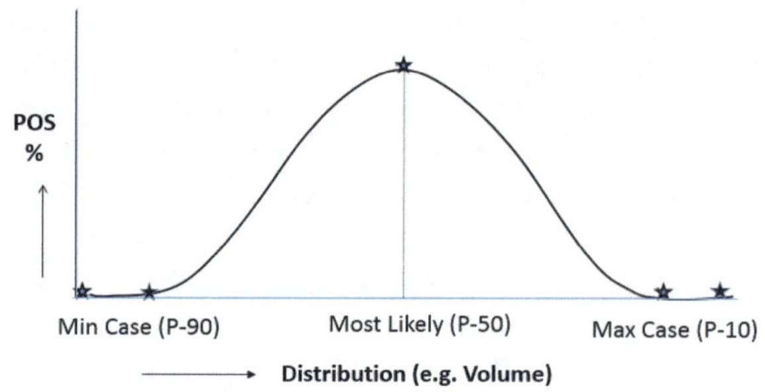

a. Probability of success versus distribution

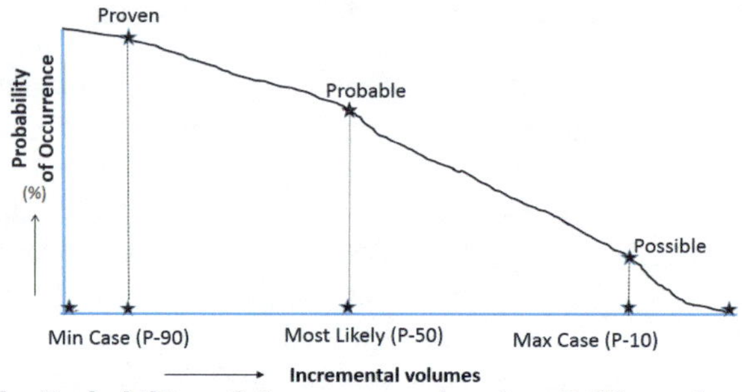

b. Probability of Occurrence showing P-90 as the proven value, over which more unproven values P-50 (probable) and P-10 (possible) may be added

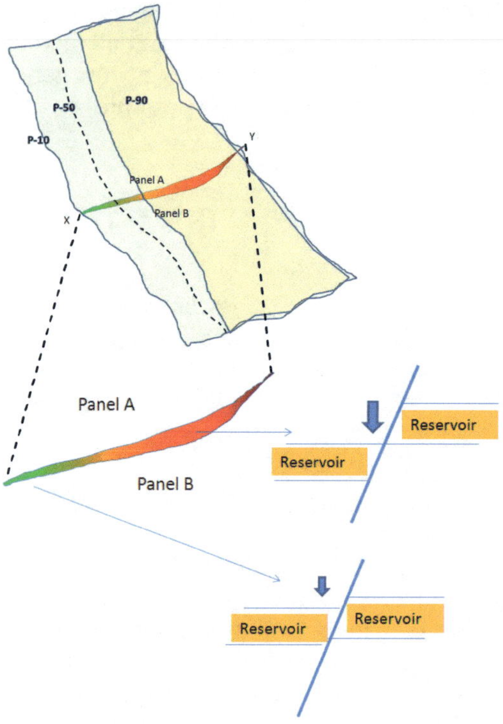

c. Mapped fairway bisected by a fault with varying throws shown in a cross-section, disconnecting the reservoir differentially.

Figure 8.5: Low, most likely and high cases related to different Probability Of Success (POS) and the Probability of Occurrence

Actual *Max case* is P100 and actual *Min case* is P0, is a success or failure case. A mapped fairway is also shown, where it is partially disconnected by a fault at point X, but fully disconnected at point Y. This is a case of **Sensitivity**, showing the analyses uncertainty.

In a., P-90 is low case while P-10 is high case, giving the less and more predicted volumes that will occur respectively; P-90 is a hope for a big surprise, while P-10 is the minimum in case of an unpleasant surprise. The possibility or POS for either case is low. It is a case where the volumes are yet unknown. P-50 is the most likely case.

In b., it is a case of known initial discovery or volume, and the other predictions (P-50 and P10) are increases that may lead to addition if they occur. The chances that P-50 occurs is 'most likely' (probable), but an additional P-10 is only a possibility; it is possible.

In **c.**, the throw is smaller and probably not able to disconnect the reservoir closer to X, as shown in the illustration to the right, which it completely disconnects closer to Y. The area covered, determining the volumes, corresponding to the *Probability of Occurrence*, and the *Possibility Of Success*, are shown in the map.

Possibility of Success focusses of percentage possibility based on quality of prospect features while probability of occurrence focusses on alternative scenarios. The chance that oil is trapped in one of the panels across the darker yellow region near Y is almost 100%, quite high (P-90); that will be the minimum hydrocarbon volume case (if a well has been drilled there, it will be a proven case).

The chances that hydrocarbon is trapped all along to the region near X are low (POS of the fault seal in the green area is too low), tending towards zero, highly unlikely (as such P-10). Although that addition would give the highest volume if it did occur. The reality is ***most likely*** somewhere between the two, where the probability is P-50. For economic evaluation, P-50 is used for being realistic while P-10 is used as being ***optimistic***. P-90 is ***pessimistic***. P-90 is the POS of minimal result and the probability of occurrence of not getting any more than the proven volumes. It gives the worst-case scenario, which must be considered even when it is assumed that the most likely will be the outcome.

For reserve booking (i.e. formally adding volumes to an oil companies published hydrocarbon reserves), P1, P2 and P3 are used to signify possible reserves (just like P90, P50 and P10, are used during prospect evaluation, respectively). In that case, P1 (or Low Case) is the minimum proven hydrocarbon volume or reserve, P2 (or Base Case) is the proven and probable reserves and P3 (High Case) is the maximum possible reserves (including the proven, probable and possible volumes). Risking and uncertainty terms related to hydrocarbon volumes are not used in the same way during exploration and when booking reserves. It is often better to use 'low case', 'base case' and 'high case' when working as geoscientist. Then specific standards, which exists as regulations, may be used for booking reserves.

Estimating the Uncertainty

The *volumetrics* of the oil pool in a prospect will in addition define the uncertainty surrounding that prospect. Despite enormous studies showing that a petroleum prospect has sufficient hydrocarbon volume, some uncertainties remain, to be defined for each parameter.

The uncertainty that a well will produce 100 million barrels of oil may be ± 10 m barrel based on studies of its *volumetrics*. If the well succeeds, it will produce for example a minimum of 90 million barrel and a maximum of 110 million barrel. Unlike the risk, the uncertainties for each parameter of interest would have to be summed together to obtain the total uncertainty of the project. Often the aim is to relate each uncertainty to how it will affect the volume. For example, if there is a 5% uncertainty in the length of a play fairway, then it could imply 5% of uncertainty in the volume. At the end, the total uncertainty is related to the volume of oil expected.

In Figure 8.6, a mapped fairway is highlighted, showing uncertainty in the width of the area and how that may affect the volume. In this case, the uncertainty in the width is (100-95/100) x 100 = 5% and these invariably affect the volume. In practice, there could be uncertainty in both the width and the length of the mapped fairway.

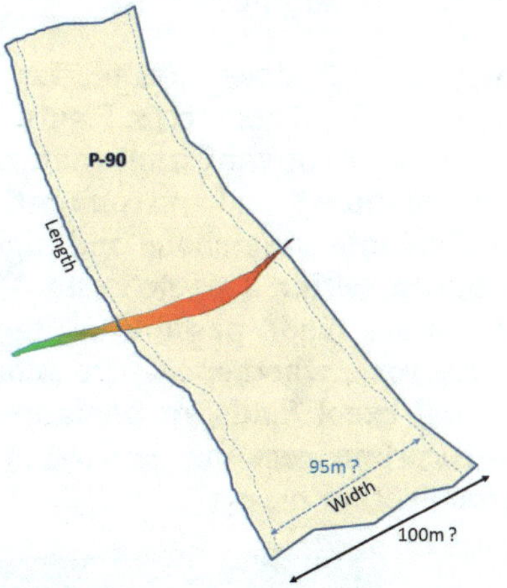

Figure 8.6: Mapped fairway showing uncertainty in the area and eventually volume

Given that the total width is 100 m, and the uncertainty is about 2.5m at each end of the fairway, 5 m uncertainty in total, the estimated uncertainty is therefore 5% of the total volume. Note that 5 m uncertainty is not equal to 5%; if the total width is 150 m and the total uncertainty remains at 5 m, the estimated uncertainty will be just over 3% i.e.[(150 - 145)/150] x 100 = 3.3%.

While probability has to do with an unknown possibility of occurrence, uncertainty has to do with an unknown precision (dimension or quantity) in the feature being evaluated. The uncertainties in petrophysical properties are much more of concern in determining oil in place and dynamic communication. They require geostatistical tool and software for modelling.

The factors affecting the prospect evaluation parameters, which could be risked and their uncertainty estimated, are described below. At this stage, answers will have to be given for each of them.

Petroleum Generation from Source Rock and Charging

This can be checked by confirming the presence of source rock, estimating its distance to the trap structure, as well as determining the available drainage surface of the source rock to the structure. The volume will be cumulative over time (the longer charging of trap continues, the more the volume accumulated). It is important to establish that the seal was there before the charging process.

Sometimes it does not matter if these are unconventional resources. Figure 8.7 shows the existence of **conventional** (natural oil and gas accumulations) and **unconventional** (immature organic rich shale and tight sand gas) resources within a single basin. These accumulations are made possible by the presence of source rock, whether mature enough to generate and expel fluids, or immature just so that hydrocarbon can be generated from them through other means.

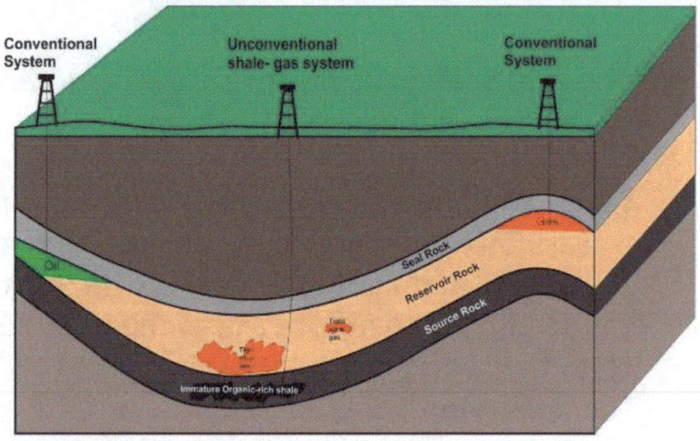

Figure 8.7: Petroleum charge system and underlying source rock leading to production of conventional and unconventional resources in the same basin

In Figure 8.7, among the questions that may be answered during the evaluation of the prospects are: Which of the deposits has more potential or more volumes? What technology is best suited for the production and cost effectiveness? What probability will be assigned to the source rock and trap system before drilling? What factors will lead to uncertainties and how can they be quantified?

Reservoir Size and Quality

Reservoirs are evaluated by first mapping its distribution to estimate its size, and by ensuring that the scales and depth conversion of the seismic data used are properly done. Amplitude reflection could help in estimating both vertical and horizontal reservoir extension. The "cleanliness of the sandstones" or the amount of "secondary porosity in the limestone" will have to be established. Porous sands with little clay or cementation content is said to be clean, while fractures and dissolution improve the permeability of limestone.

Sandstone quality could be a result of several factors, including compaction, cementation, secondary porosity, and different environment of deposition at different depth, leading to different distances of transportation, roundness of the grains and sorting. Figure 8.8 shows three cases of sandstone samples describing what their reservoir quality will be. The first sample with relatively uniform grain sizes will have better reservoir quality, because their grain are well sorted compared to the case of the third sample having poorly sorted grains. The second sample in the centre may be best, because the grains or matrix are cleaner, relatively well sorted and have better porosity.

Figure 8.8: Sandstone samples with different reservoir quality

The examples Figure 8.8 may be used to suggest what probability will be assigned to a reservoir quality, if these samples were obtained from analysis for a prospect before drilling. In addition, to be considered are what factors could lead to uncertainties and how they can be quantified.

Trap System

Trapping can be checked by seismic mapping of the structure or stratigraphic/sedimentary feature, to determine if it forms a good closure that prevents both vertical and lateral escape of hydrocarbon. There are several types of traps, ranging from *structural*, *stratigraphic* to *hydrodynamic* traps, which have been discussed in previous chapters. Structural trap evaluation is much more descriptive as it can be literally seen on seismic data, but stratigraphic traps are more difficult to ascertain; hydrodynamic traps are even much more challenging to assess. Structural traps are more effective in trapping hydrocarbons than stratigraphic traps, while hydrodynamic traps are less likely to effectively trap hydrocarbons (compared to structural and stratigraphic traps). However due to continuous charging, accumulation in hydrodynamic traps may be produced for a relatively a long time.

In the sedimentary basin shown in Figure 8.9, a shale rock lies directly on top of the basement. The shale may serve as a source rock while the sandstone overlying the shale acts as a reservoir and the marl above serves as seal. Structural and stratigraphic barriers to flow are provided by both sides of the basement. The combination of all aforementioned factors forms a potential play of interest in prospect evaluation.

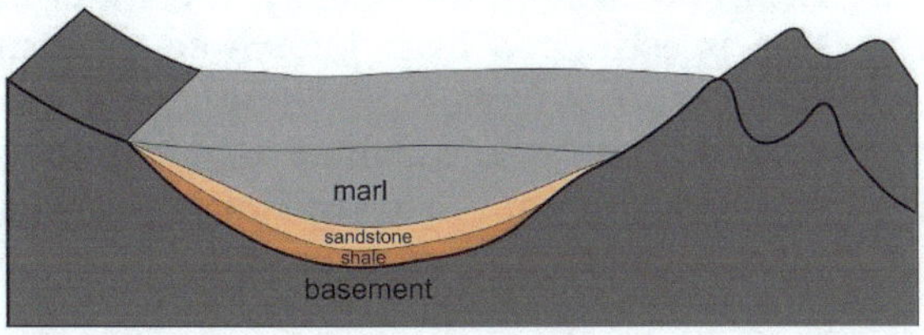

Figure 8.9: Basin with potential play, sand-shale combination and possible marl seal and (structural/stratigraphic) barrier at both sides

It is important to assess what efficiency to ascribe to each of these factors, compared to an anticlinal 4-way closure, for example. A probability will have to be assigned to the geometry and trap if this framework was suggested for a prospect before drilling. The several factors that will lead to uncertainties would as well have to be quantified.

Seal Integrity and Efficiency

The first step in seal evaluation is to determine the type of seal and its distribution. A shale bed for example, if thick enough, has a high tendency to seal. However, this is not always the case as the presence of fractures could compromise the seal. No seal is completely sealing or completely leaking. It is this degree of competency (capability to hinder fluid flow) that is tested in seal evaluation. The seal **efficiency** tests the leakage through capillary entry into pore throats in the seal, while **integrity** tests the resistance of the seal to micro-fracturing under a given pressure. *More on these tests, in the next chapter on Traps and Seal Assessment.*

Figure 8.10 shows a sedimentary basin configuration, where different sedimentary rock layers are deposited on the basement. Sandstones (yellow) and shales (dark ash) could form possible reservoir and seal respectively where hydrocarbon is generated from shale layers below the reservoir. The integrity and efficiency of the seals, as well as the presence of lateral barriers, will determine whether generated fluid will be trapped by the seals at all levels. The seals are not described and the question marks (in Figure 8.10) would require some prognosis during prospect evaluation.

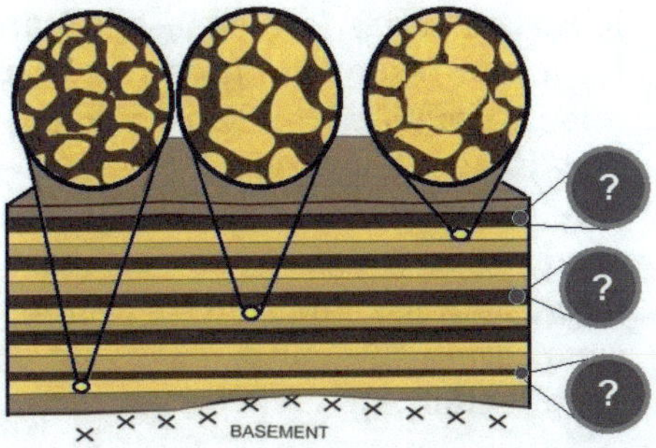

Figure 8.10: Sealing within a basin with different types of sandstone reservoir layers (yellow) and regional seals with different thicknesses (dark ash)

Assuming that lateral barriers, such as fault seal, confine the reservoirs, it will be important to assess the shale seals to know which of them will be more effective in terms of top *seal integrity* or *efficiency* (more on this below). The sandstone are the same as shown in Figure 8.8; the additional information in this figure could help in understanding what might have led to different sand samples and reservoir qualities. It then becomes possible to suggest the impact on the shaly rocks in circles with the questions marks. *The porosity will be lower, and using the idea from the sandstone sample, obviously there would be more impact of compaction for the rock in the circle at the bottom.*

Trapping and sealing mechanisms are perhaps the most important aspect of prospect evaluation in a confirmed petroleum system (Finkbeiner, et al., 2001; Bilotti and Shaw, 2005). In such oil fields where a source rock has been confirmed by offset wells, and there are interplay of seal and reservoir pairs, it is essentially the trapping and sealing mechanisms that determine if oil is found.

8.4 Reservoir Evaluation

In reservoir evaluation, there is need to not only identify the hydrocarbon reservoirs, but also to understand their vertical and lateral distributions, their oil yielding potentials (extractable to the surface) and how each unit is associated (Heid et al., 1950; McPhee and Arthur, 1991; Dembicki, 2009; Burrus, 1998). Below are the important aspects of reservoir evaluation.

Facies and facies association

Reservoir facies (e.g. sandstone) are separated from non-reservoir facies (e.g. shale) and the quality evaluated (for conventional resources) – how clean is the sand for example – what is the clay content? How massively together is the sand, or is it formed for example with shale intercalations? Will one oil well impact, on the sand, drain the entire reservoir or is it compartmentalised?

Porosity and Permeability (ø and k)

The porosity and permeability, also discussed elsewhere, are probably the most important factors for the reservoir quality. Pore space availability (**porosity**) and the interconnectivity of the pore (permeability), or presence of micro-fractures (**permeability**) make fluids to flow through them.

Net to Gross (NTG)

In the vertical seismic section, or from a well log, it is possible to establish a percentage thickness of the reservoir section with respect to the non-reservoir section. This is the **Net to Gross**, and will be the only available section permeable in commercial quantities by hydrocarbon fluid. NTG is illustrated in Figure 8.11.

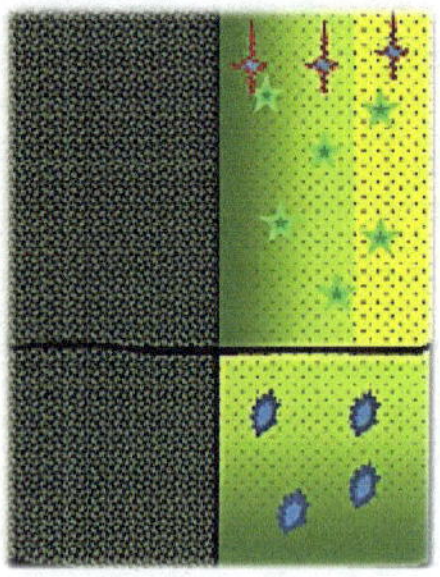

Figure 8.11: NTG illustrated as the ratio of sandy or porous (reservoir) rock to shaly or non-porous (non-reservoir) rock.

The NTG, that is, the ratio of the porous reservoir rock facies (yellow) to the non-porous reservoir rock facies (dark green) is about 50%. The Net Rock itself is not fully saturated with hydrocarbon; part of it is saturated with hydrocarbon (gas and oil in red and green stars respectively), while the lower part is saturated with water (blue stars).

Fluid Saturation (water, oil and gas: Sw, So and Sg respectively)

The fluids that saturate the reservoir - *water, oil or gas* (and other traces) are important in reservoir evaluation. Usually, the water saturation is estimated from known sources such as offset well or from logs where a well has been drilled. Then it is assumed that the remaining fraction (the upper part in Figure 8.11) is made up of hydrocarbon, oil or gas both.

Net Pay Zone (NPZ)

The **net pay zone** is the section of the net reservoir (the NTG) saturated with hydrocarbon fluid. Only the net pay zone (that part with hydrocarbon) will count in estimating the hydrocarbon in place within the reservoir; the water-filled section will not be considered even though it is part of the reservoir section of the NTG since it does not contribute to hydrocarbon volume. In Figure 8.12, two levels 1 and 2 of channels, having facies A and B are shown. Facies A for both Levels 1 and 2 is the non-reservoir section and will not contribute to hydrocarbon saturation. Facies B of Level 1 and 2 is the NTG while Facies B of Level 1 is the NPZ with hydrocarbon saturation; it is this section that will count when estimating the hydrocarbon in place within the reservoir.

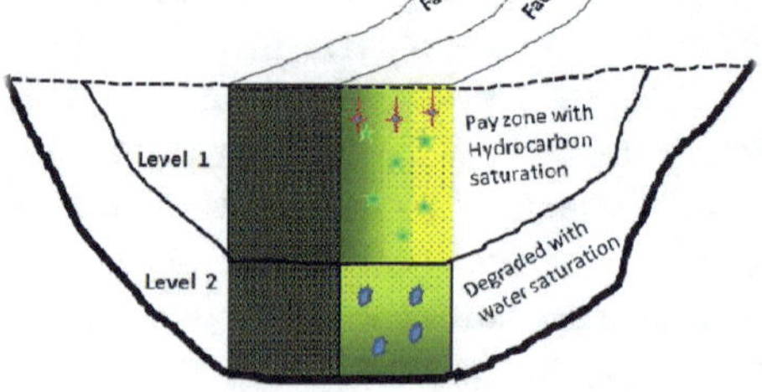

Figure 8.12: Channel with an ideal fill laterally with different facies of sediment A and B, showing a vertical fluid variation with the pay zone (good

porosity and hydrocarbon saturation) above in facies B of Level 1

The net sand in Level 2 of Facies B is in addition degraded, meaning it is not clean sand. Facies A is poor if it has some sand or simply non-reservoir if it is shaly, with low porosity and permeability, and will not contribute to the STOIIP.

The concern is on how the well will be planned to drain this pay zone. Will it be a vertical or horizontal well? Will that affect the cost and degrade or optimise the economics of the well after the prospect evaluation?

Gas Oil Ratio (GOR)

Fluid occurring as oil at depth under higher pressure will have some portion of it arrive as gas on the surface due to reducing pressure; the higher pressure at depth causes some of the gas to be dissolved in the oil, and so the gas escapes or is exsolved due to reducing pressure at the surface. ***Gas solubility (Rs)*** is a measure of how much gas remain in solution at such depth, usually more in light fluids or condensates. The ratio of produced gas to the oil is the **Gas Oil Ratio** (GOR), and it is important to evaluate this for a reservoir in order to establish its commercial importance.

Formation Volume Factor (FVF)

Similar to the effect of pressure while oil is extracted to the surface leading to some expulsion of gas, oil will reduce in size due to reducing temperature and pressure. This reduction (also contributed by expulsion of gas) is known as **formation volume factor**, and is related to the depth of the reservoir, fluid type it contains and nature of the reservoir.

Heavier oil will signify lesser reduction of the oil volume at the surface under reducing temperature and pressure. Just as for lighter fluids, the greater the depth the more tendency for gas to dissolve in oil due to higher pressures; there will be more reduction of the oil on the surface as more gas is expelled from it due to reducing temperature and pressure.

Gas Expansion

Where gas already exists at depth, it will expand on arriving at the surface due to reducing pressure. The difference in volume between the original and the final on surface is known as **gas expansion**. It also depends on the depth and nature of the reservoir and the fluid type contained in it.

Figure 8.13 shows what happens when the initial oil and gas in place is moved from higher temperatures and pressures at the subsurface to the surface at atmospheric temperature and pressure conditions. When this happens, gas expansion takes place, leading to a reduction in oil volume; they form the equivalent meter cube (m^3) of gas and Stock Tank Barrels of Oil (stb) respectively. The ratio of the expulsed volume of gas to the oil is the GOR.

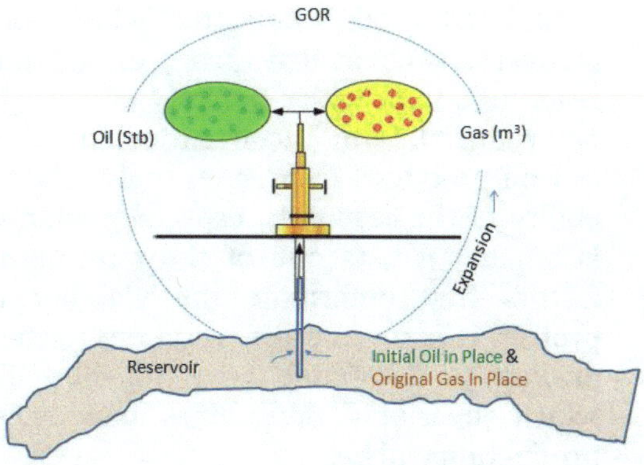

Figure 8.13: Separation of Oil and Gas on surface to the equivalent Stock Tank Barrels of Oil (stb) and meter cube (m^3) of gas respectively

When the fluid in Figure 8.13 reaches the surface, the solution (condensate for example) will separate into oil and gas due to formation volume factors, to give the **GOR.** The ratio of the volume of gas that is separated from the solution to the volume of oil at this standard conditions gives the **Gas to Oil Ratio (GOR)**.

The **Recoverable Hydrocarbon** (RH) depends on the total *Pore Volume (PV)* multiplied by the *Recovery Factor (RF)* multiplied by the inverse of the *Formation Volume Factor (FVF)* of *oil (Bo)* or *Gas (Bg)* respectively.

$$\mathbf{RH = (PV * RF) * 1 / FVF}_{(Bo \text{ or } Bg)}$$

Factors affecting the prospect evaluation parameters (*described in the section on Risk and Uncertainty*) could be risked and their uncertainty estimated. Probabilities assigned to each of the factors may be multiplied to find the POS of a prospect. *Assign an arbitrary Pore Volume of a given gross rock volume, based on the porosity of the reservoir sandstone sample; also assign a Formation Volume Factor (FVF) and Recovery Factor (RF).* Then the Recoverable Hydrocarbon (RH) for the conventional resources can be estimated.

Target Reservoirs

The overall aim of prospect evaluation is to be able to target specific reservoirs having good hydrocarbon bearing potentials. Sandstone reservoirs with lesser compaction and carbonate reservoirs with secondary porosity have higher porosity values. Reservoirs having porosity values between 20 and 40% are considered good enough. With enhanced recovery, such as fracturing or acidification, reservoirs with lesser porosity of say 10% can be produced. With unconventional methods, oil can be produced from even shale. The permeability of the reservoir, especially where there is oil and gas, is one of the most important factors that determine the viability of the project. It is usually predicted and tested after drilling to determine how the oil will flow, which invariably determines how successful production will be.

Oil saturations may be simulated by software and calculated after drilling the well when log data is available. Figure 8.14 is a log showing different sand/sandstone and shale lithologies on the left of the log; the right of the log shows porosity (Φ) and rock matrix. Three reservoir tops and bases are marked out. In Reservoir 1, different fluids are present within the pore space – gas (red), oil (green) and water (light blue). The remaining white space to the right is the rock matrix. Some residual oil was found in the water phase of Reservoir 2 and even in the shale there exists some oil.

Software computation indicates this presence of oil in even non-reservoir lithology, as it looks at actual values read off by tools such as resistivity and density or neutron-density logs. The dominant fluid, more than 50%, is often the one considered the in-situ fluid (used to define *oil-bearing* or *gas-bearing* or *water-bearing* reservoir). The total of the water-oil-gas saturations should be one hundred percent. This value should be equal to the rock porosity, since that is what the fluid occupies.

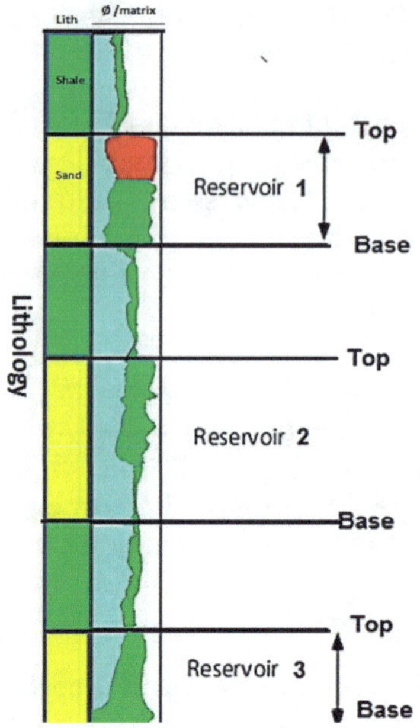

Figure 8.14: Reservoir targets showing the tops and bases of reservoir sands or sandstone

Imagine that residual oil was found in the reservoir and the shaly formation was found to contain so much immature kerogen, or that there was no residual oil but gas in the shaly silty layer. Could it be a potential unconventional resource? What method of mining will the geoscientist suggest? More on this in Chapter 10, on Unconventional Hydrocarbon Resources.

The last stage of prospect evaluation goes as far as evaluating the drillability of a given prospect. An adequate pore pressure program is planned for each prognosed interval, as shown in Table 8.1. The table shows the geologic descriptions of target reservoir formations encountered at several depths in a hypothetical well. For example, at true vertical depth with reference to the mean sea level (TVD/MSL) between 0-79 m, Formation 1 of Miocene geologic age was seen. Its dominant lithology is sandstone.

Table 8.1 suggests that to drill safely through this formation, an equivalent mud weight (EMW) of pore pressure determined to be 1.2 will be used; subsequent depths and lithology could require a different EMW. *(Well operations are discussed in the next part. the development and production of a hypothetical well, with formations depths similar to the one shown in Table 8.1, are presented in Appendix II).*

Table 8.1: Geology of the formations encountered by a hypothetical well

Depth m (TVD/MSL)	Formation	Geological Age	Dominant Lithology	EMW of Pore Pressure	Pore pressure
0 -79					
79 - 650					
650 - 1150					
1150 - 1910					
1910 - 2150	Formation 1	Miocene	Sandstone	1.2 EMW	262 bar

Before drilling a proposed prospect, target reservoirs are marked and the intervals entered in a table (as in Figure 8.14) and predicted petroleum results are plotted (as in Figure 8.14). These will be included in most pre-drill documents destined for the drilling of a prospective well. The Exploration team works on this with the Well Operations department in an oil company. These is handed over to the Drilling department as *Well Prognosis*. While drilling, the wellsite geologists send actual logs used to refine the tops and other markers below and above the reservoirs.

At the end of prospect evaluation, a static and dynamic model is built to predict the outcomes of production, not only at well locations, but also for the entire reservoir or across the field, over a given period. With each new well drilled, the model may be updated to further refine the prediction, in terms of both reservoir petrophysical results and production.

8.5 Reservoir Static Modelling and Dynamic simulation

Static Modelling

Static modelling, also known as geologic modelling, involves the use of structural grids to integrate the faulting and stratigraphic pattern of geologic bodies. It could be a final way of representing a prospect within a field or reservoir level. *Reservoir modelling*, often known as *geomodelling*, usually targets reservoir intervals in an oil field. Petrophysical and facies groups are propagated into a reservoir grid, leading to the computation of fluid volumes and later, for simulation of flow within a reservoir, in a given block. Most geologic modelling for an oil field focus on the hydrocarbon zone, above the oil water contact.

Geomodelling requires the integration of knowledge of the field, including geology, geophysics and reservoir parameters. There are two major aspects: *structural modelling* and *property modelling*.

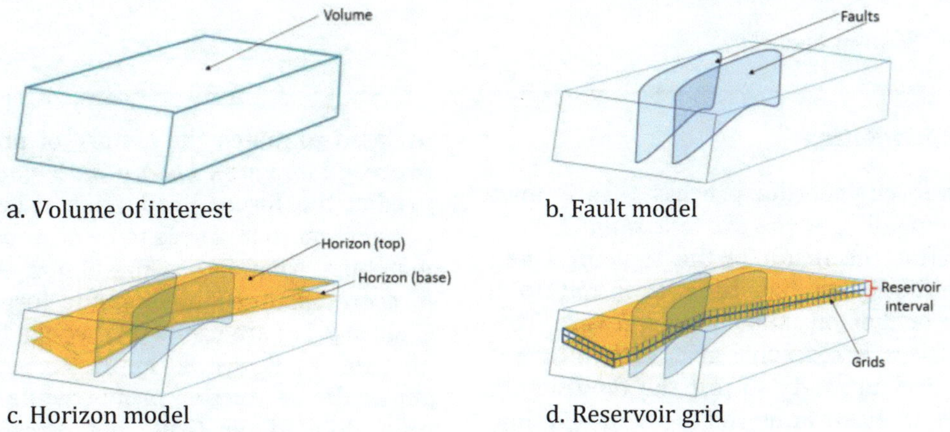

Figure 8.15: Structural modelling

Structural Modelling

This requires the initial set-up of the modelling space and features in a given software. The steps (shown in Figure 8.15) are:

a. Define the volume of interest - in the software
b. Add the faults - fault modelling
c. Add the horizons - horizon modelling
d. Do a gridding - reservoir gridding

Property Modelling

Property modelling is most successful when there is a well from which to extrapolate the properties, after defining the areas of interest. As shown in Figure 8.16, it involves:

a. Defining regions and architectures: like reservoir fairways or geologic bodies
b. Facies modelling: classifying facies
c. Petrophysical modelling: inputting parameters such as porosity, net-to-gross, and permeability
d. Fluid modelling: Inputting the fluid saturation, for oil and gas
e. Volumetrics or volume computation: plotting volumes of fluid to display maps and report figures

Structural and Property aspects of geomodelling give an output known as the *static model*. It describes the state of an oil field when the fluids are not artificially made to flow or put in production. The deliverable is a given volume of *Hydrocarbon Initial In Place (HIIP)*. This may be presented in a table that first gives the Hydrocarbon Pore Volume (HCPV), then subdivides the volumes as they occur in geobodies, structural panels, and as they are associated with different facies. It is a good practice to display the map of the bodies and facies.

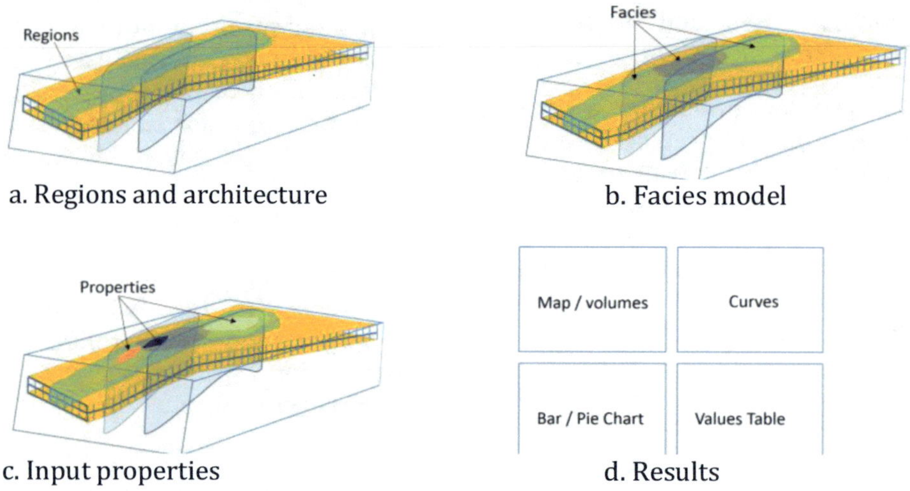

Figure 8.16: Property modelling

Dynamic Simulation

A reservoir engineering process that follows geologic modelling is the *dynamic simulation*. It is done after the result of the structural and property modelling has been exported to a simulator or dynamic simulation software. This software uses a reservoir grid with inputted petrophysical property to test the flow within the model. It shows how well the fluid will flow during production. This flow model is regularly updated to match the history of production (a process known as *history matching*) and then predict the future production using the same parameters that proves to be correct for history matching. That future prediction is known as *forecast*. History matching and forecast graphs (Figure 8.17) are useful all through the life of an oil field. In Figure 8.17, black continuous line shows the production profile, while the red and blue dots show GOR and Pressure values respectively. The trend across the dots once matched, is used to forecast future evolution of the GOR and pressure respectively *(note that the scales vary for the parameters in the vertical axis).*

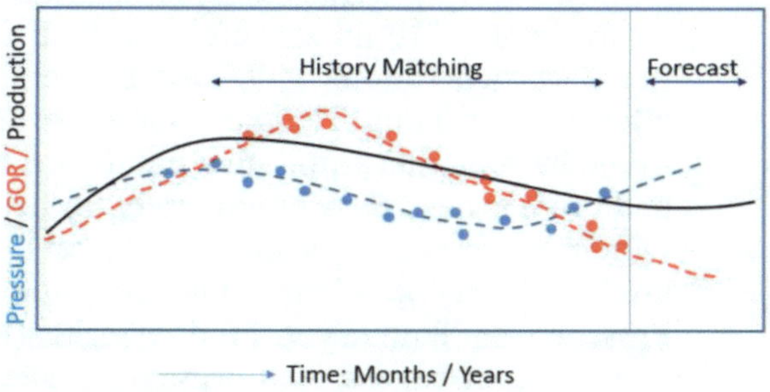

Figure 8.17: History Matching (HM) and Forecast (FC) curves in a production profile, showing Gas Oil Ratio (GOR) and pressure evolution

Dynamic simulation gives an output known as the *dynamic model.* This implies the model that involves flow or pressure differential, unlike static model that gives the result of an oil field in its natural or static condition. The result is comparable to a *production profile* that is useful for reservoir management. By updating and testing each of the parameters, such as permeabilities, fault transmissibilities and net-to-gross, the process helps to characterize (that is, know the properties of) the reservoir.

This knowledge helps in optimizing the production. The process of using local data at well level, to define the overall values at field scale, is known as *upscaling.* These are all useful in deciding where to place the injectors and producers in a field. In summary, reservoir modelling is useful for:

- Characterisation of reservoir
- Upscaling
- History matching
- Prediction / forecast for production.
- Control and optimization

Actual modelling with software takes some training and practices and the details cannot be exhausted here. A direct application of the concepts of big data, machine learning and artificial intelligence is becoming very useful in industry main interests in basin modelling, static modelling and dynamic simulations for assessing petroleum systems, estimating hydrocarbon in place and managing production.

Well Prognosis

The information obtained during prospect evaluation serves as a guide on what could be encountered in the subsurface when a well is drilled. This prediction, before drilling the well, is known as **well prognosis**. The various **well prognoses**, on the different parameters evaluated, are integrated to create an ultimate document that contains important geological and reservoir pre-drilling information, including:

- Lithology of formations
- Reservoir tops and markers
- Fluid types and contacts
- Fluid (gas-water-oil) properties
- Pore pressure predictions
- Well trajectory

It may be called *Final Well Statement Of Requirements* (FWSOR) or *Well Location Report* (WLR). This final document necessary to drill a petroleum well is handed over to the drillers and operations geologist. A very important detail is the well trajectory, stored as survey data, which contains the coordinate at every point in the well path. It is designed after careful modelling of the best well placement to target the resources, at better reservoir locations, while meeting drilling safety.

Well prognosis is similar to a doctor's prognosis based on symptoms; it is only after the tests, after drilling the well, that the encountered reservoirs and their properties are ascertained. From the results of static and reservoir modelling, the geoscientists and reservoir engineers chose best target depths and well trajectories across the field, as well as the thickness and information on the units that will be encountered when a new well is drilled.

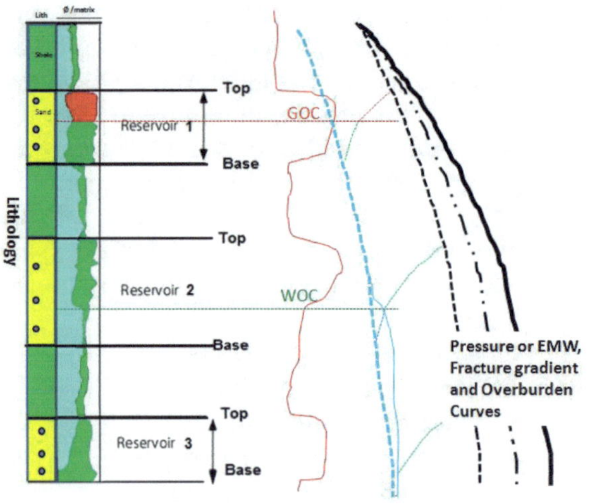

Figure 8.17: Well prognoses updated with well results (as in post-mortem) showing (expected) lithology and fluids (shown by the resistivity log curve in red)

The well prognosis may be integrated into the pre-drilling chart (Figure 8.17), which is updated with well results to show different lithologies (sand and shale), different fluid types (gas, oil and water), GOC, WOC, as well as pressure or equivalent mud weight (EMW), fracture gradients and overburden curves. A **post-mortem** is a comparison of the predicted results and the actual results after the well has been completed.

In this example (Figure 8.17), the resistivity curve may be predictive or actual, showing the Gas Oil Contact (GOC) and the Water Oil Contact (WOC). The fluid contact is sometimes indicated by flat spot - DHI on seismic, identified during prospect generation. This will be confirmed after actual drilling. The dark points within the lithology are where pressure points are taken, and there is need for a minimum of three points within a (connected) sand body to establish a gradient. This gradient could be plotted in Pressure or Equivalent Mud Weight (EMW) versus depth curves. The pressure curves and prediction shown here are for illustrative purposes *(more on pore pressure prediction in the next part on Well Geology and Operations).*

Figure 8.17 the red curve is the resistivity, which is high (to the right) in the presence of hydrocarbon, and low (to the left) in its absence. There is a sharp reduction at the WOC, but no major difference at the GOC. Differentiating between gas and oil is difficult using resistivity alone. Identification of GOC on log requires the use of Neutron-Density log (*which will also be discussed in the next part*).

EXERCISES

1. What is a petroleum prospect?

2. Identify the four features that make a good prospect.

3. Why is there need to resolve the following questions during prospect evaluation:
   i. Is the source rock mature?
   ii. What is the evidence that generated oil has migrated to the reservoir?
   iii. How good is the reservoir?
   iv. Is the trap structure fractured?
   v. Was the trap available before the charging?
   vi. Has the oil been migrated away before the formation of the structure?
   vii. Any recent tectonic activity that might have destroyed the prospect?
   viii. What is the seal efficiency and integrity?
   ix. Has the oil or gas leaked?

4. Differentiate between the different stages of petroleum migration.

5. Describe with diagrams two structural and stratigraphic traps.

6. Describe the concept of risk and uncertainty in the second stage of prospect evaluation.

7. Explain how the following factors affect prospect evaluation:
   a) Petroleum generation and charging of the trap
   b) Reservoir size and quality
   c) Trap system
   d) Seal integrity and efficiency

8. Describe the following important aspects of reservoir evaluation:
   a) Facies and facies association/compartmentalization
   b) Porosity and Permeability
   c) Net to Gross
   d) Fluid Saturation (water, oil and gas)
   e) Net Pay zone
   f) Gas Oil Ratio
   g) Formation Volume Factor
   h) Gas Expansion

9. What are target reservoirs? Why is their identification important?

10. Briefly discuss what is entailed in well prognosis.

REFERENCES & FURTHER READING

Bilotti, F. and Shaw, J. H. 2005. Deep-water Niger Delta Fold and Thrust Belt Modeled as a Critical-Taper Wedge: The Influence of Elevated Basal Fluid Pressure on Structural Styles. *Amer. Assoc. Petrol. Geol. Bull.*, 89 (11), p. 1475 – 1491.

Burrus, J., 1998. Overpressure Models for Clastic Rocks, their Relation to Hydrocarbon Expulsion: A Critical Re-Evaluation, in B. E. Law, G. F. Ulmishek and V. I. Slavin eds., Abnormal Pressures in Hydrocarbon Environments, *Amer. Assoc. Petrol. Geol. Mem.,* 70, p. 35 – 63.

Cobbold, P.R., Clarke, B.J. and Loseth, H. 2009. Structural Consequences of Fluid Overpressure and Seepage Forces in the Outer Thrust Belt of the Niger Delta. *Petroleum Geoscience*, p. 15, 3 – 15.

Dembicki Jr, H., 2009. Three Common Source Rock Evaluation Errors made by Geologists during Prospect or Play Appraisal, *AAPG Bulletin,* v. 93, no. 3, p. 241–356.

Finkbeiner, T., Mark, Z., Flemings, P. and Stump, B. 2001. Stress, Pore Pressure, and Dynamically Constrained Hydrocarbon Columns in the South Eugene Island 330 Field, Northern Gulf of Mexico, *Amer. Assoc. Petrol. Geol. Bull.*, 85 (6), p. 1007 – 1031.

Heid, J. G., McMahon, J. J., Nielson, R. F. and S. T. Yuster, 1950. Study of the Permeability of Rocks to Homogeneous Fluids. *Am. Pet. Inst. Drill. Prod. Pract.,* p. 230–244.

Kostenko, O.V., Hack, W., Naruk, S. J. and M. Paupon, 2008. Structural Evaluation of Column-Height Controls at a Toe-Thrust Discovery, Deep-Water Niger Delta. *AAPG Bulletin*, 92, p. 1615 – 1638.

Law, B. E., and Spencer, C. W. 1998. Abnormal Pressures in Hydrocarbon Environments, in B. E. Law, G. F. Ulmishek and V. I. Slavin eds., Abnormal Pressures in Hydrocarbon Environments, *Amer. Assoc. Petrol. Geol. Mem.,* 70, p. 1 – 11.

McPhee, C.A. and K. G. Arthur, 1991. Klinkenberg Permeability Measurements: Problems and Practical Solutions, in *Advances in Core Evaluation: Accuracy and Precision in Reserves Estimation*, ed. P.F. Worthington, p. 447–462.

Nederlof, M.H., 2019. Geology and Energy Analysis (GAEA) Consultancy, Software and Education for Oil and Gas Exploration. *Website,* 23 September Update. *https://www.mhnederlof.nl/emv.html*

Pochat, S., S. Castelltort, J. V. D. Driessche, K. Besnard and C. Gumiaux, 2004. A Simple Method of Determining Sand/Shale Ratios from Seismic Analysis of Growth Faults: An Example from Upper Oligocene to Lower Miocene Niger Delta Deposits. *AAPG Bulletin*, v. 88, no. 10, p. 1357–1367.

Saugy, L., and Eyer, J.A. 2003. Fifty Years of Exploration in the Niger Delta (West Africa), *in* M. T. Halbouty, ed., Giant Oil and Gas Fields of the Decade 1990-1999: *Amer. Assoc. Petrol. Geol. Mem.,* 78, p. 211 – 226.

CHAPTER NINE

TRAPS AND SEAL ASSESSMENT

9.0 Traps and Seals

The most important condition in a petroleum system is that the generated hydrocarbons have been retained. Trapping and sealing are among the most important factors of concern in prospect evaluation. The petroleum seal is any layer (or mechanism) that inhibits the flow of hydrocarbons from a reservoir in a petroleum system. In major petroleum fields, oil and gas have been generated in large quantity in past geological time. If they are not found in the fields today, or if they occur in small quantity, it is because there were no structures to trap them, or that they were eventually lost to leakage. This is why the study of seals is very important in the oil and gas industry.

An attempt is made to present some advanced concepts with simple demonstrations and figures. The main concepts will be repeated in different sections, so as to comprehend them from different perspectives. Each revisit comes with a little more depth, helping better assimilation. In this chapter, adequate attention will be given to caprock and fault sealing and how they affect trapping of hydrocarbons. Among the aspects that will be discussed are:

- Sealing process
- Indication of leakages
- Seal types
- Seal evaluation
- Laboratory studies
- Field studies
- Trap filling and modelling

Seal evaluation is carried out at all stages of the E & P value chain, during exploration and appraisal and even while developing or producing a well. The initial seal evaluation is part of prospect evaluation while the latter is for optimisation of a field and infill well placement. The examples in this chapter show how these seal assessment methods are applied to petroleum evaluation in significant fields and basins. The major focus is on how to identify and evaluate potential seals in a hydrocarbon field.

9.1 Hydrocarbon Retention

Several factors control the retention of hydrocarbons in the subsurface, and at every stage of the migration process, there is need for some kind of control on how far the hydrocarbons can disperse. Some of the factors that have been discussed include the ability of hydrocarbon fluid to migrate, the availability of source and reservoir rocks, presence of impermeable boundaries and the availability of trapping structures. A good trapping requires impermeable boundaries in different directions (Downey, 1984; Kaldi and Atkinson, 1997; Ingram, 1999). In Figure 9.1, a reservoir containing fluid (light blue) is shown in an anticlinal structure. Three directions of leakages are also shown.

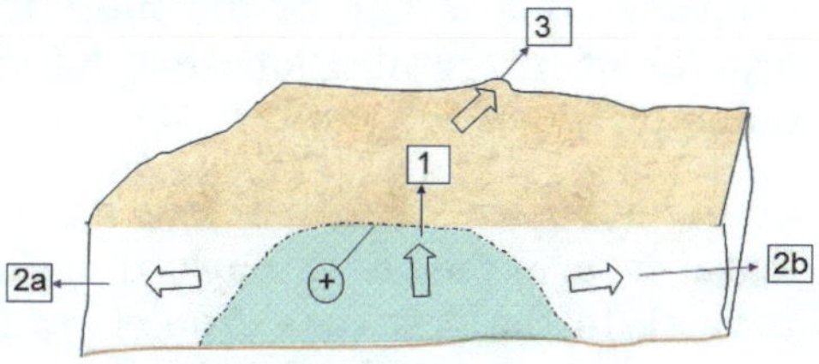

Figure 9.1: Directional leakages, which will all have to be considered in prospect evaluation

The leakages follow three axes (1, 2 and 3 as shown in Figure 9.1):

1. upwards or downwards
2. to the left or right direction
3. inward or outward direction.

Direction 1 is a vertical leakage, upwards. Directions 2a and 2b are horizontal (lateral leakages in the left and right directions respectively). Direction 3 is leakage occurring in the horizontal inward and outward directions.

These three directions correspond to the principal stresses, which play a role in leakage. In practice, fluid leaks in the direction of the minimum principal stress, where there is least resistance, or to where capillary entry pressure is minimum. To appreciate the importance of the sealing process, it is necessary to recall the various processes in hydrocarbon generation, migration and retention, in which sealing plays a role:

Generation: There is need to have a "sealed" or compacted sediment with enough organic content, limited in space by barrier, and kept under pressure and other favourable conditions of temperature and depth for the sediment to mature and generate hydrocarbon.

Migration: Migration in the first place means there is a limited part, an enclosed pathway through which the oil is migrating. These could be, for example, a sand unit "sealed" above and below by an impermeable unit.

Trapping: Hydrocarbons will continue to migrate until they get to suitable structures that allow it to accumulate. The oil therefore reaches its destination when it finds a trap, a structure that provides a closure and prevents further migration.

Retention: How long the oil or gas will remain in the trap depends on how good is the sealing property. This is one of the most important aspects of the evaluation that interests the industry.

Consider Figure 9.2 where two cases of fluid leakages are shown through the seal (dark brown). In the first case droplets of oil (light green) enters the seal, while in the second case, there is a direct flow of the oil across the seal. Different processes resulted in these two scenarios: in the first case, minute oil droplets attempts to squeeze through small pore spaces in the seal. In the second case, high-pressured oil pushes to break through the seal.

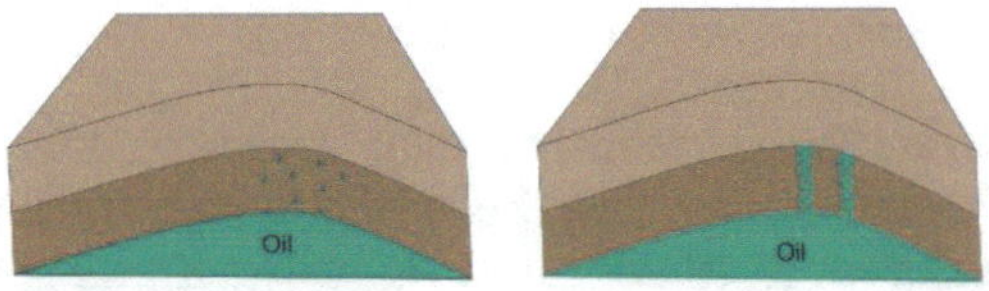

a. Minute leakage b. Breakthrough flow

Figure 9.2: Two leakage scenarios in which oil droplets permeate the seal on one hand (minute leakage) and flow through the seal (break through flow) on the other

Where the tiny drops of oil (light green) from the oil pool naturally permeate the overlying seal, it is because there are pore spaces available in the seal. In the second case, the pressure of the fluid forces its way more like a chimney through the seal by fracturing it.

9.2 Sealing Processes

A rock unit, or layer, acts as a seal if the lithology is "tight" enough to limit the flow of hydrocarbon across it substantially. This essentially involves very *low porosity* rocks such as mudstone or shale; their matrix should also be so strong that it resists micro or macro-fracturing that could lead to fluid expulsion. Fluid flow is limited in such cases, not only by the smaller pore spaces or tightness of the lithology, but also by the density and viscosity of the fluid under consideration.

Fine-grained sediments usually have *higher ductility* and have smaller pore throats/spaces through which oil and gas, depending on their *density*, may be hindered passage. *Viscous* fluids build-up high interfacial tension, at the interface between the permeable and the impermeable rock, limiting passage. In addition to the fluid type and viscosity, the inherent *pressure* and *temperature* of the hydrocarbon fluid are important in determining how a particular seal will resist hydrocarbon passage.

Leakage

When there is poor sealing (where a seal does not prevent fluid flow through it), there is said to be *leakage*. An evidence of fluid leakage is **pockmarks**. Pockmarks are concave crater-like depression often in profusion, on fine-grained sediments or muddy seabed, formed due to eruption of fluids through sediments; they are expressions of high fluid discharge or venting sometimes above reservoirs or recent channels. In Figure 9.3, a seabed isochron map is shown. To the southwestern part of the map (the red portion of the map) is an indication of major and minor pockmark alignments (dashed lines) that exist parallel to known deep anticlinal axis and recent channels.

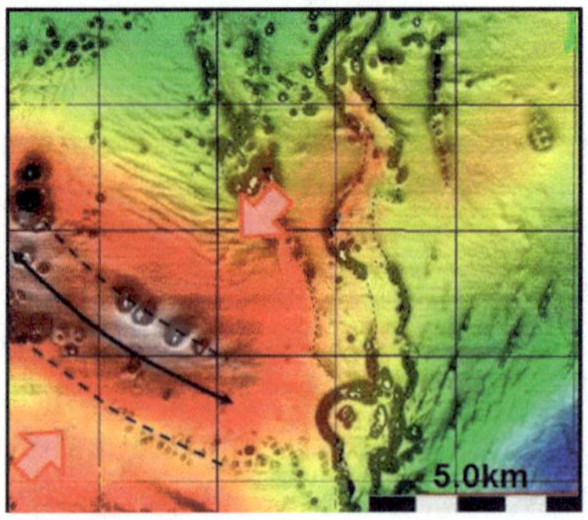

Figure 9.3: Seabed isochron showing pockmark alignments parallel to known deep anticlinal axis (major pockmarks) and a different organisation (minor pockmarks) trailing recent channels (Nosike et al., 2008a)

Pockmarks are formed by focussed fluid flow and as such indicate past and present hydraulic activity reaching the sea surface (King and MacLean, 1970; Hovland and Judd, 1988). Pockmarks have been found in the Scotian Shelf (part of the continental shelf southwest of Nova Scotia, Canada), the North Sea that connects to the Atlantic Ocean in northern Europe, the Niger Delta at the Gulf of Guinea in southern Nigeria, and elsewhere(Nosike, 2009). In places where mounds (structures of rounded elevation that form small natural hills) occur in association with *pockmarks*, they are sometimes described along with or as pockmarks.

Several sizes, shapes and volumes have been defined, including large complex pockmarks up to 300 m wide and 12 m deep in the Nyegga area of the Norwegian Sea, localised pingo-like sediments mounds up to 1 m high and 4 m wide within the Nyegga area (Hovland et al., 2005; Gay, et al., 2006; Hovland and Svensen, 2006). Oil seeps, slicks and gas venting to the surface have been highlighted as evidence of fluid leakage; other evidences of fluid leakage are mud volcanoes, gas chimneys/fluid conduits seen on seismic (Nosike, 2009, 2016). They all indicate seal breaching, but at the same time a petroleum system with possible trapping.

Efficiency and Integrity of Seals

Efficiency and *integrity* are sealing mechanisms. Oil and gas migrate upward due to buoyancy, and are hindered from reaching the surface by a seal. The two major processes that create a non-sealing scenario are:

1. *Oil leakage through the rock membrane* (*rock seal is inefficient*)

2. *Breaking down of the rock wall* (*rock seal lacks integrity*).

Seal Efficiency

When hydrocarbon is trapped because the pore spaces are too tight, this is known as "**Efficient sealing**". Rock unit acts as filter or membrane that has tiny pore spaces. When petroleum fluid molecules are able to pass through the pore spaces, then it is a seal failure or an inefficient seal. Factors affecting membrane sealing processes, apart from the nature of the fluid, discussed in details elsewhere, are:

- Capillarity
- Pressure

In lithological seal, the rock grains act as a membrane, filtering fluid across depending on the size of the pore spaces. The figure shown earlier for minute leakage is used to explain that it is a form of a membrane seal, in Figure 9.4. Pore spaces are available for petroleum fluid to flow through the seal; fluid flow occurs as shown, indicated by tiny drops of oil migrating upwards through the membrane seal. When this happens, the seal is said to be *inefficient*.

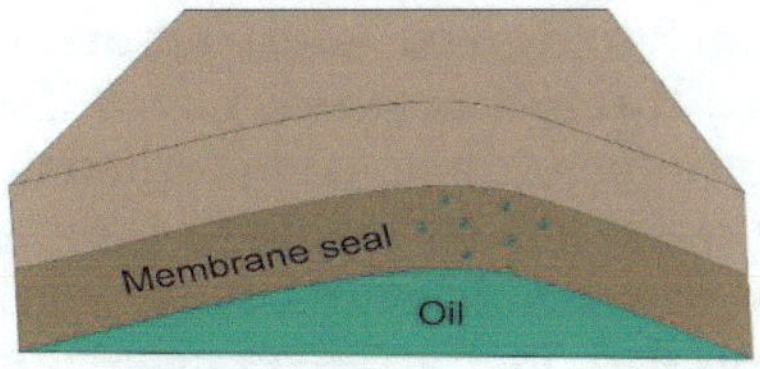

Figure 9.4: Lithological seal or membrane seal, showing seal inefficiency (compare with Figure 9.2)

Such leakage may be shown as follows:

$$\mathbf{P_w + P_h > Pe}$$

Where,

Pressure in the waterleg (P_W)
+ Pressure of hydrocarbon column P_h)
> Capillary Entry Pressure (P_e)

The pressure in water is *hydrostatic* for normally pressured reservoir. The pressure in the hydrocarbon phase is the *buoyancy*. The *capillary entry pressure* is the minimum pressure necessary to penetrate pore spaces. It depends on the sizes of the pore known as pore throat (gap between grains available for fluid to pass or flow through), on the property of the fluid in the seal (essentially the interfacial tension between the fluid and the rock matrix) and on the contact angle between the grain and the penetrating fluid. Leakage occurs in an inefficient seal when the pressures in the water and hydrocarbon are greater than that required to resist fluid flow through the pore spaces.

In Figure 9.5, a case of leakage due to capillary breaching is illustrated. On the left image, pressure of water and oil (as indicated by the purple and blue lines) is greater than the effective/capillary pressure (yellow).

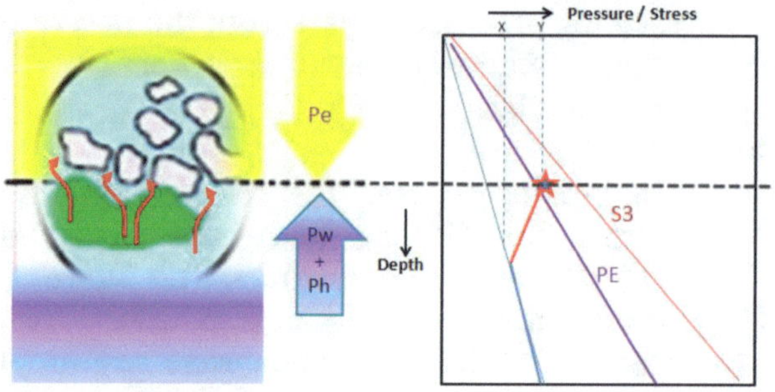

Figure 9.5: Illustration of capillary leakage where oil and water combined pressure leaks into the seal only when they exceed the capillary entry pressure (Pe or PE)

Fluid flow (green with red arrows) can take place through the rock matrix (white grains), breaching the boundary (dashed lines) between the water and oil pressures (purple arrow) and that of the rock matrix (yellow arrow). This is illustrated in the relationship between pressure/stress and depth, indicating what the water pressure would be at point x, and the pressure at point y due to additional oil pressure. The red star shows, as well, at what pressures the fluid will leak.

At this stage, it is important to begin to think of every sealing and leakage as a pressure – stress relationship between the reservoir fluid and the seal. Each type of seal or fluid leakage can be represented in a pressure or stress versus depth plot, where the rock strength or necessary pressure for fluid to sneak into it or breakthrough is plotted against the depth of the reservoir, as in Figure 9.5. There is need to be familiar with pressure-depth plot and be able to represent each case in the chart. This will be revisited in a further classification of seal types below.

Seal Integrity

There are cases where the molecules of the hydrocarbon fluid are under sufficiently high pressure to fracture the seal and flow through it. Seal integrity is possible where the rock unit acting as a plane has high enough physical strength to resist any fracture and passage of fluid through it. When hydrocarbon is trapped because the top seal has resisted fracturing, it is known as "**Seal Integrity**".
In that case the a rock unit acts as a plane, resisting any passage by sheer physical defence as fluid molecules hit the walls. If however the pressures of those fluids are high enough to break the plane, it fractures and gives way. This is known as hydraulic fracturing of the sealing structure because pressures are high enough to break the rock wall. This happens when excess buoyancy pressure and other pressures of the fluid act vertically against the cap rock, known as *top seal*.

In stress seal, the rock grains and matrix act as a stress wall blocking fluid passage until high enough pressure breaks this resistance. The figure shown earlier as breakage flow is used to explain that it is a form of stress (hydraulic) seal, in Figure 9.6. The seal has been compromised; it is a case of seal integrity failure.

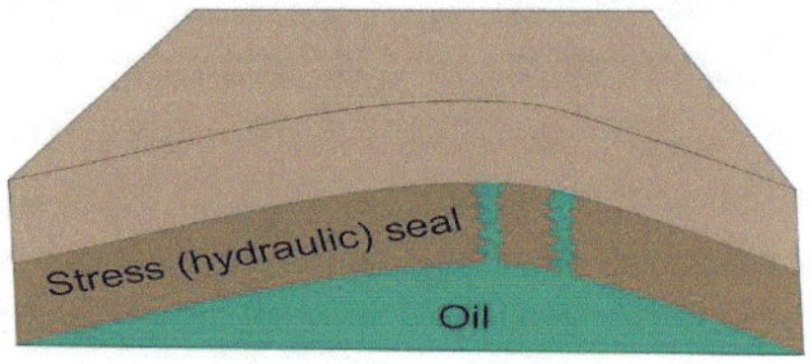

Figure 9.6: Stress seal as a kind of hydraulic seal, in this case showing lack of seal integrity

Fracture may not only necessarily be because of pressure from fluid molecules acting against the top of the seal. In reality, the plane may still fracture through its weakest points, including weak pore spaces that will offer less resistance to fluid flow. The difference in the case of seal integrity failure is that the fluid is forcefully passing through the barrier (shifting aside any barrier or rock grain) of the matrix, rather than naturally sneaking through the pore spaces (as in membrane passage). Factors affecting top hydraulic fracturing processes, apart from nature of fluid, are:

- Stress-pressure differences
- Shear and tensile strengths

On the right plot in Figure 9.7, the water pressure trend (thick blue line) is same in both reservoir. Every excess pressure (higher than hydrostatic, Hw) generated deeper is shared equally within the porous reservoir unit and the overpressure at both depths are the same. The impact is however different: at depth of 3000 m, the overpressure is below the fracture limit (due to more overburden, OBG). It is not below the fracture limit at shallow depth interval of 500 m. Therefore, the same overpressure results in hydraulic fracture near surface, where it reaches the minimum principal stress, S3, or fracture limit (dark orange curve).

In this case, hydraulic fracturing occurs because of dynamic pressure transfer from a deeper depth to a shallower depth within the same reservoir. When hydraulic fracturing takes place through the seal and smaller overburden overlying the reservoir, fluid flow to the surface may occur at the point marked by the larger red star on the left image.

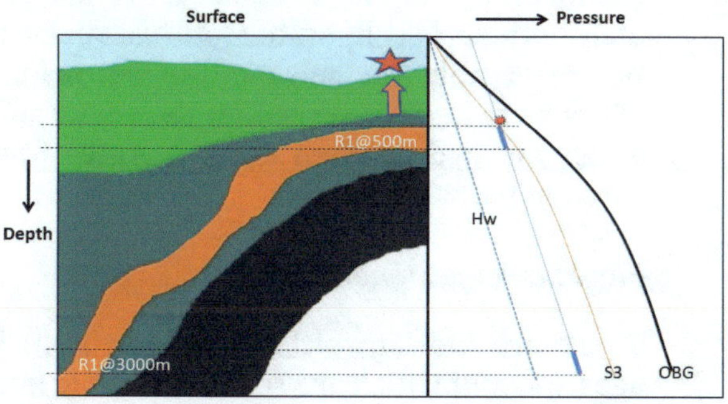

Figure 9.7: Hydraulic fracturing due to dynamic transfer of overpressure from a deeper level (3000 m) of reservoir R1 to a shallower depth of 500 m for the same reservoir

Mohr's Resistance Diagram

The tensile (resistance to pull apart forces) and shear (resistance to breakage by forces parallel to each other) strengths of the top seal can be illustrated by the **Mohr's resistance diagram** to show at which point pressure in reservoir will overcome the resistance and fracture the top seal. A more brittle seal will not behave the same way as a more ductile muddy seal. The Mohr's diagram is a two-dimensional graphical technique used to determine rock behaviour by plotting the values of principal stresses in various materials when subjected to different stresses. It is also useful for understanding rock behaviour during drilling.

Consider a situation where the stresses (or confining pressure) acting around the rock of a top seal in the y and z axes are equal (let's say $\sigma_2=\sigma_3$), then a major force can be applied directly to it (which then is σ_1) and record at which point it breaks or shears. This is simulated in the laboratory by applying a load on a rock while measuring at what point the rock breaks. The rock may be placed in a pressure chamber, so that the pressure of the chamber is recorded as σ_2 and σ_3 acting equally on the other two directions. Then the confining pressure (pressure of the chamber) is varied to obtain various σ_3 and σ_1 while changing the load (σ_1).

When the σ_3 and σ_1 are plotted on the horizontal axis, and use the centre of each ($\sigma_3 + \sigma_1 / 2$) to plot a circle (whose radius is $\sigma_3 - \sigma_1 / 2$), a diagram similar to the diagram in Figure 9.8 is obtained. A Mohr's diagram in Figure 9.8 shows the relationship between horizontal and vertical stresses (acting on a rock) and properties of the rock obtainable such as cohesion (C), minimum failure point (C_0) and shearing component or torque (τ). Fluid will leak by hydraulic fracturing where the pressure exceeds the cohesion of the seal.

This is the Mohr's diagram. Geologically, it illustrates what happens when rocks are buried and acted upon by horizontal pressures or stresses (S_2 and S_3) while another main force is applied directly on its surface (like fluid pressure acting vertically as S_1) until it reaches the breaking point. The major force could also be a drill bit during drilling, which fractures the rock. During Leak-Off Test (LOT), the mud pressure fractures the formation, such that leakage goes to the direction of the least stress.

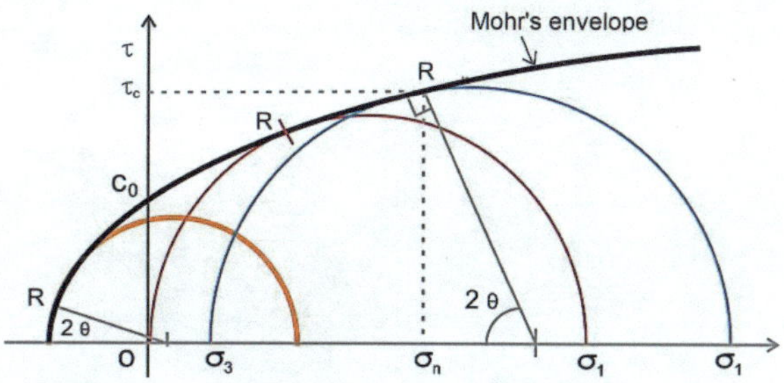

Figure 9.8: Mohr's diagram showing the cohesion of the rock C (minimum failure point at C_0) and shearing component (or torque, τ)

An envelope can be drawn by tracing a line that follows the tangent of each of the circle (thicker black line). This Mohr's envelop crosses a vertical y-axis known as the *shearing component (or torque, τ).* T gives an indication of how the rock will fail, or will be broken while drilling it. The point where the envelope bisects the τ axis gives the *cohesion* of the rock, C. (C_0 is the minimum, as shown in Figure 9.8). Any value higher than this point, or above this envelope, is in the failure domain of the rock

property. Any value within or below the envelope is in the domain where the rock has not failed. This information is vital because during drilling, it is important that the rock is not subjected to stress beyond its failure domain, to avoid formation failure, leakages or fluid losses and unnecessary drilling setbacks.
A brittle seal will have a straight envelope, whereas a ductile material tends to have a curved envelope as in the example in Figure 9.8. The angle 2Θ between the radius and the horizontal line, from the centre of the circles to the tangent line (point R) of the envelope, is related to the *internal friction* of the material. It is a measure of the ability of the rock to withstand a shear stress. *This will be revisited in the next Part, in Chapter Thirteen: Drilling and Geosteering, where the Mohr's diagram is related to rock strength and breakage while drilling.*

9.3 Indications of Seal Breaching

In the case of either seal inefficiency or integrity failure, there is leakage. In the time past, oil seepage observed at the surface has been a major way of predicting subsurface petroleum system and possible reserve. Today, the use of satellite images and remote sensing in general has improved subsurface detection of presence of hydrocarbon, and the interpretation of how they could mean traps beneath the surface. Examples of seal breaching and indication of trapped hydrocarbons are described below.

Oil Slicks

Oil slicks are identified by laser fluorescence imaging or by direct observation on the land surface or from aerial photos. Oil leaks could occur around decomposing hydrates, such as methane hydrates for example, when they are fractured by seabed instability. When associated with seabed hydrate features, slicks could be observable on seismic seabed maps. Oil slicks have been observed on the water surface when the hydrocarbon fluid migrates upward through the water body.

Fluid Leakage Signatures

There are indicators that highlight fluid leakages, most of which are seen on seismic images, such as:

1. Pockmarks (shown in Figure 9.9)
2. Pipes (like pipes on seismic section)
3. Chimneys (may have a wider funnel venting shape at the surface)
4. Change in 3D seismic (pull up and push down effects in seismic reflection)

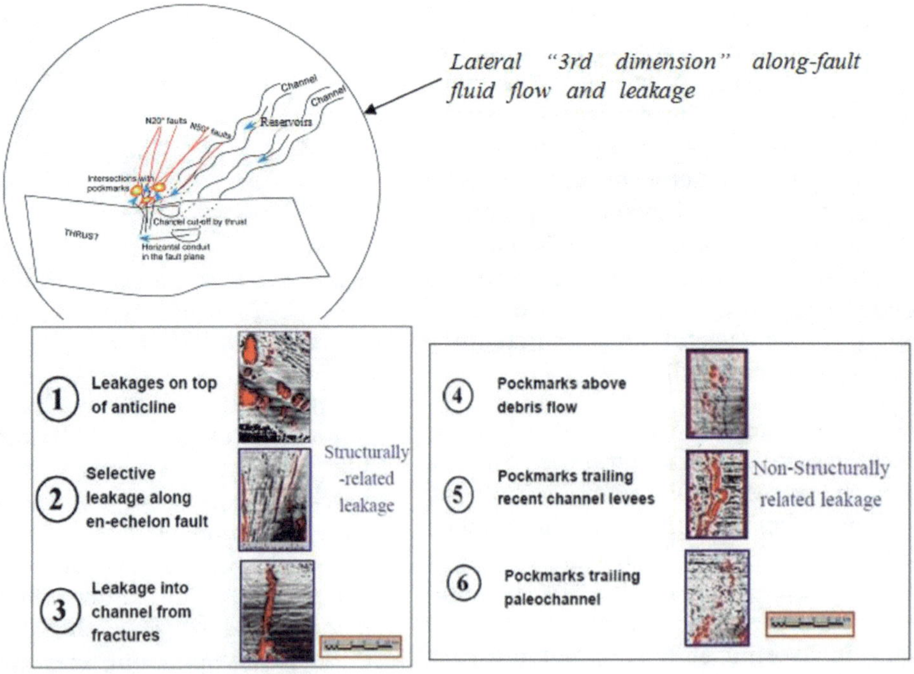

Figure 9.9: Different fluid leakage signatures showing structurally and non-structurally controlled processes (Nosike et al., 2008b) (compare with Figure 7.19)

Pull up and push down effects on seismic data occur because fluid content changes seismic velocity locally. When visible on 4D seismic data (when compared to previous seismic data of the same area), it shows changes in the velocity of seismic waves in that area. Seismic waves are faster in denser material. Pull up and push down occurs when the rock body becomes denser or lighter due to changes in its fluid content; it shows that the fluid in a reservoir has leaked away, or that oil has been produced, and replaced by water, for example.

In Figure 9.9, different fluid leakage signatures are shown, some of which are structurally related leakages while others are not. Those structurally related include leakages on top of anticline, selective leakages along en-echelon fault and leakages into channel from fractures.

Those that are not structurally related leakages are dominantly pockmarks above debris flow and those trailing recent channel levees and paleochannels. Fluid leakage indicators relating to structural activity are mostly rectilinear while those relating to sedimentary features may be sinusoidal.

Residual Oil

The presence of residual oil (found after drilling a well without economic quantity of trapped oil) indicates that there might have been an accumulation lost probably due to seepage. It can help the geoscientist in investigating the direction of this tertiary migration, and find a prospect nearby.

Empty Reservoir in a Petroleum System

Sometimes, despite all indications confirming that oil has been present, based on comparison with an oil well targeting the same reservoir unit in an offset well, a given reservoir is still be found dry. If the possibilities of degradation and lateral lithological changes are ruled out, it could mean that the oil has leaked away in a case of seal breaching.

Unlike the cases of residual oil and empty reservoir, the presence of oil slicks or fluid leakage signature does not imply that hydrocarbon is absent. Fluid flow may be occurring, but it is the degree or rate of escape that is of concern, whether it is high enough to deplete the reservoir. The aim of seal evaluation is therefore to assess stages of oil slicks and fluid signatures and avoid the residual or empty reservoir scenario. Oil slicks and fluid signatures are not enough. There is need to do other kinds of tests such as top seal integrity test, use of amplitude signatures to identify retained hydrocarbon, calculation of capillarity and fault leakages, and other seal evaluation methods shown in subsequent sections.

One of the issues to look out for is a **weak point** above reservoirs or on seals. A weak point may be created by faulting or erosion. Fault intersection creates concentric weak zones, which may be a kind of triple junction focusing migrating fluids. Change of stress direction, such as those which occur on fractured four-way closures (domal / anticlinal structure), due to compression occurring in two horizontal directions at different times, is an example. This could be caused by a lateral ramp formation as shown in Figure 9.10. A thrust sheet propagation running over a wedge could cause this scenario, leading to near perpendicular folds or ridges and fault intersections.

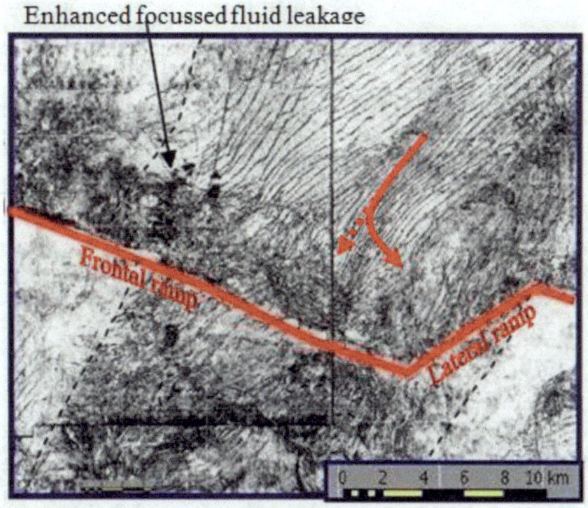

Figure 9.10: High fluid up-flux and leakage due to fault intersection in a zone of lateral ramp development (red arrow shows rotation of stress direction perpendicular to the frontal ramps) Nosike et al., 2018b

It is important to note that retention of hydrocarbon is relative; there is always some leakage and some sealing. It is the rate of sealing relative to leakage that is important. If charging is ongoing it replaces a depleted trap, else what is left depends on how much has leaked.

9.4 Types of Seals

Vertical and Lateral Seals

Just as there are two concepts of seals, integrity and efficiency, there are two types or broad groups of seals, which are both vertical and lateral seals. Most vertical seals are lithological while lateral seals are structural or stratigraphic. There is interplay of both lithology and stratigraphy in most seal types. For example, it is the lithology in fault zones, or the lithology to which the fault zone makes contact, that make them sealing or leaking. Lateral stratigraphic boundaries as well as barriers due to a sedimentary process are also common.

Sometimes there is interplay of vertical and lateral seals acting as barriers and leading to concentrated fluid leakage along an anticlinal axis. This explains the major pockmarks in Figure 9.3. The process that may lead to such scenario is illustrated in Figure 9.11. Pressure compartments created by overpressure differences across lateral fault boundaries and vertical lithological limits are known as **geopressure cells**. In Figure 9.11, a relationship between pressure and depth is shown, where different overpressures are recorded at different reservoir units (pressure cell limits) with increasing depth. Geopressure cells are created by overpressure differences.

The layer-by-layer conceptual model (Nosike, 2008) illustrated in Figure 9.11 shows different leakage constraints within various units.

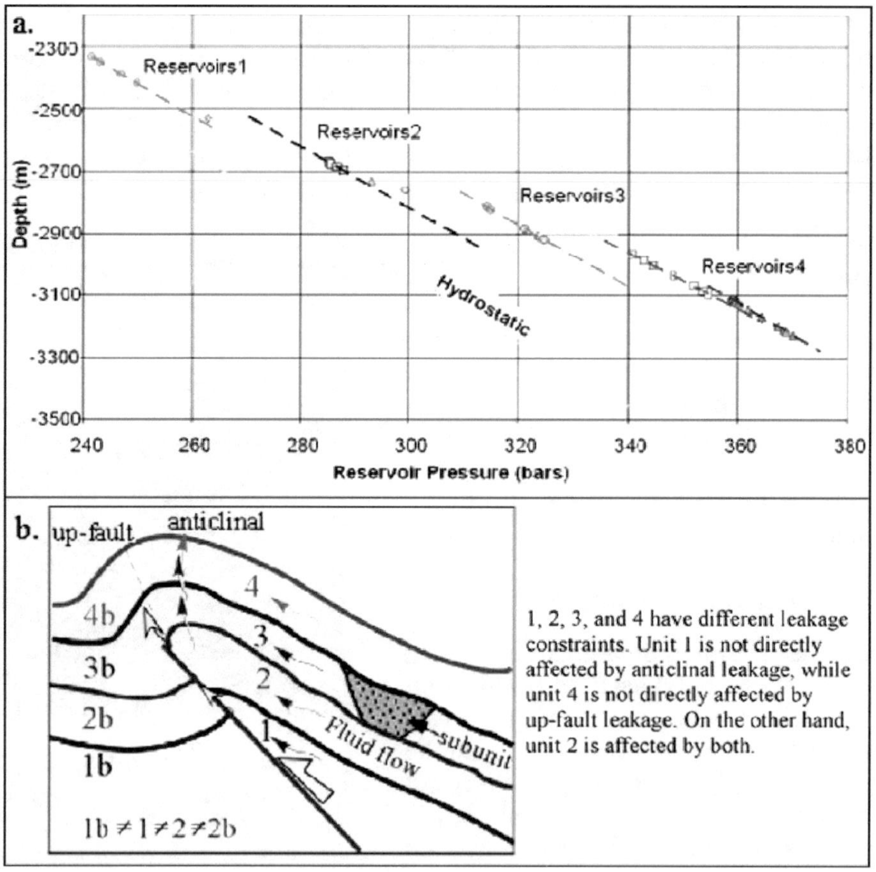

a. Pressure-depth plot showing overpressure shift at vertical pressure cell limits
b. Layer-by-layer conceptual model showing different fluid-pressure dissipation constraints on an anticline

Figure 9.11: Vertical and lateral seal combination and a pressure-depth plot to illustrate how differential leakages at a thrust-related anticline may lead to vertical and lateral reservoir and geopressure cell compartmentalisation (Nosike, 2008a)

Unit 2 is affected by both anticlinal leakage and up-fault leakage. Anticlinal leakage does not directly affect Unit 1, while up-fault leakage does not directly affect Units 3 and 4. Unit 1 retained higher overpressure. The situation highlighted is feasible for a thrust-propagation fold, which is a thrust-related anticline where the low-angle fault has not yet fully offset the folded layers.

Up-fault and Anticlinal Apex Leakages

In the schematic anticline illustrated in Figure 9.11, the major determinant leakage types above the anticlinal thrust axes were:

1. up-fault/thrust tip and
2. top seal/anticlinal apex leakages

(These two leakage types emplaced the double-row of pockmarks seen in Figure 9.3).
The layers are bounded by similar vertical lithological seals (vertical geopressure stratigraphic limits), but the combined effect of anticlinal and up-fault leakages for each stratigraphic unit varies with the nature of the thrust-propagation folds. This affects fluid-pressure dissipation differently from each unit, around the thrust fault (lateral geopressure structural limits) and associated anticline.
To discuss vertical and lateral seals further, seal types introduced above are revisited below, this time to include pressure seal and to integrate the role of lithology and faulting, as well as apply in-situ stress seal / stress components to drilling. The seals discussed below include:

a. Lithological seal
b. Stress seal
c. Pressure seal
d. Fault seal

9.5 Vertical Lithological Seal

Lithological seals include any rock layering that inhibits flow of petroleum through it; these could also include salt lithologies such as gypsum and halite. This seal type is due to rock layers acting as a membrane, where only tiny holes through the grains, known as pore throat, are available for fluid passage. Shale is the most common lithological seal; the term mudrock is used to generalise very fine-grained lithology rocks with carbonate or siliciclastic minerals. They can be very extensive, which is why they form regional lithological seal when they occur in thick sequences.
When hydrocarbon or other reservoir fluid reaches the top lithological seal, it attempts to wet the surface, or the mineral grain, and pass through the pore space. An angle formed during that wetting gives an idea of how easily the fluid will penetrate. For example, water will wet the ground more than it will wet a glass. Figure 9.12 shows the contact angle (θ) and wetting surface between fluid and mineral of rock matrix; it is used in estimating the capillary entry pressure needed for fluid molecules to flow through the grains that form the rock.

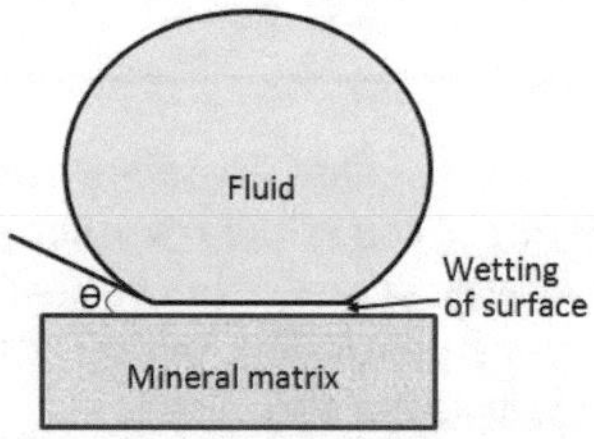

Figure 9.12: Illustration of contact angle between fluid and mineral grain, necessary for the estimation of capillary entry pressure

Within the tight seal lithology, water is often already present, and there is an interfacial tension (γ) with the water molecules. In Figure 9.12, the wetting surface has a water film on which the fluid smears as it attempts to pass on the mineral grain of the rock matrix. The pore throat size controls the angle (θ) which the hole makes with the larger reservoir lithology below it. The tightness of the seal membrane (pore throat size R) means there is need for a certain pressure (capillary entry pressure, P_e) to penetrate it. The capillary entry pressure is given as below:

$$\mathbf{P_e = 2 \cdot \gamma \cdot Cos\,\theta / R}$$

Hydrocarbon molecules would not be able to penetrate the upper seal lithology unless its own buoyancy pressure is higher than the capillary entry pressure. The buoyancy pressure of the hydrocarbon fluid depends on its column *height* (H_c) and on the difference in *density* between the underlying water (ρ_w) and the hydrocarbon (ρ_h).

The total pressure in the reservoir (P_{res}) is given by the water pressure (P_w) and the hydrocarbon buoyancy pressure (P_h). Figure 9.13 shows a case where the pressure in the reservoir due to a combination of water and hydrocarbon pressures is greater than the capillary entry pressure. This results in membrane leakage where fluid flow inhibition is impossible.

Membrane leakage occurs when,

$$P_{res} > P_e$$

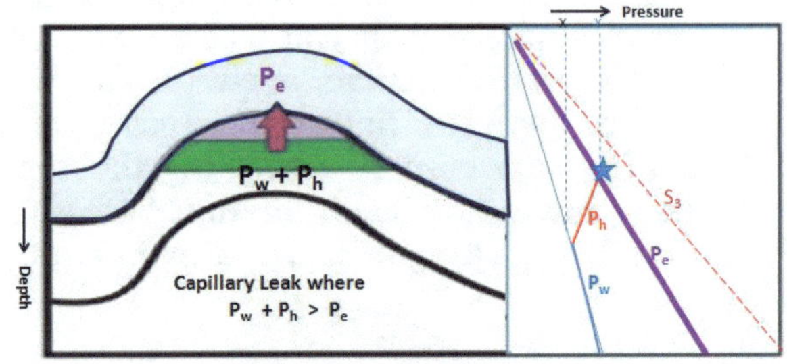

Figure 9.13: Capillary leakage requiring that total pressure at the top of the reservoir is greater than the capillary entry pressure in the seal (also represented in a pressure depth plot, where the total pressure at top of reservoir is x + y)

The pressure-depth relationship also shows this; the total pressure at the top of the reservoir is a combination of the hydrostatic pressure (P_w) and the buoyancy pressure of the hydrocarbon fluid (P_h), marked by points x and y respectively. From the above, it applies that if the capillary entry of the seal is known (that is, estimated parameters such as pore throat, and know the densities of the fluid is know, say from offset well), then the hydrocarbon column height can be calculated.

9.6 Stress Seals

As highlighted above in non-sealing mechanisms, a seal gives way when high enough hydraulic pressure, or fluid pressure, exceeds its resistive strength. This could be vertical as well, leaking the fluid to the upper unit without necessarily reaching the surface. In advanced studies, hydraulic pressure opening faults is also taken into consideration. Two types of stress related seals are:

a. Top seal hydraulic fracturing or integrity (Vertical, *discussed above*)
b. In-situ hydraulic seal or failure (vertical or lateral)

In-situ hydraulic seal or failure is much like vertical hydraulic fracturing due to lack of seal integrity. In-situ hydraulic seal is used to describe where this process occurs in the subsurface, especially in *horizontal* or *lateral directions*. To understand this, it is important to identify the three stress components:

1. The maximum of this stress is known as S_1 or maximum principal stress. It is often in the vertical direction S_v for extensive basins.
2. The intermediate stress is known as S_2 or intermediate principal stress. It is in one of the horizontal axes S_h and is lower than S_1 but higher than S_3.
3. The minimum of the stresses is known as S_3 or minimum principal stress. It is in one of the horizontal axes S_h and is the lowest of the stresses.

When under pressure, fluid will first leak in the weakest direction, which is in the direction of the minimum principal stress (S_3), be it vertical or horizontal. It could be vertical during episodic phase of compression leading to top seal hydraulic fracturing. It is usually in one of the horizontal directions in normal basin in relaxation phase. This knowledge is very useful in predicting fluid leakage during exploration and while drilling a well. Hydraulic fracturing will occur in the direction of S_3 where the strength of resistance of the rock is minimum.

In a coordinate system, the **Vertical stress,** S_V may be plotted in the y axis while the **Horizontal stresses**: S_{hz} is plotted in x direction or S_{hx} in the z direction. Figure 9.14. illustrates these stress components: the *maximum principal stress* known as S_1 (in the vertical direction; highest of the stresses as indicated by the red arrow), the *intermediate principal stress* known as S_2 (one of the horizontal axis S_h; lower than S_1 but in this case higher than S_3 as indicated by the orange arrow), and the *minimum principal stress* known as S_3 (one of the horizontal axis S_h; smallest of the stresses as indicated by the black arrow).

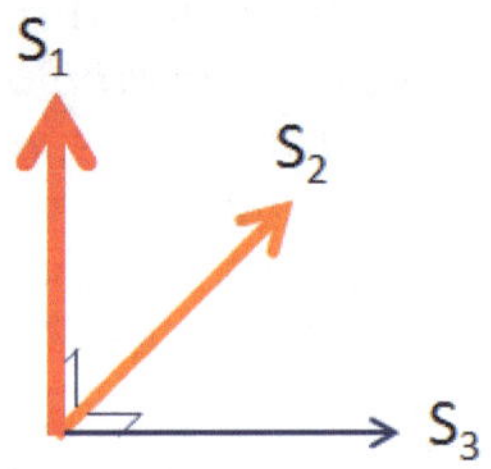

Figure 9.14: Stress components showing the principal stresses in a case where S_1 is vertical while S_2 and S_3 are horizontal

Highly overpressured fluids will only fracture the weakest point on the seal cover, usually at the crest, at the top, leading to vertical hydraulic fracturing. Hydraulic fracturing could also occur on the sides of the compartment bearing the hydrocarbon as explained above.

For example, if:

$$S_V > S_{h\text{-}xy} \text{ and } S_{hx} > S_{hy}$$

Leakage will be to S_{hy} direction.

Where,

S_V = *maximum principal stress (in the vertical direction)*

$S_{h\text{-}xy}$ = *intermediate principal stress (in the horizontal x and y direction)*

S_{hx} = *intermediate principal stress (in the horizontal x direction)*

S_{hy} = *intermediate principal stress (in the horizontal y direction)*

When there is an uplift, like an anticlinal structure where higher pressure fluid arrives at the crest, during episodic compressional phase, S_v may be smaller than S_h. These events are punctual, meaning they happen promptly or during a given relatively short period, not continually. This is the case in Figure 9.15, where:

$$S_V < S_{h\text{-}xy}$$

After such phases, the fluid is again retained. The failed top seal behaves as a valve, similar to the fault-valve model (Sibson *et al.,* 1988; Sibson 1992), where fluid pressure is relieved when it becomes too high, and then closes when pressure has dissipated. Yet during seal assessment, it is taken into consideration that the amount of fluid retained in the reservoir is controlled or limited by $\mathbf{S_3}$, through hydraulic fracturing. The fluid pressure in the reservoir is a total of pressure in the aquifer and that in the oil column.

There are cases where the horizontal stresses are momentarily higher than the vertical stress, like during punctual compressive movement. $\mathbf{S_3}$ becomes vertical. Figure 9.15 shows such a case where the pressure in the reservoir due to a combination of water and hydrocarbon pressures is greater than the top seal and fluid pressure in the reservoir forcefully fractures through the overlying vertical seal.

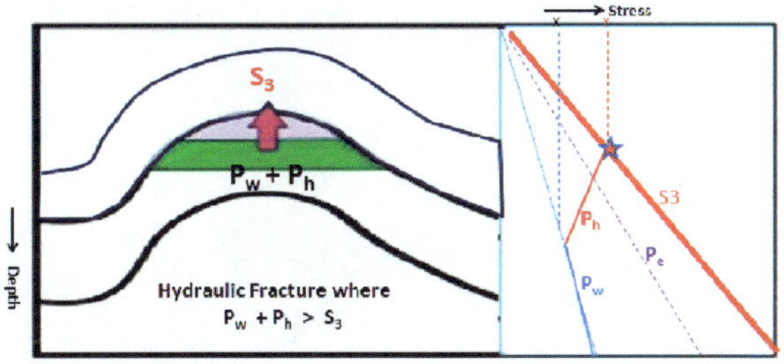

Figure 9.15: Stress seal showing that hydraulic fracturing of top seal will occur when P_{res} is higher than S_3 and the vertical stress is momentarily less than the two horizontal stresses S_{hx} and S_{hy}

Where,

P_w = Pressure in the waterleg

P_h = Pressure in the hydrocarbon column

The pressure-depth relationship also shows this; the **hydrostatic pressure (P_w)** and the **pressure of the hydrocarbon fluid (P_h)** are greater than the **minimum principal stress (S_3)**, which causes the rock to fracture at point y.

Leak Off Test (LOT) and **Formation Integrity Test (FIT)** are performed to estimate at what point the pressure of the drilling fluid with exert a stress high enough to fracture the formation. LOT and FIT give an idea of the retentive strength of a rock, at what pressure the rock will fracture. To some extent when coupled with borehole images in advanced studies it can help deduce where the minimum principal stresses are directed and at what stage leakage would occur. The principles of LOT and FIT and how they can be interpreted is shown in Figure 9.16.

Drilling mud is pumped into a cased formation leaving only an open hole in the section of the seal just above the reservoir. The pressure of leakage (maroon red line) obtained from LOT test is known as *Leak Off Pressure (LOP).* Further stress on the rock will cause it to fracture (red line), at a pressure known as *Fracture Initiation Pressure (FIP).* However, in practice, fracturing of the formation is usually avoided. Rather, the resistance of a formation wall is tested to a point where it is good enough to safely drill through with a drilling fluid of specific density just above the *Formation Pressure (FP)* (thick blue line).

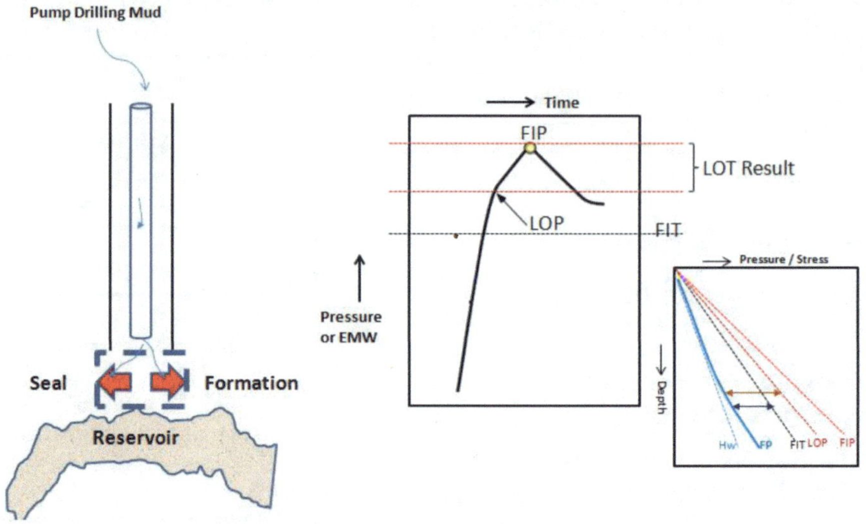

Figure 9.16: An illustration of stress test and a corresponding pressure/stress – depth plot

This test is known as *Formation Integrity Test (FIT)* (black line); it has pressure value less than the LOP.

The formation starts to fracture at the Fracture Initiation Pressure (FIP). That is the break down pressure, after which the pressure no longer holds and a drop is observed in the gauge. Usually this stabilises at about the same pressure of the LOP which is the actual resistance of the formation and is rightly the equivalent for the S_3. Different oil companies may indicate LOP as Leak Off Test (LOT) result. It is also common for some to take the FIP as LOT. What is important is to know what they indicate.

Drillers will need to drill with drilling mud pressures or equivalent mud weight higher than the formation pressure (FP in Figure 9.16) in order to balance off or counter any pressure from the formation. This is because if the formation pressure is higher than that of the drilling mud, the formation pressure will cause formation fluid to flow into the well. Therefore, the whole idea is to avoid fluid pressure influx into the well, known as having **gain**. It is dangerous when not controlled and could lead to an explosion (blowout). This is the worst nightmare of any oil company; if the situation is not properly contained, it could lead to loss of lives, damage of properties worth millions of dollars, breakdown in operations, and environmental pollution.

Where gain occurs, the weight and the Equivalent Circulating Density (ECD) of the drilling mud will have to be increased, a process known as weighting-up; mud additives such as mica and barite can be used to achieve this. Drillers will however not want to drill with mud of high pressure up to the LOP or FIP as that will fracture the reservoir, weakening the formation. This may lead to **losses** of drilling mud, shortage and eventual return of formation fluid. This will as well result in blow out. Where losses occur, measures are taken to control fluid influx, by reducing the drilling mud weight and taking measures to control the impact of gravity and the weight of the drilling bit, which may constitute additional pressures.

The idea therefore is to use an equivalent mud weight that allows the driller to safely drill through the formation, a weight higher than the formation pressure but lower than the LOP or FIT. Drilling within the **safe window** (the range of the brown double arrow line in Figure 9.16 pressure/stress - depth curve is known as drilling window) from the formation pressure to the LOP is the objective. Sometimes the FIP is reached intentionally in carefully and controlled stress test, especially in exploration wells, to know the actual resistance of the rock, which is useful for not only drilling but as well as for knowing the maximum hydrocarbon that would be retained in S_3 controlled trapping system of the field.

9.7 The Pressure Seal

Pressure seals (which may be vertical or lateral) occur because of pressure effect, where in the absence of capillary hindrance to flow, and without necessarily a resistance of a barrier wall, oil and gas are trapped because a higher pressure on the other side of the barrier prevents it from leakage. It is the fluid pressure, usually above the hydrostatic in this case, that prevents flow from one side to the other of the compartments, and not the membrane of the separating media.

An illustration of a pressure seal is shown in Figure 9.17. The oil is seen to have a high pressure (of 2 magnitudes as indicated by the arrowheads), but higher magnitudes of pressures in the seal and other overlying strata (3 magnitudes as indicated by the arrowheads) counter the pressure of the oil in the reservoir attempting to force its way through the seal.

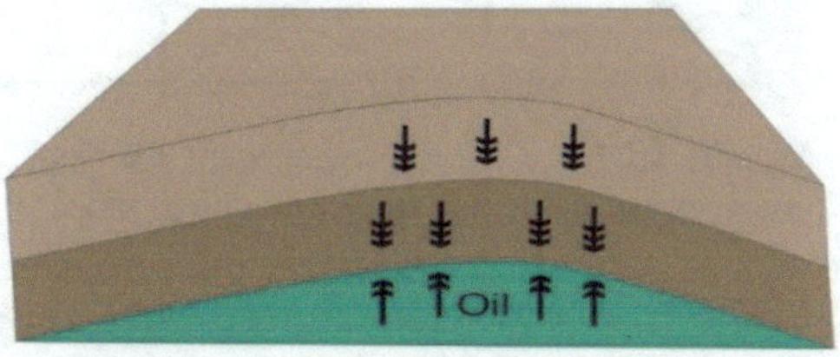

Figure 9.17: An illustration of pressure seal showing counter efforts by pressure in the seal against the pressure in the oil reservoir attempting to enter the seal

In pressure seal, the fluids act against each other and there is flow in the direction of the lowest pressure irrespective of the grain sizes of the rock matrix involved. Pressure seal is sometimes a transitional condition while the effective pressure conditions is still maintained. Figure 9.18 shows a case of a pressure seal which could only fail when the reservoir pressure (combined pressures of water and hydrocarbon, represented in the figure at x and y, respectively) becomes greater than the pressure of the seal, thereby causing fluid to flow out of the reservoir and through the seal.

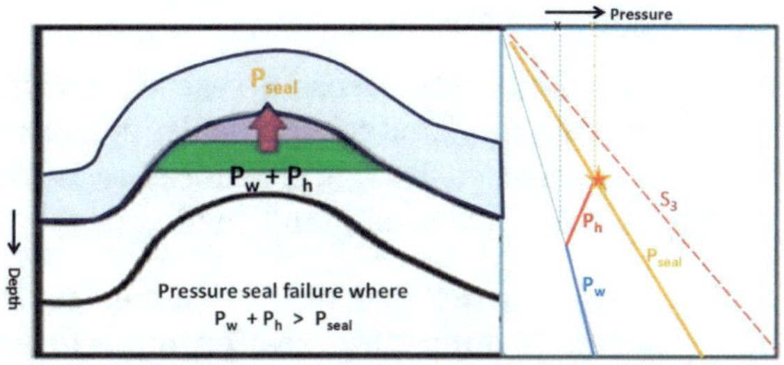

Figure 9.18: Pressure seal where the total pressure at the top of the reservoir is less than the pressure in the seal (also represented in a pressure – depth plot, where the total pressure at top of reservoir is x + y)

If the pressure seal were good enough, the counter pressure from the seal would be greater than that of the formation, resisting any form of fluid flow through it. It is also represented in a pressure – depth relationship, where sealing would occur if P_h and P_w were less than P_{seal}. This is further represented for a reservoir with hydrocarbon **column height (h_c)**, as shown in Figure 9.19.

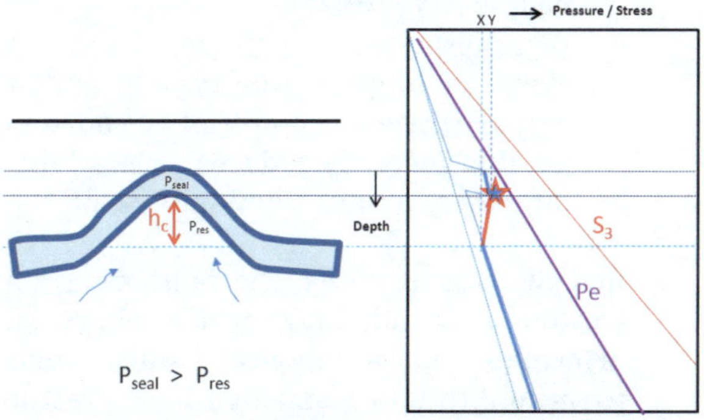

Figure 9.19: Reservoir fill to a hydrocarbon height (h_c) shown on a section as well as in a pressure/stress - depth curve

This illustration is the case of a pressure seal (P_{seal}) in an anticlinal structure, represented for a reservoir pressure (P_{res}) in a reservoir filled with hydrocarbon to a column height (h_c). The pressure in the seal is greater than that of the reservoir and so there is no fluid leakage. In the pressure/stress and depth relationship, higher water pressure (thick blue line) in the seal is observed compared to that of the reservoir; this pressure was responsible for limiting hydrocarbon fluid flow through the seal, and not the capillary entry pressure or the rock matrix.

9.8 Lateral Fault Seal

When a fault cuts across a reservoir unit, and the overlying / underlying seals, it has a tendency to seal; such seals are therefore known as **fault seals**. Two broad fault sealing process are:

1. Bringing the seal unit to juxtapose to the sand (juxtaposition seal example in Figure 9.20) or smear

- Juxtaposition, in which reservoir sands are brought against a low-permeability unit (example shale) with a high entry pressure.
- Clay smear (entrainment of clay or shale) into the fault plane, thereby giving the fault itself a high entry pressure and low permeability.

2. Introducing the seal lithology (mud or clay) in the fault zone that intersects the reservoir:

- Cataclasis, which is the crushing of wall rock to produce a fault gouge of finer grained material, giving the fault a high capillary entry pressure.
- Diagenesis, when preferential cementation along an originally permeable fault plane may partially or completely remove porosity, this generally reduces permeability and ultimately creates a hydraulic seal.

In fault seal analyses, the fault zone may be treated as a lithology, which obeys similar principles as lithological seals, including permeability entry and hydraulic fracture. In Figure 9.20, total pressure at the top of the reservoir where it makes contact with the fault will be the sum of the pressure in water phase plus the buoyancy pressure in the oil and gas overlying it.

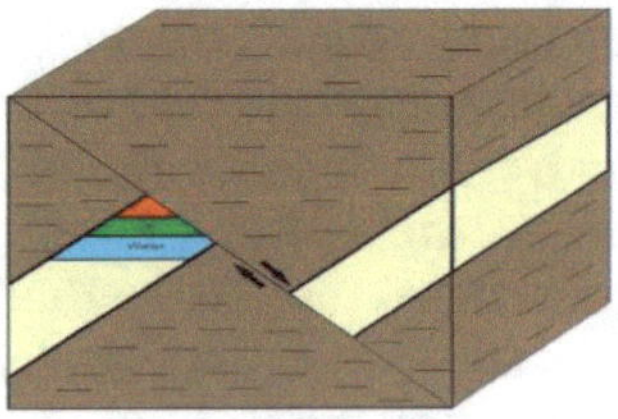

Figure 9.20: Simplest form of fault seal is by juxtaposition of reservoir to seal rock of lower permeability (example of sandstone to shale)

The fault zone has gouge properties that behave like lithological seal. Shale seal rock of very low permeability (dark brown with dashes, in Figure 9.20) is brought across a sandstone reservoir rock (dull yellow) by virtue of the normal fault that cuts across them. Subsurface fluids migrating within the reservoir rock are trapped and prevented from further migration by the seal.

Faults also seal by creating pressure differences between the petroleum bearing compartment and the opposite unit, or by providing high stress zone with a seal integrity that retains the hydrocarbon until exceeded. Figure 9.21 shows an anticlinal structure that is cut across by a fault. Where the non-reservoir or low permeability lithology is not fully juxtaposed on the reservoir lithology, or where smearing and introduction of seal lithology in the fault zone does not take place, the fault becomes a leaking fault rather than a sealing fault.

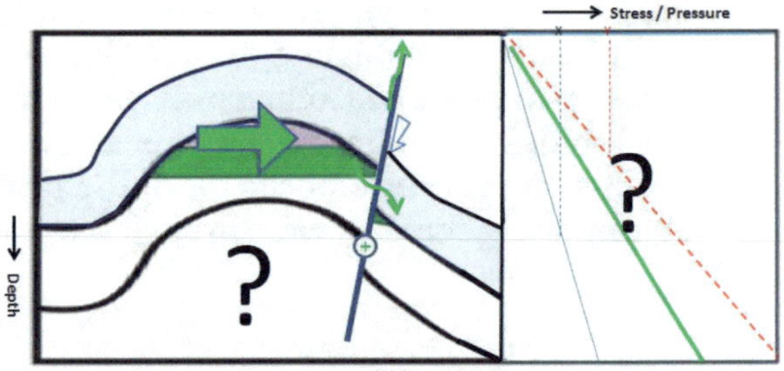

Figure 9.21: A draft schematic for fault seal and corresponding stress pressure-depth plot

From the examples above, how the fluid in Figure 9.21 may be compartmentalised, sealed across the fault by processes apart from juxtaposition, can be demonstrated. The first thing to do is find out what factors will control the ability of the fluid to penetrate into the fault, by either hydraulic fracturing or capillary entry. There is a **fault stability pressure** or **stress,** which must be exceeded to alter or open the fault. *Could there be a pressure seal across the fault? Attempt can be made to illustrate these in the stress/pressure-depth Plot.*

9.9 Seal and Hydrocarbon Column Height

As shown above, the total pressure in a reservoir determines how much fluid it will trap. To know the column of fluid contributed by the hydrocarbon phase, only the pressure contributed by the hydrocarbon (oil/gas) is taken into consideration. That pressure is the **buoyancy pressure** given by the difference between the water pressure and the hydrocarbon pressure (that displaced the water within the same column height):

Pressure of water = $g \,.\, h_{w} \,.\, \rho_{w}$ 1

and,

Pressure of hydrocarbon = $g \,.\, h_{c} \,.\, \rho_{hc}$............2

where,

g is gravity, **h** is the column thickness or the depth difference between the top and base of the oil column (that is height) and ρ_{hc} is the density of hydrocarbon. ρ_{w} is the density of water.

Buoyancy pressure, or

Buoyancy = $(g \,.\, h_{w} \,.\, \rho_{w}) - (g \,.\, h_{c} \,.\, \rho_{hc})$ 3

= $g \,.\, h \,.\, (\rho_{w} - \rho_{hc})$ 4

h_w is the same height that would have been exerted by a water pressure if it were water-bearing, within the same column. Which then is h_w for hydrocarbon-bearing.

In normal shaly top-seal, with only hydrocarbon fluid, hydrocarbon would be trapped until its buoyancy force P_h reaches equilibrium with the capillary entry pressure at the top-seal. At that equilibrium,

$P_h = P_e$ 5

Considering the formula, $P_e = 2\,\gamma \cos\theta / R$ *(see section on 9.5 Vertical Lithological Seal)*

Then,

$g \,.\, h_{c} \,.\, (\rho_{w} - \rho_{hc}) = 2 \,.\, \gamma \,.\, \cos\theta / R$................6

where,

θ is the interface angle, g is the gravitational constant, γ is the hydrocarbon-water interfacial tension and ρ_{w} and ρ_{hc} are in-situ water and hydrocarbon densities respectively. R is the pore-throat radius and h_c is the sealable hydrocarbon height.

The hydrocarbon height can therefore be obtained from the equation:

$\mathbf{h_c = 2 \,.\, \gamma \,.\, \cos\theta / R \,.\, g \,.\, (\rho_w - \rho_{hc})}$ 7

In cases where h is already known, the equation can be rearranged to find R:

$R = 2 \,.\, \gamma / h \,.\, g\,(\rho_{w} - \rho_{hc})$ 8

In practice, after obtaining R as shown above, it is plotted against other properties such as permeability and porosity (estimated from logs where it cannot be measured on seal rocks). The relationship is then used to predict hydrocarbon column in offset wells or in similar field. R can also be measured in the lab. Where there is a waterleg, it is the total pressure exerted by the underlying water column and by overlying hydrocarbon buoyancy that is taken into consideration.

9.10 Evaluation of Sealing and Leaking Faults

In-depth fault gouge study requires advanced techniques. In many oil companies, especially in operational settings where attention is most given to quick look methods on operational demands, structural analyses and fault characterisation is limited. Quantitative fault connectivity analyses are some of the less mastered approaches in assessing trapping and seal evaluation until recently.

In addition to qualitative approaches, such as visual evaluation of fault segmentation and clay smearing from seismic for example, two major fault analytical approaches may be used for advanced in-depth fault connectivity assessment *(which are programmable for automation)*. This is based on fault evaluation principles of juxtaposition and fault zone clay content. The two methods are:

1. Fault Throw Mapping
2. Shale Gouge Ratio (SGR) Computation

These are complementary and no single evaluation method is conclusive without a knowledge and application of other structural analyses and general geologic understanding of the field.

Fault Throw Mapping

Fault throws (vertical displacement of rock due to faulting) can be plotted on a map using colour codes to highlight big and small throws. For example, big throws may be indicated in red and smaller throws in green. This is simplified by the use of industry software. In the example in Figure 9.22, the throw varies along the fault and the higher throws are in the centre. The architecture of the faults is common on anticlinal structure; they are known as crestal collapse faults. Throw mapping can help to estimate compartmentalisation when bigger throw offset reservoir and juxtapose them on seal rocks.

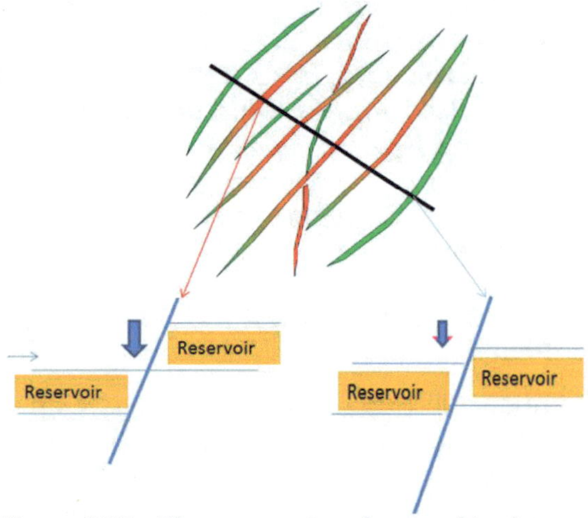

Figure 9.22: Throw mapping showing big throws in red and smaller throws in green (compare with Figure 8.5)

In Figure 9.22, a throw map is shown that highlights different throw sizes; big (red colour) and small (green colour) throws are highlighted with the bigger throws mainly at the centre along the fault. Two different normal fault sections are also shown, one on the left where the fault juxtaposes the reservoir section on non-reservoir (full compartmentalisation of units takes place; fault could be sealing), while the one on the right shows partial juxtaposition (units not compartmentalised; communication between reservoirs means fault is not sealing).

In Figure 9.23, a reservoir fairway is shown on a throw map, where it is cut across or intersected by faults. Fault 1 has a smaller throw (green), while fault 2 has a bigger throw (red). Similar colours for subsequent faults on the fairway illustrate magnitude of throw. Usually, the colour code should be applied in such a way as to respect the reservoir thickness, where there is a cut-off indicating where the fault throw is bigger than the reservoir thickness. Fault may seal where its throw is higher than the reservoir thickness, like in a narrow sand bed alternating with a thick shale layer.

Based on the above guide, in Figure 9.23, assuming that the red indicates a throw of 40 m and the green a throw of 10 m, the assessment needs to suggest which of the fault would likely seal by juxtaposition. If the reservoir thickness is 20 m, a suggestion can be made as to which of the faults 1 and 2 will compartmentalise the reservoir, that is, will act as a sealing fault and stop fluid from flowing across.

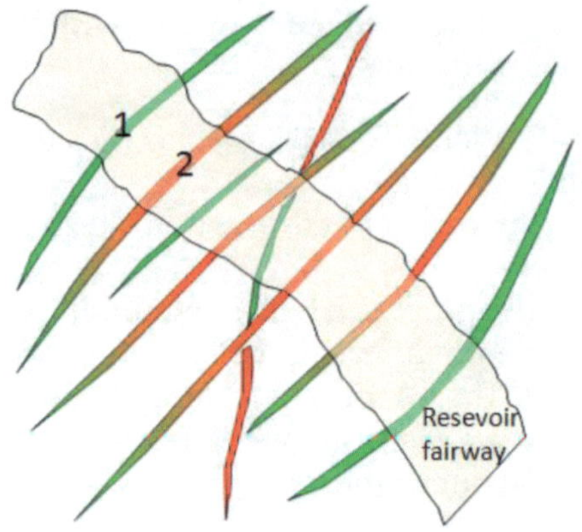

Figure 9.23: Reservoir fairway across a throw map showing places where the faults intersect the reservoir (compare with Figure 9.22)

In addition to the above, use of 4D seismic in comparing the difference in amplitude and other properties at time T1 (initial acquisition) and T2 (second seismic was taken) may show fluid limits across a fault. Such limits to flow indicate fault boundaries or sealing faults. Where 4D seismic shows that the fluid passes despite the presence of fault, it is established that the fault is leaking. There is communication between the panels (compartment or blocks) separated by the fault.

Shale Gouge Ratio (SGR) Computation

The SGR algorithm is an important way of estimating the average proportion of clays within fault zones (Yielding and Needham, 1997, Fristad et al., 1997). The Shale Gauge Ratio (SGR) can be estimated by adding together the sum of each of the percentage of the clay for each unit A to D, and then dividing the result obtained with the throw (vertical distance/displacement) of the fault, as shown in Figure 9.24a. The fault zone property in the red star is an average of the mix that the brown circle gouge material experienced as the fault slid across the throw.

An adapted approach to SGR computation (which is adaptable to *automation* in fault seal assessment) is illustrated in Figure 9.24b. It is a summation and averaging of well log derived *Volume of Clay* or *Volume of Shale* data (*Vclay* or *Vshale*) over different throw distances. It is then applied on a throw map for different faults or fault areas with different throw. In this actual computation of SGR for different throws (indicated in Figure 9.24b), the SGR was initially high, but began to reduce at the reservoir (top and bottom of reservoir denoted by X and Y respectively), where there was more sand presence in addition to clay.

The SGR calculation was done by using Vclay from *.Las file* (*Log ASCII Standard* file-format used in the oil industry to store well log information) obtained from well logs. Gamma ray log can as well be used to estimate the shale and eventually clay content of fault wall rock lithologies. Often the higher the shale, the higher the clay content of a given rock sample. Lab measurement of quantitative mineralogy also provides clay contents by summing the clay minerals in the rock. The percentage of clay mineral in a shale rock may be used to calibrate an oil field after several measurement; for example, it can be established that 100% Vshale gives 70% Vclay.

In Figure 9.24b, there is an example where the fault gouge had relatively high SGR, which reduces with depth at the zone of reservoirs, where there is more introduction of sandy material into the fault zone. X and Y indicate a fault zone across a reservoir top and bottom where the SGR is about 0.4 (40%) and 0.2 (20%) respectively. Most faults would leak for SGR less than 20% and seal for SGR higher than 40%. However, this varies depending on how the SGR was estimated and other factors affecting the fault zone. When Vshale is used instead of Vclay, the SGR values are usually higher. Another way to solve this problem is to compute the SGR for known sealing and leaking faults, then use that as cut off for similar faults in the field. It is the point of lowest SGR along a fault that determines if that fault will leak.

Sometimes a fault zone is slightly sandy and the clay content is not high enough, leading to partial leakages. When a fault is leaking partially, it is said to be *baffling* or having a role in *fluid retardation*. It is as well important to say if the sealing is in *static condition* (before oil production) or in *dynamic condition* (during production and well injection). A fault may connect two panels over geologic time but will not allow fluid communication by natural depletion during oil production. Likewise, a fault may communicate with high water injection but will not have active fluid flow across it in static condition. This understanding will be useful in *machine learning* for fault seal analysis.

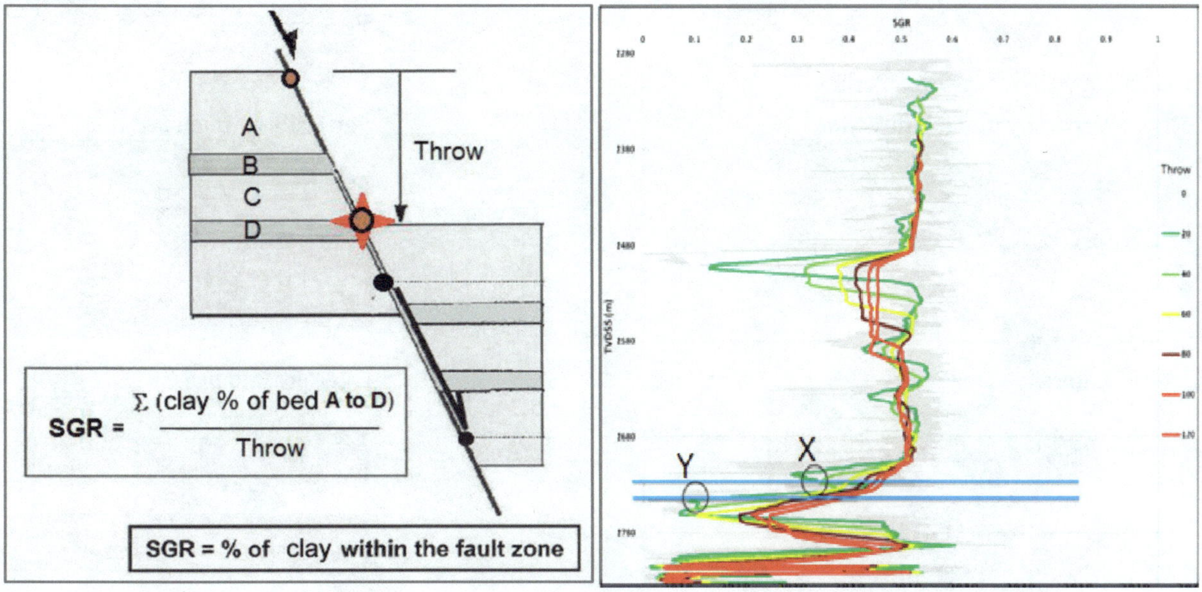

a. Principle of SGR computation b. Actual computation of SGR for different throws

Figure 9.24: Shale Gauge Ratio (SGR) computation principle and an actual computation for different throws

9.11 Field Study of Fault Sealing

Fieldwork offers a broad understanding of fault zones, as observation shows that what may be represented as a single line on seismic is actually a zone of sometimes up to 50 meters. Relying on information from seismic alone is not sufficient for understanding fault zones; there is need for up-close studies of faults in the field to understand their sealing capacities better. Characterisation of the fault structures in the field suggests a relationship between the kinematics and the fault zone fabrics, which in turn affect the fault zone hydraulic property. Fieldwork can help to understand how the character of the fault structure, as well as its movement and the nature of the lithology in the fault zone will determine whether a fault would seal or leak.

The effects of throw and local tilting related to episodic tectonic movement, along with rheological contrasts between the different sedimentary layers, can result in infill or destruction of fault gouge, which determines leakage and sealing. In Figure 9.25, core samples are collected from different sample points and in different directions within the fault after an initial fault zone field study. The idea is to be able to collect representative samples of the materials in the fault zone so that extensive laboratory studies can be carried out to reveal more information.

Figure 9.25: Coring of sample points

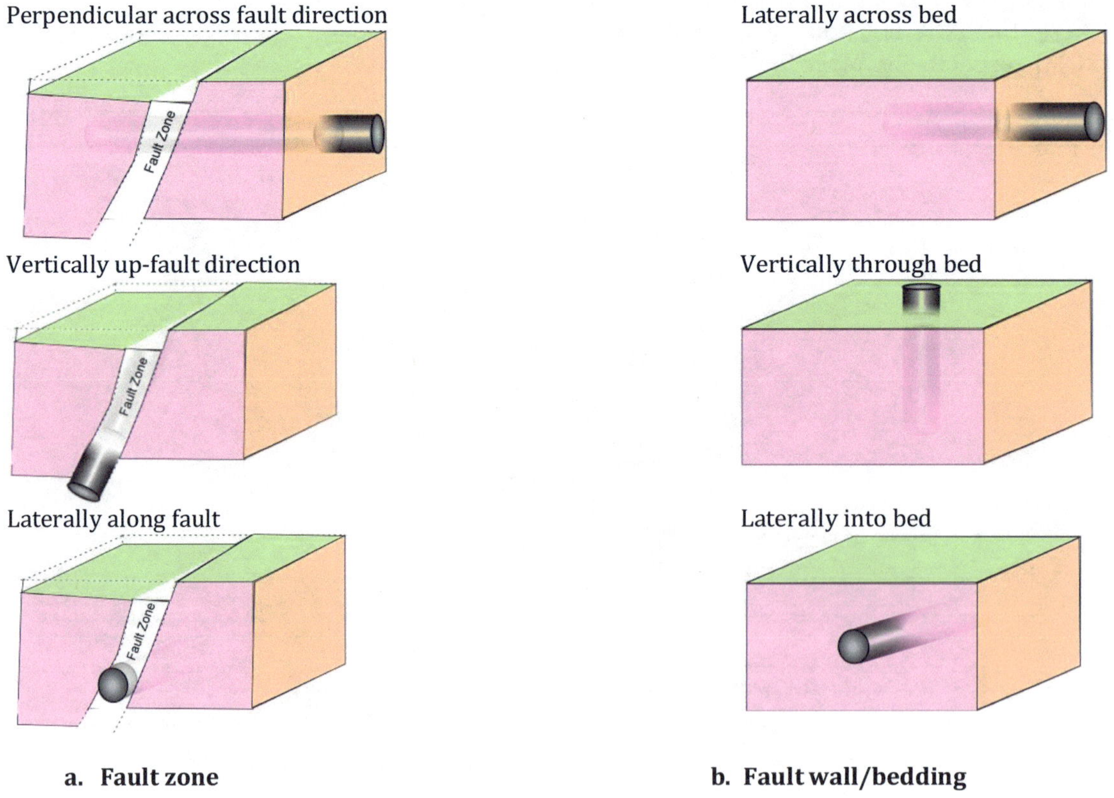

Figure 9.26: Directional coring of fault zones and walls

Layered gouge materials are smeared in the zone of widening during slip across the points of refraction or inflexion along the fault, leading to local fault zone heterogeneities and gouge fabrics that can be described in the field (Knipe, 1992; Gibson, 1994; Knipe et al., 1997; Van der Zee, 2002; Wibberley et al., 2008). After an initial fault zone field study, a useful next step is to collect samples (Figure 9.25). For fault zone, the samples may be cored in different directions within the fault zone, as illustrated in Figure 9.26 for laboratory studies. This is done only after measuring and logging the fault wall so their thickness is known.

Figure 9.26 a. and b. shows how the directional coring of fault zones (a.) and walls (b.) is achieved. Core samples can be collected perpendicular to fault direction, in vertically up-fault direction, and laterally along fault. It can also be collected laterally across a bed, vertically through the bed or laterally into the bed. In the example shown in Figure 9.27, the hydraulic properties vary across the major fault contacts as shown in Figure 9.27 a and b. The distinctive major fault contact and layering allowed for the comparison of permeabilities at different distances from the contact, which is plotted against the drawing of the faults.

In these examples, the permeability of the cores was measured (details in the next sections); permeability increases with distance below the fault contact, with fluctuation of high and low permeability due to local heterogeneities and localised shear or lithological lenses. The higher permeability seemed to be related to the fragmenting and presence of micro-clasts (clastic rock fragments not visible to the eye) in the reworked gouge sediments, and due to shear permeability enhancement close to not only the major, but also minor fault contacts and foliations.

In Figure 9.27, hydraulic property variation across major fault contacts is shown, where permeability comparison at different distances from the contact was made and plotted against the fault.

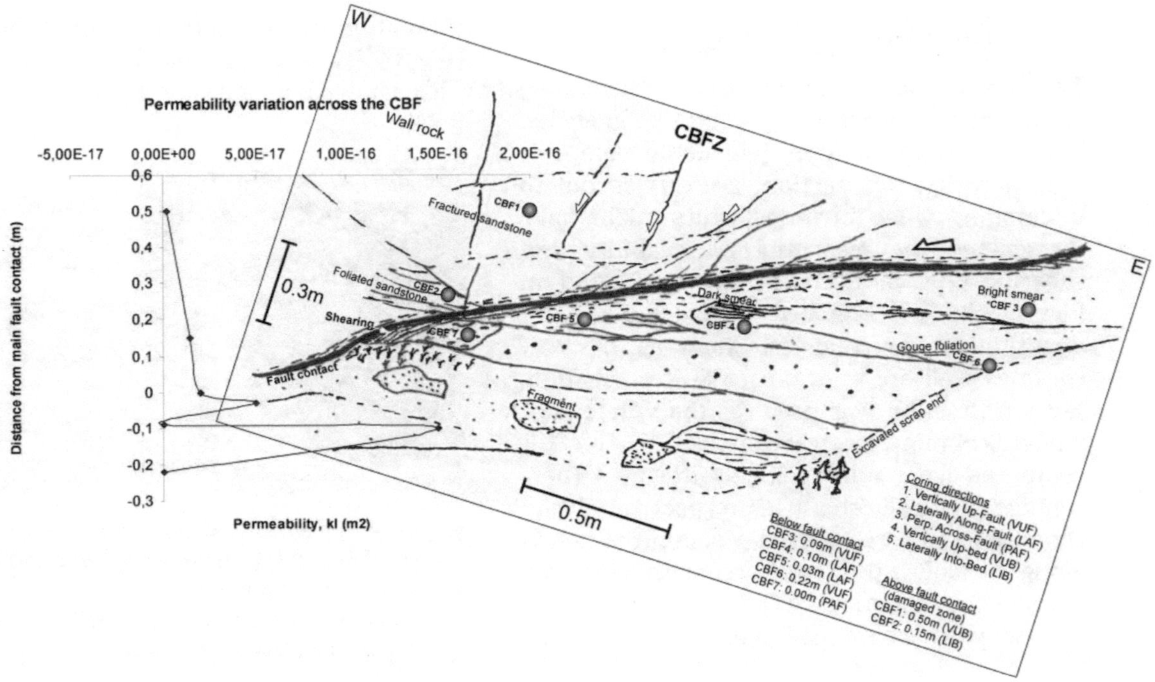

a.

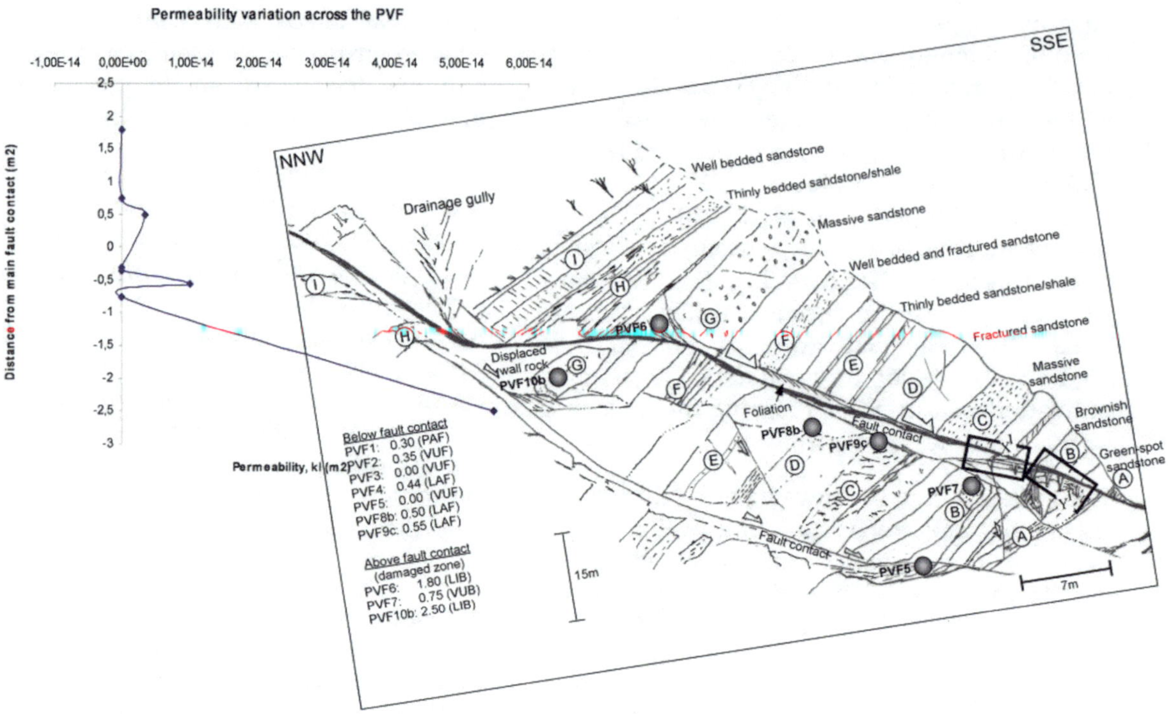

b.

Figure 9.27: Variation of permeability across two fault zones whose distinctive contacts and gouge properties were characterised in the field (Nosike, 2009)

9.12 Laboratory Study on Fault Seal

Two types of laboratory analysis of a fault zone are done: qualitative and quantitative analyses. The qualitative analyses (elemental composition, porosity, thin section) are carried out to understand what intrinsic factors might have determined the measured permeability and shearing type of the fault example cored in Figure 9.26 and sketched in Figure 9.27. The quantitative mineralogical analyses of wall rocks as well are relevant for SGR estimation. X-ray diffraction was used for the whole rock analyses of mineralogical composition. The SGR computed from wall rock clay fractions (percentage phyllosilicates with respect to silicate contents) and percentage clay content at points along the fault is then compared with that given by SGR computation. There was usually a similarity in such comparisons.

The porosity of samples from the fault zones sampled by the author was quite low, and not measurable by classical methods (such as Hg-injection or Archimedes method), especially for the clay-rich gouges. Visual porosity (Φ_v) estimation was attempted on the thin sections (Figure 9.28) using visual porosity software. This showed a porosity of between 1.0 and 3.3%.

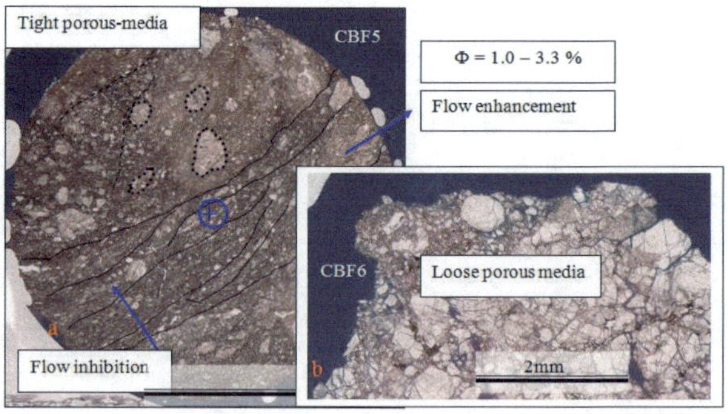

a. Very tight microfabric with possible flow parallel and into shear alignment and inhibition across the alignment.
b. Relatively loose microfabric with inter-granular and micro-fracture porosity

Figure 9.28: Example of thin sections illustrating types of porosity and flow properties of fault zones (Nosike, 2009)

In Figure 9.28, two examples of thin sections are shown: the first one on the left being tight porous media which could inhibit flow due to relatively lower porosity; the second on the right being loose porous media which could support flow due to relatively higher porosity. Thin section studies are carried out using petrographic microscopes, which allow for properties not observable with the naked eyes to be seen.

The presence of little combined illite/smectite and smectite clay minerals indicate possible minor diagenetic levels; the breakdown of some smectite being converted to illite; likewise the effect of superficial weathering on the production of calcite cements by the release of calcium transferred to pores and micro-fractures in the shear zones. Such minor weathering is often inevitable in outcrops and is a major limitation of analogue studies when applied to the case of syn-sedimentary faults affecting rapidly buried Tertiary sediments with little or no diagenesis affecting the fault zones, as is suggested for the Niger Delta Basin.

9.13 Permeability Measurement

Intrinsic permeability is an important factor for evaluating the hydraulic property or fluid flow in porous media (Zoback and Byerlee, 1975, 1976; Zhang and Cox, 2000). Its measurement in the laboratory at different confining pressure is one way to predict the flow of hydrocarbons in real geological systems. Ideally, intrinsic permeability is a measure of the mobility of fluid within a porous material and as such is related only to the pore spaces or geometry of the rock. This is therefore independent of the fluid type. If an incompressible fluid or liquid were to be used for the measurement, the permeability k is determined from the linear form of Darcy's law:

$$Q = -\frac{k\,A}{\eta}\frac{dP}{dx}$$

where Q is the rate of fluid flow (fluid volume measured per unit time), A is the cross-sectional area of the specimen, η is the viscosity of the pore fluid, and dP/dx is the pressure gradient across the sample (change in pressure across the length of the sample).

If the pressure gradient through the specimen is constant, then:

$$Q = -\frac{kA}{\eta L}(P_{up} - P_{down})$$

where L is the specimen length and P_{up} and P_{down} are pore pressures at the upstream and downstream ends respectively.

However, where a compressible pore fluid is used, such as N_2 in the present experiment, the density of the fluid differs within the specimen, and the average gas permeability k_{gas} is given as (Tanikawa and Shimamoto, 2008):

$$Q = -\frac{k_{gas}\,A}{2\,\eta\,L}\left[\frac{P_{up}^2 - P_{down}^2}{P_{down}}\right]$$

In this example by the author, the set-up for the permeability measurement is shown in Figure 9.29.

The equipment used for permeability measurement by the author shown in Figure 9.29 consists of an oil pump for confining pressure, pump gauge, confining pressure vessel, upstream pressure gauge, flow meters vented to atmosphere, gas regulator and gas cylinder (numbered 1 – 7), respectively.

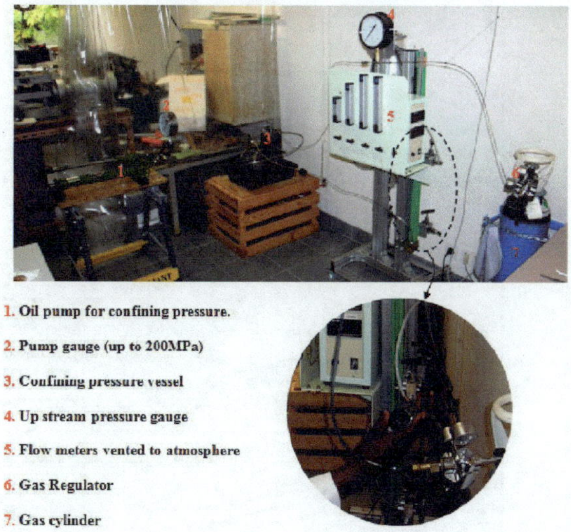

Figure 9.29: Equipment setup for permeability measurement

The permeabilities in the cored sample described above were measured using these methods in Figure 9.29. The range of measured permeability values are:

Max. = 10^{-14} m² (10 mD)
Min. = 10^{-22} m² (10^{-7} mD)

(1 milliDarcy = 10^{-15} m²)

The permeabilities varied for different flow directions for both *fault zones* and for *fault walls* as shown in Figure 9.30 a. and b. respectively. The results were useful for flow modelling discussed in subsequent sections. For the fault zone, along-fault permeability was the highest and this lateral migration along a fault plane laterally is often ignored in fault seal analysis.

In Figure 9.30, a clear variation in permeability for different flow directions for both fault zones and for fault walls is observable. While Figure 9.30a shows permeability plots for along-fault, across-fault and up-fault flow directions in heterogeneous fault zones, Figure 9.30b shows that for across-bed, into-bed and up-bed flow directions in fault walls.

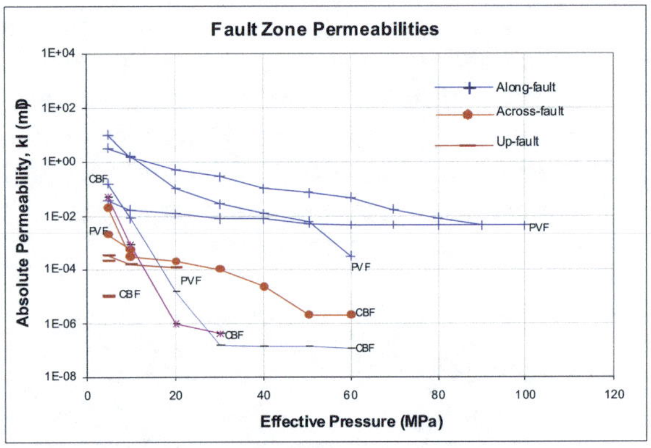

a. Permeability plots for different flow directions in heterogeneous faults zones

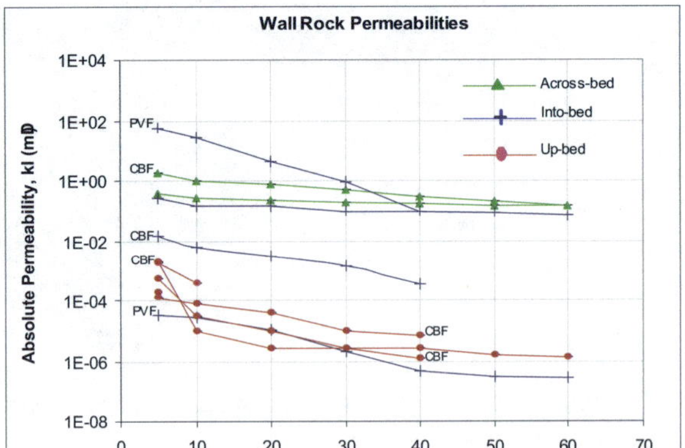

b. Permeability plots for different flow directions in fault walls

Figure 9.30: Fault zone and wall rock permeabilities plotted against effective pressure, showing the impact of heterogeneity and anisotropy (Nosike, 2009)

The permeability was measured for different confining pressures while increasing pressure (simulating burial) as well as while reducing the pressure (simulating uplift), to highlight the impact of burial depth, erosion and uplift on fault zone sealing property, and the changes for the same fault zone when uplifted. Such uplifts and burial affect minimum principal stresses as well as compaction, porosity and permeability. These factors control the types of seal mentioned above. The observation helped the author to explain episodic expulsion of fluid in gravity controlled thrust fault zone (shown in Figure 9.39).

9.14 Trap Filling

Filling and Underfilling

The final aim of seal evaluation is to know if the petroleum trap has oil or not, to be drilled and the amount in place. A trap is *filled* if the seal is integral and efficient. Sometime a trap is *underfilled*, not because there is leakage but because there was never enough hydrocarbons to charge it in the first place. If on the other hand it is filled and yet being charged, then it is *filled to spill.*

A reservoir is good enough to store hydrocarbon if:

1. It has a relatively high porous lithology (carbonate/sand) net to shale (or other non-reservoir lithology) ratio. This is known as Net To Gross (NTG);
2. The pore spaces are connected and fluid can flow through it (permeability);
3. The water saturation (S_w) is low, meaning the hydrocarbon saturation is high (S_h).

A reservoir is good enough if it has high porous rock lithology compared to non-reservoir lithology, high permeability, and low water saturation. The portion of a good lithology reservoir filled with hydrocarbons determines the trap filling. This is estimated using several methods, and used to estimate the initial oil in place and the reserve. The factors affecting the seal is shown in Figure 9.31, including the seal efficiency / integrity and the load or overburden (overburden gradient when represented in a pressure or stress-depth curve). The filling process of an anticlinal trap relates to pressure in the reservoir and the nature of the seal, and can be represented in a pressure-depth curve.

Imagine there was need to drill an exploration well in the structure shown in Figure 9.31. If the pressures are known, it is possible to know what will be the expected hydrocarbon column height.

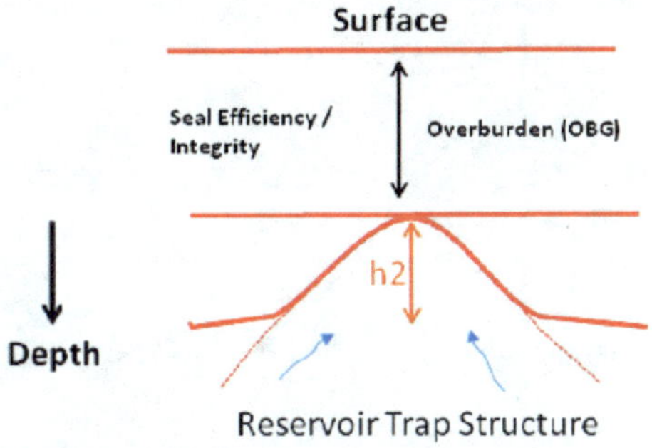

Figure 9.31: A reservoir and trap structure showing the factors that affect the seal (which indirectly affect the fluid retention and the hydrocarbon column height, h2)

The water phase is represented in Figure 9.32, first the normal hydrostatic pressure (broken blue line) and the actual water pressure in the waterleg (thick blue line, P_w). Whatever is the limiting seal is shown to the right before the overburden gradient, OBG. Next, how the trap fill process is represented in a pressure- or EMW-depth curve needs to be shown in stages as below.

The top and base of the reservoir is added to show the possible fillable hydrocarbon height (Figure 9.33) as the base serves as spill point and starts to spill the fluid with progressive charging.

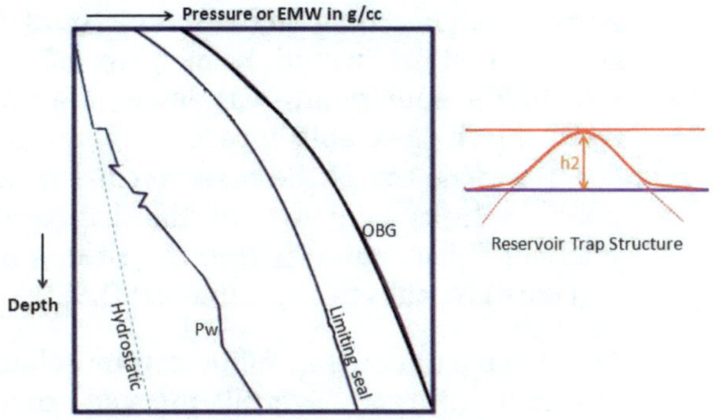

Figure 9.32: A schematic pressure-depth plot showing the waterleg and the limiting seal (indicating the window where the hydrocarbon column could fit following charging within a reservoir)

In Figure 9.33a, it is noted that more column would have been filled if there were more available structure, that is, if the flanks were lower. This is because the limiting seal (LS) was still higher, had higher pressure which was not yet reached by the column height. This is because the top of the reservoir below the leak point, where the hydrocarbon gradient (black curved line from the water leg, P_w) reached LS curve at Ty. The limiting factor (which could have been any of the seal constraints: S_3, P_e, P_{seal}, fault) could have trapped more fluid column (h3) down to Tz.

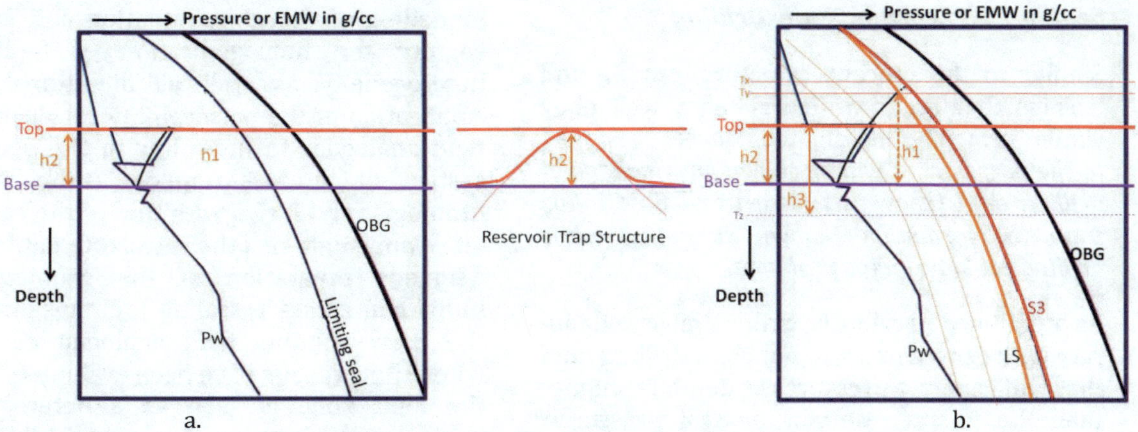

a. b.

a. Trap filling showing hydrocarbon gradient up to the top of the reservoir
b. Trap filling with the hydrocarbon gradient extended (conceptually) above the top of the reservoir

Figure 9.33 *Trap fill to spill illustrated on a pressure-depth curve*

The *conceptual* extension of the hydrocarbon gradient, to highlight the potential for further filling in Figure 9.33, shows that the limiting stress is higher than the reservoir fluid pressure. Actual **h1** would have been **h3** if the structure's spill point was lower, and there were more available space and continued charging. The top of the reservoir is below the possible leak point where the hydrocarbon gradient (black curve within the reservoir top and base) reaches the limiting seal (LS).

The illustration of trap filling can be related to underfill, fill and overfill to spill charging scenario highlighted in a previous chapter. The small inset image of the different fill scenarios shown in Figure 9.34, with oil in green under the traps, can be linked to the pressure-depth curve. *The scenario not properly represented in the curve can be attempted to improve the plot.*

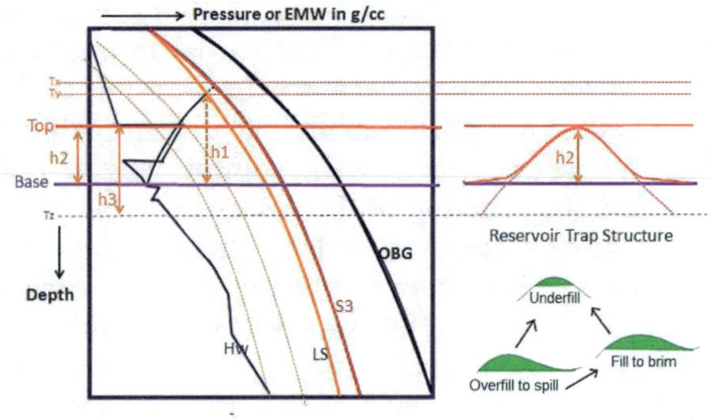

Figure 9.34 Pressure-depth curve with different trap filling scenarios and an inset map to the bottom right showing different oil (in green) fill scenarios

Sealing and Leakage While Drilling

Similar to the concept in nature, sealing and leakage that occur while drilling a well obey similar principles highlighted above.
In the next part, it will be shown how leakage of drilling mud (known as losses) and fluid influx from the formation (known as gains) while drilling could be major problems.

As mentioned previously, drillers must collaborate with geoscientists to not use a drilling mud that will generate pressures at depth far higher than the in-situ reservoir or seal pressures. They will as well not use too low-density fluid that will have very low pressure, far lower than that of the reservoir or seal formation, thereby allowing fluid influx). The acceptable drilling condition (Figure 9.35) is where the mud weight is higher than the formation pressure but less than the Limiting Seal (LS) constraint.

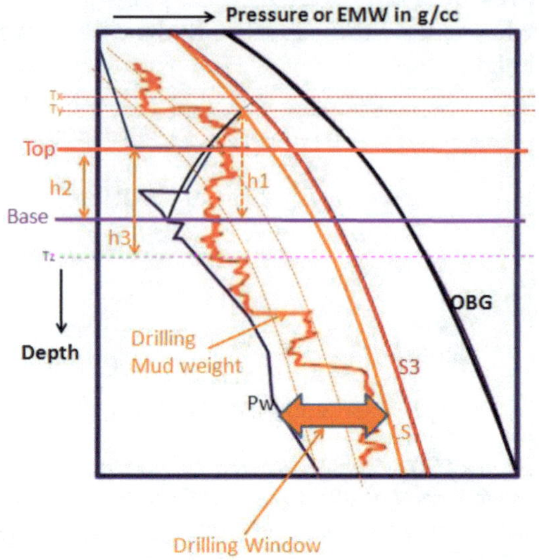

Figure 9.35: Pressure/EMW-depth plot showing formation pressure (P_w and gradient showing hydrocarbon buoyancy in the reservoir layer)

The safe *drilling window* is the range between the formation pressure and the limiting stress (often the minimum principal stress given by LOT or a little less in the area of the LS or FIT explained previously) as shown in Figure 9.35. The figure also shows the right drilling mud weight for the exemplary well in the trap filling schematics. These concepts and techniques form the major discussion in the next *Part on Well Geology and Petrophysics.*

9.15 Modelling of Charging and Leakages

Modelling of charging, retention and leakages require the homogenisation of fault zone heterogeneity as well as directional flows. Application of the permeability result from the field analogues to fluid flow in the petroleum system of interest requires comparing the lithologies and facies with that of the case fault on seismic data or other available data type. It demands separation of the geometries of individual zones, rescaling their magnitude to the corresponding field analogue equivalent whose flow results have been evaluated. Where the fault zone is seen as a heterogeneous porous media with varying permeabilities, just like layered bedding units, it is possible to resolve the layered geometries, then the permeabilities can be homogenised for flows through the layers.

The results of permeability studies and field work show that overall averaging out of fault petrophysical properties is not sufficient for accurate fluid flow modelling and that there is a tendency of differential fluid migration in the direction of the fault plane (Nosike, 2009). The fluid would reach other sedimentary bodies, leaking away or leading to reservoir connectivity by faults perpendicular to the reservoir, even when the reservoirs are far apart. The classical averaging of permeability (e.g. transmissibility multipliers, parallel and series flows) needs to be adapted for a distinctly layered fault zone as shown in Figure 9.36.

To estimate the rate of flow along this case fault, the measured permeabilities at 5 – 60 MPa pressure of confinement (equivalent simulated burial of the reservoir targets of case oil field level, on which the flows were modelled) were used for the calculations. It was assumed that the fault zone became saturated before the generation of overpressure that occurred around 4.2 million years (Myrs) ago to limit the potential risks, it is better to use the lowest permeability when estimating reservoir charging and the highest permeability when estimating reservoir leakage.

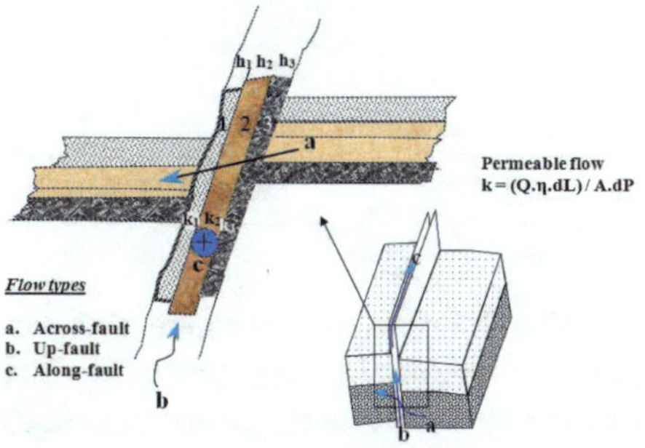

Figure 9.36: Distinctly layered fault zone with different average permeabilities in the across-, up- and along-fault directions (Nosike, 2009)

The result of the model shows that while cumulative along-fault leakage is highest because of higher permeability and pressure difference between the reservoir compartment and its exterior, cumulative across-fault leakage is higher than cumulative up-fault leakage despite far lower pressure difference. This is because of a higher contact surface area of leakage for across-fault flow.

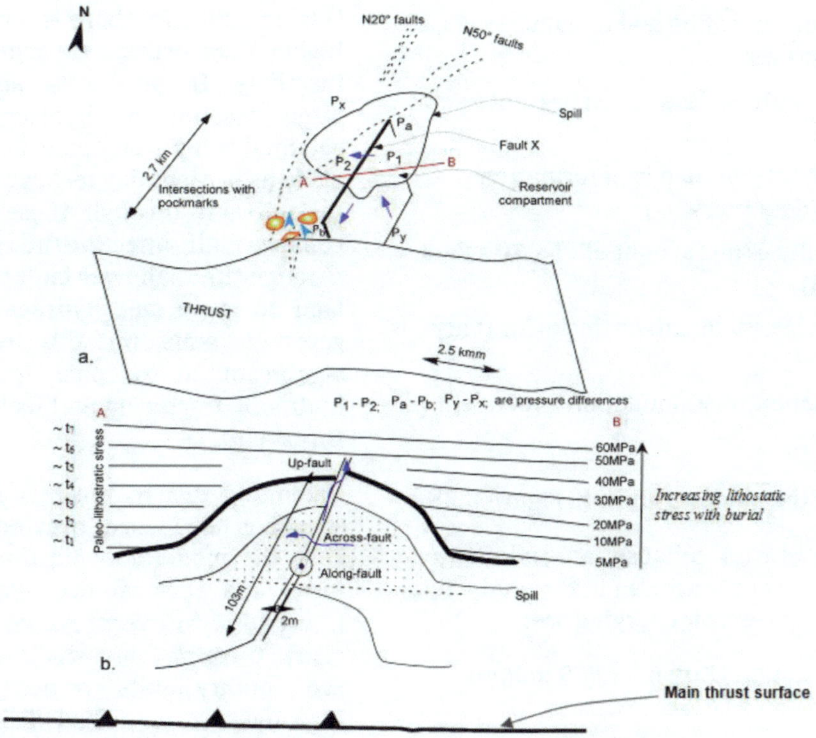

a. Reservoir compartment leaking laterally to the zone of lateral ramp development
b. Transect A-B across the fault showing the impact of the three leakage directions

Figure 9.37: Modelled case of fault leaking a major southern reservoir compartment (Nosike, 2009)

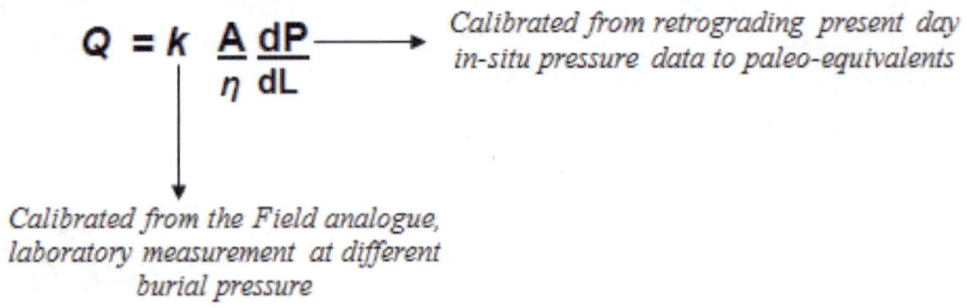

Despite increasing pressure differences, there was decreasing leakage per year during inactivity. Present day leakage during fault inactivity for a single crestal collapse fault is only about **0.013 m³ (≈ 0.08 barrels) per year** while it was up to **4.8 m³ (≈ 30 barrels) per year** 7 million years ago. This shows that before the reactivation of the syn-sedimentary fault, the effect of burial remains more important in limiting permeability flow than increasing the overpressure enhancing it. During such period of inactivity, leakage by other means, such as hydraulic fracturing, would be triggered.

The input parameters (which vary in some cases for different directions and charging history) were:

- ☐ Timing of initial charging ≈ 7.0 Myrs ago
- ☐ Timing of saturation and dynamic leakage < 4.2 Myrs ago
- ☐ A: from fault surface, thickness and height (varying)
- ☐ dP: from retrograded well formation pressure data (varying)
- ☐ η: viscosity of oil (of light oil, 1×10^{-3} Pa-s was used).
- ☐ dL: fault lateral length or thickness (varying)
- ☐ Burial: laboratory simulation equivalent (5 – 60 MPa)

The result of the model is given in Figure 9.38.

For a single crestal collapse syn-sedimentary fault of the type characterised in the study, total leakages since reservoir charging are:

Along-fault = 4,060,501 m³ (25.5 million barrels)

Up-fault = 430,696 m³ (2.7 million barrels)

Across-fault = 2,849,864 m³ (17.9 million barrels)

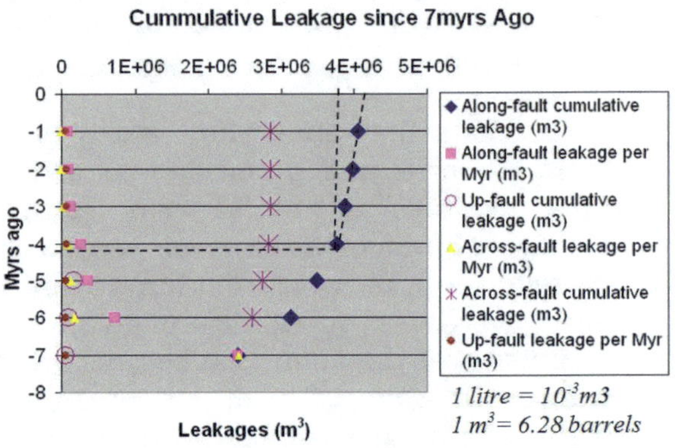

Figure 9.38: Fluid flow model result of case syn-sedimentary fault (Nosike, 2009)

During fault slip, there is a major expulsion far higher than leakage through the faults during inactivity. Despite these significant leakages, large amount of hydrocarbons would be retained where charging is continuous (especially as present day leakage during inactivity is insignificant through these shallower crestal collapse faults affecting the reservoir). Leakages through the shallower fault zones modelled can lead to important hydrocarbon loss at local reservoir scale, but this leakage is not very significant at regional levels compared to hydraulic fracturing and leakages during major thrust slip.

Extending this to the broader picture, taking what has been learnt previously into consideration, the model showed that the static (capillarity and permeability related) and dynamic flows (due to overpressure gradient or variation) through the shallower multiple syn-sedimentary faults are not the major cause of high overpressure fluid dissipation and reservoir depletion, that is, compared to the more violent expulsions of giant pockmarks and mud volcanoes at the threshold of pressure build up.

Episodic Leakages in Gravity Controlled Tectonics in the Niger Delta

Although the compartments in the extensional zones onshore of the Niger Delta, with normal growth/antithetic faults and grabens, could show very high pressure profile in the highly subsiding areas (high sedimentation and low Net To Gross ratio of the sandy facies with respect to the shale), fluid pressures most often remain lower than the vertical confining stress and the least principal stress. The chance of top-seal failure or sudden breakthrough to the surface across the overburden is less.

The overpressure generated by the sediment load (including contribution by thick seawater layer) is transferred dynamically to the distal ends uplifted by thrusting, reaching top-seal constraining stress and breaking through. A similar scenario, as between the extensional and compressional zones, is produced in the outer fold belts where the imbricate thrust faults act as structural barriers for each pressure cell and create differential folding and local accommodation space for sediment deposits. Table 9.1 summarises the effect of gravity movement in the proximal and distal offshore fronts. How this forms and controls the petroleum system in the case study is illustrated in Figure 9.39.

The above workflows and examples illustrate the role of geoscientist the training, quality checks and definition of the standards for integrating flow analysis in machine aided modelling. The role of geoscientists in artificial intelligence is the provision of definitions of geologic bodies in languages that computers understand. They have to provide structures for the discretization and modelling of fluid flow in reservoirs, faults and other geologic features. This is possible when quantitative and qualitative approaches are adopted for the assessment and representation of geologic information. Another important role of the geoscientist is in providing conceptual model based on observations and experience, for example where empirical data is unavailable or do not apply.

Table 9.1: Effect of gravity movement in the proximal and distal offshore fronts

Proximal front	Distal front
Detachment/shortening	Thrusting
Local accommodation space	Folding
Sediment loading	Compartmentalisation
Geopressure generation	Top-seal stress changes/hydraulic fracturing
Fluid pressure transmission	Fluid pressure dissipation/hydrocarbon leakage

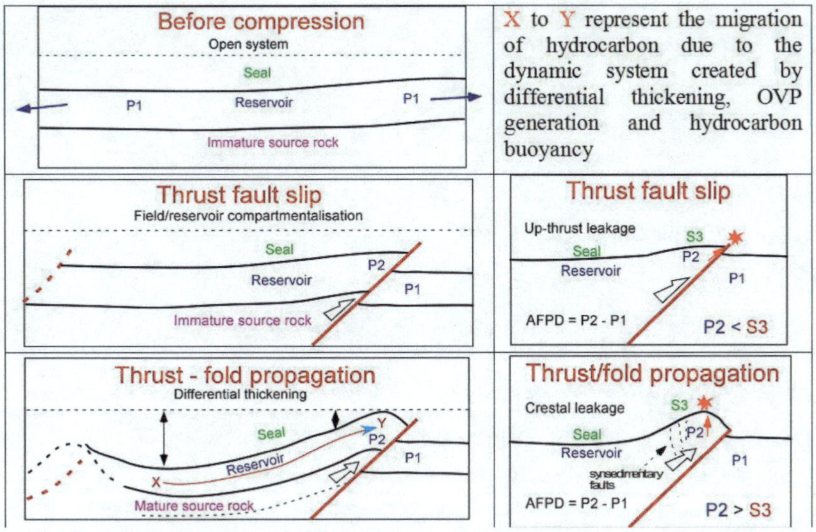

Figure 9.39: Formation and control of petroleum system by gravity tectonics (Nosike, 2009)
P1 and P2 are pressures at either side of the fault; S3 is the least principal stress in the top-seal; red star signifies leakage

EXERCISES

1. a) What is a petroleum seal?
   b) Describe the role of a seal in a petroleum field.
   c) Describe the sealing process, explaining how the properties make one rock a better seal than the other.

2. a) What are the two major processes that create a non-sealing scenario?
   b) What is seal efficiency?
   c) What is seal integrity?

3. a) What are the indicators of seal breaching?
   b) Describe the two major types of seals.

4. a) Describe the concept of filling and underfilling of a trap.
   b) What roles do net-to-gross, permeability and saturation play in determining the quality of a reservoir?

5. a) Using a diagram, describe ways in which directional leakages could take place in a prospect.
   b) What roles do pockmarks play in the identification of hydrocarbon potentials?

6. Explain how the relationship between pressure effects of water, hydrocarbon and capillary entry affect the sealing capacity of a rock.

7. Explain how Mohr's envelope is obtained and used to determine failure domain of a rock.

8. Use a diagram to describe the three major stress components. Under which of these stresses will fluid leakage occur?

9. Describe in detail the principle of LOT and FIT, and show how they can be used to determine a safe drilling window.

10. Explain how fault throw mapping and SGR computation are used to evaluate sealing and leaking faults.

REFERENCES & FURTHER READING

Downey, M. W., 1984. Evaluating Seals for Hydrocarbon Accumulations. *AAPG Bulletin*, 68, p. 1752–1763.

Fristad, T., A. Groth, G. Yielding and B. Freeman, 1997. Quantitative Fault Seal Prediction: A Case Study from Oseberg syd, in P. Møller-Pedersen and A. G. Koestler, *Hydrocarbon Seals: Importance for Exploration and Production. NPF Special Publication 7*, p. 107–124.

Gay, A., M. Lopez, P. Cochonat, D. Levaché, G. Sermondadaz and M. Séranne, 2006. Evidences of early to late fluid migration from an upper Miocene turbiditic channel revealed by 3D seismic coupled to geochemical sampling within seafloor pockmarks, Lower Congo Basin. *Marine and Petroleum Geology 23*, p. 387–399.

Gibson, R. G., 1994. Fault-Zone Seals in Siliciclastic Strata of the Columbus Basin, Offshore Trinidad. *AAPG Bulletin*, v. 78, no. 9, p. 1372–1385.

Grauls, D., F. Pascaud and T. Rives, 2002. Quantitative Fault Seal Assessment in Hydrocarbon-Compartmentalised Structures Using Fluid Pressure Data. *NPF Special Publication 11*, p. 141–156.

Hovland, M. and A.G. Judd, 1988. Seabed Pockmarks and Seepages. Impact on Geology, *Biology and the Marine Environment.* Graham & Trotman Ltd., London. 293 p.

Hovland, M., Svensen, H., Forsberg, C.F., Johansen, H., Fichler, C., Fosså, J.H., Jonsson, R. and H. Rueslåtten, 2005. Complex Pockmarks with Carbonate-Ridges off Mid-Norway: Products of Sediment Degassing. *Marine Geology* 218, 191–206.

Hovland, M. and H. Svensen, 2006. Submarine Pingoes: Indicators of Shallow Gas Hydrates in a Pockmark at Nyegga, Norwegian Sea. *Marine Geology* 228, p 15 - 23.

Ingram, G. M. and J. L. Urai, 1999. Top-seal Leakage through Faults and Fractures: The Role of Mudrock Properties, in A. C. Aplin, A. J. Fleet and J. H. S. Macquaker, eds., Muds and Mudstones: Physical and Fluid Flow Properties: *Geological Society of London Special Publications* 158, p. 125–135.

Kaldi, J. G. and C. D. Atkinson, 1997. Evaluating Seal Potential: Example from the Talang Akar Formation, Offshore Northwest Java, Indonesia, *in* R. C. Surdam, ed., Seals, Traps and the Petroleum System. *AAPG Memoir 67*, p. 85–101.

King, L.H. and B. MacLean, 1970. Pockmarks on the Scotian Shelf. *Geological Society of America Bulletin*, 81, p. 3141–3148.

Knipe, R.J. 1992. Faulting Processes and Fault Seal. In: R.M. Larsen, H. Brekke, B.T. Larsen and E. Talleraas (Editors), *Structural and Tectonic Modelling and its Application to Petroleum Geology*, NPF Special Publication 1. Elsevier, Amsterdam, p. 325–342.

Knipe, R. J., Q. J. Fisher, G. Jones, M. B. Clennell, A. B. Farmer, A. Harrison, B. Kidd, E. McAllister, J. R. Porter and E. A. White, 1997. Fault Seal Analysis: Successful Methodologies, Applications and Future Directions, In: P. Moller-Pederson and A.G. Koestler (Ed) *Hydrocarbon Seals: Importance for Exploration and Production*, Norwegian Petroleum Society Special Publication, 7, p. 15–40.

Ligtenberg, J. H., 2005. Detection of Fluid Migration Pathways in Seismic Data: Implications for Fault Seal Analysis. *Basin Research*, 17, p. 141–153.

Logan, J. T., M. Friedman, N. Higgs, C. A. Dengo and T. Shimamoto, 1979. Experimental Studies of Simulated Gouge and their Application to Studies of Natural Fault Zones. Proc. Congr. VIII – Analysis of Actual Fault Zones in Bedrock. *US Geological Survey.* Open-file Report 79-1239, p. 34–67.

Nosike, L., C. Wibberley, C. Jardiné, and Pacalin, Y. 2008a. The Role of Tectonics in Geopressure Compartmentalization and Hydrocarbon Leakage in the Deep Offshore Niger Delta: *YORSGET Extended Abstracts, Trabajos de Geología,* 6p.

Nosike, L., C. Wibberley, J-L. Montenat, C. Jardiné and Y. Pacalin, 2008b. Relationship between Thrust Fault Architecture, Pressure Cell Stacking and Hydrocarbon Leakage: Deep Offshore Niger Delta. *AAPG International Conference and Exhibition*, 26-29 October, Cape Town, South Africa.

Nosike, L., 2009. Relationship between Tectonics and Vertical Hydrocarbon Leakage: A Case Study of the Deep Offshore Niger Delta. *PhD Thesis*, University of Nice-Sophia Antipolis, France, 281 p.

Sibson, R. H., Robert, F., and K. H. Poulsen, 1988. High-Angle Reverse Faults, Fluid-Pressure Cycling, and Mesothermal Gold-Quartz Deposits. Geology, 16(6), p. 551-555.

Sibson, R. H., 1992. Implications of Fault-Valve Behaviour for Rupture Nucleation and Recurrence. Tectonophysics, 211(1), p. 283-293.

Tanikawa, W. and T. Shimamoto, 2008. Comparison of Klinkenberg-corrected gas permeability and water permeability in sedimentary rocks, International *Journal of Rock Mechanics & Mining Sciences* (in press) 10 p.

Van der Zee, W., 2002. Dynamics of Fault Gouge Development in Layered Sand-Clay Sequences. *Shaker Verlag Aachen* p. 155.

Wibberley, C.A.J., W. Kurz, J. Imber, R. E. Holdsworth and C. Collettini, 2008. The Internal Structure of Fault Zones: Implications for Mechanical and Fluid-Flow Properties. *Geological Society of London,* Special Publication V. 299. 373p.

Yielding, G., B. Freeman, and D. T. Needham, 1997, Quantitative Fault Seal Prediction. *AAPG Bulletin*, v. 81, p. 897–917.**Yielding, G., 2002.** Shale Gouge Ratio – Calibration by Geohistroy, *in* A. G. Koestler and R. Hunsdale, *Hydrocarbon Seal Quantification.* Norwegian Petroleum Society (NPF) Special Publication 11, p. 1–15.

Zhang, S. and S. F. Cox, 2000. Enhancement of fluid permeability during shear deformation of a synthetic mud. *Journal of Structural Geology,* 22. p. 1385–1393.

Zoback, M. D. and J. D. Byerlee, 1976. Effect of high-pressure deformation on permeability of Ottawa Sand, *AAPG Bulletin,* 60, p. 1531–1542.

Zoback, M. D. and J. D. Byerlee, 1975. The effect of microcrack dilatancy on the permeability of westerly granite. *J. Geophys Res.*, 80, p. 752–755.

CHAPTER TEN

UNCONVENTIONAL HYDROCARBON RESOURCES

10.0 Concept of a Hydrocarbon System

As shown in the previous chapter, *Chapter Eight, on Prospect Evaluation*, conventional prospect evaluation assesses risk of success and assigns Probabilities Of Success (POS) to each of the four elements of the play concept, which include source rock, reservoir, trap and seal. In this case the Probability of Success (POS) defines the risk, the percentage chances that there are a good source, reservoir, trap and seal that will generate and accumulate hydrocarbons. This as well includes assessing the timing of migration relative to the formation of the trap structure. It is not the same in unconventional resources. Yet concepts of the play and hydrocarbon system are not useless in the evaluation of unconventional hydrocarbons.

In an unconventional prospect, the source rock may be the reservoir, irrespective of trap, seal or timing; oil may be generated from the immature source rock, making such a rock both the source and the reservoir rock. Although knowing possible positions of trap or seal rock and migration history of a basin may be useful in assessing some unconventional deposits, the main evaluation is that of quantifying the source rock / reservoir and the volumes. It is this source / reservoir rock volume, if relatively large enough, that will constitute one of the factors determining the profitability of extracting the deposit.

The factors considered in the evaluation of the various types of unconventional deposits are:

- Deposit quality
- Available budget
- Required technology
- Production type and distribution
- Human capital and organizational unit
- Geologic risk and environmental risk
- Probabilistic determination of geometry and resource extent
- Final prospect economics and profitability

How each of these is integrated into the development of the resource is described in this chapter, after looking at the various unconventional resources.

Risk assessment involving pore-pressure is not always necessary, since there is no impact of buoyancy in such unconventional resources as oil shale or other resources quarried directly from the surface. The effect of pressure is limited where overpressure in the deposit is nearly absent, due to the absence of buoyancy in such resources. It is other factors such as cost of engineering, budget, production and more discussed below that determine the viability of the resource.

10.1 Unconventional Hydrocarbons

When hydrocarbons occur in basins, at their right maturity for exploitation with the most popular or conventional tools, they are known as *conventional* hydrocarbons. Hydrocarbons have been generated from other methods such as laboratory experiments under controlled environments, which mimic their generation in the subsurface (Speight, 1999, 2008). When they are generated from artificial decomposition of present day immature fields, plant and other organic materials, whether embedded in rocks or not, they are known as *unconventional* hydrocarbons. The oil shales, seafloor hydrates and biogenic mounds are examples of unconventional hydrocarbon resources.

Therefore, unconventional hydrocarbon is petroleum produced through methods that are not considered conventional (that is drilling an oil well and allowing the oil to flow into the well with minimal effort). Conventional petroleum production works for crude oil generated under normal conditions, natural gas and condensates that are less viscous (which enable flow with minimal effort when a well is drilled) and stored in porous sandstone or similar permeable rocks known as reservoirs.

Unconventional hydrocarbons include *Coal-To-Liquid (CTL)*, *biofuel* (from plant and animal, *biomass)*, laboratory-based chemical reaction of carbon and hydrogen yielding substances (Mabro, 2006; Mantripragada and Rubin, 2011; Choi and Lee, 2013.). They also include natural *bitumen*, which may occur in *oil sand* and *kerogen oil* from high organic matter shale rock. When mined in an open excavation, the deposit may be cut in different terraces, providing steps for stability which is maintained as the mining site is widened from the top. This is common in strip mining where the *overburden* is removed to reach the deposit near surface. Best practices and land use policies require that the site is restored as much as possible to its pre-mining state after exploitation (Mills, 2008; Kelly et al., 2010).

Mining of the unconventional resources is sometimes done by the extraction of the resource through its removal from an *open pit*. Figure 10.1 shows an open pit mine site, where an unconventional resource (dark green) is excavated step by step, from steps 1 to 4, after the overburden (green/yellow) is stripped or removed during widening. Excavation commences and is first completed at the step closer to the surface, at step 1, after which subsequent steps are excavated with increasing depth until the resource is exhausted or becomes uneconomical or unsafe to produce. The deepest level 4 is the base or bottom, also known as *bench*.

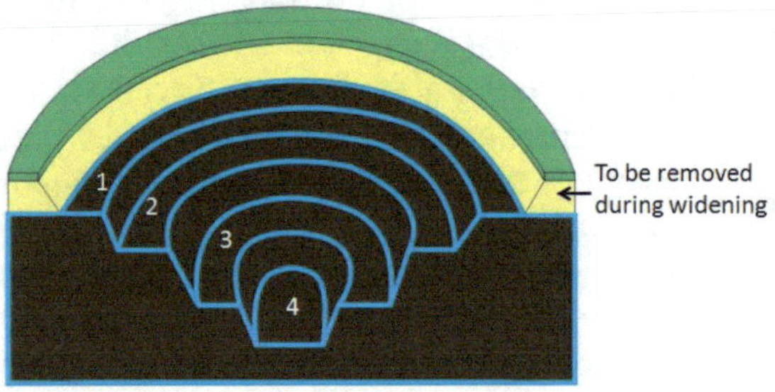

Figure 10.1: Illustration of mining site for unconventional resources cut in terraces, showing steps 1 -4 that maintain stability

10.2 Evaluation of Unconventional Deposits

Unconventional hydrocarbons occur in the same or basins similar to conventional basins. While some are exposed or associated with mountains, others are exposed by valleys. They are generated in the same way as the conventional, except for processes and conditions which did not allow unconventional resources to be transformed to conventional. Most of them require some form of fracturing of the strata before they can be extracted by unconventional recovery processes. Unconventional hydrocarbon reserves have added up to world petroleum resources and its utility and availability is limited only by the risks its exploitation poses to the environment. It is hoped that the risk must continue to reduce since the technology for its extraction is continually advancing.

The first step in the evaluation of unconventional resources is to have a geologic framework and understanding of the basic petroleum system. Among the questions to ask are:

- What stage of consolidation is the unconventional deposit?
- What is the level of maturity of the hydrocarbon content?
- How is the hydrocarbon impregnated (chemically bonded or saturated liquid)
- What is the quantity or volume of the petroleum rocks?

A major part of the evaluation of unconventional resources is similar to that of conventional. However, unconventional resources require several laboratory tests on collected samples. It is the properties of the rocks, sandstone or shale, or other rock lithology with imbued hydrocarbons, that is key. The sample may look like regular samples that are collected or cored in conventional petroleum fields (Figure 10.2). Quantifying and ascertaining the minimum amount of petroleum content per volume of rock is important for assessing the economic viability.

Figure 10.2: Oil shale and other samples displayed for analyses

Risk assessment of unconventional hydrocarbon deposits involving pore-pressure is different, as there is no impact of buoyancy in unconventional resources such as oil shale, or others quarried directly from the surface. As production starts, there is continuous feedback, and post-mortem to compare prediction with actual result.

The factors considered in the evaluation of the various types of unconventional deposits, from deposit quality, through several concerns such as technology and environmental risks, to final prospect economics, is discussed in a later section (10.6 Development of Unconventional

Resources).

10.3 Types of Unconventional Resources

Biomass and Biofuel

Biomass consists of plant and animal materials, which is essentially organic matter, animal dungs, terrestrial forests and herbs. Many sources for biomass exist and biomass may be used to produce biofuel. The most common biofuel worldwide is ethanol, which is derived from fermentation of sugar. It can be adapted for use as fuel in petrol engines or mixed as a blend with gasoline (Hammerschlag, 2006).
Conversion of cassava chips and similar raw material to ethanol and reacting with hydrogen to produce hydrocarbons have been proven to be feasible. Figure 10.3 illustrates how petrol is generated from cassava tubers. It is first cut into chips, from which ethanol is generated. Ethanol is then produced by the action of micro-organisms and enzymes from sugar /starch or cellulose fermentation, the latter being more difficult. The enzymes are used to release sugar from stored starch, after which it is fermented, distilled using significant energy and dried. This product is then subjected to processes that eventually generate petrol. Assessing the relative benefit and economic value of such petroleum as compared to direct use of cassava for staple food is important.

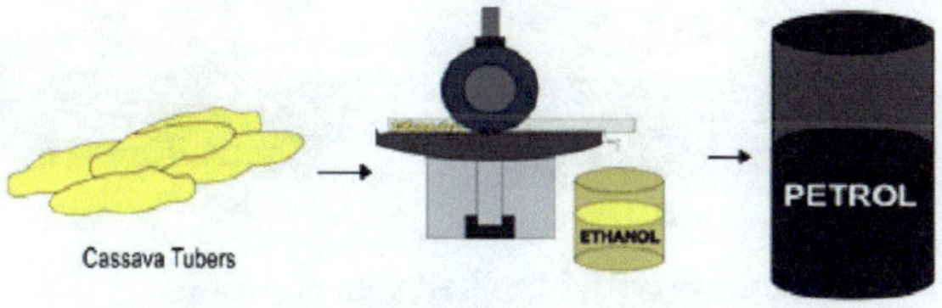

Figure 10.3: Cassava tubers as raw material for generating ethanol, which is then converted to petrol

Other forms of biofuel include bio-alcohols (such as methanol and butanol), biodiesel (produced from oils or fats), biogas (produced by digestion of organic matter in the absence of oxygen by bacterial organisms), and solid biomass fuels (such as charcoal and wood).

Oil Sands

This is the closest to conventional reservoirs. They are sand, except that they contain very heavy crude oil or bitumen that is too viscous to be produced by normal drilling and production processes (Dusseault, 2002; Smandych and Kueneman, 2010). *Oil sand* is also commonly known as **tar sand**, because of the viscous crude known as tar. This high viscosity is the reason it cannot easily flow into wells unless enhanced recovery methods such as thermal cracking or dilution is utilized. Steam is used to reduce their viscosity if drilled and drained by gravity or they are mined by surface extraction and then refined under special conditions (Butler, 1991; Century, 2008).

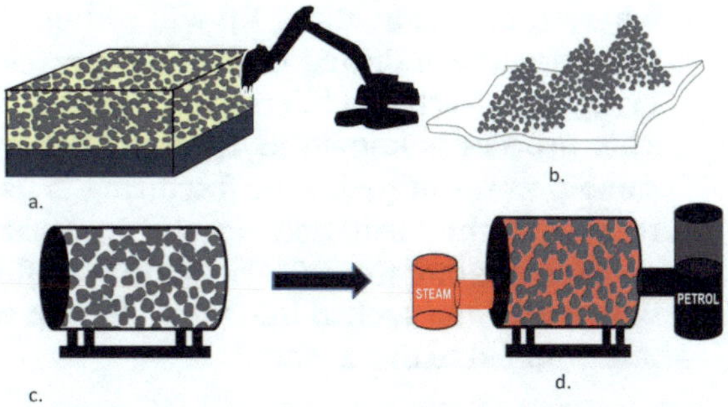

a) Oil sand deposit underlain by a dark source rock
b) Extraction of oil sand to the surface using an excavator
c) Drum serving as container or reactor for oil sand processing
d) Steam injection into drum at one end and recovery of oil at the other end

Figure 10.4: Local setup to show oil sand extraction and processing, using drum in lieu of modern standard refinery

In Figure 10.4, an oil sand deposit overlying a source rock (as shown in a.) is excavated, since the use of an excavator is possible where the oil sand deposit is near the surface. After its excavation, it is transferred into a drum where reactions that will generate oil will take place. Steam is injected at one end of the drum, so that heat reduces the viscosity of the hydrocarbon fluid, enabling it to flow and be collected at the other end of the drum.

Oil sand deposits that are relatively shallow can be exploited using surface mining methods, where the overburden can be removed to access the deposit. Alternatively, steam injection is used to recover heavy oil in-situ (on site). Two horizontal wells are drilled to the top and bottom of the oil sand formation, after which steam is injected into the upper well, which melts the bitumen that flows into the lower well and to the surface. In addition, hydrocarbon solvents can be used for the separation of bitumen from sand, which may be more economical and produce more oil. These varying costs and methods are part of what is considered during the prospect evaluation.

Tight Oil

Tight oil occurs in normal reservoir type rocks but with very low permeability, or it may occur in silt, tight sandstone or shale rocks. *Tight oil* is known as *shale oil* (different from *oil shale*) where it occurs in shale. Oil will ordinarily not flow into a well drilled in such tight reservoirs, except it is fractured to create artificial opening. This process is known as **fracking**, where the same process of hydraulic fracturing is used to fracture tight formation usually in horizontal wells to liberate the oil. The oil is then flooded with water and sucked into the well from where it is pumped to the surface.

The conventional reservoir, with high porosity before compaction or cementing, may undergo a reduction in pore volume to become tight. In Figure 10.5, a reservoir condition with high porosity and permeability is shown (in **a.**). Where this exists, hydrocarbons can flow conventionally when a well is drilled to target the reservoir. However (as shown in **b.**), burial load and other geologic processes may compact the reservoir, resulting in reduction of porosity and permeability. Pore spaces become tight, so much that the oil is not recoverable by conventional means. The reservoir rock will need to undergo fracking (as shown in **c.**), where the rock is hydraulically fractured to create secondary porosity and permeability, allowing the fluid to flow and be recovered at the surface.

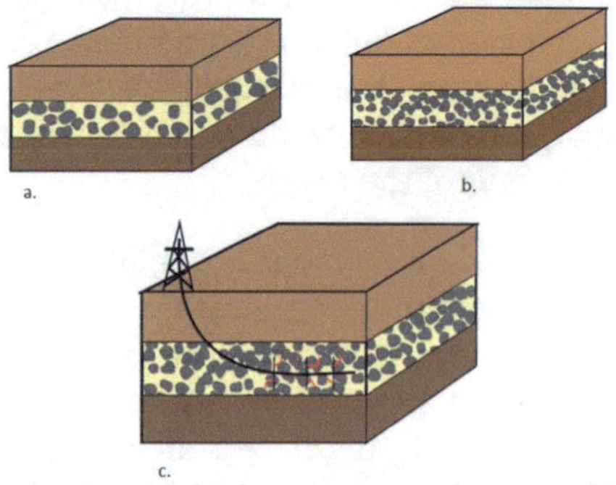

a) Reservoir with high porosity and permeability
b) Reservoir with low porosity and permeability with tight oil
c) Extraction of tight oil by hydraulic fracturing

Figure 10.5: Production of tight oil by fracking from an oil deposit in very tight, low permeability reservoir

Because of its occurrence in low permeability rock, usually silty shale or tight sandstone, compaction or reduction in its pore space volume may have occurred or continue to occur over time, leading to the oil being further imbibed in the sediments. It will therefore need to undergo processes of hydraulically fracturing it, again mostly through horizontal wells.

Oil Shale

Most known **oil shales** are mainly of marine/lacustrine origin. Oil shale deposits are rich in organic content capable of generating oil but for its immaturity, due to insufficient temperature and pressure required to transform the kerogen in the oil shale to oil. Oil shale is of kerogen and bituminous origins; organic sources include remains of algae, pollen and spores as well as woody plant materials (Cane, 1976; Urov and Sumberg, 1999; Hutton, 1994; Altun et al., 2006). It is produced by surface mining at shallow depth, or by drilling and heating the kerogen to maturity at deeper layers. Figure 10.6 shows high kerogen content (oil) shale and the conversion of the oil shale content to crude oil by heating before extraction using semi-conventional wells.

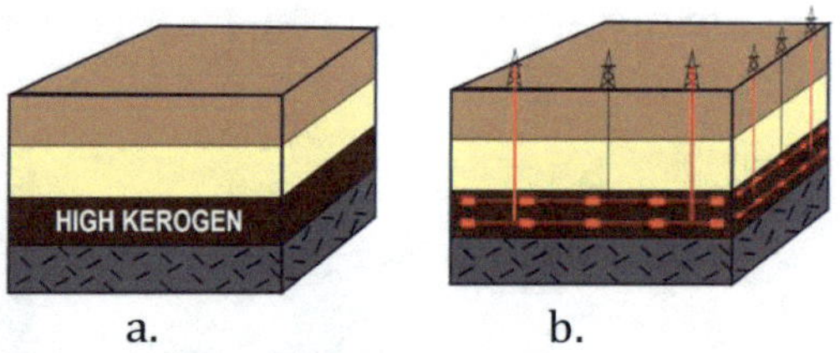

a. Oil shale with high kerogen content
b. In-situ transformation of oil shale to crude oil through heat

Figure 10.6: Production of petroleum from oil shale in-situ

The red wells in Figure 1.06 b. indicate heater wells, while the dark wells are production wells. Once heated, they subject the oil shale to temperature conditions that would have generated conventional crude, generating oil in the process. Oil shales would have been source rock for conventional hydrocarbon, as they have high kerogen content. By heating them either in-situ or after extraction, the oil is expelled by maturing the kerogen artificially, by pyrolysis or thermal dissolution.

Coal Liquefaction

Coal is a sedimentary rock that is often addressed as a mineral because of its appearance and direct use once mined. It is a fossil fuel generated from plant matter (which initially forms peat, from which lignite, sub-bituminous, bituminous and anthracite coal deposits are formed), consisting mainly 65 to 95% carbon, together with hydrogen, oxygen, nitrogen and sulphur. Unlike shale, it does not show thin fissility though it may behave as if slightly flaky like charcoal when broken; it is therefore hardly capable of being split or divided in a particular direction compared to shale.

As a sedimentary rock, its deposit could have a large extension, which is why it can be used en masse for energy generation. Coal can serve as source rock and as seal for conventional resources, but is a direct source of hydrocarbons in unconventional resources. The derivation of petroleum products from coal resources uses old traditional and more recent sophisticated methods.

The process through which solid coal is converted to synthetic liquid fuels is known as **coal liquefaction;** it is a Coal-To-Liquid (CTL) conversion. This involves the conversion of coal to oil and to gas by heating and subjecting it to reaction with other solvents or gases in the presence of a catalyst (any substance that initiates or speeds up a chemical reaction without itself being affected or forming part of the end product of the reaction). One major method is known as the *Fischer-Tropsch* process. Different types of coal deposits exist for this purpose, with anthracite and sub-bituminous coals thought to have higher yields.

Direct Method

Step 1: Coal is dissolved in a solvent and heated under high pressure

Step 2: The produced liquid is hydrogenated in the presence of a catalyst

The *direct method* of petroleum production from coal involves liquefying coal at high temperature and pressure, with addition of hydrogen and reaction to form petrol. In Figure 10.7, a sedimentary deposit with different coal seams is shown. The coal seams, being relatively close or exposed at the surface are excavated with an excavator and transported to the coal processing plant. Coal is dissolved in a solvent and heated under high pressure to produce a liquid that is hydrogenated in the presence of a catalyst to generate oil.

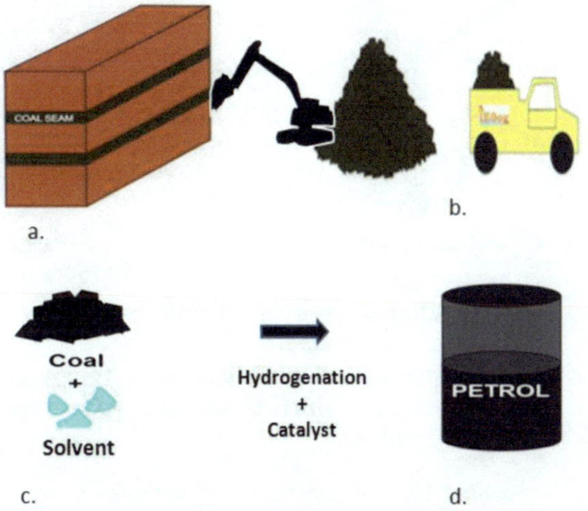

a) Sedimentary rock showing coal seam
b) Coal transportation from site after extraction from seam
c) Dissolution of coal in solvent for transformation of coal to liquid
d) Production of petrol by hydrogenation in the presence of a catalyst

Figure 10.7: Production of petroleum from coal by direct method

Indirect Method

Step 1: Coal is gasified to produce syngas (CO and H_2 gas mixture)

Step 2: The carbon monoxide and hydrogen are reacted to produce petrol

This is illustrated by the equation:

$$nCO + (2n+1)H_2 = C_nH_{2n+2} + nH_2O$$

This involves the Fischer-Tropsch synthesis (shown in the above equation) but may also involve a separate phase of methanol synthesis or conversion of methanol to hydrocarbon by hydrogenation of the CO component of the reaction. One of such processes is known as ***methanation***, in which carbon oxides and hydrogen are converted to methane and water in the presence of nickel catalyst, as shown in the equation below:

$$CO + 3H_2 = CH_4 + H_2O$$

In the production of petroleum using the indirect method as shown in Figure 10.8, the process of methanation is applied.

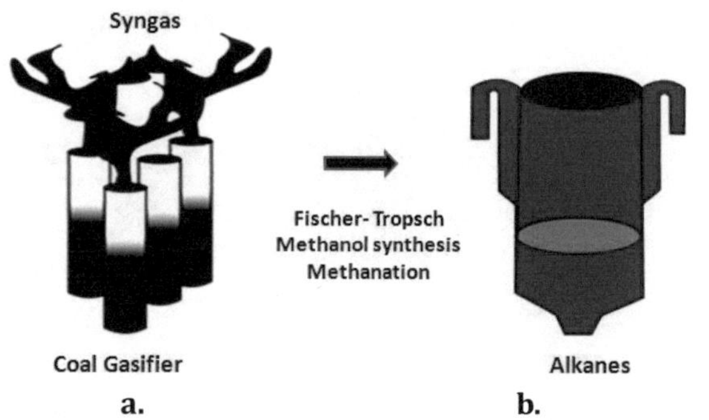

a. Heating of coal in gasifier to generate fumes known as syngas
b. Formation of methanol after removal of gas impurities, which is subsequently transformed to petrol

Figure 10.8: Production of petroleum by indirect method involving gasification of coal and using an intermediary step to convert the CO to essentially alkanes

Coal is made to undergo gasification (conversion of coal to gas rather than liquid as in the direct method) that creates a gas known as *syngas*, which is subjected to methanation to produce methanol after the removal of gas impurities, and subsequently used to generate oil. Coal gasification is achieved by different means, such as by blowing oxygen and steam while subjecting coal under pressure. These require the use of so much water compared to conventional hydrocarbons and biofuel.

Thermal Depolymerisation

Chemical processes similar to geologic processes that form hydrocarbons are used to break down long chain hy*drocarbons into smaller chains under heat and pressure, a process known as thermal depolymerisation.* Hydrocarbons so formed are similar to those formed by fossil fuels under favourable conditions. Thermal depolymerisation is used to describe heating that converts solid waste or synthetic material with potential to generate hydrocarbons, including those that occur in nature, like petroleum coke, to car tyres and plastics.

In Figure 10.9, tyres are used as raw material for the thermal depolymerisation process. The tyres are melted and processed in the presence of a catalyst, from which petrol is generated as the main product, and sticky melt residue is generated as a by-product or waste.

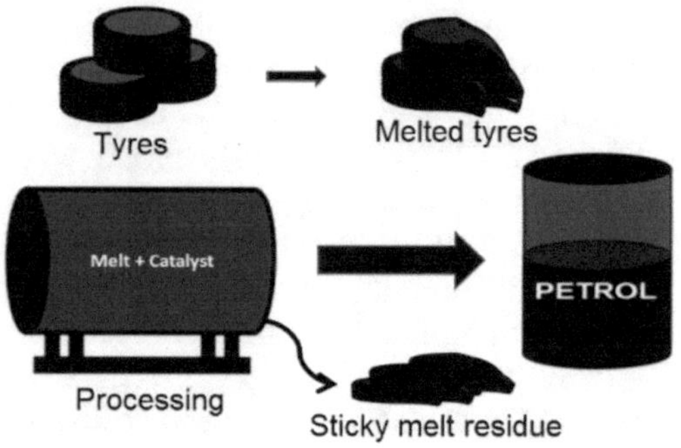

Figure 10.9: An illustration of thermal depolymerisation of melted tyres in the presence of a catalyst to form petrol while leaving sticky melt residue

10.4 Unconventional Technologies and Benefits

Biomass Technology

The use of **biomass** for ethanol production involves a number of methods, which include adding enzymes to digest sugars in starch, sugar fermentation, and distillation processes requiring significant heat, and drying. Biogas is produced and recovered from biodegradable treatment of waste materials (through processes explained above, in section *10.3 Types of Unconventional Resources*).

A biomass processing centre could act as a biorefinery (facilities and equipment that allow for processes that convert the biomass into fuels), including the use of farms and forests material to produce fuel at small and large scales. A cellulosic biomass and animals dung serving biofuel production will be profitable if achieved in a centre where the biomass is converted to oil and gas before transport. Though there might be less pollution due to decomposable organic matter (compared to synthetic polymers), the danger with this is again the depletion of biomass in nature. The use of waste and dung makes this danger less, perhaps, rather to become a positive derivative or second value.

Oil Sand Technology

As already introduced above (*10.3 Types of Unconventional Resources*), several technologies and methods are available for exploitation of oil sand deposits. **Cyclic Steam Stimulation (CSS)** also known as "huff and puff" method is achieved through the use of steam injection to recover heavy oil; the well is injected with steam, and after a while, when the heat must have reduced viscosity, the well is produced. The **Steam Assisted Gravity Drainage (SAGD)** allows two horizontal wells to be drilled to the top and bottom of the oil sand formation, after which steam is injected into the upper well, which melts the bitumen that flows into the lower well and eventually extracted to the surface (Attanasi and Meyer, 2010; Alvarez and Han, 2013).

Issues of compartmentalisation are as well important in the production of oil sand through wells. In the example shown in Figure 10.10, the importance of adequate geologic study of compartmentalisation is highlighted, where a single vertical well would not have produced a sand compartment. In addition to the need to reach and fracture every area mapped out for draining oil sand, horizontal well help to abate issues of compartmentalisation, which are major issues of concern in heavy oil that will not flow across the minimal less permeable boundary.

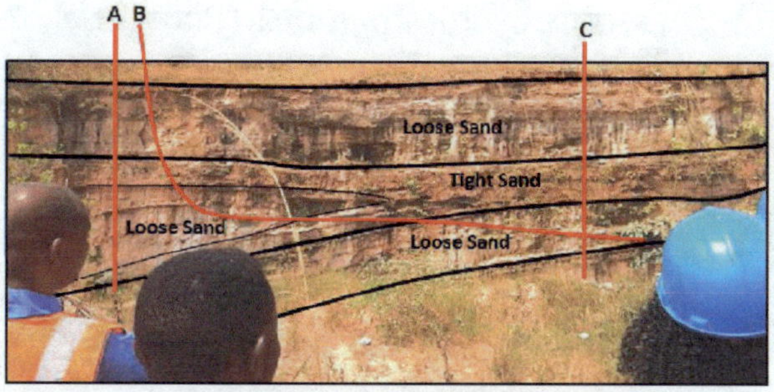

Figure 10.10: Illustration of compartmentalisation from an outcrop in the Anambra Basin

Figure 10.10 shows how minor and less permeable boundary (including sandstone), could pose a challenge in the extraction of heavy oil. The lithology in the region is predominantly sand, and compartmentalisation issues among the various sand deposits will affect the nature of drilling or well planning to extract as much hydrocarbons as possible. This problem is usually abated by the use of horizontal wells. In this example, Wells A and C may leave unswept hydrocarbons because of their high viscosity and therefore their inability to fully flow conventionally into the well.

If this were a lighter oil produced by conventional method, the impact of the separation among the different units might affect compartmentalisation less. Note that Well A cuts across all the layers and would drain all of them if the oil were less viscous. Well B not only cuts the layers vertically, but also laterally. Well B is most adapted to drain the entire compartments with assisted drainage such as steam injection.

In non-thermal **Vapour Extraction (VAPEX)** technique for production of oil sand, hydrocarbon solvents rather than steam is used for the separation of bitumen from sand by reducing the viscosity of the heavy oil (diluting the heavy oil so that it can be drained by gravity to lower horizontal wells from where it is extracted); sometimes it is more economical and may produce more oil. **Cold Heavy Oil Production with Sand (CHOPS)** is another method used, where sand filters are removed from heavy oil wells and sand is produced with the oil, together with water and gas, which are separated from the oil before the oil itself is upgraded to a synthetic crude. Newer experimental methods such as **Toe to Heel Air Injection (THAI)** and **Combustion Overhead Gravity Drainage (COGD)** make use of both vertical air injection wells and horizontal production wells to achieve separation of bitumen from sands.

Tight Oil Technology

Due to the heterogeneous nature of some formations with **tight oil**, with low porosity and permeability, they can undergo a process of fracking. This involves injecting high-pressure liquid in order to create fractures in reservoirs that will generate fluid flow paths, leading to the eventual extraction of oil. The amount of oil recovered between wells varies, due to the heterogeneous complexities of the reservoir's intrinsic characteristics. Tight reservoirs containing only oil are difficult to produce, and so natural gas may be required for enhanced recovery purposes in the pore spaces to allow the oil flow towards the well.

If the tight sand in Figure 10.10 contained oil, it will be important to know how differently Well

B can be optimised to ensure it is produced. From an initial observation point of view, this depends on the heterogeneity of the layer. If the interest were only in the oil in the tight sand, then the horizontal well within that layer would be maintained to remain in the tight sand as much as possible. That would be a scenario where oil has been produced from the top and bottom layers, but the tight sand was not producible by conventional method since it requires fracking.

Oil Shale Technology

Most **oil shale** will have to be mined using surface *open pit* and *strip mining* methods, where the overburden is removed to expose the oil shale deposit, after which it can be mined or subjected to heat on site in the absence of oxygen or further processing to generate oil. Oil shale may look like normal shale or may be darker due to higher kerogen content. Therefore, its texture, fissility, compaction and possible fracture affect the methods used for production, likewise its depth of occurrence. The fractures in the oil shale shown in Figure 10.11 suggests that would make production possible if it occurred at depth, due to ease of flow when subjected to heat.

Figure 10.11: Slightly compacted fissile oil shale with fractures (the hammer for scale)

If the oil shale deposit is at greater depths, surface mining becomes uneconomical; underground mining methods such as *room-and-pillar* (where rooms and pillars are created as the deposit is mined horizontally) will have to be employed. If processing will be done on site for oil shale underground, the oil shale is heated before extraction, allowing for possibility of higher recovery.

The fissility of shale helps oil produced from shale to flow in a given direction, sometimes horizontally, to a collection point near surface, or to a well where it is heated for in-situ production. Slightly fractured shale, as in Figure 10.11 also makes for easier extraction of oil when processing is done in-situ.

Coal liquefaction Technology

The first challenge is to extract coal in a safe and sustainable process. Coal deposits are available in many parts of the world and occur mainly in the southeastern region in Nigeria. Coal looks like charcoal in nature and may occur as an interbed within other sedimentary beds, mainly shale. The major difference in terms of occurrence between coal and charcoal is that while coal is naturally formed due to the prolonged action of burial, heat and pressure on plant remains mainly in an anaerobic decay, charcoal is formed by manmade or fire-related quick action of heat on woody materials in the presence of limited oxygen on the earth surface. Figure 10.12 shows a coal exposure, where coal could be easily broken out and mined. Even with a small sample, it is possible to use coal to produce hydrocarbons.

The two types of coal liquefaction methods have been discussed: the direct liquefaction method and the indirect method. Both requires the transformation of coal into liquid or gaseous phase composed of hydrogen and carbon monoxide, before converting it to liquid or gaseous hydrocarbon fuel (Höök, et al., 2013). This transformation requires heat, lots of water, pressure and catalysts, which may lead to the construction of heavy machineries if production is at an industrial scale.

Figure 10.12: Exposed coal seam intercalation between shaly and silty rocks, hidden below weathered sediments (hammer used for scale)

Thermal Depolymerisation Technology

There are different but similar ways of achieving **thermal depolymerisation** (example in section *10.3 Types of Unconventional Resources*), with all requiring water, and high temperature and pressure (to undergo a process known as *hydrous pyrolysis*). The raw material used as feedstocks for the industrial process of thermal depolymerisation are usually solid waste or synthetic material, such as tyres and plastic bottles. In manufacturing plants where the processing of a large quantity of raw material is required, there is need to put in place facilities for maintaining high temperature and pressure.

The feedstock is usually ground into smaller masses and mixed with water. It is then put into a first stage reaction chamber where it is pressured to about 40 to 50 bar and heated to about 200 to 300°C for about 15 to 30 minutes, depending on the nature of the original feedstock. A mixture of crude hydrocarbons products and solid leftovers or melt residues are formed from this first stage. The residues are separated after which the crude hydrocarbons are transferred to a second stage reaction chamber where they are heated, sometimes with the use of catalyst, to higher temperatures of about 400 to 500°C to further break down the long chain hydrocarbons. Different fractions of the hydrocarbons are obtained by fractional distillation. *(An illustration of thermal depolymerisation was shown above in Figure 10.9).*

10.5 Basic Measurements and Chemical Tests

Before considering the development, the unconventional resource would have to be identified first. While drilling tight oil and oil sand, continuous tests similar to those in conventional drilling are carried out to ascertain the nature of the formation and to identify the oil levels. Samples may as well be collected and analysed onsite or in the lab. Some of the tests are illustrated in Figure 10.13.

Tests such as flame test, limestone test, lithological description, porosity, cement test and permeability test all aid in the processes associated with hydrocarbon exploration and production. Few basic tests can be used to determine if a deposit contains immature or mature hydrocarbons. Immature hydrocarbons from oil shale may require pyrolysis (transformation of a substance by the action of heat) to measure Total Organic Carbon (TOC).

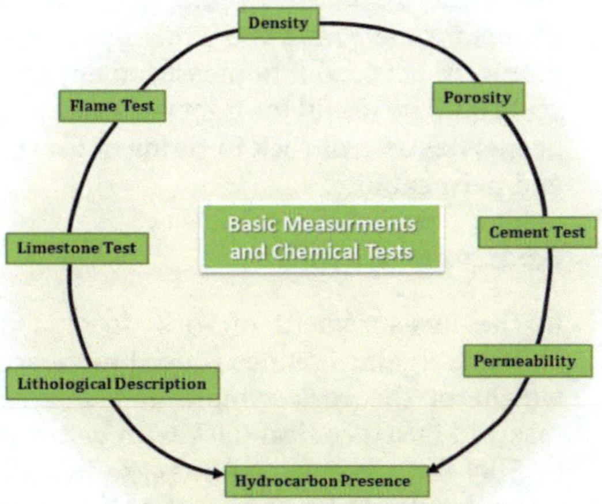

Figure 10.13: Schematic showing few simple measurements and chemical tests applied in conventional and unconventional resources (iESog Laboratory, 2019)

TOC is a measure of organic carbon in the source rock indicative of its hydrocarbon generative potentials. Oil from oil sand can be identified by fluorescence test. Other simple tests include density, flame test, lithological description, limestone identification, hydrocarbon fluorescence, cement test and more that can be carry out in the laboratory.

Porosity Measurement

In the measurement of porosity, a conical flask with clear graduation is used. A rock sample is inserted in it and water is added up to say 50cl and left for up to 24hrs. The next day, there will be a reduction in the water volume due to the absorption of the water by the pore spaces of the rock. This reduction in water volume is the volume of the pore spaces. This is a rough estimation as pore permeability may impede actual absorption of the water as well; the water may permeate the rock matrix slightly increasing the porosity. The porosity of the rock sample is key in deciding production method for heavy oil and tight sand and expected volumes of oil in place. Advanced porosity measurement methods are available.

Permeability Measurement

Permeability is the ability of the fluid to flow within the rock, meaning the pore spaces are connected and fluid can flow through them. *This has been discussed in the previous chapter, Chapter Nine: Traps and Seals Assessment.* The nature of permeability measurement applied in seal could be useful for tight oil, as it has similar properties as seal rock in terms of low porosity and permeability.

Density Measurement

In the measurement of rock density, a small digital weighing balance is used to measure the weight of the rock sample. It is immediately inserted into a conical flask with water in it, up to 50cl for example. An increase in volume of water level will be observed. If it increases up to 55cl then the volume of the rock is 5cl, which is equal to the volume of water displaced. Density is given by the ratio of the mass to the volume. The mass is obtained by dividing the weight obtained from the digital weighing balance by the gravitational constant. Rock density is important in drilling as well as in the processing of many unconventional petroleum deposits.

Coal Identification and Flame Test

Testing for coal involves colour (usually dark) identification, texture, streak (colour produced when any mineral surface is scratched on a surface known as a streak plate; colour may be different from the obvious colour noticeable with the naked eyes before scratching) and other geological rock tests. Two additional simple tests carried out on coal are density and flame test. Coal has low density of about 1.5 - 1.7g/cm^3 compared to other similar dark rocks. It burns with flame when lit, which is a confirmatory test when other tests indicate the same. Other similar dark rocks may not burn due to the absence of flammable organic matter as found in coal. The nature of the emitted soot and the burning style may indicate the type of coal.

Lithological Description

While this is not a test on its own, it can be used for the identification of rock type, especially in granular form under the microscope. During such description information such as colour, sorting, roundness and hardness are recorded. When the grains are about the same size, they are *well sorted*. When they are of different sizes, they are *poorly sorted*, which is the case for most rock samples with porosity and permeability. These may give an indication of tight sediments, as in tight oil. Roundness is an indication of provenance (the origin of the sediments; where they come from), as sediments that have been transported from a distant (distal) would be more rounded than sediments near (proximal) provenance.

Limestone Identification Test

Limestone may occur alongside unconventional deposits and it is often important to identify them. In addition, cementation from limestone may result in changing intrinsic properties of the rock, affecting the conservation, quantity and production of unconventional hydrocarbons. A rock sample is tested for limestone by adding 10% dilute HCl. If there is bubbling and emission of CO_2 gas then it most likely contains limestone.

Cement Test

While drilling tight oil, casing may be necessary as in conventional wells. Such casings are cemented (explained in the next Part, in the *Chapter on Well Operations*). It is common for cement to muffle information on lithology and formation, and sometimes confusion arises as to whether the cutting sample obtained at the surface is cement or carbonate rock, and so attention is paid to when the cement is drilled out before assessing new formation. To check if there is cement in a sample, phenolphthalein is add to it; if the sample colour changes to pink then there is cement in the sample.

Hydrocarbon Fluorescence Test

Hydrocarbon test is the ultimate, as it eventually identifies the presence or absence of hydrocarbons. In this test, rock sample is placed on a petri-dish or spot tray and observed under a fluorescence machine (under UV light). Propanol may be added to the rock cuttings and it is observed for fluorescence. If fluorescence results in blooming bluish or milky brightening colour, then there is presence of hydrocarbons in it. Hydrocarbon content in a rock sample, which may be seen under fluoroscope, is known as *shows*.

In Figure 10.14, the samples on the left and right have fluorescence while the sample in the centre does not, as the sample in the centre does not have hydrocarbons.

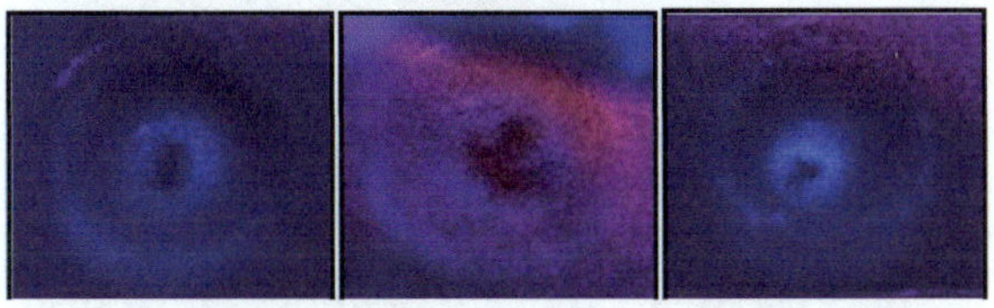

Figure 10.14: Three samples observed under a fluoroscope (from iESog Laboratory, 2019)

10.6 Development of Unconventional Resources

Deposit Quality

The measurements and laboratory tests described above are essential for determining the quality of unconventional resource deposits. Attention is given to coal because it is one of the most used secondary sources of power and unconventional hydrocarbons readily available, and which has been studied immensely during the course of writing this book. Other aspects of quality for considerations are listed below.

Ethanol as biofuel requires no natural deposits. It is produced from agricultural plants, such as cassava, wheat, corn and sugarcane. Other forms of biofuel include other bioalcohols (such as methanol and butanol), biodiesel (produced from oils or fats), biogas (produced by digestion of organic matter in the absence of oxygen by bacterial organisms); solid biomass fuels (such as charcoal and wood) occur in nature.

Oil sand or tar sand is sandstone deposits, sometimes loose sands that were not fully consolidated, containing extremely heavy and viscous crude difficult to produce conventionally. It is cracked in-situ. Canada and Venezuela have some of the largest deposits.

Tight oil or shale oil is found in lithologies of low permeability, usually shale or sandstone, where compaction or eventual reduction in its pore space volume may have occurred. Due to this low permeability, it will need to undergo processes that hydraulically fracture it through horizontal wells. Countries with the highest volumes of tight oil include United States, Russia, China, Argentina and Libya.

Oil shale deposits occur in geologic marine or lacustrine environment or other paleoenvironment of high organic content shale deposition, having potential to generate oil, but are today immature source rocks. Oil shale occurs in many places, abundant in North America; it exists as well in Nigeria and has been sampled in the Anambra Basin (Figure 10.11).

Coal deposits are found in many parts of the world, and in Nigeria it is very abundant in formations of the Anambra Basin. Different types of coal deposits are used for unconventional resources, with anthracite (highest carbon content among all coal ranks, between 92 – 98%) and bituminous coals (of less quality than anthracite but more quality than lignite coal, between 60 – 80%) thought to have higher yields of CTLs. While considering the exploitation of coal and related gas, it must be brought to mind that surface and underground mining are not the only options.

Drilling for Unconventional Gas in Coal

Drilling for associated gas in coal deposits needs to be considered, especially now that gas has become a major need for power generation. Associated gas is common in fractured coal bed as have been observed in the Anambra Basin. The gas in the Anambra Basin has quality that can be used as Liquefied Natural Gas (LNG) or for directly generating power, as well as for CTL. In Figure 10.15, water is injected into the formation through an inner tubing, using a submersible pump. The gas is driven into the outer bore or through an outer tubing and collected at the surface, while the water is processed (filtered) and re-injected for continuous production after separation from gas.

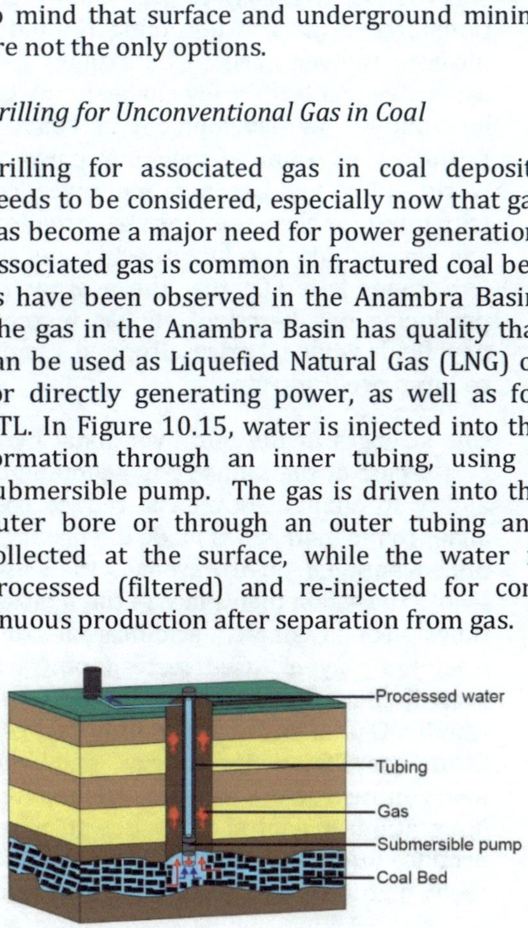

Figure 10.15: Basic technique for drilling coal associated gas

Artificial Deposits and Synthetic Remains

Some waste product (artificial) deposits may be used to produce unconventional resources. These organic and synthetic remains as well as natural petroleum related substances can be used as source material. As has been explained in previous sections, thermal depolymerisation employs chemical processes to breakdown long chain hydrocarbons (from such synthetic remains) into smaller chains, under heat and pressure, similar to the geologic processes that form hydrocarbons in nature. This is sometimes considered as recycling and may help to clear or reuse unwanted petroleum by-products.

Available Budget

Oil prices were higher a few years ago and huge profits were made by governments, oil companies and other parties involved. With fluctuating oil price, challenges faced by conventional crude oil with respect to the unconventional, and the fact that easy to get onshore conventional resources are becoming limited (deep offshore regions need to be exploited - more expensive and technically challenging), there is less budget for development of conventional resources in some developing parts of the world. Additional expenses like taxes, coupled with quest for cleaner energy, have made the oil and gas business less lucrative in some regions. Sometimes it is just that the average cost of producing one barrel of oil has increased in new fields while oil prices dropped, resulting in reduced profit margins.

One strength of the unconventional hydrocarbon sector of the industry is adaptability. The ability to adapt new means, reduce cost and adapt to the market has made the production of unconventional hydrocarbons to show the ability to sustain their place in the market at all times. For example, acidification to open fractures may be used instead of fracturing, even if this takes longer or yields result less rapidly. One of such optimisation tactics, apart from simplifying technology and reducing heavy infrastructure based development, is the integration of natural processes to reach and produce larger deposits with less input. One of such methods is subsurface coal microbial metabolism known as *in-situ methanogenesis* (Figure 10.16).

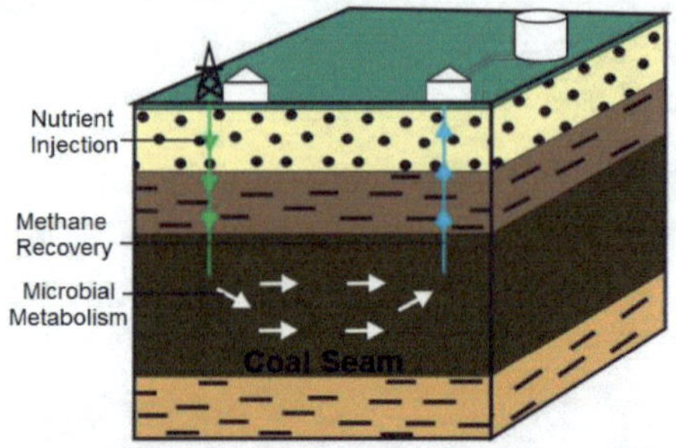

Figure 10.16: In-situ methanogenesis as a cost optimisation method

In Figure 10.16, a combined technical and natural-assisted process to reach larger coal deposits and generate larger quantity of methane gas. The process simulates natural microbial activity that produces biogenic gas (methane) in the subsurface at shallower depths. Methanogenesis requires injection of nutrient and sometimes anaerobic microbes known as methanogens into coal seams that occur near the surface (at a depth where the temperature is not yet too hot to kill the microbes). The nutrient may be injected through a well that facilitates coal microbial metabolism (methanogenesis), upon which methane is recovered at the surface. An evidence of degraded hydrocarbons by existing microbes may inspire this method. Methanogenesis requires technological know and a good budget, to yield substantial results.

Unconventional resources are much more expensive to produce, and unless there are substantial amounts of the deposits that will last a considerable period of time, money spent to recover such resources may not yield return relative to oil price, especially with lack of investment, high taxes and tough government policies. Despite this, the need for unconventional sources of hydrocarbon remains necessary. Availability of these resources allows for production growth, and less reliance on conventional crude oil production.

Production Type and Distribution

Biofuels, such as biodiesels, are biodegradable and not toxic and as such are safe to handle and transport in comparison with conventional diesel fuels. For oil sand resource, a network of pipelines will be used to gather and collect crude bitumen from oil sand deposits, from where they are sent to collection points that allows for easier access to proper refining, marketing and exportation. Railway transport can also be used to move the crude bitumen, especially when pipelines are unavailable. Before it is deemed fit for use in conventional refineries, there is need for pre-processing, which may as well be onsite or near site, involving the removal of sand, water, waste and other impurities.

With modern drilling making it possible to deviate in the subsurface, sometimes only a single site is visible on the surface while underground wells spread to distant lateral positions at depth. A community may have wells below their buildings and farms several kilometres away from the unconventional drilling site. Terrain instability, gas leakages and even earthquakes related to horizontal well drilling far from surface site has been reported. Adequate HSE and environmental impact assessment are therefore necessary for drilling into shallow coal seams. While the impact of surface mining hazards is obvious, drilling underground is as well not exempted in possible environmental degradation, as shown in Figure 10.17.

In Figure 10.17, the impact of not only lateral, but as well vertical distribution on production mechanism is indicated with increasing impact on the environment. Both cases of single and multi-layered coal seam could lead to possible environmental degradation if not properly managed.

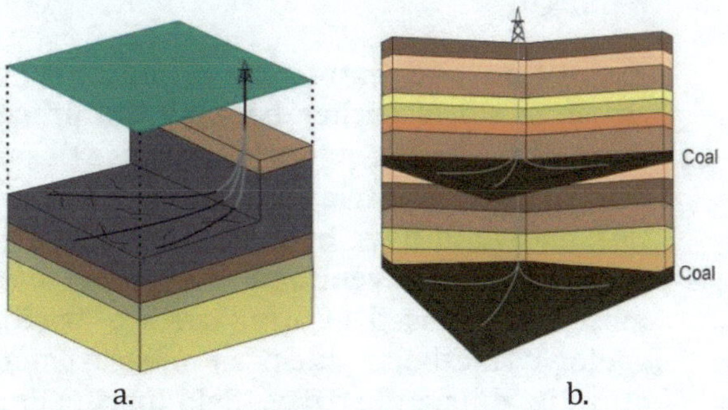

a. Production of widespread deep deposit with multiple lateral horizontal wells
b. Production of widespread shallow and deep deposits with multiple lateral wells

Figure 10.17: Single versus multi-layered coal seam

The upper coal seam in Figure 10.17b may be at the level of the fresh water table, and so the impact on the environment is more in this case. Where fracking is necessary, suitable drilling rigs are used, for example in production purposes for tight oil. Such rigs are supposed to have modules to curtail high pressure on surface to avoid risks to personnel and environment. Infrastructures are put in place for the collection and transportation of produced oil as well as the water needed for fracking. Water usage may affect natural ecosystem or deplete streams available for the local community (Nosike, 2019). Production will be hindered by the unavailability of water in rural areas and appropriate drilling rigs.

Human Capital and Organizational Unit

Lack of new skilled and motivated personnel is one of the challenges facing the petroleum industry (Nosike, 2019). Sometimes those who are motivated have no skill and those who have the skills are few. Open mining is not as coveted as conventional drilling sites, sophisticated drillships and office spaces of oil and gas companies. The government policies and general societal apprehension of unconventional petroleum and fossil fuel in general do not make things easier.

It will be difficult to exploit unconventional resources to sustain its growth over time without the required technical expertise. There is need for rapid learning; companies are seeking to improve effectiveness of their existing workforce while positioning themselves to ensure continuous innovation and collaboration with diverse disciplines required to take unconventional hydrocarbon resources to the required level of sophistication. With appropriate workforce in place, exploitation of unconventional deposits would be more viable.

Geologic Risk and Environmental Risk

Health Safety and Environment (HSE) and general efforts against pollution and environmental degradation are now a watchword in the petroleum industry. One of the major challenges faced with the use of biofuels is air pollution, the same faced with the use of conventional petroleum resources. Pollution results from increases in the production of nitrous oxides (N_2O), carbon dioxide (CO_2) and carbon monoxide (CO). The need for crops to generate biofuels has also affected agricultural land use, requiring heavy use of fertilizers and insecticides. These tend to work against the tenets of HSE, which traditional oil companies attempt to uphold.

In addition to geologic risks, such as overburden failure, environmental risks of pollution as well as immediate accidents are also a challenge for the workforce in the unconventional resources sector. Fracking poses a risk of underground water pollution and land instability. Single point recovery of heavy oil from oil sand when properly done is similar to conventional oil and gas wells. Fracking from single point may however not be effective and its impact on the environment depends on several factors (example shown in Figure 10.18), especially where there are multiple layers of unconventional deposits. Done minimally however (as in steam injection), it might cover less extent and limit impact laterally. Compared to multiple horizontal wells (as in Figure 10.17), the reason fracking from a single vertical well point would have more impact on subsurface is the creation of unguided fractures, as shown in Figure 10.18.

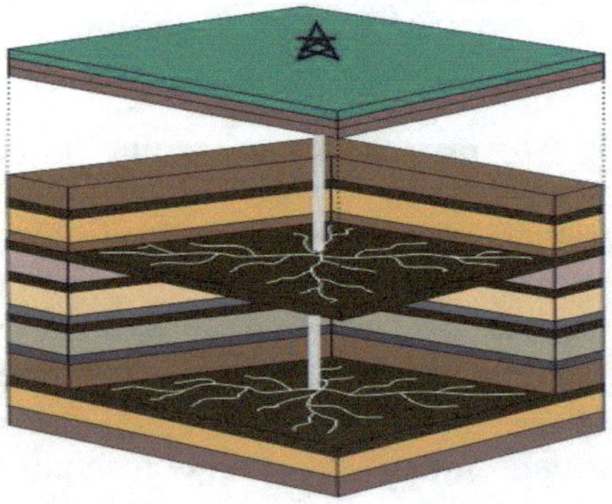

Figure 10.18: Production with fracking from a single vertical well point

Many health concerns are involved regarding oil sands. Disintegration of wild life and aquatic habitat, increase in soil erosion due to overburden removal and deforestation during surface mining are some of the concerns. There could be pollution and depletion of water resources during production and contamination of land by heavy metals such as lead, mercury and arsenic during extraction. This may not always be the case as these metals may not necessarily be present. Similar contamination occurs as well in mining mineral resources and other major engineering construction or excavation sites.

It sometimes seems that there are health hazards in unconventional as much as there are in the conventional, if not more in the unconventional. This is a major hindrance to the development of unconventional resources. In Nigeria and Africa, as advocated by concerned platform such as the *Integrated Elvee Services*, many canvass for only a minimum exploitation of unconventional hydrocarbons as an energy mix to abate or reduce any hazards to the minimum (Nosike, 2017). As well for oil shale, carbon dioxide and methane emissions may occur. In other cases, like for CTL, considerable heat is associated with coal liquefaction. One solution is to know these hazards and abate them, instead of turning a blind eye or not understanding what goes on in the industry.

Geometry and Resource Extent

Affirming the distribution, volumes and net-to-gross of an unconventional petroleum deposit is sometimes a challenge. The process of determining the geometry and resource extent is similar to that of conventional resources (discussed in the chapters on *oil field assessment and prospect evaluation*), however, there might not have been enough studies, and as detailed as that in the conventional oil and gas field appraisal before the beginning of production. An unconventional deposit might not require as detailed porosity studies and pore pressure prediction, which limits the knowledge of formation properties. Likewise, trap assessment and seal analysis are not always required.

Excavation and extraction of unconventional resource starts earlier because the process is more direct and requires less surface to subsurface probe for intrinsic petrophysical properties, as would have been required for efficient recovery by conventional rigs. Also, because most are exposed or near surface, traditional geologic methods such as measurement of attitude: dips and strikes, field logging and use of trigonometry for estimating geometry and extents of deposits may be employed, as shown in Figure 10.19.

Biofuel geometry and extent are only limited by technology to produce them, as the raw materials can be made very much readily available with good agricultural practices. Biofuels are easy to produce through modern-day biological processes from plants and breakdown or digestion of wastes. On the other hand other resources like oil shale have a wide geometric and resource extent; they are found in all oil provinces of the world as they serve as source rock for most conventional oil resources. A majority however occur at great depths that do not allow them to be economically extracted with present technology.

Figure 10.19: Oil shale stratification, with dipping beds which allowed for the measurement of the attitude and as such vertical thickness of the strata

Prospect Economics and Profitability

Unconventional hydrocarbon resources could be more profitable to produce than conventional hydrocarbons in the long-term; they are sometimes closer to the surface and they have a longer depletion span. The large increase in the production of tight oil led to significant amount of available oil and the subsequent drop in oil prices. If the quality of the kerogen in the oil shale is good, and if it can be efficiently extracted, production of oil shale would be even more profitable if the energy used to mine and process it is brought to the minimum. The factors in the costing are shown in Table 10.1.

Table 10.1: Simplified costing chart and economics

Costing	Economics
Mechanical	*Size of deposit*
Human capital	*Availability of water*
Government tax	*Cost of gas or barrel of oil*
Environment	*Deposit Extent and duration*

Actual economics involves subtracting the cost and the gross sales in monetary value of the oil and gas production. The factors listed in Table 10.1 and shown in Figure 10.20 highlight how recycling and use of locally available tools and human capital could help in profitability, by reducing spending cost and expenditure. Other profits may not be monetary, but availability of energy for the community sponsored by government or private company through Corporate Social Responsibility (CSR) provisions.

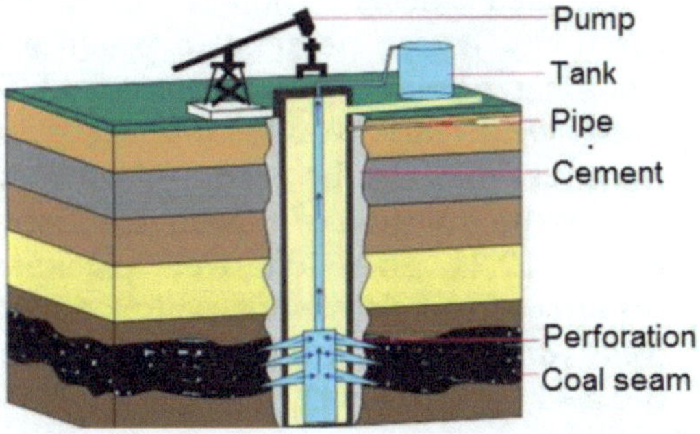

Figure 10.20: Set up of gas production from coal through fracking and perforations (compare with Figure 10.15)

Figure 10.20 shows an example of cost consideration with a development system for the production of gas from coal. Question to consider for such development may include utility near the production site, available Liquefied Natural Gas (LNG) plant, pipeline, nature of the gas/purity level, market price or demand and original gas in place. These are similar questions to ask in the case of other unconventional resources such as tight oil and oil shale. Water recycling helps in reducing cost as well as locally made or sourced pumps, pipeline, cement and completion tubings. Most importantly is nearby utility and storage facility, which contributes profitably by limiting downtime and allowing non-stop productivity.

Refer to the section on Petroleum Economics in Chapter 8: Prospect Evaluation, for more on the general process of economic assessment.

Together with conventional sources of hydrocarbon, oil shale will be large enough to meet the growing worldwide demands for these resources which have been witnessed in recent past, even though there is a higher cost of production and transportation of the unconventional resource. Where conventional oil prices are higher, it may become more economical to produce and market unconventional resources if the ill effects of such activities are drastically reduced.

Re-Evaluation of Abandoned Fields

In practice, while major oil companies divest and marginal fields are sold for conventional hydrocarbons, it may turn out that some abandoned fields contain unconventional resources. It also possible to produce unconventional resources from abandoned conventional wells. Unconventional and conventional resources may also be produced concomitantly during the life time of a field, as illustrated in Figure 10.21. Such left over unconventional resources in a field may be exploited using *infill wells* (drilled after the initial wells planned in the field development).

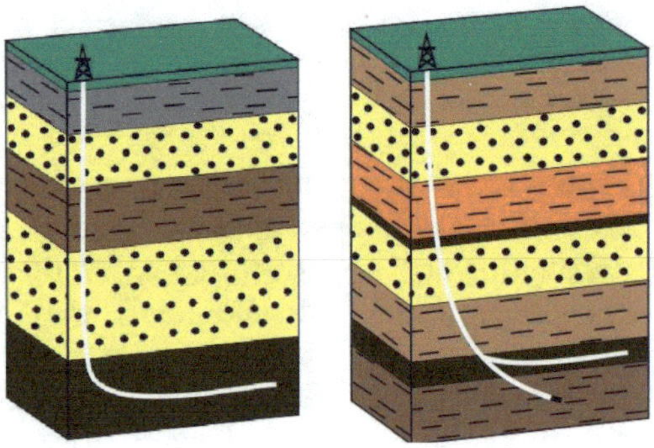

a. b.

Figure 10.21: Unconventional resources drilling below an already depleted sand-bearing reservoir

In Figure 10.21, two sandstone reservoir layers – yellow with dots, having had their conventional hydrocarbon resources tapped in a. overlay the dark green unconventional layer below also tapped by the same well. In **b.** a sidetrack well is drilled to deviate horizontally into an unconventional layer (dark green), with the same well used to drill the conventional oil in the sandstone layers above.

The drilling or reuse of conventional wells for 'infill' opportunity (i.e., for resources not taken into consideration in the original *Field Development Plan – FDP*), applying combined exploitation methods (involving conventional and unconventional resources), and use of oil and gas knowledge base or heritage, will help in maintaining an energy mix while transiting to new sustainable energy.

The next part deals with well operations techniques necessary to produce conventional hydrocarbons, which is as well the base technology for the production of unconventional resources.

EXERCISES

1. a) What are unconventional hydrocarbons?
   b) List 5 examples of unconventional resources.

2. What roles do source rocks, reservoir rocks, traps and seals play in the formation and accumulation of unconventional hydrocarbon resources?

3. How do the following factors determine the viability of unconventional hydrocarbon resources?
   i. Deposit quality
   ii. Available budget
   iii. Technology required
   iv. Production type and distribution
   v. Human capital and organizational unit
   vi. Geology risk and environment
   vii. Geometry and resource extent
   viii. Final prospect economics and profitability

4. a) What is the most common source of biofuel worldwide?
   b) List three other sources you know.

5. Differentiate between oil sand and oil shale with respect to the formation of each resource.

6. a) What role or effect does porosity and permeability play in the formation of tight oil?
   b) List three countries with high volumes of tight oil.

7. How do prices for commercial exploitation of unconventional hydrocarbon resources compare to that of conventional? What reasons could contribute to your answer?

8. Briefly describe 5 methods applied in the exploitation of oil sand.

9. a) Which unconventional hydrocarbon resource is fracking most applicable?
   b) Describe how the method is used to achieve exploitation of the resource.

10. a) Describe the two coal liquefaction methods.
    b) Describe the processes of thermal depolymerisation.

REFERENCES & FURTHER READING

Altun, N. E., C. Hiçyilmaz, J. Y. Hwang, A. Suat Bağci, M. V. Kök, 2006. Oil Shales in the World and Turkey; Reserves, Current Situation and Future Prospects: A Review. Oil Shale. *A Scientific-Technical Journal. Estonian Academy Publishers.* 23 (3), p. 211–227.

Alvarez, J. and S. Han, 2013. Current Overview of Cyclic Steam Injection Process. *Journal of Petroleum Science Research.* Science and Engineering Publishing Company.

Attanasi, E. D. and R. F. Meyer, 2010. Natural Bitumen and Extra-Heavy Oil. *Survey of Energy Resources, eds., J Trinnaman and A. Clark: World Energy Council*, p. 123-150.

Butler, R., 1991. Thermal Recovery of Oil and Bitumen. *Englewood Cliffs: Prentice-Hall.* p. 104.

Cane, R. F., 1976. The Origin and Formation of Oil Shale. In Teh Fu Yen; Chilingar, George V. Oil Shale. Amsterdam: *Elsevier.* p. 1–12; 56.

Century, J. R., 2008. Tar Sands: Key Geologic Risks and Opportunities. The Leading Edge, 27(9), p. 1202-1204.

Choi, Y. J. and S. Y. Lee, 2013. Microbial Production of Short-Chain Alkanes. *Nature.* 502, p. 571–4.

Dusseault, M., 2002. Cold Heavy Oil Production with Sand in the Canadian Heavy Oil Industry. Alberta Energy.

Hammerschlag, R., 2006. *Ethanol's Energy Return on Investment: A Survey of the Literature 1999-Present. Environ. Sci. Technol. 40, p. 1744–50.*

Höök, M., D. Fantazzini, A. Angelantoni, and S. Snowden, 2013. *Hydrocarbon Liquefaction: Viability as a Peak Oil Mitigation Strategy. Philosophical Transactions of the Royal Society A. 372 p.*

Hutton, A. C., 1994. Organic Petrography and Oil Shales. Energeia. University of Kentucky. 5 (5).

iESog, 2019. Integrated Elvee Services Laboratory Analyses, *Unpublished Practical Manuals*, Enugu. 66 p.

Kelly, E. N., D. W. Schindler, P. V. Hodson, J. W. Short, R. Radmanovich, and C. C. Nielsen, 2010. Oil Sands Development Contributes Elements Toxic at Low Concentrations to the Athabasca River and its Tributaries. PNAS. **107** (37), p. 16178–83.

Mabro, R., 2006. Oil in the Twenty-First Century: Issues, Challenges and Opportunities. Oxford University Press. 368 p.

Mantripragada, H. and E. Rubin, 2011. Techno-Economic Evaluation of Coal-To-Liquids (CTL) Plants with Carbon Capture and Sequestration. *Energy Policy.* 39 (5), p. 2808–2816.

Mills, R. M., 2008. The Myth of the Oil Crisis: Overcoming the Challenges of Depletion, Geopolitics, and Global Warming. *Greenwood Publishing Group.* p. 158–159.

Nosike, L., 2017. Overview of the Petroleum Industry: A New Perspective in the Saga of the Oil and Gas Struggle. *Delizon Publishers,* 191 p.

Nosike, L., 2019. *A Career in the Petroleum Industry: In a Time of Energy Transition.* Delizon publishers. 253 p.

Smandych R. and R. Kueneman, 2010. The Canadian-Alberta Tar Sands: A Case Study of State-Corporate Environmental Crime" in R. White (ed.) *Global Environmental Harm.* Cullompton: Willan.

Speight, J. G., 1999. The Chemistry and Technology of Petroleum. CRC Press. 953 p.

Speight, J. G., 2008. Synthetic Fuels Handbook: Properties, Process, and Performance. McGraw-Hill Professional. p. 9–10.

Urov, K. and A. Sumberg, 1999. Characteristics of Oil Shales and Shale-Like Rocks of Known Deposits and Outcrops. Oil Shale. *A Scientific-Technical Journal. Estonian Academy Publishers. 16 (3 Special),* p. 1–64.

PART THREE

WELL GEOLOGY AND OPERATIONS

Well geology and field development turn focus away from the office exploration studies to rig site operations, and back again to base or office, where the results are analysed and interpreted. A rigsite is a location for drilling, completion and workover operations, where massive structures exist to drill hydrocarbon wells. Different types of well geometries are drilled for several kilometres to tap oil pools and gas caps. Likewise, different types of rigs are used depending on the nature of the field and a lot of rig site personnel are involved, working together to ensure successful drilling.

When planning a well, safety is the first thing to consider, in order to carefully drill through pressured formations, mobile shale and zones with fractures and faults. The personnel involved work together to isolate, avoid and minimize or manage any hazards. Before drilling commences, there is need to predict the abnormal pressures at depths, especially within the shales overlying the reservoir, in order to determine the drilling fluid mud weight necessary to safely drill to the desired targets. There is direct pressure measurements in the reservoir while and after drilling.

The monitoring team onsite, known as mud loggers, participate in the drilling for a proper follow up of the well. They do this by estimating and recording the depths, the mud properties, the drilling parameters and processes, and the gases arriving at the surface. They collaborate with the wellsite geologist to collect and describe cutting samples, carry out laboratory measurements, and prepare reports to be transmitted to the base in town. Gas monitoring while drilling is important as it can provide invaluable information on hydrocarbon presence, character of fluid, nature of seal, compartmentalization, and nature of the lithology in the subsurface.

CHAPTER ELEVEN

WELL OPERATIONS

11.0 Essentials of Well Operations

The wellsite geologist focuses essentially on well operations. The management in oil companies is most concerned about what goes on in the billion-dollar development and production facilities at the operations location. Whether a young geoscientist or a reservoir engineer, or the cost controller checking expense notes, understanding the well operations onshore and offshore will be the major task on the table.

Specific survival training courses are given to rig site personnel who must collaborate to ensure successful drilling in safe working conditions and in an environment-friendly way. The geologist, geophysicist, drillers and reservoir engineers work together to isolate, avoid and minimize or manage any hazards. Today, all eyes are on them to get it right regarding health, safety and environment conservation.

Wells could be vertical or deviated in which case they are near or horizontal at the drain sections. As such well planning involves using maps, different types of well geometries, studying the locations and considering the technicalities of drilling several kilometres in the subsurface. A drilling program contains well operations procedures involving drilling activities, casing sections and completion.

All the above form the bases of well geology, transiting from exploration and appraisal to development and production, which are treated in the following chapters.

11.1 Teamwork and Preparation

The first step in working in the well operations team in an oil company is to get used to relating with field personnel. Wellsite and operations geologists learn to work in an integrated team and people with field drive. They face people with different experience, likely from different countries and speaking in different languages and accents, but with same objective of carrying out the well operations successfully. They work on a rig as rotational staff (personnel that work sometimes at the rig and sometimes off work for weeks). Even if a member of staff visits the rig only occasionally, he or she needs to understand the psyche of those who work there for the major part of their career (Nosike, 2019).

In the wellsite and in the camps (onshore) or residential areas (offshore) where workers stay, there are different levels of hierarchy which are strictly respected. Among them are:

- *Company man*: who happens to be the oil company's representative on a drilling location;
- *Camp manager*: in charge of camp and staff management while overseeing the work of subcontractors working on the rig;
- *Camp supervisor*: responsible for fulfilling all duties as the Company representative for the camp, such as setting up camp facilities and its daily inspection;
- *Camp boss*: responsible for all hotel and catering services;
- *Tool pusher*: a senior administrative personnel in the drilling crew responsible for providing materials and skilled personnel for efficient operations to be carried out;
- *Material supervisor*: who has a knowledge of drilling materials, coordinating and controlling all activities associated with the materials.

In general, people are friendly on the rig site, due to the fact that they are isolated; everyone wants the operations to progress and to live in harmony. All the same, there are these orders and relationships. The new site geologist has to start adjusting to camp or remote life in preparation for oil field work. As the machines on the rig look bigger than the individuals there (Figure 11.1), there is need for a stronger bond with the people to adapt to the environment.

Synergy is created via teamwork, which allows for greater achievement by working together. Individuals focusing on machine while communicating with people require good relationship and experience to adapt to rig site conditions. While in office, personnel focuses so much on small machines such as computers and on individual colleagues. At the rig, personnel focus on big machines and few people – that creates the need for an extended teamwork.

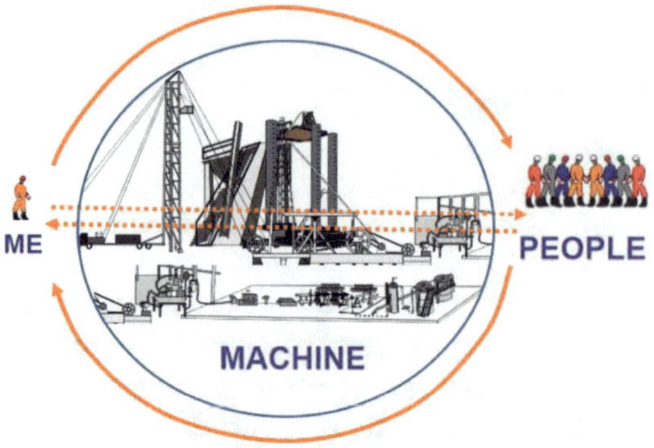

Figure 11.1: Extended teamwork emphasized when there is a relationship between people and machines

There is need to be knowledgeable and able to discuss, to study books covering the wide topic of geosciences and operations and undertake training courses that cut across domains and departments in the oil industry (Nosike, 2019). The geologist should be able to cooperate with others in order to understand what they do, and to serve beyond just a helping hand in well operations.

Types of Rigs

Preparations for well operation depend on the type of rig in question. The type of rig depends on the nature of the drilling intended (Gerwick, 1986; Hyne, 2001). Below are rig types and how they work.

1. **Heli rigs and truck mounted rigs** – light and made up of parts that can be transported by helicopter or trucks and coupled onsite.
2. **Fixed platform rigs** – stand on metal or concrete legs fitted at the base of the sea in near offshore locations.
3. **Drill barges** – used in shallow offshore locations or in swamp, where ballast (inflatables used for stability) keep them firm on the water bed.
4. **Jack-up rigs** –set in place by jetting the legs out until they reach the sea bottom while the platform itself remains above the sea level.
5. **Semi-submersible rigs** – partially inserted in the sea body by a combination of floatation and slight immersion by their weight. The pontoons (airtight hollow float structures) lower down the rig and several anchors provide buoyancy and maintain it in place.
6. **Drill ships** – ships adapted for drilling purposes in deep offshore where dynamic positioning systems control the thrusters that keep the ship in a fixed position. They counter the effects of water and wind currents, so as to stabilize the ship.

The fixed platform rig is suitable for shallow offshore setting. Figure 11.2 shows close-by landmass and vessels including helicopter on the helipad and supply boat on the open sea.

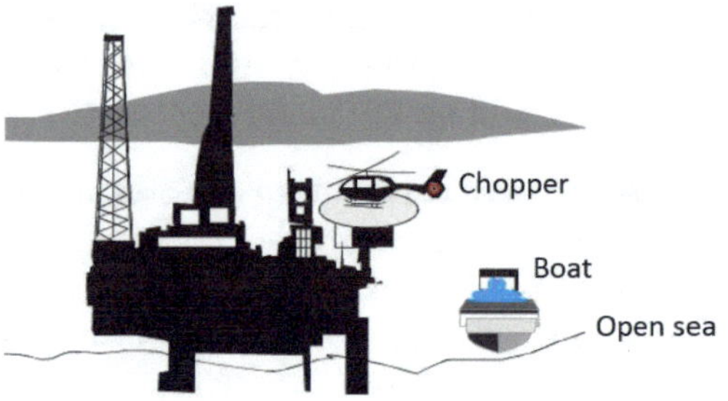

Figure 11.2: Illustration of rig operations by a fixed platform rig

Other types of rigs are shown in Figure 11.3. Figure 11.3 a. to c. shows a semi-submersible rig, a jack-up rig and a drillship respectively. A semi-submersible rig consists of an operating deck located above the sea level, with its other end submersed below. This provides more resistance to wave action than normal ships due to the good stability of its design. They are suited for offshore drilling operations, just as the drillship, where fixed structures such as the jack-up rig and near-onshore rigs are not suitable. The jack-up rig can self-elevate by raising either of its long support legs upon getting to the drilling location; the legs are jetted out until they rest on the sea bottom, after which the structure is raised to a particular height above the water to prevent tidal and wave actions on the structure.

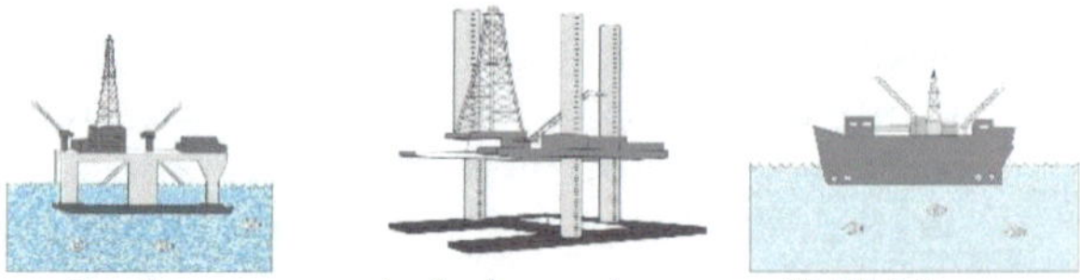

a. Semi-submersible rig b. Jack-up rig c. Drillship

Figure 11.3: Types of rigs (modified from Exploration Logging Inc., 1979)

A drillship is a conventional ship with a drilling rig mounted in the centre. The ship sails to the drill site and then passes its drill strings through a hole in the *hull* (the frame or body of a ship), called the *moon pool*, through the water or *marine riser* to drill down the seafloor. The ship floats over the drill site until drilling is completed. Going to drill in a drill ship requires the most intense preparation, as it implies going offshore for several days. The personnel will have to adapt for sailing in a ship as well as for survival in case of helicopter crash, if going by helicopter.

Rig Jobs

Another factor to take into consideration in preparing for a field mission is the type of job. Apart from the actual drilling, rig site work involves a period of cementing and waiting on the cement to set. There are also **workover** operations aimed at servicing and maintaining production of oil and gas wells. *Workovers*, otherwise known as well interventions, are done later or years after a well has been drilled. In some cases, a smaller equipment known as *service unit* or *workover unit* is taken to the rig site and mounted for the minimal operations.

A service, well servicing, or pulling unit uses hoisting equipment mounted on a truck body or on a trailer to raise and lower equipment into the well during a workover. A winch and mast system is mounted on the truck or trailer. The winch is driven off the truck engine. The hoisting only ensures that tools are lowered and brought back to the surface. This particular type of well intervention is also known as **rigless operation** (Figure 11.4).

Figure 11.4: Rig site service unit

In a rig, the hoisting and drill string form the main rotary system that is used to effect the torque and pressure necessary to break rocks while drilling, as well as carry out the circulation with drilling fluid. This is a process the geologist monitors closely. The drill string is the lower part, which carries the bit and drill parts to the bottom of the wells. The upper hoisting carries the swivel (a mechanical device responsible for suspending the weight of the drill string while it rotates) and the travelling block above the kelly (a base level on the drilling floor used to transmit rotary motion from the rotary table to the drill string as it is moved up and down). Rig hands work essentially around the hosting system and they are trained to handle equipment and have to be safety conscious.

The rotary system (Figure 11.5) is the rotating parts, used to produce torque and pressure necessary to break rocks while drilling. The hoisting system consists of crown and travelling blocks, hook, swivel and draw works responsible for the up and down movement of the drill string. The drill string consists of the drill pipe, drill collar and drill bit, primarily responsible for actual drilling into formations

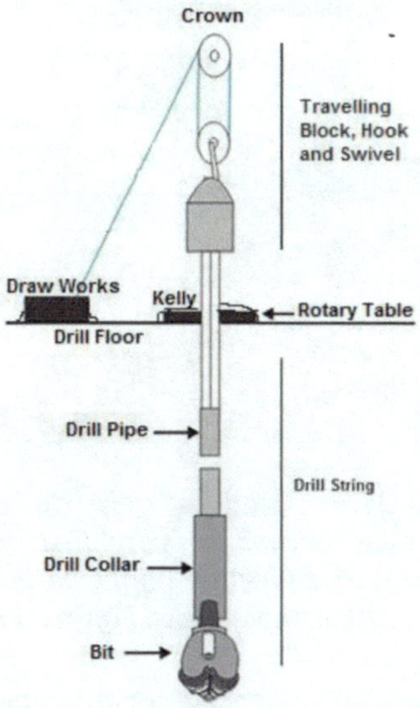

Figure 11.5: Hoisting and drill string showing the rotary system

A full rig is made up of several components, which can be grouped under:

- Power generating system
- Hoisting system
- Drilling tools
- Rotary systems
- Sensors and computing systems
- Drill floor and platform
- Circulation system

Only a few of these are important in the duties of the geoscientist to understand (some are more important to drillers, some to the mud loggers, and others to other personnel working on the rig site), and even fewer are shown in the figure below in order to focus on the circulation system. The circulation system is important for monitoring the well and the formation, which is a core task of geologists.

Rig components shown in Figure 11.6 include the derrick, top drive, drill pipes laid out for serial attachment to the drill string, power generator, and a blowout preventer to shut-in the well in case of a kick that may result in a blowout. Others include the well casing, drill string, drill collar and drill bit, all located below the surface

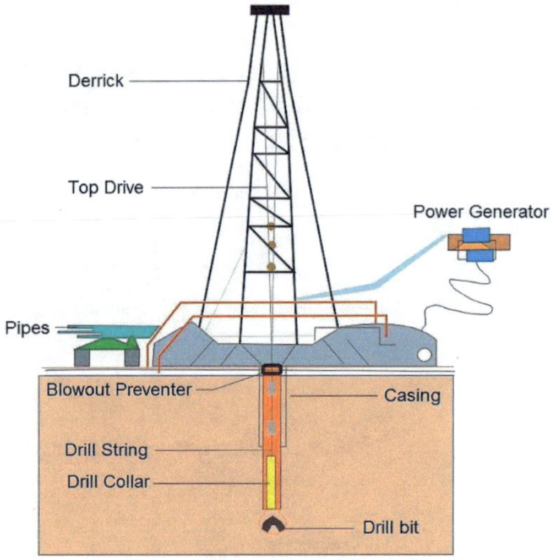

Figure 11.6: Sketch of few rig components and subsurface assembly (simplified to show mainly aspects whose function might be in the scope of the geoscientist; compare with Figures 11.5 and 11.7)

The power system generates energy to operate the rig; the circulation system ensures that the well is fluidified, including the surrounding of the drill string, so that drilling tools can rotate in safe conditions. The mud that flows in and out, among other functions, lubricates the drill bit and helps to bring cuttings to the surface. Most logging systems used to evaluate petrophysical properties of formations operate within the drill fluid (mud) and send impulses through them. Figure 11.7 is a sketch of the mud system, simplified to show essentially the closed system flow. This closed system helps the geologist and drillers to monitor *losses* and *gains* while drilling. The mud pit is continuously monitored for gains or losses by constantly monitoring the volume of drilling mud in circulation; where the volume is higher or lower than that circulated, it could indicate possible gains and losses respectively.

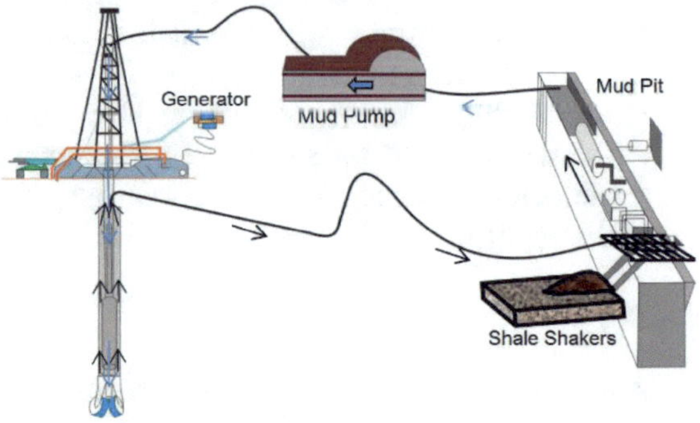

Figure 11.7: Flow pipes and closed circulation showing the principle of the circulation system on the rig

The mud is a heavy viscous drilling fluid mixture used to facilitate the drilling process by stabilising the formation during drilling, while cooling the bit and sending rock cuttings to the surface. The mud pump is used to send mud into the formation through the hollow drilling pipe to the nozzles at the drill bit from where it is circulated. The drilling fluid from the well on getting to the surface is first filtered at the shale shakers (then goes through other filters known as desilters and desanders, to remove silt and sand fractions respectively) before it is pumped back to the hole by the mud pump(as shown in Figure 11.7). A degassing system removes excess gases.

The drill cuttings collected over the shale shakers are washed and analysed under the microscope by mud loggers and wellsite geologists. A *mud logger* is the personnel responsible for monitoring drilling operations and the accurate collection of samples at defined intervals during drilling. The samples would be subjected to different detailed onsite and laboratory analyses. This requires major preparation of bags, chemicals and lab equipment. Somewhere near the shaker is a gas sampling system, a degasser, to collect gas samples that would be analysed at the mudlogging unit to gather additional information about the well as well as the geologic formations encountered. The nature of the work and environment comes with some risks, physical and emotional.

Working offshore is probably one of the most challenging aspect of oil field work, in terms of separation from city life, but the conditions in a drillship may be more comfortable than on land rigs. Land rigs serve essentially for drilling purpose (sometimes without integrated residential accommodation), since the personnel there are on land and have some access to normal life on land (often in a close by residential camp). Whether on a land rig or drillship, the many responsibilities conferred on the personnel running it are enormous and demand sensitive attention to work, life and environment.

11.2 Health Safety and Environment (HSE)

The oil and gas industry is very competitive and faced with a continuously and now threatening demand for transparency and accountability for the impact of its activity on the environment. In an international setting where different people, cultures and technical background collaborate to carry out this activity, it is important that there are laid down specific and strong checks on heath safety and environment.

Onsite, an HSE supervisor will induct the new personnel on arrival and provide a time or roaster "T-card", and other information needed while on site. A T-card is a time card used to indicate in- and out-ward bound movements of personnel to and from the rig or campsite. It is easy to use, requiring little or no training. This is important for ensuring that everyone is accounted for in case of an emergency or need to evacuate the site. During drills, everyone has to remove, turn or displace their T-card to another board to show they are all safe and sound.

Safety culture is one thing every personnel needs to fully imbibe to work in the oil producing environment. There is need to be conscious of hazards and risks. Many trainings would be provided to the personnel in this regard before going onshore or offshore to work. Each personnel will have to obtain, or would be provided with Personal Protective Equipment (PPE). This is usually a hard hat / safety helmet, coverall, boots, earplugs, gloves and goggle. The wearing of PPE is strictly enforced on the rig.

Hazards, Risk Control and Maintenance

Hazards refer to situation that has the tendency to cause harm. The well operation site is a hazardous environment with many risks (Schempf, 2007; Krah, and Unger, 2013; Nosike, 2017). These risks range from those causing personal injuries to those brought about by environmental hazards such as bad weather, heat and noise stress. Among the common hazards on the rig are:

- Jack-up rig leg failure
- Lifting operation with a risk of falls
- Blow Out Preventer (BOP) failure
- Simultaneous Operations (SIMOPs) with risk of interference
- Rig floor operations requiring manipulation of heavy moving metals.

Hazards specific to the offshore wellsite include helicopter crash, high water waves, and the consequences of accidents such as fire or explosion having higher gravity (Silcox et al., 1989; Leffler et al., 2011). Safety training for offshore rig work should include a basic safety induction for personnel new to the offshore oil industry, such as the *Basic Offshore Safety Induction and Emergency Trainings (BOSIET)*, which includes what is probably the most important training for going offshore: the Helicopter Underwater Escape Technique (HUET). This popular technique involves escaping from a capsizing helicopter underwater.

Part of the training involves each trainee jumping into water and eventually regrouping in a huddle, to simulate what is required of them to do while waiting for rescue in a case where they abandoned ship and lifeboats are insufficient. The huddle posture helps to lessen loss of body heat, and enables collective rescue of the group rather than searching for individuals. Figure 17.8 shows the part of the training requiring that personnel are dipped into the water in a helicopter-shaped cabin, simulating the ditching of a helicopter. The cabin is then rotated under water while the trainees are required to unbuckle, jack-off the cabin window, leave cabin, inflate the life jacket and swim to escape.

Also included sometimes for offshore workers is the *Totally Enclosed Motor Propelled Survival Craft (TEMPSC)* training, aimed at mass evacuation of personnel on board in case of an emergency. Other training courses are related to firefighting, first aid and self-rescue. Another general training is the *Minimum Industry Safety Training (MIST)* aimed at introducing workers to fundamental safety elements required to function effectively in the oil industry, especially for those new to the industry.

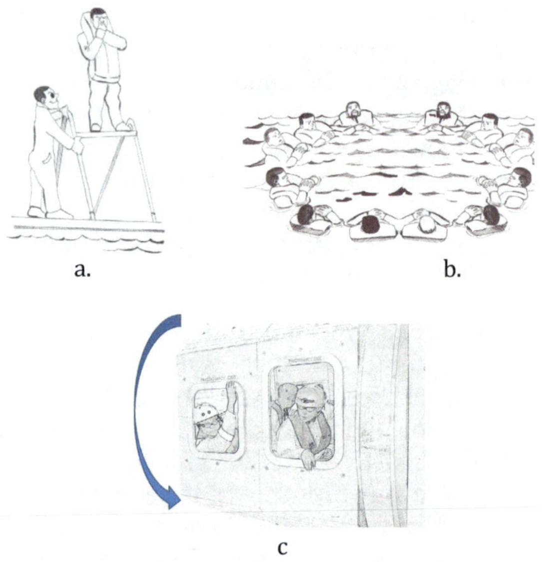

a. A trainee jumping from a height into the water
b. Trainees performing the huddle posture in a water pool
c. Helicopter Underwater Escape Technique (HUET)

Figure 17.8: Offshore survival training

Targeted training is provided for those who may work with H_2S or other dangerous gases. There are guides on these trainings, locations and their certifications in different regions (Nosike, 2017; iESog, 2019). The aims of most of these training events are to:

- Assess the Risk (awareness)
- Prevent it (put measures in place)
- Control it (where it occurs or is inevitable)

Risk awareness campaign cover:

- Chemical
- Gravity
- Electrical
- Noise
- Atmosphere
- Confined space

The personnel could contact the personnel *Responsible for Safety and Environment on Site (RSES)* on arrival at the rig for more information and on-the-job rig training on each of the mentioned risk areas. Despite the presence of trained safety officers, ultimately everyone is responsible for own safety while working on the rig site. The employer can do only so much to provide a safe working environment, but cannot necessarily ensure protection from hazards, and so the employee is responsible for his or her own safety by observing safe culture and safe work practices, one that personnel can follow to improve work safety and performance.

To control the hazards, there are step by step priorities, starting from controlling to maintaining to managing a hazardous situation. The steps are:

- Controlling
- Eliminating
- Reducing
- Isolating
- Use of engineering control
- Communicating safety measures
- Rescuing in case of accident
- Maintaining the preventive measures
- Managing risk and learning some lessons when it occurs

The staff is expected to follow the *Employee Reporting Structure* (ERS) to communicate safety observations on the rig, in a hierarchical order from the employee to top management. However, it is more important to ensure that an accident is prevented than to wait to report it through the employee's hierarchy. Other immediate reporting systems on the rig are the **STOP** and the **SET** cards. The *STOP: Safety Training Observation Programme* uses a card in which what is observed and the solution the observer has proposed are stated. This is stacked in boxes and reviewed daily by the HSE personnel. They are read out in morning safety meetings. Awards are sometimes put in place to encourage participation in regular filling of the STOP card. The *SET: Safety Effective Trouble-free* operations, is a variant of the STOP, and there are others.

In addition to the safety measures mentioned above, a basic knowledge of First Aid is necessary for working on a rig site. Whether onshore or offshore, learning *Cardiopulmonary Resuscitation (CPR)* procedures, treating of basic injuries and getting certifications to practice such could come in handy in emergency situations, as illustrated in Figure 11.9. Personnel should learn the basics of firefighting, types of fire extinguishers and how they work. Climbing skills and use of protective equipment require training as well. Working at heights requires protection and use of harness (that is, a kind of support consisting of an arrangement of straps used to attach the body to an object) to prevent falls. These are common challenges on a rig.

Figure 11.9: Man working at height, requiring protection and use of a harness to prevent fall

It is important that personnel are trained to offer emergency CPR procedures in case of emergency and the victim is unable to breathe (example following a fall as illustrated in Figure 11.9). CPR is given to injured personnel, which is an emergency procedure involving heart compressions and sometimes mouth-to-mouth resuscitation to allow the injured regain consciousness or preserve brain functions until further measures are taken to stabilise the condition of the injured.

Personnel that will work only onshore will get onshore safety training on such drills as forest driving on SUVs or on surviving insect and snakebites. The experience gained during these trainings will not be regretted, and the lessons learnt from them will help with safety in personal day to day activities. Each personnel will need to undergo an annual medical aptitude test to check hearing, sight and general health.

Environmental Challenges Onshore and Offshore

While developing wells onshore and offshore, environmental issues that could arise include:

- Accidental discharge of waste onshore or offshore
- Sea life segregation and agglomeration (destabilization of natural habitats)
- Regulations of the administrative authorities and government

Local inhabitants near an onshore rig may benefit by increase in sales at the local market, hotel or logistical services. However, the forests, wildlife, and subsurface water and other resources may be adversely affected. Regulations set by government may not be fully adhered to, especially where agencies or authorities responsible for ensuring that such are fully enforced neglect their importance.

11.3 Well Planning and Development

After appraising a prospect, the well is developed in order to start producing hydrocarbons. Several technical and logistics meetings known as *Well Statement of Requirements* (SOR) are held before drilling the well. Well planning and development are the aspects of oil business that take into consideration the oil and gas market trends, including oil price and improving recovery technology that can make a given well either economical or not, for a given time. A secondary recovery method is any method that helps enhance the production. An example is the injection of water just below the oil water contact, in order to add pressure to drive the hydrocarbon upwards (Figure 11.10). Where a well is not supported in any way, it is said to produce by *natural depletion*.

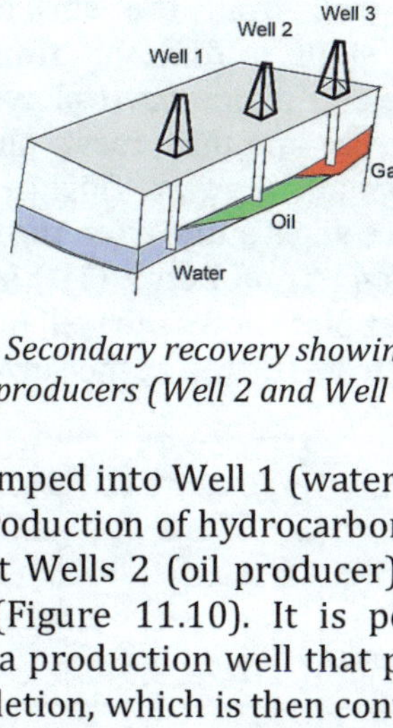

Figure 11.10: Secondary recovery showing the injector (Well 1) and producers (Well 2 and Well 3)

Water is pumped into Well 1 (water injector) to drive the production of hydrocarbons overlying the water at Wells 2 (oil producer) and 3 (gas producer) (Figure 11.10). It is possible that Well 1 was a production well that produced by natural depletion, which is then converted to an injector and flooded with water after initial hydrocarbon recovery, to drive hydrocarbon production in other wells. Water support is so common that it is considered a basic production requirement in most oil fields and not an enhanced recovery mechanism.

Predictions or conclusions reached in the exploration phase will be confirmed or invalidated during appraisal and production; information obtained at the exploration stage will likely not be sufficient to understand the entire oil field (Baars et al., 1989; Economides et al., 1997; Heshelow, 2010). Additional results at the appraisal and production stages can help redefine subsurface parameters and rock properties, for better understanding of the field. One of the reasons for this, apart from discrepancy between model and real life scenario, is that most exploration wells are vertical. Once the oil has been discovered, several sub-vertical and near horizontal drain sections will be used to produce the field (Gatlin, 1960; Schempf, 2007; Li and Dai, 2008.). It is necessary to ensure that the well is properly positioned to meet the target, following the most effective and yet cost effective trajectory. This requires a proper mapping of the development area, and understanding of the map by the well operation personnel. A guide to achieving this is in the well program that will be provided to the geoscientist on board the rig.

The initial **Well Profile (WP)** is decided during the well SOR and included in the **Well Location Report (WLR)**; the final one will be generated as the well is drilled. WP is a plot of the well trajectory from the surface closure to the true vertical depth (the vertical distance of the drilled well from the surface to the point drilling stops is different from the measured depth along a non-vertical well). The graph usually has another inset plot, showing the *northings* and *eastings* (Figure 11.11). The WP therefore gives a line from the Well Head (WH) to the final Total Depth (TD) in a section view. The inset plot shows the well in a plan view as if looking from the top at the surface.

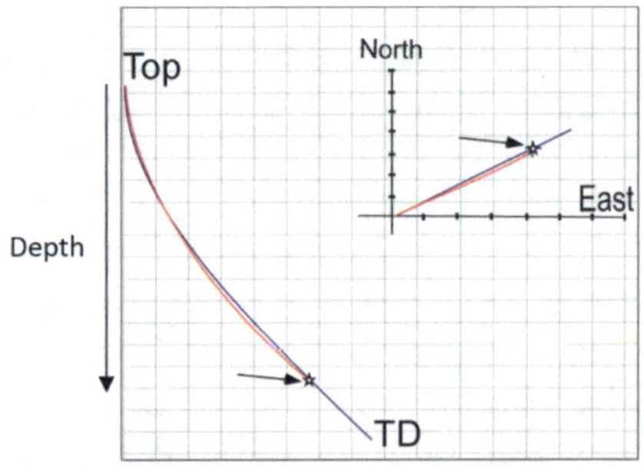

Figure 11.11: Well profile and inset plot showing trajectory

The WLR contains the necessary information for drilling the well, such as the elevation (of the rotary table) to use for drilling, expected depth and maximum inclination in the case of a deviated well (see Appendix I for more information on WLR and general wellsite reporting).

The larger curve shows the well from top to the intended final TD. The external axes should be depth for this larger curve. The inset curve shows the sample well trajectory from a plan view, plotted on the northings and eastings axes. On the curves, the star shows the point of actual drilling. The driller is not doing so badly, as the red line (actual trajectory) is close to the blue (prognosed trajectory). The wellsite geologist obtains these curves by plotting the actual surveys against the prognosed survey already on a spreadsheet. It gives a good indication of the fidelity to the intended well path. A wrongly followed well path could lead to missing the reservoir target, which in turn delays drilling activities and increases operation cost.

Types of Well

There are essentially four types of wells, depending on the primary and secondary objectives of the well. These are:

1. *Production well:* for producing hydrocarbons
2. *Injection well*: for injecting fluid as in Enhanced Oil Recovery (EOR)
3. *Service or utility well*: for observation, waste treatment or disposal
4. *Relief well* : in case of kicks and other down hole–pressure related hazards

In Figure 11.12, a story can be narrated of how Well 1 was drilled and supported by Well 2. However, water injection in Well 2 and perhaps charging from the left / hydrodynamics led to overpressure and spill to Well 3 (dotted blue line shows possible overpressured aquifer spilling from the anticline on the left to the right). When this was observed in Well 3 as a kick, it was too sudden and discovered late, making it impossible to control. A decision was made to plug and abandon the well as soon as the well was brought under control. A relief Well 4 was drilled to reduce the pressure in the overpressured formation before controlling and plugging Well 3 for abandonment. These scenarios depending on field development strategies are foreseen and possible mitigation measures planned before field development.

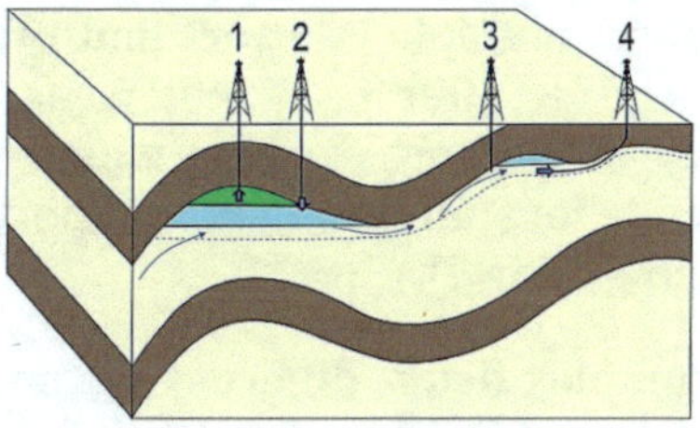

1-Production well; 2-Injection well; 3-Utility well; 4-Relief well

Figure 11.12: Exemplified types of well with a conceptual relationship among them

The type of well and the objectives will determine the various well operations and personnel. Oil and gas services companies propose their services to the oil company who hires them for casing operations, cementing, mudlogging, petrophysical logging, coring, waste management, and many more services, including catering and recreation. The most important of these is the drilling service company, whose rig and drilling equipment are used. The operations geologist, with management teams and the local authorities may call for tenders (in this case a formal proposal to service firms and vendors to bid for the job) and will meet with the contractors, whom after inspection and negotiation will be selected for the contracts. Well planning is an extensive project that goes from the technical to logistics and wellbeing.

Well Information and Headers

A name is given to each field and numbers of wells planned out to drill the various reservoirs. Field A will have wells A1 to A100 for example, depending on the number of wells planned. There is a **Well General Information** section on the document, together with the Well Location map, which gives the exact *well head coordinates* (X = and Y= in meters). The *Total Depth (TD)* is also given on that section. These are also included on the Well Headers, found in most documents used during the well development, as shown in Table 11.1. Other information in the header is the survey measurements, which will be discussed, in a later chapter on drilling and geosteering.

The **well map**, usually an ***isochron*** map with interpretation of structural and sedimentary features, presents the reservoir level of interest, highlighting previous well positions (if this is not the first well), faults, compartments and the targeted oil pool zone. An isochron map is a map that shows different seismic transit time (usually with depth equivalents and contours) by colour variation in which the same time / depth shows the same colour.

11.4 Well Clusters and Surface Geometries

When a field is developed, the wells are planned in such a way that some will serve as producers and others as a support system. The supporting wells are used for secondary recovery, which becomes important after the beginning or later in the life of the field. Provision has to be made for this right from the onset. There are essentially three types of arrangements:

1. Central support system
2. Line support system
3. Array of support systems

There are different kinds of well clusters used to achieve this. They form different geometries, some of which are shown below. In **central support**, injectors are placed at the peripheral water contacts to drive oil to a central producer. In a **line support**, producers and injectors are lined opposite each other. The **direct support** is a kind of line support used in deep offshore turbidite where the injector is placed within the channel fairway of the producer for example. Figure 11.13 shows the different types of supports.

Table 11.1: Well header information

Wellhead relative coordinates Easting(m) Northing(m)	TD:	Depature at TD: inc at TD: azim at TD:		TD coordinates X: Y:		COORDINATES X	Y	Eastg	Northg
Calculation Method:						SURVEY DATA			
Surv N	Survey Tool Type	Survey Depth (m)	Inclin Angle (deg)	Hole Azimuth (deg)	Course Length (m)	TVD RKB (m)	TVD subsea (m)		TVD subsea (m)

The nature of support will depend on the ease of the hydrocarbons to flow – compartmentalization of the reservoirs, the fairway trend and the available reserve. In the example in Figure 11.13 c and e, the oil producers are in green while the injectors are in blue. The array of support systems must respect the minimum spacing and be drilled in designated well slots (Figure 11.13e).

The geometries in Figure 11.13d is an **array of support systems** showing the different options, a combination of the central support, line support and direct support. The blue line/geometry shows a multi-injector well support system for the well at the centre (but it may as well support other wells beside it). The red and yellow lines/geometries show several wells, some of which are injecting while others are producing. Controlled injection and production are practiced when (production / injection) rate monitoring is important to avoid flooding the wells, in situation where the producers are drilled alongside the injectors or drilled as infills later in the life of the field.

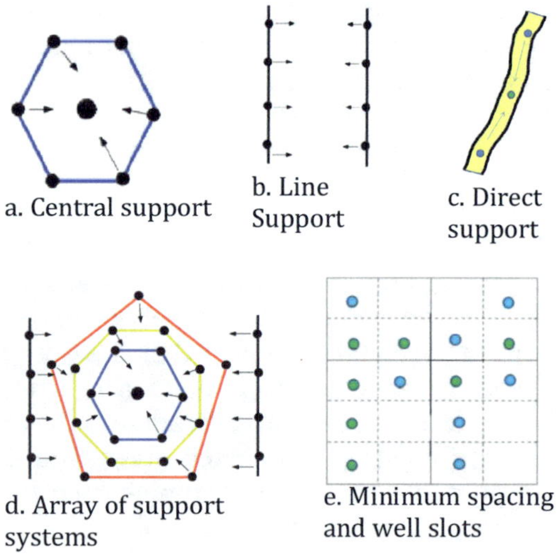

Figure 11.13: Well clusters showing different geometries to enhance drive

It is possible to convert a producer well to an injector, when it becomes flooded with water and drill an attic well (closer to the crest of the structure, to target the *attic oil*) as producer. It is sometimes easier to move to a new slot where available, or to the closest and suitable location, and drill a new well, after considering minimum authorized spacing (Figure 11.13e.). Drilling manifolds are made available in those slots to be used when needed. Manifolds are junctions with chambers where many pipes for drilling and production purposes meet. In old oil field, multiple injectors and producers have been added after many infill wells, leading to an array of support systems where the injectors provide overall pressure support and sweep (Figure 11.13d).

In complex fields, different system clusters are used to form an array with several spots. The more complex and closed up a cluster, especially in closely compartmentalised reservoirs with several well penetrations, the more the need to use strict geosteering techniques to ensure that there is no collision of wells. When more wells are drilled while some are already producing, a special precaution is used in what is called a Simultaneous Operation (SIMOP). Sometimes a nearby well will have to be shut while drilling another. Crane, boat and helicopter operations offshore become more demanding in terms of operations monitoring and management (Silcox et al., 1989; Tanaka et al., 2005).

Once the well location has been surveyed and confirmed, an initial excavation is made, known as *cellar,* usually dug out and cemented below the rig for major wells (Figure 11.14). The cellar serves as container for mud returns while drilling the initial hole before surface equipment and the wellhead are implanted on it. It may also serve to retain the BOP stack.

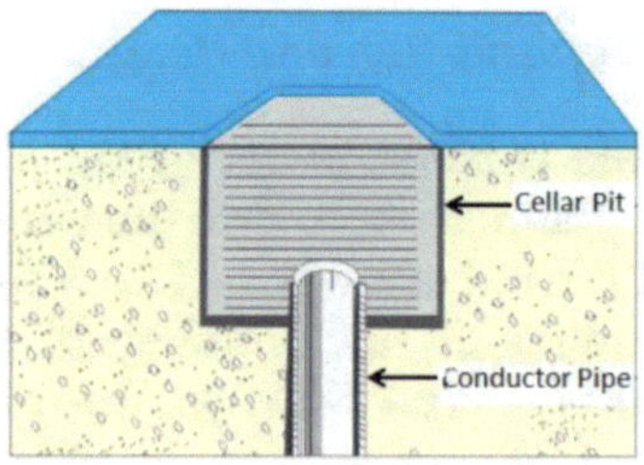

Figure 11.14: Cellar pit showing a conductor pipe serving as the first casing of the well

The conductor pipe, the short length but largest and first casing, is sometimes driven into the ground and may be anchored on the cellar deck. At this stage, the geoscience and engineering teams are ready to drill, except the pore-pressure of the formation for safe drilling is yet to be determined.

EXERCISES

1. a) Differentiate between "primary" and secondary recovery techniques.
   b) Describe briefly the various enhanced recovery methods.
   c) What is a Well Statement of Requirements (SOR)?

2. At what stage or stages are the predictions or conclusions reached in the exploration phase confirmed or invalidated?

3. a) There are essentially four types of wells. Briefly describe them.
   b) What roles do supporting wells play during production?
   c) List the three types of arrangements for a support system.
   d) Use diagrams to explain the roles of an array of supporting systems.

4. How do personnel prepare to carry out well operations? What is the importance of teamwork?

5. What is a rig? Differentiate between the following types of rigs:
   a) Heli rigs and truck mounted rigs.
   b) Fixed platform rigs.
   c) Drill barges.
   d) Jack-up rigs.
   e) Semi-submersible rigs.
   f) Drill ships.

6. Use diagrams to show the various components of a drill rig.

7. a) Why is it important to carry out Health, Safety and Environment (HSE) activities in the industry?
   b) List the various items that make up Personal Protective Equipment (PPE).

8. a) What is a hazard?
   b) List common hazards that could occur on the rig.
   c) What are the necessary steps taken to control or manage a hazardous situation?

9. a) Give two examples of safety training courses for offshore rig work.
   b) What is the importance of these training courses?

10. How is safety observations communicate on the rig? What are the environmental challenges faced both onshore and offshore?

REFERENCES & FURTHER READING

Baars, D. L., W. L. Watney, D.W. Steeples, and E. A. Brostuen, 1989. Petroleum; a Primer for Kansas (Educational Series, no. 7 ed.). Kansas Geological Survey. p. 40.

Bourgoyne Jr., A. T., K. K. Millheim, M. E. Chenevert, and F. S. Young Jr., 1986. *Applied Drilling Engineering*. Richardson, Tx, USA: Society of Petroleum Engineering.

Economides, M. J., L. T. Watters, and S. Dunn-Norman, 1997. Petroleum Well Construction, John Wiley & Sons. Duncan, Oklahoma: Halliburton.

Exploration Logging Inc., 1979. Field Geologist's Training Guide: An Introduction to Oil Field Geology, Mud Logging and Formation Evaluation, Sacramento.

Gatlin, C., 1960. Petroleum Engineering, Drilling and Well Completions. Englewood Clifs, N.J., USA: Prentice-Hall.

Gerwick, Jr. B.C., 1986. *Construction of Offshore Structures*. Baffins Lane, Chichester, UK: John Wiley & Sons.

Heshelow, K., 2010. Investing in Oil and Gas: The ABC's of DPPs. Iuniverse. p. 52.

Hyne, N. J., 2001. *Nontechnical Guide to Petroleum Geology, Exploration, Drilling, and Production*, PennWell, 598 p.

iESog, 2019. Integrated Elvee Services Laboratory Analyses, *Unpublished Practical Manuals*, Enugu. 66 p.

Krah, J., and R. L. Unger, 2013. The Importance of Occupational Safety and Health: Making for a "Super" Workplace. National Institute for Occupational Safety and Health.

Leffler, W., G. Sterling, and R. Pattarozzi, 2011. Deepwater Petroleum Exploration & Production: A Nontechnical Guide. Tulsa, Okla.: PennWell Corp.

Li, Z. and J, Dai, 2008. Controlling Technique on Navigating Path of Horizontal Butt Well and Cross-Well. Natural Gas Industry; 21(2), p. 70-72.

Nosike, L., 2017. *Overview of the Petroleum Industry: A New Perspective in the Saga of the Oil and Gas Struggle*. Delizon publishers. 191 p.

Nosike, L., 2019. *A Career in the Petroleum Industry: In a Time of Energy Transition*. Delizon publishers. 253 p.

Schempf, F., 2007. Pioneering Offshore: The Early Years. [Tulsa, OK]: PennWell Custom Pub.

Silcox W. H., J. A. Bodine, G. E. Burns, C. B. Reeds, D. L. Wilson and E. R. Sauve, 1989. Offshore Operations. *Petroleum Engineering Handbook – Chapter 18,* USA: Society of Petroleum Engineers. p. 18-1 – 18-52.

Tanaka, S., Y. Okada, and Y. Ichikawa, 2005. Offshore Drilling and Production Equipment, in *Civil Engineering*. In *Encyclopaedia of Life Support Systems (EOLSS)*, EOLSS Publishers, Oxford, UK.

CHAPTER TWELVE

PORE PRESSURE PREDICTION AND INTERPRETATION

12.0 Pressure Prediction

As part of well planning, there is need to predict the subsurface pressures that will be encountered along the well paths during drilling operations, across strata and within formations. Drilling commences after geopressures have been predicted. The wellsite, operations, development and asset geologists need to understand pressure relations, how to plot pressures with depth and represent them graphically. Pressure prediction requires the understanding of basin temperature gradient and burial history, and the use of offset well logs where these are available. Pressures in shale or non-reservoir sections are estimated using different algorithms, but pressures in the sandstone or reservoir sections are measured or predicted based on pressures calculated for the overlying and underlying strata.

Pressure estimation starts before, continues while and after, drilling a well. The predicted pressures for each formation are used to prepare the well architecture and to plan the drilling mud densities. Other data obtained while drilling, such as gas values, drilling parameters, and logs, help to confirm prognosed pressures. The most accurate pressure value is however the one measured using Well Formation Tester (WFT) tools, but these require proper interpretation as there may be cases of no sealing, tight formation, probe plugging, gas trapping and supercharging, which are all related problems discussed in this chapter. These tools are also able to sample formation fluids to give precise petroleum results of a well. The next sections look at how the pressures in the non-reservoir and in the reservoir sections behave, and how they are estimated. Pore pressure application in drilling is discussed in this chapter, alongside related technologies, hydraulic fracturing, stress tests such as Formation Integrity Test (FIT) / Leak Off Test (LOT), and well stability.

12.1 Pressure Basics

Pressures in the subsurface, at the depth of petroleum generation and storage, are sometimes higher than normal and usually referred to as *abnormal pressures* (Swarbrick and Osborne, 1996; 1996). **Geopressure** refers to abnormal pressures, usually *overpressures*, occurring in deep geologic units in the subsurface. An **overpressure** is any pressure that exceeds the **hydrostatic pressure**. The **hydrostatic pressure** is the pressure that a column of water (or salty water known as brine in the oil industry) will exert at any depth. Sometimes a pressures below normal, or below the hydrostatic, occurs due to depletion and this is known as *underpressure* or *subnormal pressure* (Whittaker, 1991; Gluyas, and Swarbrick, 2003).

Pressure (P) is the application of Force (F) over a given Area (A), as shown in Figure 12.1. When acting on a plane, or deforming a rigid surface, the term *stress* is used.

$$P = F / A \quad \text{.......................................} 1$$

Force is given by Mass (m) * Acceleration due to gravity (g)

So,

$$P = F / A = (m * g) / A \quad \text{....................} 2$$

To relate this to Density (ρ) which is a common property of interest in fluids, consider that density is the ratio of the Mass (m) to its Volume (v),

$$\rho = m / v \quad \text{.......................................} 3$$

As such Mass = density * volume

Or,

$$m = \rho * v$$

Replacing m in equation 2 above,

$$P = \rho * v * g / A \quad \text{..............................} 4$$

(where Area is *length (l) * width (w)* and Volume is *length (l) * width (w) * height (h); l*w in Area cancels out that in Volume to leaves only height h*)

That is to say that,

$$P = \rho * g * \cancel{lw}h / \cancel{lw} \quad \text{..........................} 5$$

Therefore,

$P = \rho * g * h$ 6

The main expression needed to start and understand pressure studies is:

$$\mathbf{P = \rho * g * h}$$

This means that pressure is a product of the effect of gravity (g) on the density of the fluid (ρ) with depth (h). The ρ symbol is pronounced "Rho".

Gravity is the force of attraction of the Earth's mass for bodies near its surface, while density is the measurement that compares the mass of a substance to its volume. The depth is the distance of interest below the surface at which pressure is to be determined.

When acting in fluids (liquid or gas) this force (blue arrows in Figure 12.1 a.) is referred to as pressure; it is referred to as stress (black arrows in b.) when acting on a plane or having an impact on solids, where the property of the solid may be affected by the pressure.

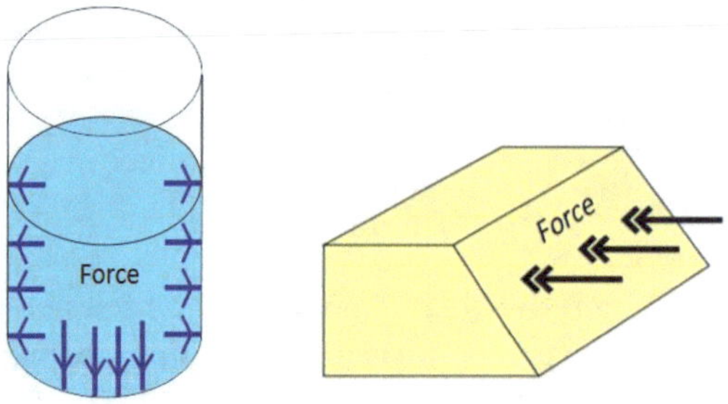

a. Pressure in liquid   b. Stress on a plane

Figure 12.1: Force acting over a given area to produce pressure and stress

Pressure gradient is the change of pressure over a given depth. This is the effect of gravity (g) on the fluid density (ρ). It can be obtained by multiplying the density of the fluid (ρ) with the gravitational constant (g)

Pressure gradient (PG) = $\rho_{\_fluid} * g$

$= P/\Delta Z$

(compare with: $P = \rho * g * h$, where h is Z)

Where Δ is a change (*which may also represented by "d" elsewhere*). The density (of drilling fluid / mud weight) equivalent of pressure expressed for a given depth is known as the Equivalent Mud Weight (EMW). That is, the EMW is the total amount of pressure at any given depth expressed in terms of mud density exerted at a true vertical depth.

Pressure gradients may indicate the type of fluid at a given depth based on its density (Figure 12.2). Density of gas is about 0.1 - 0.3 g/cc; density of oil is about 0.4 - 0.9 g/cc; while that of water is about 1.0 - 1.2 g/cc.

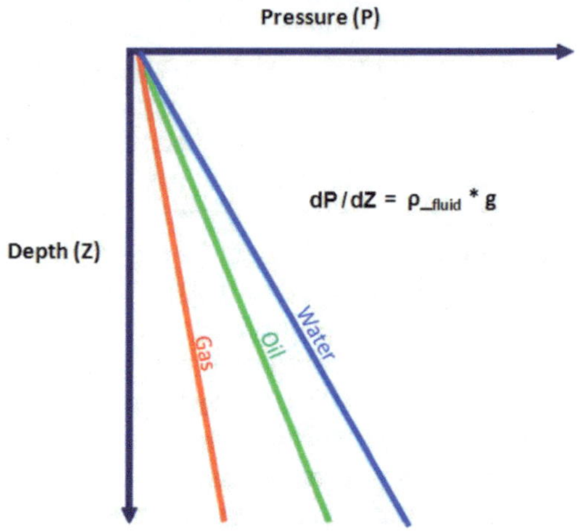

Figure 12.2: Pressure gradient as a function of fluid density, showing the pressure-depth relationship between gas, oil and water

It is a common practice in the oil industry to express oil field pressure values in *bar* or *pounds per square Inch (psi)* due to the gross amount of pressure involved. *Most pressure measurement unit used in this section is in bar.* The bar is a unit of pressure equal to 100,000 Pascal (100 kilopascal), and is slightly less or about equal to the average atmospheric pressure (one atmosphere) on Earth at sea level. To be more specific, measured pressure in psi atmosphere, psia, takes into consideration the additional atmospheric pressure of one atmosphere on the surface. This atmospheric pressure should be subtracted from psia to obtain burial only related pressure (pounds per square inch gauge - psig). *(See end of chapter for pressure unit conversions).*

1 atm (at sea level) = 14.7 psi

1 psia = 1 psig + 14.7

Expressed in bar,

P = EMW ($\rho_{\_fluid\ equivalent}$) * g * depth)

(compare with: $P = \rho * g * h$, where h is depth)

EMW = P (bar) / [TVDSS (m) * 0.09806]

*(compare with: P = 0.09806 * g * h, where h is given in TVDSS and g is 0.09806 while ρ is in EMW)*

TVD is the *True Vertical Depth*, referenced to the Mean Sea Level / Subsea (MSL/SS).

The three major fluid of interest is oil, gas and water, in petroleum exploration and production. Identifying their contacts help in quantifying the fluid (column height) to be extracted and in estimating the pressure exerted by each fluid, with known densities. The main possible contacts are:

- Gas - Oil Contact (GOC)
- Gas - Water Contact (GWC)
- Water-Oil Contact (WOC) *or Oil–Water Contact (OWC)*

The gradient of fluid in pressure depth plot gives the contacts, separating the gas from the oil and the oil from water, as shown in Figure 12.3 below.

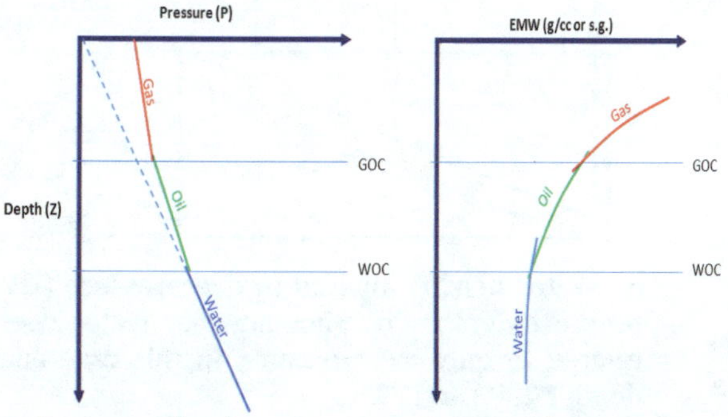

a. Pressure-depth curve

b. Equivalent Mud Weight–depth curve

Figure 12.3: Pressure and EMW-depth curves showing gradients, Gas Oil Contact (GOC) and Water Oil Contact (WOC)

These are the points at which the gradients of gas and oil (GOC) as well as water and oil (WOC) meet respectively. The dashed line in a. denotes hydrostatic pressure. The EMW curves show gradients in hyperbola or curvy shape.

With projection, a contact can still be found even in non-drilled formation by projecting a trend of overlying and underlying fluids. Where the actual contact is not known, because either the well ended or that section is not logged, or the formation pressure was not measured, then the fluid is "Up-To" a given depth. Likewise, the fluid could be "Down-To" a certain depth. As such there could be:

- Gas Up–To (GUT)
- Oil Up–To (OUT)
- Water Up–TO (WUT)
- Gas Down-To (GDT)
- Oil Down-To (ODT)

Fluid types separated at contacts may be identified in logs using a combination of logging responses, such as resistivity and neutron-density. By plotting the pressure points against depth, the gradients can indicate which fluid is present. There could be a contact between the same types of fluid (e.g. water - water contact) where fluid from different origins interface, but this is generally not considered as a contact; rather it is seen as ***fluid segregation*** or ***differentiation*** along the same gradient.

Water Head / Potential Plot

The *potential* or *water head* is the column of water in height that will rise (or descend) above the mean sea level from an open hole, if the hole were to be created above a given fluid at depth. In Figure 12.3, the pressure-depth curve is the most standard curve. However, the **potential** or **water head** curve is used to highlight fluid contacts. The EMW-depth plot is used mainly by drillers because it helps them to work with drilling fluid whose impact on pressure is adapted by varying the mud weight, weighting up or down by adding such material as barite and additives to the drilling fluid. When plotted using a water density of 1 g/cc as in freshwater (blue dashed line in Figure 12.4a.), the water head plot is known as true potential. Else, it is known as pseudo potential. **Pseudo Potential** curve is used to exaggerate the gradient and to show water contact sharply, by expressing pressure in its potential to raise water head from a higher to a lower gradient. This distance (that can be raised in meters) is plotted against depth.

To understand *pseudo* and *true potential plot,* consider the potential of reservoir pressure P1-P2 to raise water above the mean sea level (MSL) as shown in Figure 12.4b. If the reservoir is normally pressured at W1, then it will not raise any water above the Mean Sea Level

(MSL). Because the reservoir is connected, the higher pressure generated in P1 is transmitted uniformly within the reservoir such that the excess pressure in P1 is the same in P2 and P3. The overpressure is transmitted away from P1 through P2 to P3, as the pressure is higher than what the load above it would have generated naturally. This means that though at different depths, and different absolute pressures, they have the same water heads.

Only overpressure is transmitted and shared in connected unit; the water rising above sea level is same to compensate for that overpressure. Gas will rise most, second to oil and then water. The pseudo potential plots shows this (increasing rise in meters) with depth. The height, "h", is the total vertical height in the main pressure equation: $\mathbf{P = \rho * g * h}$. To get the water head , which is the extra height above sea level, from a given downhole pressure value, the True Vertical Depth down from mean sea level (TVD SS) is subtracted from the ***pressure head***. This is shown below:

Pressure head (m) = $\mathbf{h = P_a / (\rho_w * g)}$

True-Potential (m) = $\mathbf{h}$

$$= [\mathbf{P_a / (\rho_w * g)}] - \mathbf{TVD\ SS\ (m)}$$

where,

ρ_w is density of water in g/cc and P_a is absolute pressure (from which atmospheric pressure would be subtracted).

For Pseudo-Potential, the density of water is left as 1 g/cc and ρ_w is removed from the equation.

The potential for water to rise above the mean sea level is due to overpressure, not natural increase in water pressure with depth. That water rise (above sea level) is represented as water head or potential versus depth curve. In Figure 12.4c, it is seen that H1 to H3 has the same water rise because the reservoir is connected; the rise is related to the excess pressure shared equally and not due to the normal hydrostatic pressures, as shown in **c.** In a depleted reservoir for example, with underpressure, H will be lower than the MSL. Parallel overpressure line shown in the pressure-depth plot suggest reservoir connectivity or communication.

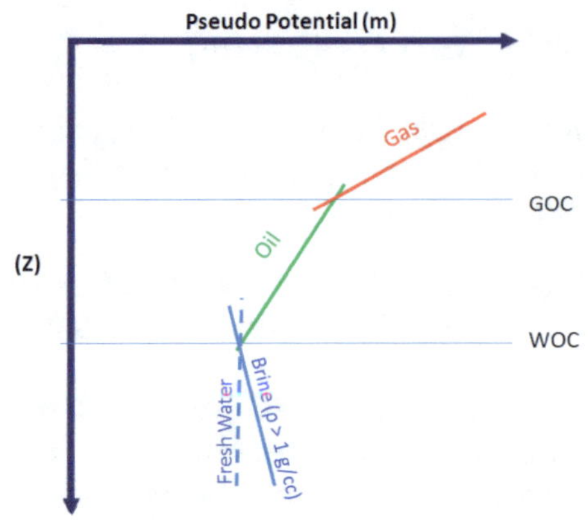

a. (Pseudo) Potential curve

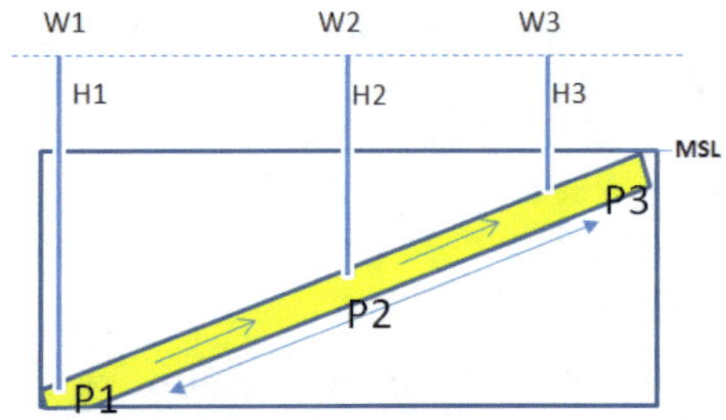

b. Illustration of potential or water head

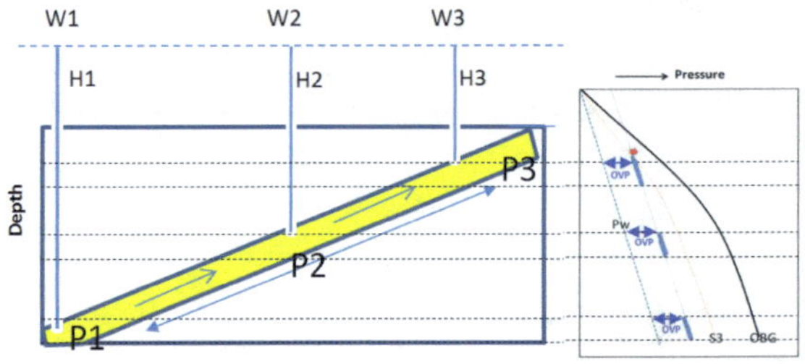

c. Water head compared to overpressure (OVP) in pressure-depth plot, showing that water rise H is related to only overpressure, in this case equal at depth P1, P2 and P3

Figure 12.4: Pseudo and true potential curves and illustration of water head

12.2 Oil Field Geopressures and Representation

Pore spaces in the subsurface are filled with water, even when oil or gas is trapped above this water. The lower part of a reservoir or other geologic structure containing pore water is known as the **waterleg**. The pressure in the waterleg could be represented as $P_{wl.}$ or just $P_{w.}$; if it is a normal water pressure, it will plot as hydrostatic (dotted line) on a pressure-depth curve, as shown in Figure 12.5. If plotted in EMW, water pressures becomes vertical lines (dotted vertical lines as shown in Figure 12.6

b.) for the various possible water pressures close to the hydrostatic (normal hydrostatic water pressure has an EMW=1).

When overpressured, the water pressures, usually in shale, known as P_{shale} or shale pressures in *Pore-Pressure-Prediction* terminology, will shift to the right increasingly with depth, to give the plots in Figure 12.5 and Figure 12.6b labelled as **Pw** and **P_{shale}**, respectively.

If the water pressure is undisturbed, it is said to be in *hydrostatic* condition, meaning the normal pressure that water (or formation fluid / brine) will exert at that depth. This normal or hydrostatic pressure can be represented as P_{nl}. It takes the same gradient of water shown in Figure 12.6a. As shown in this figure, the gradient of a *rock unit's* density is as well shown as lithostatic gradient or overburden in pressure-depth curve, as well as in Equivalent Mud Weight – depth curve.

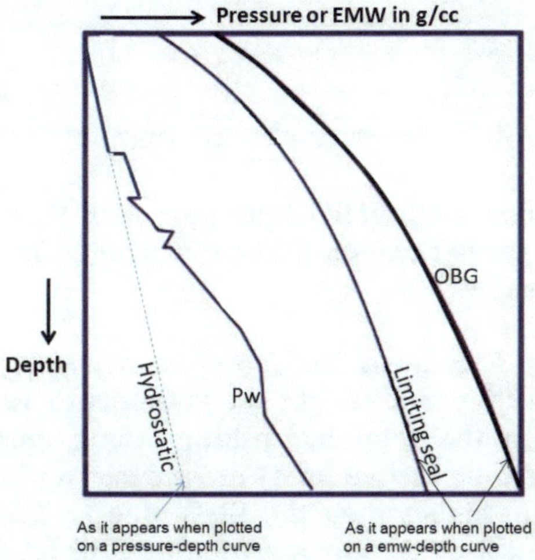

Figure 12.5: A mix of pressure and equivalent mud weight plot (with hydrostatic / water pressure, Pw, plotted in the pressure-depth curve and Limiting seal, as well as OBG, plotted in the EMW curve)

The overburden is the pressure due to a column of rock above any given depth (different rock layers piled on top of each other to a certain depth of interest); as such, it is a cumulation of the various densities that is plotted against depth.

To plot in Equivalent Mud Weight, the formula enunciated above is used, where the EMW is the equivalent mud or fluid density that will produce the same pressure at that depth.

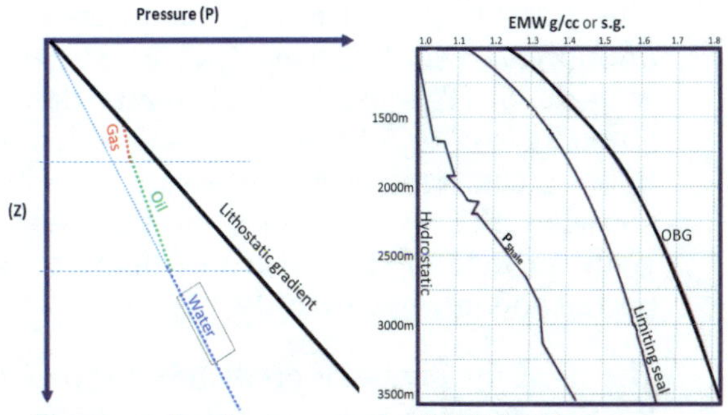

a. Pressure depth curve b. EMW-depth curve

Figure 12.6: Pressure-depth and EMW-depth curves showing fluid gradients and lithostatic / overburden gradients

With the pressurisation of a waterleg (by say burial load), the in-situ pressure becomes higher than the normal; this new pressure ($P_{in-situ}$) is above the normal pressure (P_{nl}) because of the additional pressure acting on the interval of interest. The additional pressure is known as overpressure (OVP). It is obtained by subtracting the normal pressure from the in-situ pressure.

$$P_{in-situ} = P_{nl} + OVP$$

Therefore,

$$OVP = P_{in-situ} - P_{nl}$$

This can be represented graphically as shown in Figure 12.7

The transition zone (shown in Figure 12.7) is the region marked by a change in the normal trend of the fluid gradient due to abnormal pressure. The $P_{in-situ}$ is what is measured with instruments while drilling.

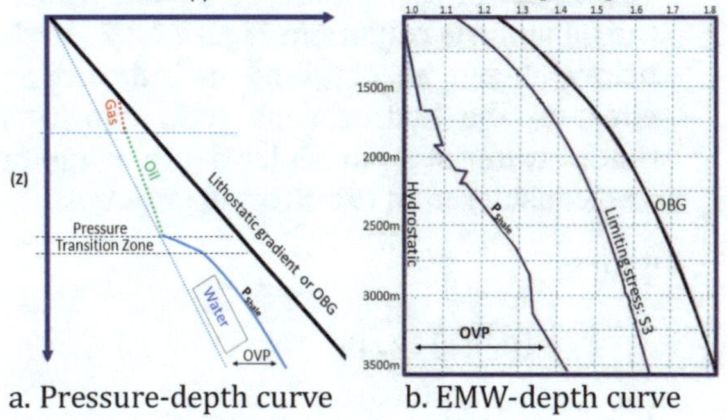

a. Pressure-depth curve b. EMW-depth curve

Figure 12.7: Pressure-depth and EMW-depth curve showing hydrostatic gradient and overpressure (P_{shale}), the fracture gradient or limiting stress (S_3), and the lithostatic gradient (OBG)

The normal pressure can be calculated by multiplying water density (ρ_w) by gravity and with depth ($P_{nl} = \rho_w * g * h$). Water density is about 1g/cm^3 or 1 kg /m^3 for fresh water and a little more for brine. Pressure prediction, knowing the overpressure in the reservoir, gives an idea of the nature of fluid to be expected and in what quantity.

The need for pressure prediction in shale or in the reservoir is necessary because the measurable pressure before drilling is not known. This must be estimated in order to determine the drilling fluid (mud) weight, to balance off formation pressure and avoid formation fluid gains (kick) or drilling fluid losses into the formation. Both of these affect borehole stability and are dangerous scenarios that can cause blowout (uncontrolled fluid gush out from the well) and rig explosion in case of fire.

The pressure in the waterleg of a reservoir will be:

$$P_w = \rho_w \cdot g \cdot h_w$$

$$(p_w = p_{nl} \text{ if hydrostatic})$$

Where:

ρ_w is water or brine density and h_w is the water vertical height

The pressure of hydrocarbon is dependent on the effect of gravity (g) on its density (ρ_h) and thickness (h_t). If the hydrocarbon (oil or gas) is floating above the waterleg, an additional pressure known as *buoyancy* is exerted within the reservoir structure. It is a kind of overpressure due to oil, as is different from overpressured water due to burial. That additional buoyancy pressure could be represented as P_b. This is the pressure reached by the gradient of the oil and gas column in Figure 12.7. As shown below, it can be obtained by subtracting the effect of the hydrocarbon fluid density from that of water at a particular depth, while taking into consideration the effect of gravity.

That is,

$$P_b = (\rho_w - \rho_h) \cdot g \cdot h_t$$

Where:

ρ_w is the density of water and ρ_h is the density of the hydrocarbon phase replacing it up to the vertical height h_t.

In Figure 12.8, a lithological column is shown with mainly shale (green with dashes) and sandstone (in yellow with dots). The sandstone layer is isolated, leading to high overpressure in that reservoir section (which may affect immediate overlying and underlying shale pressures). The gradient plotted above is used for the fluid within the reservoir while that of the lithostatic column or OBG is used to show the maximum burial pressures and as well the minimum principal stress S_3.

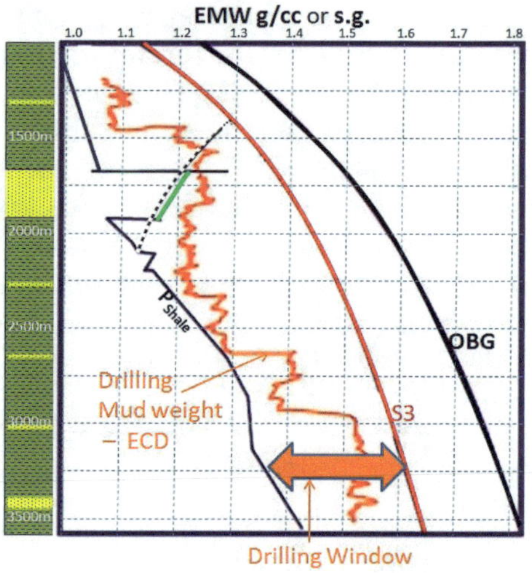

Figure 12.8: EMW-depth plot, with lithocolumn to the left and an ECD curve within the drilling window

The *Equivalent Circulating Density (ECD)* is the effective mud weight at a given depth, resulting from the total hydrostatic (static mud) and dynamic (active mud) pressures. The ECD is a little higher than the EMW due to additional impact of the moving/rotating drill bit in mud while drilling. During non-drilling phase, the well is in static **Equivalent Static Density (ESD) Mode**, while it is in active **Equivalent Circulating Density (ECD) mode** during drilling. The drilling window is attained by determining pressure values above the abnormal pressure and below that required to fracture the formation (that is, above P_{shale} and below S_3). This determines the safe *drilling window* or *margin.*

In Figure 12.8, consider the lithological column (lithocolumn) to the left and the sandstone or reservoir regions. The main reservoir (yellow with dark dots) is shown between 1500 and 2000 m depth. Note the shoot up of pressure value in the reservoir section; at that area, the pressure is not easily predictable by empirical calculation as it might be a transmission from other depths. Sometimes reservoir pressure may be inferred from the values computed in the immediate overlying and underlying shale in conventional Pore Pressure Prediction (PPP) methods, but in other cases the reservoir pressure is different and can only be ascertained by direct pressure measurement during or after drilling.

The green gradient in Figure 12.8, extending in the emw-depth plot, shows oil gradient in the sandstone reservoir. The limiting factor for the trap is not the S_3, as the S_3 curve is still further to the right. The black dotted line trailing this green gradient was extended beyond the reservoir thickness, to suggest that the trap would have filled further below with oil if spill point was lower and charge sufficient.

12.3 Pore Pressure Prediction (PPP)

The aim of pore pressure prediction is to prognose the pressure in the shale. That in the sandy reservoir is assumed to be close to the overlying and underlying shale (where the sand pressure is in equilibrium with that of the shale). Regional knowledge and comparison with offset well is used to predict shale in the sand or carbonate reservoir. Once the pressure in this shaly unit is known, it is possible to drill safely through them and hopefully, through the reservoirs whose pressures are relatively known from neighbouring well and measured while drilling (Mann and MacKenze, 1990; Alixant and Desbrandes, 1991; Stump et al., 1998).

Pressures in shale can only be calculated, or predicted, not properly measured with a tool because most tools require good permeability to measure. Pressure can be predicted before drilling (Pressure prognosis) and / or while drilling (Pressure While Drilling – PWD). Logs measured while drilling, including resistivity, density and sonic logs, give indication of how compacted the shales are, thereby giving an indication of what their pore pressures could be. That is when compaction or undercompaction is the main cause of overpressure, common in tertiary rocks (Grauls, 1999, 1994; Grauls and Baleix, 1994, Harrold et al., 1999).

The normal compaction shown in Figure 12.9 is similar to what happens when a soaked sponge is pressed. Overpressure is not generated in this case and water pressure remains hydrostatic even with increasing supply of water, which is easily calculated and predicted.

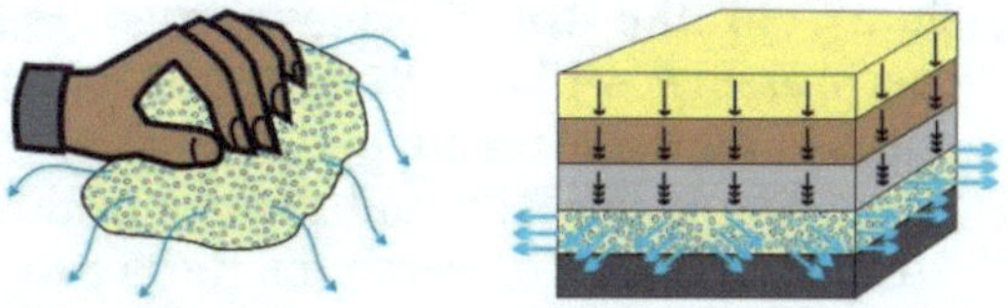

a. Water expulsion from a sponge

b. Water expulsion from porous sediment

Figure 12.9: Normal compaction where there is no barrier and compression of sediments leads to expulsion of water

A *Normal Compaction Trend (NCT)* of porosity or similar indicator curve will be as in Figure 12.10 down to the zone of transition. For porosity, the NCT will ideally decrease as porosity reduces with depth. For density, the NCT will ideally increase as density increases with depth. Most basins will give a NCT in the upper units and it is recognizable once any of the overpressure indicator parameters such as Resistivity (R), Density (D), or Time (sound or sonic transit time, log T) is plotted.

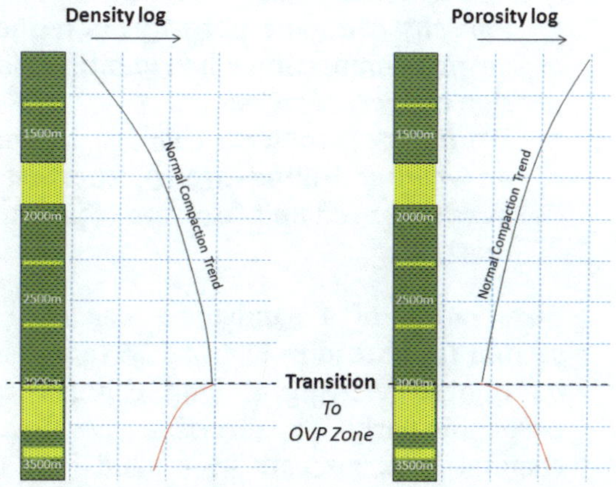

Figure 12.10: Upper zone showing normal compaction trend down to a transition to overpressure zone

NCTs have been established in most petroleum sedimentary basins and oil fields. Once the curves enters an overpressure zone at depth (lower zone in Figure 12.10, below the transition), there is no longer a Normal Compaction Trend (NCT). Parameters whose value reduces with normal compaction (e.g. porosity) will increase at the overpressure transition, while parameters whose value increases with normal compaction (e.g. density) will reduce at the overpressure transition.

This shows that the indicator (e.g. porosity, density) will increase or reduce respectively as shown in the trend. This change indicates overpressure – like a litmus test, which produces a decisive result by measuring relative indicators as porosity and density in this case. Attempt is made to search for these properties or indicators in logs and other drilling parameters and use them to predict overpressure where undercompaction is the major cause. Other causes of overpressure include chemical reaction, temperature increase, dynamic transfer of pressure from deeper level, hydrocarbon generation and more.

The use of compaction and undercompaction to predict overpressure is based on the establishment of normal compaction trend and finding where it contrasts in undercompaction trend. Such contrasts indicate overpressure as pressure is trapped in undercompaction situations. The tools for pore pressure prediction are based on:

- ***Undercompaction:*** means that pore spaces bear load in the water and not only on the matrix, due to water not being expelled.
- ***Porosity changes:*** porosity not reducing despite compaction when identified is a sign of overpressure.
- ***Drilling parameters:*** changes depending on whether drilling through a normal or overpressured unit (e.g. rate of penetration).

Observation of a sandstone lens on seismic section (as shown in Figure 12.11a and b) may be due to viewing a long channel fairway perpendicularly to direction of flow. The channel in *a.* appears as a sand lens on the cross-section in *b.* showing tendency to be overpressured due to inefficient dissipation of fluid or poor dewatering when buried.

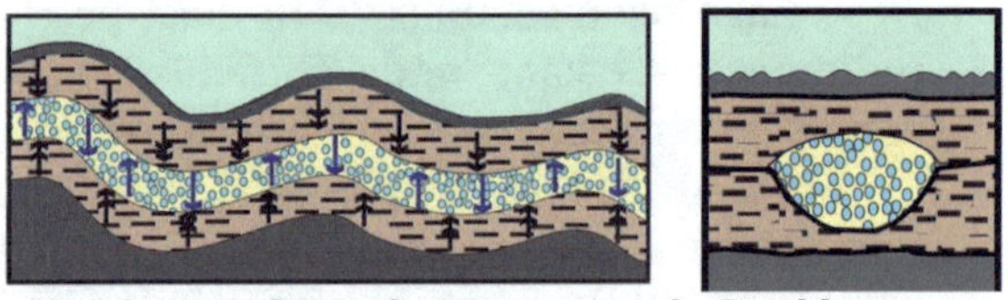

a. Sandstone channel b. Sand lens

Figure 12.11: Isolated sandstone channel (yellow with blue dots indicating trapped water molecules)

Main methods for predicting pore pressure include:

- ***Ratio Method:*** the ratio of normal compaction indicator to undercompaction indicator could be used to get the overpressure based on established relations
- ***Equivalent Depth Method:*** suggests that if the porosity above is known, which is equivalent to an unreduced porosity at depth, it can be used to estimate the overpressure trapped at depth
- ***Dexponent and drilling break:*** combination of drilling responses gives indication of overpressure, as well as direct observation of drill cuttings and fluid flow while drilling

(Each of these will be discussed in the next sections).

Other causes of overpressure include chemical reaction, temperature increase, dynamic transfer of pressure from deeper level, hydrocarbon generation and more. These other causes have less impact in rapidly subsiding basins with petroleum deposits, and are not fully taken into consideration by the conventional methods of pore pressure prediction.

12.4 Pressure Before Drilling (PBD)

Eaton (1975) provided an equation used for pore pressure prediction in shales that have been pressurized by burial or loading by overlying sediments, in a process known as *disequilibrium compaction*. That means a compaction happened so quickly (often in a quickly buried, subsiding basins) that there was no time for the water to escape (Terzaghi, 1923; Terzaghi, 1968; Walder and Nur, 1984). The equation is based on rock mechanics (Terzaghi, 1943) that relates burial load to pressures borne by the in-situ fluid and the rock matrix.

From Terzaghi's understanding of soil mechanics, the equation can be presented as,

Vertical Stress = Fluid pressure + Effective Stress

The sedimentary package (known as overburden or **OBG**) above a given depth of interest pressurizes the (fluid in the) shale beds at depth. The pressure in the shale resulting from this is known as **Pshale** or **Psh**. The pressure that is not borne by the fluid in the shale is borne by the matrix, that is, actual grains at that depth. The part of the load borne by the matrix is known as *effective stress* (**Peff**). When solid grains (grain-grain contact) are considered, it is seen in terms of *stress*; when the fluid in the shale is considered, it is seen in terms of *pressure*. When sediments compact so quickly that the fluids were not expelled, it is known as disequilibrium compaction. The fluid pressure becomes higher than normal, and it becomes overpressured.

Overpressure occur in rapidly subsiding basins, deltas for example, as shown in Figure 12.12a where quick subsidence increases accommodation space for sediment supply, having its provenance from mountains or highlands subjected to weathering and erosion.

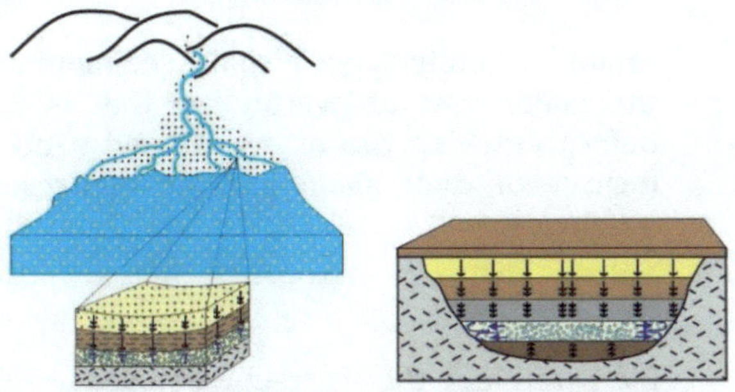

a. Rapid sedimentation b. Inefficient dewatering

Figure 12.12: Disequilibrium compaction and fluid retention in sand

In the presence of fault or lateral barrier as in Figure 12.12b inability of water to escape leads to overpressure in shale as well as in the reservoir such as sandstone.

The overburden (OBG) load is borne by the **Psh** and the **Peff**,

$$OBG = Psh + Peff$$

This above can be rearranged such that the pressure in the shale can be obtained by removing the effective stress from the overlying load:

$$Psh = OBG - Peff$$

The bulk rock densities (density of rock and the fluid it contains) are estimated by sonic values, from seismic data, and other geophysical prospecting methods used during exploration. In oil fields, the average densities of the rock are known. In which case, the OBG can be calculated from the fundamental equation:

$$OBG = \rho * g * h$$

(where ρ is $\rho_{\_bulk}$ for the density of both rock and fluid in the rock)

The calculation of OBG is done for each rock unit above the depth of interest, and then summed up at each stage to obtain the cumulative bulk densities. Software packages are available to help in this calculation. The OBG is practically the vertical load or stress, usually known as Sigma_V, S_V, Sigma 1 or S_1. This would be easy to calculate if the rock densities were the same in all the overburden, as shown in Figure 12.13.

In practice though, there is need to calculate the lithostatic stress for each unit with different bulk density (density of rock matrix and fluid). Once the OBG is obtained, all that is needed is the effective stress in order to know the Psh at a given depth. But the effective stress (which is the difference between the OBG and the Psh, as shown in Figure 12.13), is also not known when without the Psh. Nearby well data (from offset wells) and other methods, shown below, are used to go around this.

The seismic image in Figure 12.13 shows different rock units overlying a reservoir of interest; each of these units has varying densities, and as such the total OBG will be a sum of the total vertical load of each rock unit to the depth or reservoir of interest. If the rock or individual layers had the same densities, then it would have been easier to compute.

For example, for a depth of 3000 m and a rock of density 2.3 g/cc,

$$\begin{aligned} OBG &= 2.3 * 3000 * 9.8 \text{ m/s}^2 \\ &= 67620 * 0.01 \\ &= 676.2 \text{ bar} \end{aligned}$$

(The answer was multiplied by 0.01 to get the result in bar).

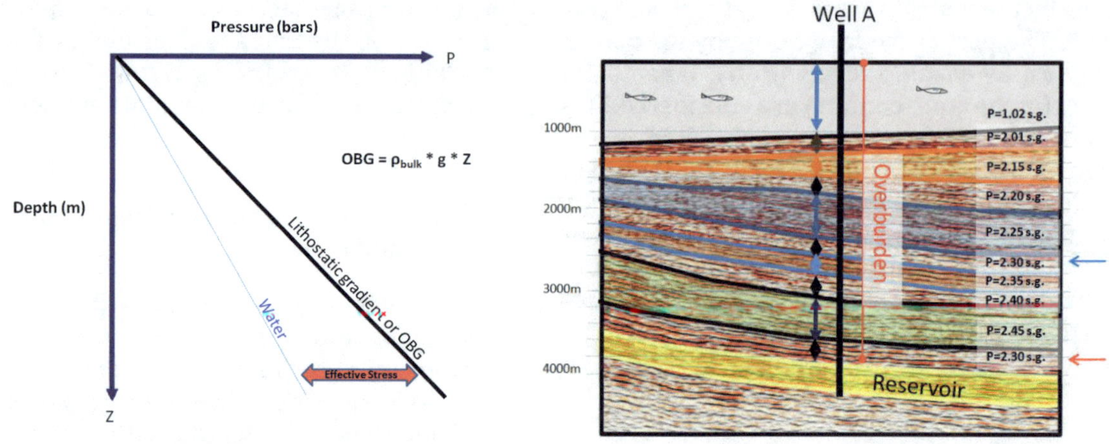

a. Lithostatic gradient

b. Seismic image with different density layers

Figure 12.13: Simplified calculation/representation of overburden, OBG or lithostatic gradient, and water pressure (Psh), showing the effective stress as the difference

However, because bulk density would vary for several units, there is need to compute OBG for each unit with their corresponding thickness and represent them incrementally with depth, as below:

$$OBG = \rho_1 * g * z_1 + \rho_2 * g * z_2 + \rho_3 * g * z_3 \ldots.$$

(where, 1 to 3... represent the various densities and thicknesses of the units)

In Figure 12.13b., consider calculating the overburden gradient for the units above the reservoir, using the densities in the right of the seismic image. Use the provided depth calibration approximately. Could these densities be represented in a log form? Attempt to plot the density against depth and construct a normal compaction trend. What happened at the depth where there is a red arrow? Compare it with the value of density at the blue arrow.

*Refer to the next sections and then predict the overpressure in the reservoir using both the **ratio** and the **equivalent depth methods**.*

In exploration scenario or frontier zone, different estimations are made before drilling based on the geologic knowledge of the basin. This estimations are then refined once the first exploration well is drilled. In some cases, once the OBG is obtained, S_3 is estimated as a fraction of it, like 75% or so depending on the predicted relationship between the vertical and the horizontal stresses. In the deep offshore for example, hemipelagic / shaly sediment forms the major part of overburden that is drilled before reaching the overpressured sand. PPP focuses on that shale part or non-reservoir overloads.

Ratio Method

By extending the Terzaghi equation, Eaton's formula explains that the effective stress can be obtained by dividing the observed response on certain logs with what would have been the response at that depth if it were under a **normal** or **hydrostatic** condition, and multiplying by the difference between OBG and P_{nl} (remember, P_{nl} is the normal or hydrostatic pressure). Example of logs that can be used for this is the Resistivity (R), the Density (D), or the Time (sound or sonic transit time) log (T).

Let's add "n" to show a normal or hydrostatic condition. Example, a normal T will be T_n while an overpressured T will be T_o.

From Psh = OBG – Peff, it means that there is only need to multiply the normal values with this additional fraction that gives the overpressure, such that

$$Peff_{\mathbf{n}} = [(OBG - P_{nl}) * \mathbf{(\Delta T_n/\Delta T_o)^a}]$$

Where:

P_{nl} is the Psh for a hydrostatic condition; 'a' is a parameter calibrated within a given basin; for some basins, the exponent a = '1.2'

may be used for resistivity log and d-exponent while '3' may be used for sonic log)

Therefore, from Psh = OBG – Peff, replacing Peff with $Peff_n$ above,

$$Psh = OBG - [(OBG - P_{nl}) * (dTn/dTo)^a)]$$

All the parameters on the right of this formula, OBG and P_{nl} can be calculated. The log values are obtained from offset wells, or the sonic/transit time data from seismic is used. The calculation can be done over a 2D or 3D seismic image to provide a pressure prognoses cube usually coloured to show zones of high or low pressure in shale. A good practice is to provide pressure prognoses curves in EMW, using Pressure Prognoses software.

To establish the *Normal Compaction Trend (NCT)* and before using any indicator of over-pressure generation such as density, it is important to first identify those that occur in the shale or non-reservoir sections. This is because fluid pressure in sand does not represent the bulk impact of lithology forming the overburden. In fact sand fluid pressure may be higher or lower, transferred from elsewhere in porous connected units. This process of first isolating only lithologies affected incrementally by compaction, is known as *picking the shale points.*

In some software, there is need to literally click on the area or depth points to be used for the shale or non-reservoir lithology. Then the software identifies the shale points on the gamma ray log, then filters and calculates the shale resistivity points on the resistivity log. It establishes the NCT on the filtered resistivity and then calculates the *pore pressure.*

Similarly, pore pressure may be estimated using the density log. After establishing the NCT on the density log, it is extrapolated to get the normal indicator (D_n) at a depth after deviation, as shown in Figure 12.14. The deviation with the observed (D_o) gives the ratio (D_n/D_o). X is the transition from normal compaction to overpressure.

The above method is based on the Ratio Method. In the above formula, the "dT_n/dT_o" part is inverted when using Resistivity log, where it becomes dR_0/dR_n. This is because while speed or seismic velocity reduces in overpressured unit (taking more time and making T_n to be higher than T_o), resistivity reduces in overpressured unit, such that R_o is higher than R_n. Sound travels faster in solid (the denser the material the faster the speed), and overpressured unit is less compact due to more fluids (the less the compaction the more the salty fluid or brine and the less the resistivity). In overpressured units, water is more and salinity increases.

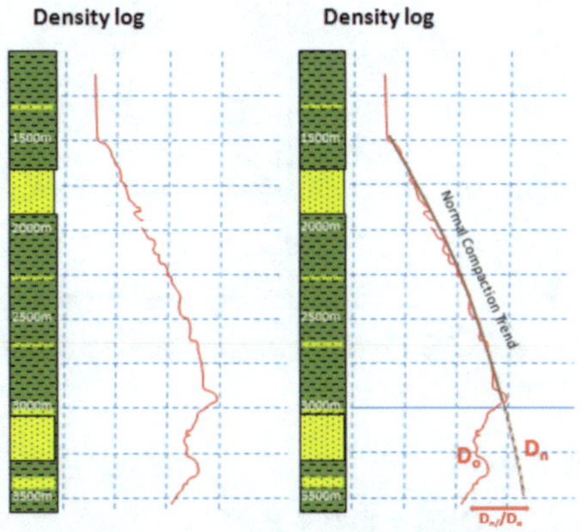

a. Density curve   b. Curve, NCT and D_n/D_o

Figure 12.14: Pressure prognoses: ratio method

Equivalent Depth Method

This method identifies a point in the upper normal compaction trend that has the same indicator of porosity or effective stress as an abnormal compaction trend at depth, and uses that effective stress above to estimate pressure at the point at depth. This is where the normally compacted shale for example at a shallower depth has the same compaction (due to under-compaction with burial) as a deeper shale unit. That is, in a case where Peff remains the same at depth as in an upper normal compaction region, while OBG had increased at that depth. Overpressure increases when OBG increases but Peff remains constant.

The point (D_n) in the upper normal compaction trend on the density log (Figure 12.15) is used to estimate density at the point (D_o) at the lower and abnormal compaction trend at depth. Note that although the value of D_n and D_o is the same at Y as shown in **a.**, the OBG at A and B in **b.** is different; it has increased in B (ignore every other parameter in the juxtaposition of the Density and NCT curve in the OBG curve in b., made just for illustration).

The accepted relationship in *Terzaghi's principle* is that the downward load of the *overburden* is equal to the sum of the upward pressure of the *pore fluid* (pore-pressure) and the *grains* (effective stress).

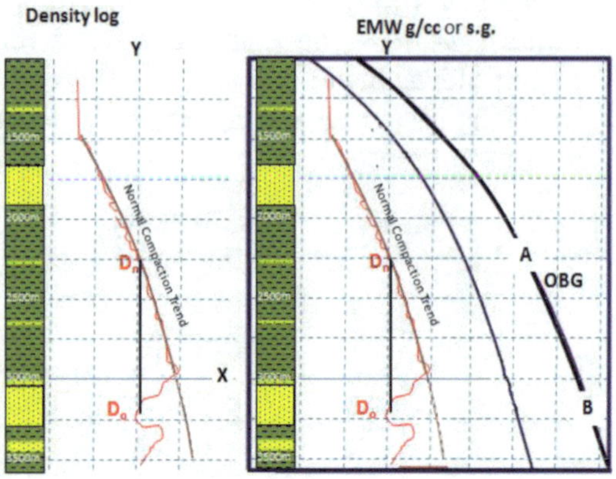

Figure 12.15: Pressure prognoses: equivalent depth method

Such that,

OBG = Psh + Peff

Or

Psh = OBG – Peff

Where,

Psh = pressure in shale or pore pressure,

Peff = grain matrix or effective stress

Then it is possible to always deduce *pore pressure* when *overburden* changes (e.g. with increasing burial at the depth of overpressure), since the *effective stress* remains the same.

Example,

If,

Psh = 40 bar when OBG = 100 bar and Peff = 60 bar

Then,

Psh = 65 bar when OBG is 125 bar and Peff is maintained as 60 bar.

The OVP is 25 bar if Psh at 40 bar was hydrostatic.

The effective stress at that upper point (D_n) would be known, if the hydrostatic pressure in the normal compaction trend section has been computed. It is also important to establish porosity versus effective stress curve or a similar curve with the other indicators of compaction to be able to use the equivalent depth method.

Refer to Figure 12.13 b; an attempt can be made to predict the overpressure for drilling the reservoir in that figure, using both the ratio method and the equivalent depth method.

Pore Pressure Prediction and Well Architecture

Predicted pressures are not only useful, but a must before drilling complex wells. It helps the drillers to know which mud weight to use to avoid kicks and losses. Armed with this overpressure value for each unit in a proposed well, the driller is ready to start drilling. He will need to plan the well and the well architecture though. This is the design and construction of the well. The well architecture depends on the number of levels with varying pore pressures. Each level requires its own casing so a specified mud weight can be used at that level.

Consider the plot and drilling window in Figure 12.8. In the uppermost section the driller cannot drill with a mud weight less than 1.1 s.g. down to 2000 m as he will have a kick when the pressure shoots up in the reservoir section. It is better to stop way above around 1500 m with that mud weight. The driller must stop at that depth, case the well with a metal pipe known as ***casing (csg)*** and cement it before drilling further with a heavier mud weight of say 1.28 s.g. If not, the higher mud weight which is injected right from top will still fracture the formation and result in losses in that upper section.

In Figure 12.16, for an EMW of 1.12 s.g., a 26" diameter casing is used, while 20 and 12.25" diameter casings are used for EMW of 1.28 and 1.44 s.g. respectively. The last depth is an open hole eventually run by a liner – a production tube with screens similar to perforation of a casing pipe. The overpressured sections therefore could be divided with the vertical bar as shown in **a.** and **b.**, then the right mud weights / ECD are used to drill each of them and case before proceeding. This division therefore, as well, determines the casing plan known as well architecture as shown in Figure 12.16 c. and d.

While drilling, there is need to remain vigilant and check that the predicted pressures are correct. PPP is continuous while drilling.

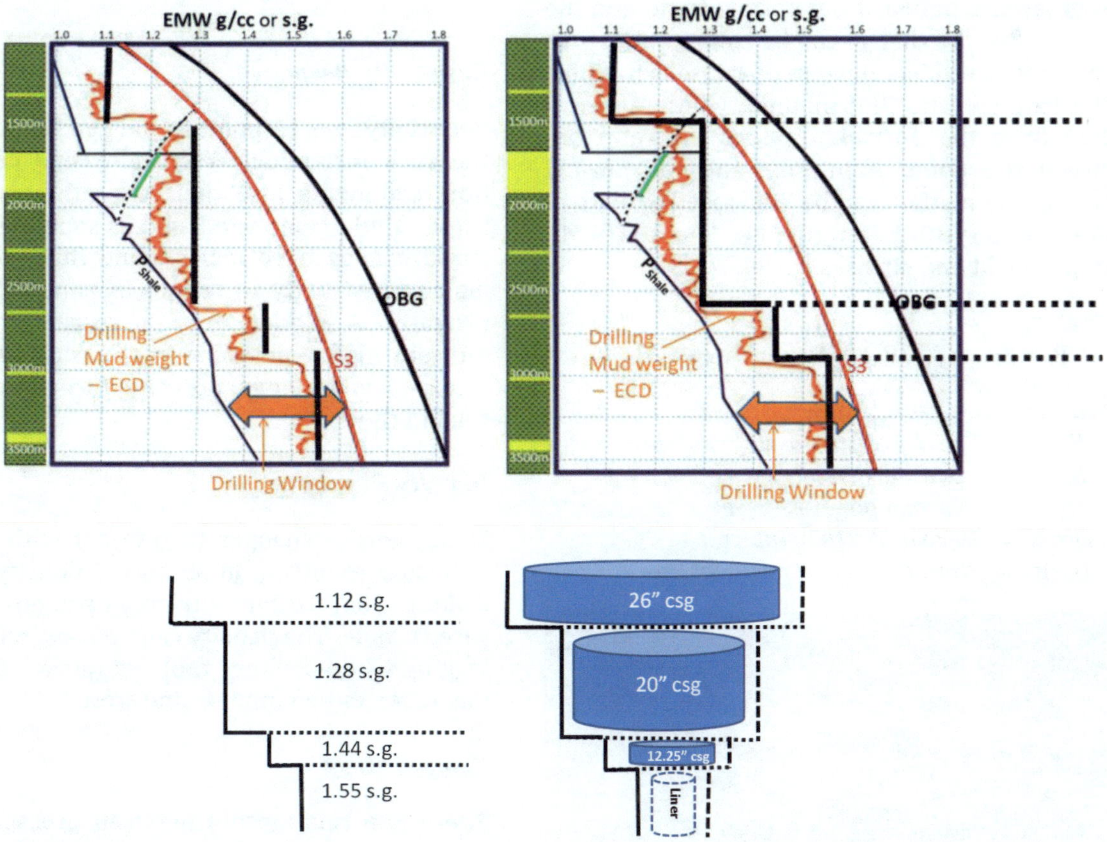

Figure 12.16: Pore pressure prediction and well architecture showing how the drilling window determines the casing plan

12.5 Pressure While Drilling (PWD)

Despite predictions, pressure still remains a challenge and must be monitored while drilling (Whittaker, 1991; Law and Spencer, 1998). The importance of this is to validate the pre-drill pore pressure model, which most time is full of uncertainties. Reservoir pressures may also be measured while drilling, giving the most reliable value. Some of the ways and tools used to achieve PWD, or *real-time pore pressure monitoring*, are described below.

Gas While Drilling (GMWD)

Total gas measurement shows values that could have a relationship with pore pressure. Over-pressured zones make gas to be liberated from the formation into the well bore that had been masked by a drilling fluid of certain mud weight, overbalanced in zones with normal pressure. This comes up and is recorded at the gas detectors in the mudlogging unit. Increasing gas shows when drilling with the same mud weight, if porosity, rate of penetration (ROP), and lithology remain the same, could mean that the formation pressure is increasing.

Different types of gas shows and trends, such as *Connection Gas (CG)* and *Background Gas (BG)* trend increase, may correspondingly show an increase in pressure (if other parameters remain the same). It is the trend that matters, as shown in Figure 12.17.

Dexponent (Dexp) and Drilling Break

This algorithm implies that if certain drilling parameters (e.g. ROP, drilling break) change while every other condition remains the same, then it is likely due to changing pressure condition. Dexp is more reliable when associ-

ated with other indicators of pressure changes such as gas reading, as shown in Figure 12.17.
A difference between the normal trend and the trend after the change can be calibrated to give the pressure value. Dexp in that case is used for the Eaton's ratio PPP method, where a deviation from the normal provides the ratio between the point of interest and the normal trend. This method can be influenced by Rate of Penetration (ROP), Rotation Per Minute (RPM), mud weight and bit sizes.

That is,

$$P_o = S_1 - [(S_1 - P_n) * (Dexp_i/Dexp_n)^{1.2}]$$

Where:

P_o = *Pore pressure at depth of interest*
S_1 = *Overburden at depth of interest*
P_n = *Normal pore pressure*
$Dexp_i$ = *Dexp at depth of interest*
$Dexp_n$ = *Dexp on normal gradient line*

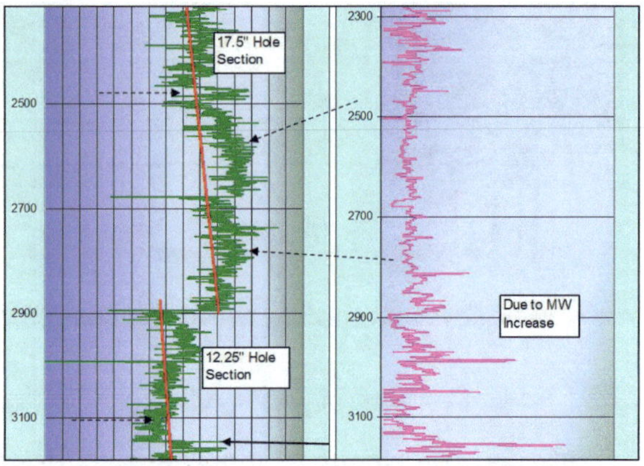

a. Dexp plot b. Total gas plot

Figure 12.17: Dexp showing possible zones of overpressure while drilling (in this case, it is a change of trend that is significant)

In Figure 12.17, the green curve marks the main trend and the red denotes the average trend within a specified hole section in the dexp plot, while the pink line marks the trend in the total gas plot due to mud weight increase. Dexp plot deviation from the normal below 3100 m may be indicative of balance problem, since the gas curve also presented an important anomaly.

Before the use of more modern techniques such as resistivity logging while drilling, Dexp was one of the major indicators of overpressure while drilling. Apart from indicating overpressure, it indicates balance issue, such as *underbalance* (drilling with a lower mud weight than is necessary to stay above the formation pressure).

Resistivity Response

Resistivity logs respond to the geochemistry of the rock formation. With increasing compaction, sediments lose their pore fluid, usually brine. Undercompacted, and as such overpressured, shales have more saline fluid, thereby making resistivity to reduce in zones of overpressure. A sudden drop in resistivity could indicate higher pressure in the rock (if other factors, e.g. lithology, porosity, formation fluid, remain the same).

Density Response

Shale density changes (increases) with depth, as such a reduction in measured density while drilling may mean undercompacted (pressured) shale. The density can be acquired using logging-while-drilling tool or shale density measurement by mudlogging crew.

Temperature

There is a relationship between pressure and temperature. Increasing temperature may occur where pressure increases.

Direct Measurement

Modern pressure while drilling measurement tools, such as the Schlumberger's Stethoscope™ probe for example, can be used to measure pressure while drilling. In practice, there is need to stop rotating or drilling temporarily, then carry out the pressure test in the reservoir, before resuming drilling. To save time, the measurement is often done when pulling out of hole with the sensor in the drilling string or Bottom Hole Assembly (BHA).

Most pressure measurements in reservoir are now done while drilling; the tools are coupled with logging while drilling tools. Pressure measurement allows for direct measurement of pressure using **Well Formation Tester** (WFT) in sand or reservoir. With direct measurement, fluid sampling and pressure values are recorded while drilling and compared with prognoses as shown in Table 12.1. More on direct measurement in the next sections, as it is also performed after drilling.

Table 12.1: Drilling and acquisition plan after pore pressure prediction and well architecture

Depth (m)	Drilling section	Petrophysical logs and seismic	Fluid Sampling	Pressure / EMW	Stress
0 -79	26 inches casing	Shallow seismic check shot	None	NCT	NCT equivalent
79 - 650	23 inches casing	Resistivity, Gamma Ray, Neutron-Density	Gas While Drilling	NCT	NCT equivalent
650 - 1150	16 inches casing	Resistivity, Gamma Ray, Neutron-Density	Gas While Drilling	NCT	FIT
1150 - 1910	12 ¼ inches casing	Resistivity, Gamma Ray and Cement Bond Log	Gas While Drilling	WFT / (*Predicted 1.2 s.g.)*	LOT
1910 - 2150	8 ½ inches casing	Resistivity, Gamma Ray, Neutron-Density and Sonic	WFT and fluid sampling	WFT / (*Predicted 1.25 s.g.)*	None

There might be an initial large casing of diameter up to 36 inches, known as *conductor pipe*, which is driven into the ground. The drill bit and later casings are placed inside the conductor pipe.

12.6 Pressure After Drilling (PAD)

Pressure measurement after drilling is achieved by wireline method, by descending a testing tool on a cable to the depth of interest. Wireline tools are lowered into the open hole after drilling. A formation pressure tester using a "probe", a nozzle, is pressed onto the formation, allowing fluid to flow in at a given flow rate, which indicates the formation pressure. Some WFT tools have a fluid sample collection and analyses option.

In pressure after drilling, the focus is on reservoir pressure. PAD gives the most reliable value since pressure measurement is taken after drilling (under static conditions), but has its disadvantage since the rock environment has been disturbed by the drilling process and in-situ pressure modified to some extent; inherent reservoir conditions have been altered and so measured values may slightly differ from the original conditions. An example of Wireline Formation Testers is Schlumberger's **Modular Tester** (**MDT**). Seismic surveys (e.g. Vertical Seismic Profiling – VSP) can also be used to identify pressure conditions post well drilling.

Pressure Test (Pretest)

Direct pressure measurements, either while or after drilling, are known as *Pretests*. The aim is to obtain direct measurements of formation pressure. It is achieved by using Well formation Testers (WFT) in relatively permeable units (usually the reservoir section) where fluid can flow easily to the tools.

A probe is plugged to the formation (yellow) as shown in Figure 12.18, then the pressure tool is used to draw fluid thereby cleaning out the mud cake and to some extent the invaded zone (green and orange respectively) that forms in the reservoir section (porous unit, like sandstone) before suction of the reservoir fluid.

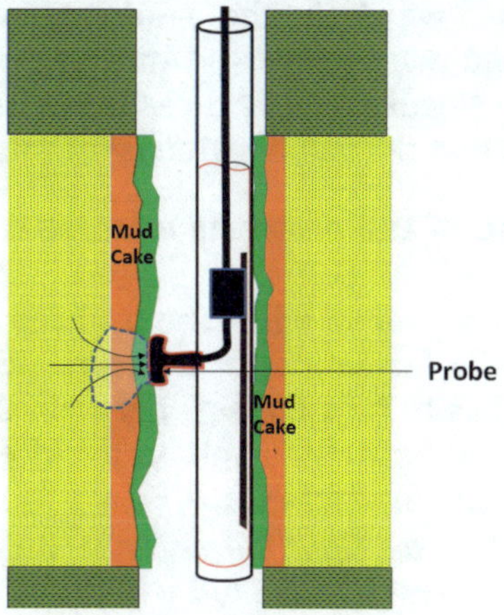

Figure 12.18: Principle of pressure measurement using a Well Formation Tester

The first pressure recorded during *pretest* is the drilling mud pressure (which drillers call hydrostatic mud pressure). This pressure is usually higher than the reservoir pressure (except the well is underbalanced, which is a risk) since higher drilling mud pressure is used to counter that of the reservoir to avoid kicks. The higher mud weight pressure drops after the drawdown and then stabilizes to be flat before going up again to the mud weight pressure. This complete signature is important to ascertain a good and full measurement (Figure 12.19). Else, an error is suspected, for example where the initial pressure keeps going down without ever stabilizing, it could be that the reservoir pressure is too low.

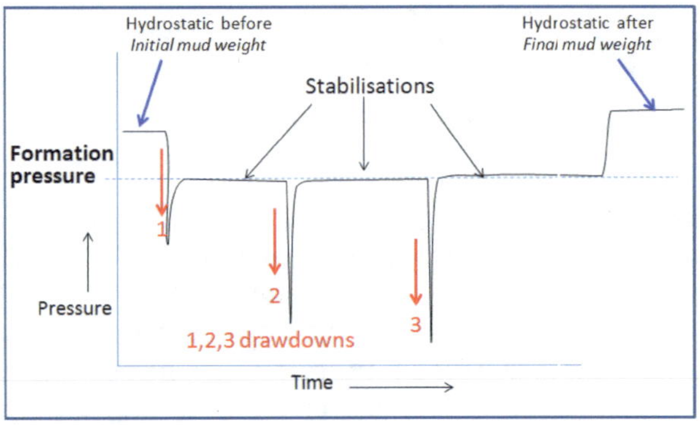

Figure 12.19: Pretest curve and interpretation highlighting the initial and final hydrostatic or mud weight pressures and stabilization plateaus showing the formation pressure

The suction is known as drawdown, which shows an increase in pressure until the end of drawdown when the pressure is allowed to recede into the formation up to a stabilization value, which is the reservoir pressure. The drawdown is repeated during the test, in this case (Figure 12.19) three times as a confirmation that the stabilization is repeatable.

Some of the problems associated with pretest are:

- *No sealing:* when the probe is plugged onto the formation, to measure pressure, there has to be a full seal between the probe and the formation wall, to avoid external pressure interference.
- *Dry or tight formations:* the principle of pretest is that fluid will flow into the probe, which is not possible if formation is too tight.
- *Probe plugging:* probe may not be properly plugged due to not being well aligned on the well wall. Proper alignment should be tested before each pretest.
- *Gas Trapping:* gas trapped in probe or within the tool will impair reading.
- *Supercharging:* this occurs when pressure value is too low and the drawdown, suction necessary for the measurement, is not sufficient.

There are situations where the probe is unable to protrude when signal has been sent from the surface, or where drawdown is unable to commence. Before the pressure test tool is lowered into the hole, the wellsite geologist will need to witness the protrusion and protraction of the probe. This ensures the probe is functional before use.

Mobility

Rock formation which is permeable, and allows flow with respect to the viscosity of the fluid it contains, is said to have high **mobility**. It is easier to perform pretest on formations with high mobility, compared to those with low mobility. Fractured reservoirs could present high mobility, but fracturing over a long reservoir unit may make "draw down", fluid flow into probe, difficult. In that case a dual packer may be used.

Dual Packer

In highly fractured reservoirs, an arrangement is made to isolate a portion of the reservoir to be tested. The tool for this is known as **dual packer;** contrary to a probe, it uses an upper and lower ring-like bumper to provide an enclosure along the well for the draw down (Figure 12.20).

The zone of interest is packed at its top and bottom before pretest is done. This process, similar to a drill stem test, allows for pressure reading in a gauge, similar to pretest with well formation tester. The advantages of the dual packer are that it can be used:

- In fractured reservoir, as mentioned above
- For Leak Off Test (LOT), a test of the strength of the formation
- As a mini Drill Stem Test (DST), which is a temporary production of an isolated section of a well for well test purposes
- In tight formation, to increase surface of suction or drawdown

- In laminated reservoir, to isolate lamina gaps and focus on rock

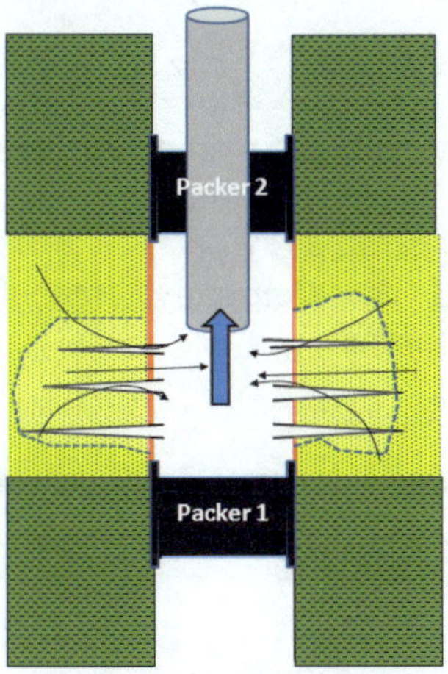

Figure 12.20: Dual packer, illustrating the use of an upper and lower packers to isolate a zone of interest (yellow) for pressure measurement

During the pressure measurement on the rig, a wellsite geologist follows up and validates the process. This is known as **witnessing**. The worksheet for witnessing pretest should include:

- Depths: Measured Depth (MD) and True Vertical Depth (TVD), each referenced to the Rotary Table (RT) and to the Mean Sea Level (MSL) or Subsea (SS)
- Hydrostatic Pressures (in this case, it is the mud weight of the drilling fluid)
- Formation Pressures (recorded after each measurement)
- Temperature (to be read off on the pretest system)
- Quality Indicator (provided by Logging Engineer based on readings on his system)
- Operational Comments (observations)
- Pressure versus depth/TVD plot (plotted on the wellsite sheet or available software)

The example in Table 12.2 shows pressure data for a reservoir level, from time of test to comment on the quality control. The value of the gradient (approximately 1) suggests that the reservoir is water-bearing. It is often important to represent the pressure in EMW. *The depth reference is not shown in the figure, but usually depth is reported in TVD/MSL or TVD/SS.* Hydrostatic before and hydrostatic after are the drilling fluid mud weight before and after the test (they have to be closely the same, as a check that the WFT reading is correct).

From the above, pressure depth curves are plotted and other pressure related estimations are made; fluid contacts are established and models could be updated following pressure results. Fluid sampling, provided by modern pretest tools, gives the most reliable information. Downhole fluid analysers give the initial fluid type and quality, and then further analyses on surface provide the actual densities and other fluid properties. As they are collected in bottles designed to retain the downhole conditions, fluid sampling can give the fluid properties, including ratio of water to oil in both reservoir and surface conditions. Pressure-Volume-Temperature (PVT) tests give the ratio of the gas to the oil (Gas Oil Ratio - GOR) and actual volumes with respect to the initial hydrocarbon in place.

Fluid Sampling

Sampling the fluid in a well is important for ascertaining petroleum result and for installation of production facilities. The fluid downhole affects the well completion and production requirements. The WFT wireline logging tools make it possible to lower special sampling tubes or bottles into the well to collect the fluids and bring them to the surface without much change to the downhole conditions. This is known as *fluid sampling*, using the same WFT tool during pretest. As the fluid is extracted and brought to the surface in bottles within the tool, that attempts to retain the pressure, volume and temperature (PVT) conditions.

Table 12.2 Pressure test recording table

Time	Test no.	Depth	Reservoir Level	Hydrostatic Before	Hydrostatic After	Draw down	Formation Pressure	Formation Pressure	Potential	Gradient	Mobility	Tempe rature	Quality Control
03:00	1	2300 m	R800b	4791 PSI	4790 PSI	3700 PSI	4484 PSI	309 bar	255 m	1.02 g/cc)	430	60 ºC	Good

The fluids are then extracted in controlled laboratory condition and pressure-volume-temperature (PVT) tests carried out on them. One physical test is to separate them through an apparatus to release the gas from the oil and further release the water, thereby obtaining their exact percentages. Fluids are also extracted and identified from cores or in cuttings while drilling, but in that case the changing pressure and temperature of extracted fluids at the surface, in atmospheric conditions, makes it difficult to know what the state of the fluid downhole was..

Nature of sampled fluid in the reservoir is used for identifying the presence or not of oil or gas. When water is sampled in a reservoir, the salinity of reservoir water is used to identify its conductivity or resistivity, and eventually in the calculation of water saturation versus oil saturation (as explained in Chapter Sixteen, on Log Interpretation). There are other reasons why formation water sampling is useful. For example, varying properties of formation water could be an indication of reservoir compartmentalization. Formation water analyses are used to identify the possible reaction with drilling fluid.

Precautions

Due to the important and delicate nature of pressure measurement and fluid sampling, there is need to ensure that every procedure is working properly during pressure jobs. It is necessary to:

- Prepare the worksheets and other documents in the logging unit.
- Check the tools and test modules.
- Determine operational procedures.
- Organise safety briefing.
- Check packer for cracks.
- Ensure rig-up, installation, is efficient.
- Label bottles for fluid sample collection.
- Add serial numbers and address for proper shipping.
- Determine tool string velocity and speed of descent or pulling (up) out.
- Remember to carry out correlation passes (to determine depths by matching peaks on the gamma ray logs).
- Call the Company Man if there is any problem.

12.7 Pore Pressure Application in Drilling

Pore Pressure Dependent Drilling Window

Understanding pore pressure is important to drilling as much as it is to exploration. In drilling, it is the ability not to fracture the formation, and not to allow formation fluid influx, that is important (Whittaker, 1991; Traugott and Heppard, 1994; Traugott, M., 1996). There is a maximum stress that the geologic units, well bore walls, can withstand before the fluid pressure within the well breaks through it. When the fluid pressure in the borehole reaches the minimum principal stress (S_3) of the well wall, it fractures the wall. This must be avoided, as fluid will flow into the formation. This is call ***loss circulation***. *(How a balance is made between drilling fluid and formation pressure, and applied in well architecture, has been discussed above: Figure 12.16).*

Figure 12.21 shows that it is the mud column weighted after PPP based on predicted formation pressure that prevents kick. Pressures from the drilling fluid presses on that of the formation fluid, preventing a well *kick*. The mud weight should be high enough just to counterbalance pressure effect of the formation; if too high, it can fracture the formation, resulting to losses.

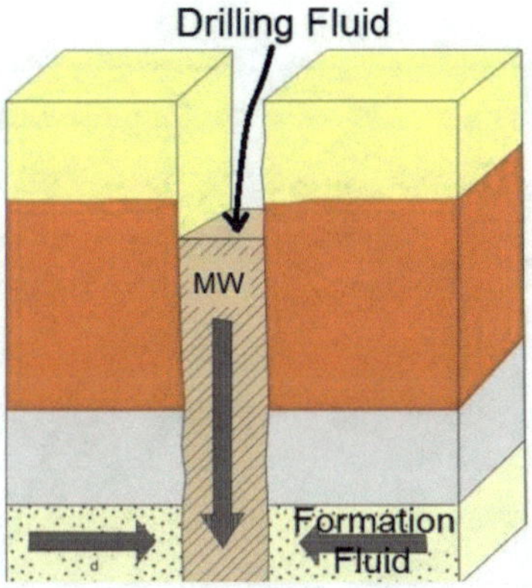

Figure 12.21: Pressures on the bottom of a well balanced by an interplay of drilling mud and formation fluid pressure

The well bore slightly rearranges the stress downhole while drilling, making the reality more complex than the basic principle of top seal hydraulic fracturing when in-situ pressure attains or exceeds the overlying confining stress. Presence of fracture, pump rate of the drilling fluid, ROP and many other conditions increase the actual effect of borehole fluid density while circulating the drilling mud. Observation of drilling parameters and responses while drilling remains key.

Below is a reminder of the impact of hydraulic fracturing as well as vertical and horizontal stresses on well stability. Hopefully this becomes clearer with the above details on pore pressure prediction.

Hydraulic Fracturing

The process of fluid breaking their way through a sealing layer and escaping is known as Hydraulic Fracturing (HF). As explained in previous chapters, borehole fluids break through when their pressure exceeds that of the formation wall resistance. Since formation wall will break in the weakest direction (minimum confining stress), Leak of Test (LOT) gives an indication of the in-situ minimum stress (S_3).

In earlier chapters, in *part two*, an explanation was given on how explorationists use this to estimate how much column of hydrocarbon a reservoir could trap before leaking or hydrofracturing at its weakest point. In drilling, it is the degree of fracturing and how it affects the drilling window (that is the difference between the well bore drilling fluid mud weight and the stress value/strength of the well wall) that is of interest. This window is provided by LOT or LOP and other fracture constraints described below.

Actually, fluid leaks at the onset of injection into the formation, before the formation breaks. Such point of leakage when found from LOT is known as *Leak Off Pressure (LOP)* point. There are instances though where fluid leaks before LOP, such as capillary entry pressure and presence of fractures. There are two fracture-related reasons why the well bore pressure may leak through the formation wall:

1. *Reopening an Existing Fracture*

   An existing fracture or micro-fracture can be reopen and make fluids to leak when the formation is subjected to excess pressure from the well bore. The pressure needed to achieve this is known as **Fracture Reopening Pressure** (FRP), and it is usually smaller than the Fracture Initiation Pressure.
2. *Creating a New Fracture*

   A new fracture or micro-fracture may be formed and make fluids to leak when the formation is subjected to excess pressure from the well bore. The pressure necessary to achieve this is known as *Fracture Initiation Pressure (FIP)*, and it is usually higher than the Fracture Reopening Pressure

In fact the FRP and FIP is where leakage begins, before attaining the LOP and LOT. The **FIP** also depends on the principal stresses (S_1, S_2 and S_3), affected by increased pressure due to the Equivalent Circulating Density (ECD) or active mud pressure while drilling. The effect of borehole pressure on vertical and horizontal wells are not the same, due to the impact of gravity and drilling bit weight. A too low mud pressure can lead to the well walls collapsing in (compressive strength failure - where the well walls compress due to inadequate support of mud pressure until they eventually collapse).

Likewise, a high mud pressure can lead to the formation walls breaking out (tensile strength failure - where mud pressure is so high that it stresses the well walls until they eventually give in). Theoretically, fluid will flow into the borehole during compressive strength failure, while it will flow from the borehole into the formation during tensile strength failure. In real case scenario, based on field observations, fluid could flow either in or out in either cases.

In Figure 12.22, leakage will occur earlier on *Side 2* due to reduced *Fracture Reopening Pressure (FRP)* than the *Fracture Initiation Pressure (FIP)* required on *Side 1*. In this case a leak off test at that location may not represent the minimum principal stress of the formation rock matrix.

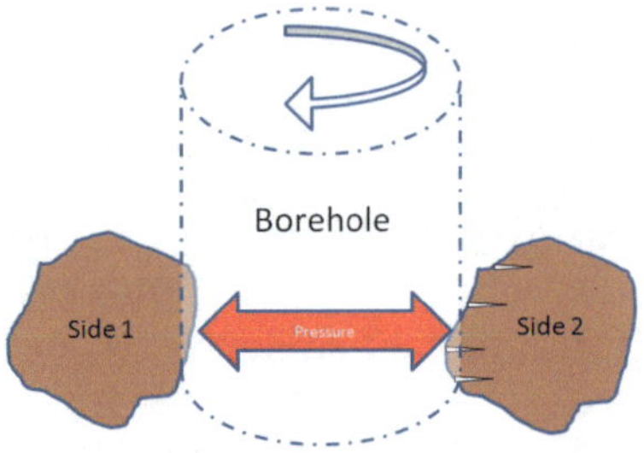

Figure 12.22: Borehole across formation with intact (side 1) and existing fracture (side 2)

12.8 Stress Tests and Well Stability

The borehole behaviour while drilling is controlled by the geopressure and stress conditions of the geologic unit (Yassir and 1994; Grauls, 1994, 1997). The Leak Off Test (LOT) is performed below the *casing shoe* (short, hollow steel rounded bottom of the casing string together with the cement around it) to estimate at what point the pressure of the drilling fluid will exert a stress enough to fracture the formation. That breaking pressure or stress at different depths gives the fracture gradient of the formation. Matthews and Kelly (1967) provided a calculation for estimating the fracture gradient in the absence of LOT data.

The formula has been integrated in most PPP software and the curve is often abbreviated as **Frac (M&K).** While doing pore pressure prediction, few LOT values from offset wells are used to adjust the curves calculated from the M&K Fracture algorithm. As shown in Figure 12.23, the existing LOT values are plotted and compared with the calculated Fracture gradient from M&K to give a minimum (FIP_{min}) and maximum (FIP_{max}) range of where the fracture would occur.

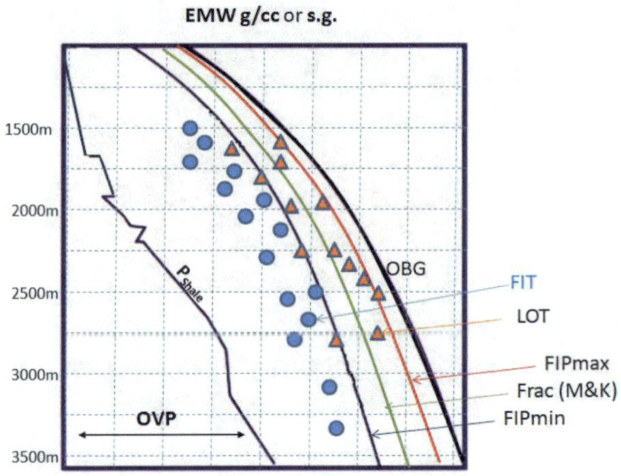

Figure 12.23: Fracture gradients showing stress test points and fracture curves

In practice, drillers often avoid fracturing a formation or even leaking off at all. They test the resistance of a formation wall to a point where they feel it is good enough to drill with a drilling fluid of specific density they want to use. Such limited test is known as Formation Integrity Test (FIT) and is a pressure value less than the LOP, shown in Figure 12.23. Sometimes accidental LOT occurs while attempting to carry out FIT; the pressure suddenly drops indicating that the formation has been fractured. In such case, the drilling window is too narrow, and insufficient for drilling the next section.

In Figure 12.23, the blue circular and orange triangular points indicate FIT and LOT values respectively. The range from FIP_{min} to FIP_{max} is also shown, where the *Frac (M&K)* is centred to give the expected fracture curve or a median.

Standard Leak Off Test (SLOT)

As discussed previously, the test carried out to know the strength of the unit is known as Leak Off Test (LOT). It gives the pressure at which the well fluid starts to leak into the formation. So the LOT actually gives the leak off pressure at first, then further gives the point after, if continued, where there is opening of the microscopic walls. This could have been due to overcoming the pressure in the formation. However, since any pressure in the formation would be less than the strength of the matrix in that formation, it means that the LOT is only accurately fracturing if the formation strength is in equilibrium with the in-situ pressure or if the test is continued until full breakdown of the formation. This is seen as a major fall in the pressure curve (Figure 12.24), as in Hooke's Law when the curves show a broken deformation of the string.

Coupled with borehole image, more can be said about LOT values. Borehole images show drilling induced fracture which can be used to understand the effectiveness of applying the right parameters in the drilling phase. For example, the orientation/direction of the fracture gives the weakest direction and consequently the direction of the minimum principal stress (S_3).

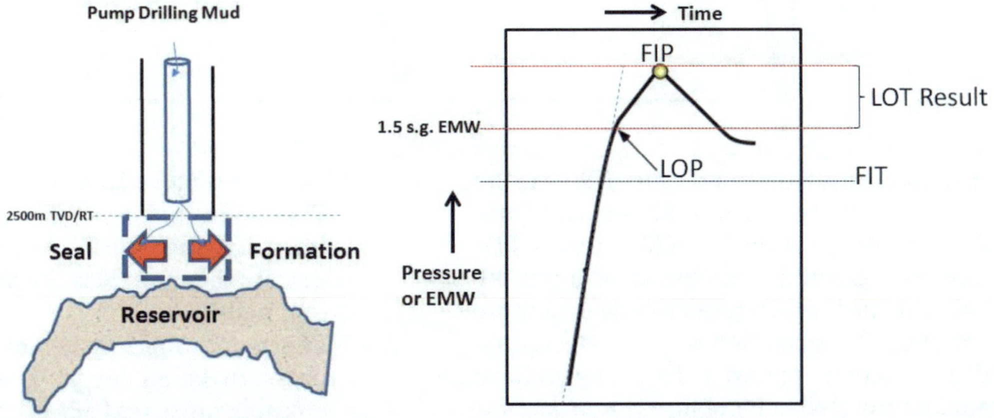

a. Well bore showing LOT b. Graph of LOT

Figure 12.24: Standard LOT @ 2500 m True Vertical Depth / Referenced to the Rotary Table and showing early detection of LOP/LOT at 1.5 s.g. equivalent mud weight

Extended Leak Off Test (ELOT)

If risk of flow into or out of the reservoir is not a major issue, then a type of extended Leak Off Test known as **Step Volume Test** (SVT) can be used to determine the fracturation point and to estimate the formation pressure. This advanced method involves gradually pumping the fluid such that it is clear when fracture starts and when the fracture closes after pumping stops. From this, the FIP_{max} and FIP_{min} can be determined, as well as the FRP or Fracture Closure Pressure (FCP), all of which are important for the driller to ensure a stable drilling phase. With comparable parameters, such data can be used elsewhere in offset wells.

During an extended LOT, several fracture points could be identified (including for example Fracture Closure Pressure – FCP), which provides more in-depth and multi-variant assessment of pressures while drilling. The FCP gives an idea of how a fractured formation (while drilling) will behave (afterwards), when it has lost its initial wall resistance. This is achieved by testing out the Fracture Reopening Pressure (FRP which is the same order as the FCP) for a fracture caused during standard leak off test (on many occasions though this would be avoided due to loss of formation resistance leading to losses from well after the test). When done, such a fractured interval would have to be cemented properly before drilling is continued. The prolonged test providing more information (Figure 12.25) is known as **Extended Leak Off Test** (ELOT).

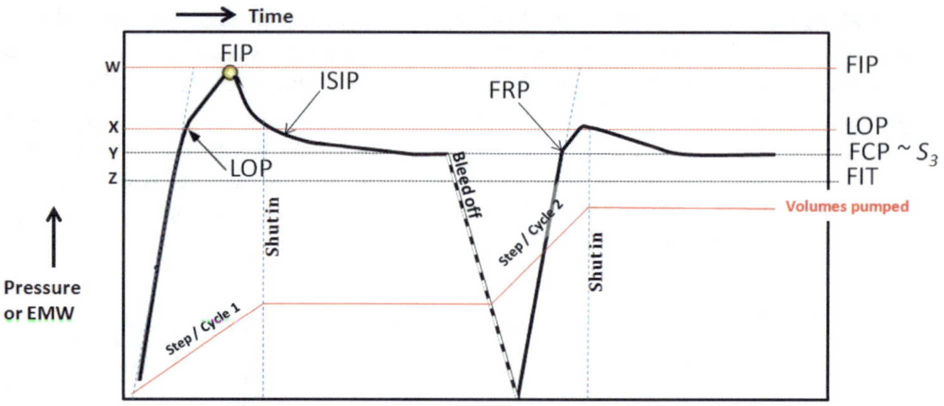

Figure 12.25: Extended Leak Off Test Step showing LOP, FIP, ISIP and FRP and FCP

The Instantaneous Shut In Pressure (ISIP) is the pressure drop at the end of injection. In Figure 12.25, W = FIP, X = LOP, Y = FCP and Z = FIT. The red plot below is the cycles of injection, showing the increase in rate at cycle 1 and the shut in after the formation breaks, and then a bleed off (action required to safely relieve pressure or normalize it) before the process is repeated in cycle 2. In cycle 2 there is an earlier leak off due to already broken resistance of the wall in the first cycle – this gives the fracture reopening or closure pressure (ideally equivalent to S_3)

Leak Off Test (LOT) Procedure

The procedures for performing LOT are:

1. At a well section Total Depth (TD or final TD, drill a little further after casing (usually 10 feet “rat hole” of fresh formation), or create some barrier using a dual packer at a depth in a well so pressure can be pumped directly against a limited part of the rock formation.
2. Homogenise the mud to have equal weight in and out.
3. Pull back into the casing shoe.
4. Shut-in well.
5. Pump continuously or in steps and record the behaviour of the formation. The pump rate will correspond to increasing pressure in the well, until a point where further pumping leads to stabilization and then drop in pressure. This is because the formation has been breached and fluid leaking from the well and flowing into the formation.
6. Plot this (see Figure 12.24 b.) and highlight:
   - The point before leakage (Formation Integrity Test – FIT, which reassures drillers they can raise the mud weight to that point).
   - The point where it begins to leak and pressure is no longer linear with increasing pressure (this is the LOP)
   - If pumping is continued, then record the Fracture Point, where the well wall is fully burst and leaks rapidly – like the failing point in a Hooke’s Law diagram. This is the LOT point of fracture.
7. If the test is continued further (Figure 12.25), what happens to the pressure after pumping has stopped may be seen; the gauge will build up again as the pressure returns. That is the *Fracture Closing Pressure*, as the formation tends to re-establish equilibrium.

This build up or coming back (in no. 7) is the restoration of the formation integrity. Ideally, it should again give the minimum principal stress (S_3), closing in on the weakest direction onto which a fracture had been initiated (if fracture was fully initiated by the pumping). This is not necessarily the precise situation in real scenario and minimum confining stress after borehole destabilisation is not exactly equivalent to original in-situ minimum stress.

LOT Correction

It has been established that the horizontal stresses depend on the effects of vertical loads or overburden (S_1) on the lateral or horizontal direction. This means that LOT measured offshore is dependent on burial depth (overburden from ground level to the subsurface) and on water depth (overburden from overlying water pressure, from water surface to sea bottom). The minimum principal stress (S_3) or confining pressures (LOP) can be a fraction of S_1, making LOT comparable in similar basins once the S_1 or overburden is similar. As such

LOT must be plotted in burial depth of geologic rocks. That way, geoscientist can apply the same value elsewhere, at offset wells. Where the LOT is measured in a well offshore, with burial starting underwater, a correction has to be applied before applying the LOT result or LOP to other offset wells with different water depths.

$$\mathbf{LOP_{\_corrected} = LOP - WP}$$

Where:

WP is the Water Pressure of the sea column, from the former well with measured LOT result

There will therefore be need to add the actual water pressure when estimating LOT for the new offset well.

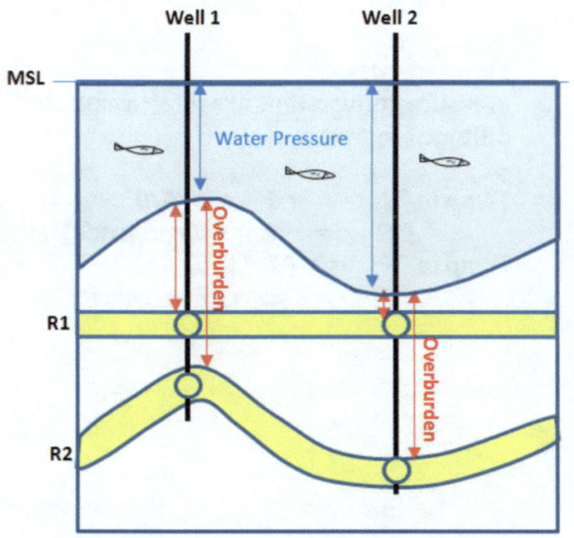

Figure 12.26: LOT correction when applied to a new offset well in a field or basin

Once LOT values that are not water depth dependent has been obtained, it is possible to apply them elsewhere by adding the new water pressure of a longer or shorter sea column at the offset well, as shown in Figure 12.26. That is on the assumption that the lithology, stratigraphic correlation and burial history are similar. In Reservoir 1 (R1), the burial related OBG in Well 1 is different from that of Well 2 and a new PPP as well as LOT estimation is necessary. However in Reservoir 2 (R2), the burial related OBG is the same and as such the Pore and Fracture Pressure is transferrable from Well 1 to Well 2. But the water pressure in Well 1 is different and would have to be subtracted before adding the water pressure overlying Well 2.

This does not mean the process of drilling the different wells or their architecture will be the same. Pressure measurement and interpretation combine several aspects of the well which cannot be discussed without understanding the processes involved in drilling the well. *The next chapter looksclosely at drilling and geosteering.*

PRESSURE AND TEMPERATURE UNITS CONVERSION

Pressure Units Conversion

The standard unit of Pressure is in Pascal or N/m^2. Because of the large amount of pressure and to adapt to locally calibrated values, the oil and gas industry uses other units and expressions to represent pressure and pressure gradients. These are given below.

General Pressure Units

1 kg/cm^2	=	*14.223 psi*
1 Mpa	=	*145.038 psi*
(MPa	=	*Mega Pascal)*
1 g/cc	=	*8.34540 lb/gal*
(g/cc	=	*gram per cubic centimetre)*
1 lb/gal	=	*0.119826 g/cc*
[lb/gal	=	*pound per gallon (ppg)]*

Pressure in Bar

1 bar	=	*1.01972 kg/cm^2*
1 bar	=	*0.10 MPa*
1 bar	=	*14.50377 psi*

Pressure in Atmosphere

1 atm	=	*14.696 psi*

Pressure in Gradients

1 g/cc	=	*0.4335 psi/ft*
1 psi/ft	=	*2.3066 g/cc*
1 psi/ft	=	*0.226206 bar/m*
1 psi/ft	=	*22.6206 Mpa/km*

Length

1 m	=	*3.2808 ft*
1 ft	=	*30.480 cm*

Volume

1 bbl	=	*158.987 lit*
(bbl	=	*barrel)*
1 bbl	=	*42.000 US gal*
(gal	=	*gallon)*
1 bbl	=	*5.61458 cubic feet*
1 gal	=	*3.7854 lit*
1 m3	=	*35.315 cubic feet*
1 m3	=	*6.290 US bbl*
1 cubic feet	=	*7.485 US gal*
1 cubic feet	=	*28.317 lit*

Mass

1 lb	=	*0.4536 kg*
(lb	=	*Pound)*
1 kg	=	*2.2046 lb*

Area

1 acre	=	*4046.873 m2*
1 acre	=	*43560 ft2*
320 acres	=	*1.295 km2*
1 ha	=	*10000 m2*
(ha	=	*hectare)*
1 ha	=	*107639.104 sq. ft.*
1 ha	=	*0.01 km2*
1 ha	=	*2.471 acres*

Temperature

Pressure conversions are often associated with different temperatures.

Temp in °C	=	*(°F - 32) * 5/9*
(°C	=	*degree Centigrade)*
Temp in °F	=	*(°C * 9/5) + 32*
(°F	=	*degree Fahrenheit)*

EXERCISES

1. What is geopressure? Mention five causes of overpressure. How is each of the causes identified? Differentiate between overpressure and underpressure.

2. What is the basic expression for starting pressure studies? Demonstrate the relationship between the parameters, starting from showing pressure as force acting per unit area.

3. What is a waterleg? What brings about changes to the pressure in it? How does the pressure in the waterleg affect the hydrocarbon column?

4. What does EMW stand for? How is it determined? Why is it very important in drilling? Give examples.

5. Define fluid pressure gradient and provide an expression to explain its variation. What is the difference between water, oil and gas gradients?

6. What are the various ways of expressing pressure measurement in an oil field? How is the information from pressure obtained and used before, during and after drilling?

7. Differentiate between the hydrostatic and thee in-situ pressure. How can overpressure be determined from them?

8. Why is there need for pore pressure prediction? Give the different formulas used for PPP.

9. Differentiate between the following borehole stability tests
   a) Formation Integrity Test (FIT)
   b) Leak Off Test (LOT) / Leak Off Pressure (LOP)
   c) What is the difference between SLOT and ELOT?

10. Determine the overpressure (or underpressure) for an in-situ pressure measurement of 510 bar (51.0 MPa) at a depth of 4000 m in the aquifer. *Assume water density of 1020 kg/m³.*

REFERENCES & FURTHER READING

Alixant, J. L. and R. Desbrandes, 1991. Explicit Pore Pressure Evaluation: Concept and Application. *Society of Petroleum Engineers Paper* 19336, p. 50–56.

Eaton, A.B., 1975, the Equation for Geopressure Prediction from Well Logs. *SPE 50th Annual Meeting*, Paper no. 5544.

Gluyas, J. and R. *Swarbrick*, 2003. *Petroleum Geoscience,* Blackwell Publishing, 359 p.

Grauls, D. J., 1997. Minimum Principal Stress as Control of Overpressures in Sedimentary Basins. *Geofluids* II'97 – Extended Abstracts, Hendry, J. P. (ed.), p. 219–222.

Grauls, D., 1999. Overpressures: Causal Mechanisms, Conventional and Hydromechanical Approaches. *Oil & Gas Sciences and Technology – Rev. IFP*, v. 54, no. 6.

Grauls, D. J. and J. M. Baleix, 1994. Role of Overpressures and In-Situ Stresses in Fault-Controlled Hydrocarbon Migration: A Case Study. *Marine and Petroleum Geology*, v. 11, p. 734–742.

Harrold, T. W. D., R. E. Swarbrick and N. R. Goulty, 1999. Pore Pressure Estimation from Mudstone Porosities in Tertiary Basins, Southeast Asia. *AAPG Bulletin*, v. 83, p. 1057–1067.

Law, B. E. and C. W. Spencer, 1998. Abnormal Pressures in Hydrocarbon Environments, in B. E. Law, G. F. Ulmishek and V. I. Slavin (eds.), Abnormal Pressures in Hydrocarbon Environments. *AAPG Memoir 70*, p. 1–11.

Mann, D. M. and A. A. MacKenze, 1990. Prediction of Pore Fluid Pressures in Sedimentary Basins. *Marine and Petroleum Geology 7*, p. 55–68.

Matthews, W. R. and J. Kelly, 1967. How to Predict Formation Pressure and Fracture Gradient: *The Oil & Gas Journal*, February 20 Issue, p. 92–106.

Pennebaker, E. S., 1968. Seismic Data Indicate Depth and Magnitude of Abnormal Pressure. *World oil*, 166, p. 73–82.

Stump, B. B., P. B. Flemings, T. Finkbeiner and M. D. Zoback, 1998. Pressure Differences between Overpressured Sands and Bounding Shales of the Eugene Island 330 Field (Offshore Louisiana, USA) with Implications for Fluid Flow Induced by Sediment Loading, *in* A. Mitchell and D. Grauls, eds., Overpressures in petroleum exploration. *Workshop proceedings. Elf Exploration Production Memoir 22*, p. 83–92.

Swarbrick, R. E. and M. Osborne, 1996. The Nature and Diversity of Pressure Transition Zones. *Petroleum Geoscience,* 2, p. 111–116.

Swarbrick, R. E. and M. J. Osborne, 1998. Mechanisms that Generate Abnormal Pressures: An Overview, in B. E. Law, G. F. Ulmishek and V. I. Slavin eds., Abnormal Pressures in Hydrocarbon Environments. *AAPG Memoir 70*, p. 13–34.

Terzaghi, K., 1923, Die Berechnung der Durchlass igkeitsziffer des Tones aws dem Verlanf der Hydrodynamischen Spannungserscheinungen (English extract). *Sb. Akad. Wiss. Wien*, p. 132–135.

Terzaghi, K., 1943. *Theoretical Soil Mechanics.* New York, John Wiley, 510 p.

Terzaghi, K. and R. B. Peck, 1968. *Soil Mechanics in Engineering Practice.* New York, John Wiley.

Traugott, M. O. and P. D. Heppard, 1994. Prediction of Pore Pressure Before and After Drilling – Taking the Risk Out of Drilling Overpressured Prospects. *AAPG Hedberg Research Conference*, v. 70, p. 215–246.

Traugott, M., 1996. The Pore Pressure Centroid Concept: Reducing Drilling Risks. *Abstract, Compaction and Overpressure Current Research,* 9-10 December, IFP, Paris.

Walder, J. and A. Nur, 1984. Porosity Reduction and Crustal Pore Pressure Development. *Journal of Geophysical Research. 89*, p. 11539–11548.

Whittaker, A., 1991. Mudlogging Handbook. Englewood Cliffs, NJ: Prentice Hall, 531 p.

Yassir, N. A. and Bell, J. S., 1994. Relationships between Pore Pressure, Stresses, and Present Day Geodynamics in the Scotian Shelf, Offshore Eastern Canada. *AAPG Bulletin*, v. 78, p.1863–1880.

CHAPTER THIRTEEN

DRILLING AND GEOSTEERING

13.0 Drilling Process

Drilling is the act of creating a cylindrical hole into the Earth using a drill bit and other drilling tools for the purpose of reaching a desired target in the subsurface. Modern drilling techniques take several factors into consideration. Important factors to consider in this chapter are:

- Drilling hazards
- Drilling equipment
- Well structure and types
- Casing and cementing
- Drilling fluid or mud
- Rock mechanics and stresses
- Effect of temperature and pressure
- Directional drilling and geosteering
- Well trajectory and downhole surveys

Most of these are included in the document that serves as guide to drillers and geoscientists working in a drilling site. This document, known as drilling program, contains the well operations procedures involving drilling activities, mud or drilling fluids to be used, casing sections, cementing and completion, and sometimes production options and facilities.

13.1 Drilling Hazards

The first concern before planning any well is safety. Safety issues may include:

- **Shallow gas** (methane), which can be identified from high resolution seismic and avoided. Shallow gas may gush out violently if penetrated and could lead to gas kicks or blowout if not effectively managed.
- **Mobile shales**, which can cause drill string sticking and other similar problems; can be identified from seismic and by correlating with offset wells.
- **Fractured or brittle (metamorphic) levels**, where drilling parameters such as Weight On Bit (WOB), Rotation Per Minute (RPM) and Rate Of Penetration (ROP) need to be adapted by the drillers.
- **Active faults**, which could create unexpected fluid flow when the fault zone is penetrated, leading to displacement of blocks in severe cases. Wells are planned to avoid fault, by deviating the trajectory for example, where necessary.

In Figure 13.1, the second well offshore shows potential production from unconventional resources on the left of the oil pool. The Geologist and Geophysicist, the Drillers, the Reservoir and the Operations Geologists, work together to isolate and avoid any hazards. For example a new trajectory will have to be set to avoid crossing active faults, shallow reflecting bodies on seismic (which could denote presence of shallow gas), major seabed pockmarks, and the like.

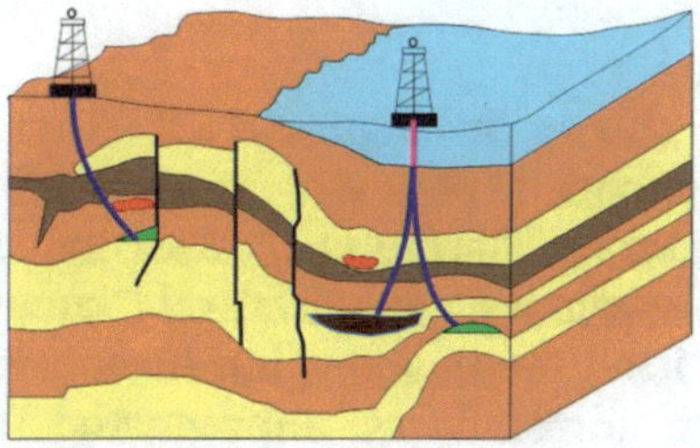

Figure 13.1: Well paths (blue lines) avoiding faults (black lines) and other drilling hazards (red lens).

13.2 Drilling Program

Irrespective of a personnel's affiliation, there is need to lay hands on the drilling program if one is going to participate in a well operation. It covers what is happening there on the rig – drilling, drilling, drilling. Many times drilling will stop, but between then and the next drilling phase, every action is geared towards that next drilling.

The well program presents different stages in the operation, input from the geosciences department, including the important factors to take into consideration such as:

- Lithology (rock type) to be encountered
- Sand tops and base
- Presence of faults
- Tentative final depth to be reached, also known as final Total Depth (TD)
- Type of casing to be used for isolating / "casing off" the well from the surrounding formation
- Health, Safety and Environmental policies

The program will include a description of the *well structure*, the tools that will be used to achieve the desired results and the methods, some of which are discussed below. The rock type to be encountered will determine the type of bit used.

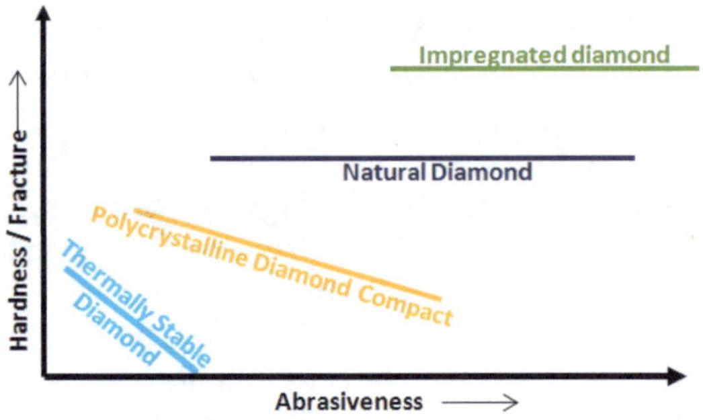

Figure 13.2: Example of plot to show decision making for choice of bit types

As shown in Figure 13.2, Thermally Stable Diamond (TSD) will work well for low to moderate hardness / fracture and medium abrasiveness, while Natural Diamond (ND) may be used for medium hardness and medium to high abrasiveness. Impregnated Diamond (ID) is perhaps most suitable for hard, fractured and abrasive rock. Polycrystalline Diamond Compact (PDC) is one of the most used bits, as it fits in the median position for different hardness, fracture and abrasive rocks.

Spudding a Well

The start of drilling a well is referred to as "spudding", which involves the removal of soft sediments, rock fragments or dirt, and the initiation of jetting of the Conductor Pipe (CP) (Gatlin, 1960; Bourgoyne et al., 1986; Economides et al., 1997; Hyne, 2001). Onshore and offshore wells are slightly structured differently. The essential difference is the need to pass through a tube or casing known as marine ***riser*** in the water column offshore, which is not necessary onshore. The riser is a large diameter pipe connecting the BOP stack at the seafloor to the rig on the water surface; it serves as storage for drill pipes and conduit for mud circulation, as well as transfer of other materials from the seafloor to drilling and sometimes production facilities on the water surface.

For an onshore well, a ***cellar*** is dug to keep well head equipment below the ground level while the final *Christmas tree* for production is at reachable height after completion. From the top to the bottom, the upper structure of a petroleum well (Figure 13.3) include:

- **Cellar:** the initial shallow hole dug so as to install the well head or the floor from which the well will be drilled.
- **Well Head (WH):** this is the point from where the well is drilled, fitted with a strong steel holding the casing and tubing equipment. It is on the surface when on land and at the bottom of the marine riser landing on the seafloor when drilling offshore.
- **Conductor Pipe (CP):** the pipe used, which eventually forms the upper part of a well, which is not drilled but pushed/driven down soft sediments.

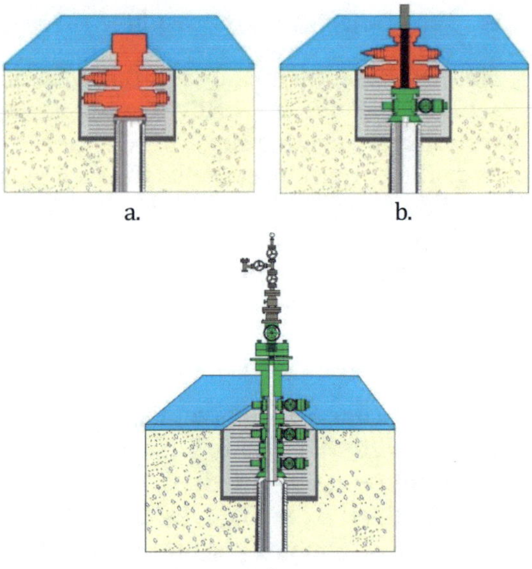

a. BOP installed above the CP
b. BOP installed above well head equipment after initial casing
c. Cellar with well head equipment above the CP and Christmas tree installed above the tubing spools for production

Figure 13.3: Cellar with well head installations

After jetting or driving in the CP, a *Blowout Preventer (BOP)* or a *Diverter* is installed before continuing drilling. The BOP is shut in emergency in case of influx into the well; the diverter in a similar situation diverts the upcoming fluid / gas away from the rig for safety. The Christmas tree, also known as production tree, is installed after drilling the well and removing the BOP; it contains chokes, a kind of control valves, through which the produced hydrocarbons pass into the flowline for storage or transportation. The Christmas tree also has a shutdown system in case of emergency.

A **marine riser** is the main casing or tube with auxiliary lines such as the choke, kill lines for circulating drilling fluid and other control lines to the BOP. The riser contains and protects the drill stem as it goes from the drillship and platform rig to the sea floor (Figure 13.4). The use of riser offshore is one of the major difference between onshore and offshore drilling.

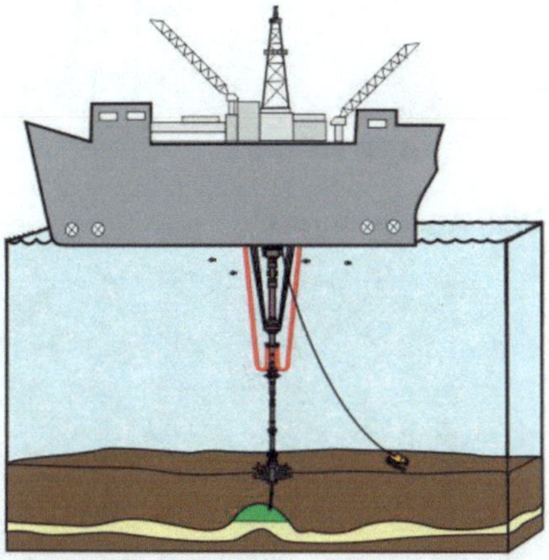

Figure 13.4: Drilling riser showing the flowlines (in red) and a Remote Operated Vehicle (ROV) in yellow for visualization of subsea installations

The riser is anchored to the drillship at an open hole (covered sides to avoid water inflow) in the ship known as Moon Pool and connected to the BOP at the bottom of the sea. A flexible and protractible system makes it adapt to the swinging of sea waves.

Rig floor

This is the small work area in which rig personnel carry out operations, adding and removing drill pipes while drilling. The rig floor is the most dangerous place on the rig due to movement of heavy equipment and possible blowout. New personnel on the rig are usually told to avoid the area or go with a supervisor. On the rig floor, the rotary system is placed on a rotating platform known as rotary table (Figure 13.5); the actual rotary table is a circular table on the rig floor that is rotated clockwise to drill. The Kelly bushing is placed above and grips the rotary table, turning it during drilling.

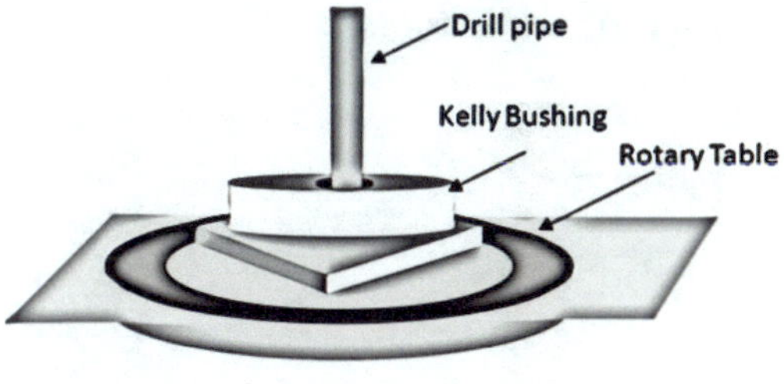

Figure 13.5: Rotary system platform on the rig floor

These levels on the rig floor are important because depths are referenced to the Rotary Table (RT) or Kelly Bushing (KB) by drillers. Though the Kelly Bushing is an older technology.

Drill bit

The *drill bit* is the cutting tool attached to the end of the system that drives the bit and any other tools attached to it. The ensemble that goes down is known as a Bottom Hole Assembly or BHA. The expression *drill string* is also used for this drill stem, when it is considered as a string used to lower other tools to the subsurface. While drilling, the drill bit is constantly within the drilling fluid to ensure good lubricity. Figure 13.6a. is a *tricone bit* showing the nozzle from which mud is injected into the well for circulation. , also known as **rock bit**, has hard rotating rollers. There are different types of bits (Figure 13.6 b. and c.) and the nature of the well and rock formation to be drilled determines the one employed.

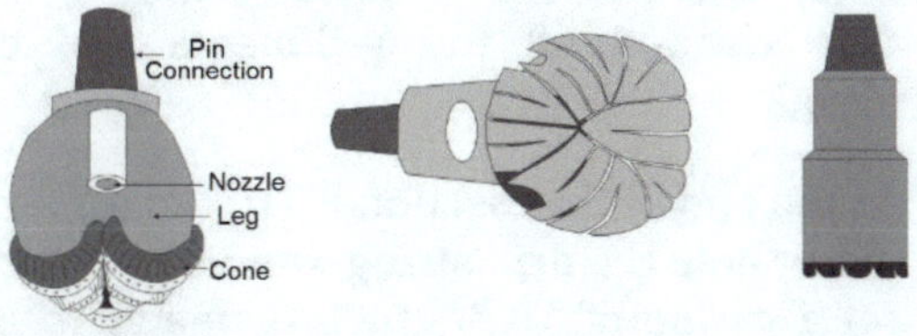

a. Tricone bit / rock bit b. Diamond bit c. PDC bit

Figure 13.6: Examples of drilling bits, showing the jet nozzle from which drilling fluid is injected into the formation

The drill bit is attached to the BHA (Figure 13.7), which is a stack of several pipes joined at the collar. Different tools are attached to the

BHA depending on the operation going on at the time, such as logging (with logging while drilling tools) or pressure test using well formation tester tools.

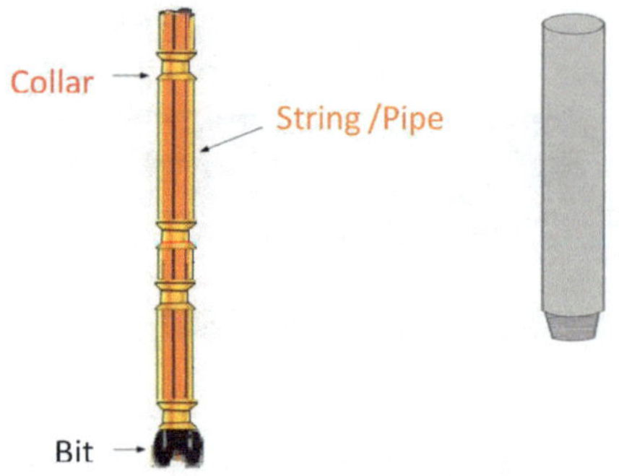

a. Bottom hole assemble

b. Joint of a drill collar

Figure 13.7: BHA and drill collar showing its joint

13.3 Well Structure and Casing

Even before drilling, the structure of the well is planned in order to use the right equipment and make good decisions. The well is drilled in phases, from the larger hole size which is cased off and cemented before another smaller diameter hole is started, by using a smaller drill pipe and bits inserted into the previous hole. This is highlighted by the well architecture below.

Sometimes the tubing is not hanged from the surface, but measured and lowered to the final Total Depth (TD), such that the top is just above the last casing. In the example of the well structure in Figure 13.8, the conductor pipe or upper casing was hammered down to a **refusal depth** of 12 m, before the drillers found resistance and decided to start drilling within it; they cleaned off the sediments and drilled ahead.

At the end of that sectional TD, the drillers pull out of hole the drill string, run in hole with and set a casing of 18 $^5/_8$ inch diameter to a *shoe depth* of 122 m. They cemented behind the casing and lowered a smaller drill string to drill to 345 m, then lowered a casing of 9 $^5/_8$ inch diameter. They then drilled to the final TD of 361 m (according to the logger) or 362 m (according to the driller). No more casing will be lowered at this depth because a smaller tube known as liner, adapted for perforation and extracting oil from the formation was used instead.

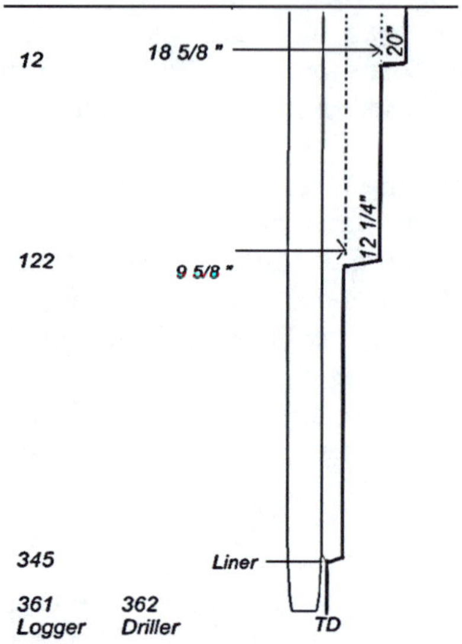

a. Well Architecture

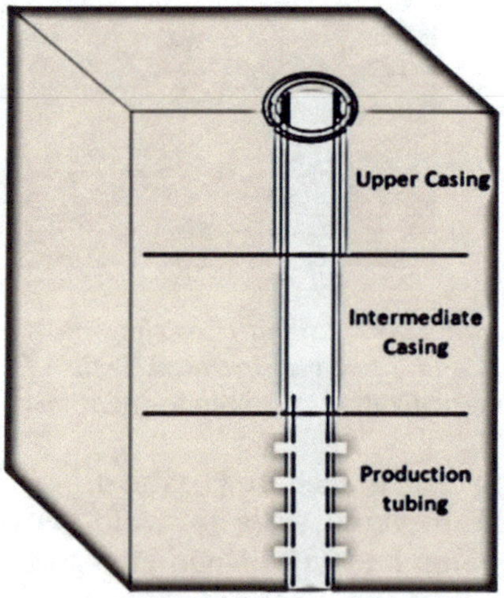

b. Casing and cementing sections

Figure 13.8: Well structure and sections, showing casings and production tubing

This architecture is planned before drilling the well, but some adjustments are made while drilling depending on real-time results.

The driller and the logging team usually have slight difference in depth, both of which are recorded. The difference stems from the minor extension that drill string or logging tool can experience at a depth with high temperature and pressure. The loggers have corrective measures for this, and often their depth is preferred by the geoscience team.

Table 13.1: Drilling fluid composition and additives

Material	Function	Specific Gravity (s.g.)	Properties
Water	Main component	1.0	High surface tension, and superior ability to dissolve different substances
Base oil (diesel, gas oil, kerosene)	Main component	< 1.0	Base for addition of dispersed water, agents and gels, also acting as emulsifier to control fluid loss
Brine (calcium and sodium chlorides)	Weighting material	> 1.0	Higher densities than fresh water and absence of solid particles that might damage producible formations
Bentonite Clay	Mud additive	2.6	Fine-grained clay used for increasing viscosity and reducing fluid loss in brine
Calcium Carbonate	Weighting material	2.7	Soluble in acid, making it suitable for drilling, completion and workover
Barite	Weighting material	4.2 – 4.5	Virtually insoluble in water and does not react with other components of the mud; used to raise mud density to stabilize formation
Carboxymethylcellulose (CMC)	Mud additive	1.59	Anionic polymer adsorbed on clay particles; for reducing filtration and raising viscosity
Lignite	Thinning material	1.1 – 1.4	Organic thinner serving water muds for reducing filtration, oil emulsification and stabilizing properties against high temperature effects
Ferrochrome lignosulfonates	Thinning material	1.1	Effective dispersing agent at high pH; used to prevent salt flocculation (clumping together to form cloggy filtrate)
Caustic soda (NaOH)	Mud additive	1.1 – 1.51	Sodium hydroxide, used to raise pH in water-based mud, counteract corrosion, and neutralize hydrogen sulphide
Soda ash (Na_2CO_3)	Mud additive	2.53	Sodium carbonate used to remove soluble salt from water and mud to enhance clay yield
Lime $Ca(OH_2)$	Mud additive	2.2	White powder used to furnish sufficient calcium so as to prevent hydration and dispersion of drilled clay and shale
Galena (PbS)	Weighting material	6.7 – 7.0	High specific gravity, used to control formation pressure, prevent formation caving, and facilitate pulling of dry pipe
Hematite (Fe_2O_3)	Weighting material	5.0	High specific gravity, used in both water and oil based fluids to control formation pressure, prevent formation caving, and facilitate pulling of dry pipe
Attapulgite Clay	Mud additive	2.4	Polymers used to increase viscosity for hole cleaning, suspension of solids and fluid loss reducer
Graphite	Lubrification	1.9 – 2.3	Agent used to reduce drilling fluid coefficient

	slurry		of friction, to decrease torque and drag

Drilling Fluid

The fluid used for drilling, usually a mix of water and different additives, also called "mud" help to:

1. cool the drill bit
2. avoid sticking of the drill string down hole
3. recover the drill cuttings from the formation when the fluid returns to the surface
4. stabilize the formation by controlling formation pressure
5. serve as a channel for data communication, induction of response, interaction with the formation and tool, and eventually transmission of signal to the surface

There are two main types of drilling fluid:

1. Oil Based Mud (OBM), or (Non-Aqueous-Based Mud (NABM)
2. Water-Based Mud (WBM)

Table 13.1 details the different composition and uses of drilling fluid. Barite with a specific gravity of about 4.2 s.g. is one of the most commonly used minerals for controlling the weight of drilling fluid (weighting up or weighting down), because it is suitable for both oil based mud and water-based mud.

Cementing

The casing is *run*, lowered down the hole to ensure the stability of the well; it is slightly smaller in diameter than the well so that it could be cemented onto the wall of the well. The cement is injected into the casing pipe and pressured so it comes up from behind the casing and solidifies between the pipe and well wall in a few hours. A type of cement used for this purpose is known as *Portland cement,* which is adapted for bottom hole temperature and pressure.
There are two types of cementing job described below: primary and secondary cementing.

Primary cementing, where a separating adjunct known as ***spacer*** is used to displace the cement, overlying the mud, from inside the well to behind the casing where the cement sets. This could take about 12 to 24 hours in normal operations, depending on the length of the well section cemented. In Figure 13.9, cement is injected into the casing during primary cementing and forced behind it where it solidifies. Secondary cementing is carried out due to improper cementing job at the primary stage or in longer casing sections.

Secondary cementing, which is necessary for long casing section, and where there is a weak formation leading to a non-proper cementing from primary cementing. In that case there might be a need to do a "top up" from above, or "squeeze in" from the side. In either case, cement is injected directly behind the casing to fill up outstanding section. In extreme cases, the casing may be pierced so as to fill the poorly cemented regions.

There are two common logs used in assessing the efficiency of a cementing job. These are:

1. *Cement Bond Log (CBL):* This evaluates the bonding between the casing and cement
2. *Variable Density Log (VDL):* This evaluates the bonding between the cement and well wall

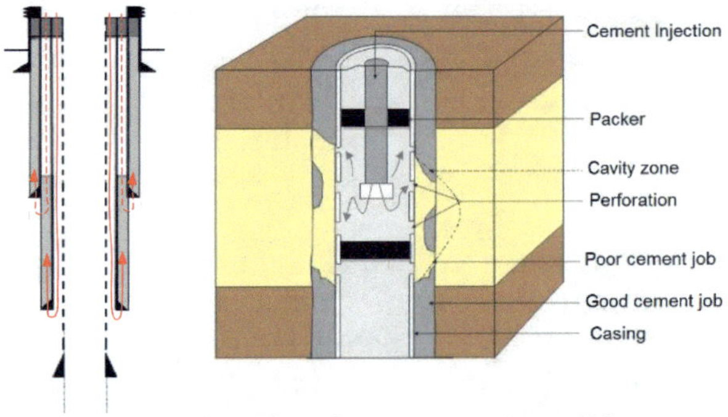

a. Primary cementing   b. Secondary cementing with squeeze job

Figure 13.9: Cementing operation.

In Figure 13.10 the quality of the gamma ray (GR) is poor due to casing. The transit time is lower in good casing and higher in poor or absence of casing. The principle of casing and cement evaluation logs, such as Cement Bond Log, is that good casing reduces the amount of waves that get to the receptors after reflection, giving lower amplitudes. They work on the basis of a standard calibration of sonic waves based on empty casing; when measured it gives a base value of amplitudes as will be read by a normal casing pipes.

Any increase in amplitude to the right implies poor casing (lower response in Figure 13.10) and reduction to the left implies good casing (upper response in Figures 13.10).

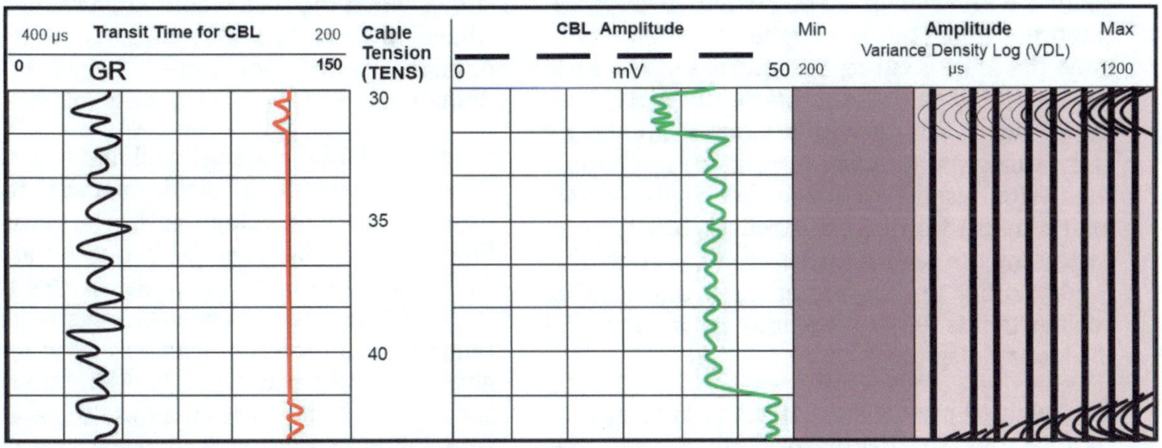

Figures 13.10: Illustration of casing and cement (as well as cement to well wall) evaluation logs: CBL-VDL with GR and sonic transit time within casing

For the Variance Density Log, increased attenuation (fading of sound waves) implies good cement job. On the VDL shown in Figure 13.10, the weak signals on the upper response imply good contact with formation wall whereas stronger signals into the formation on the lower response imply poor cement job. The amplitude presented in shades of vertical grey bar show the level of penetration of the waves in VDL.

13.4 Drilling and Rock Mechanics

Before any drilling takes place, it is important to understand the nature of the sediment or rock unit to be drilled. This is known from the result of geological work, use of seismic and other geophysical methods and from previous offset wells in the region. Once there are samples from a given well, further measurement is made in the laboratory to understand the petrology and collaborate it with nearby or regional knowledge. For drillers, the most important information is the resistance of the rock, how hard it will be when the drill bit tries to penetrate it.

The resistance of a rock can be measured by applying a force directly to it and noting when it breaks (Jaeger, 1979; Breckels et al., 1982; Atkinson, 1987). At the point it breaks, the applied force is known as S_1. If applied at an angle of about 45°, the rock breaks in a way that not all the effect of the force has affected it. That is, some of the rock has been directed to break away from the direction of the force. This type of breakage or fracture is known as *shearing* (Figure 13.11b). The force at the point of breakage, in the second major direction, is known as S_2.

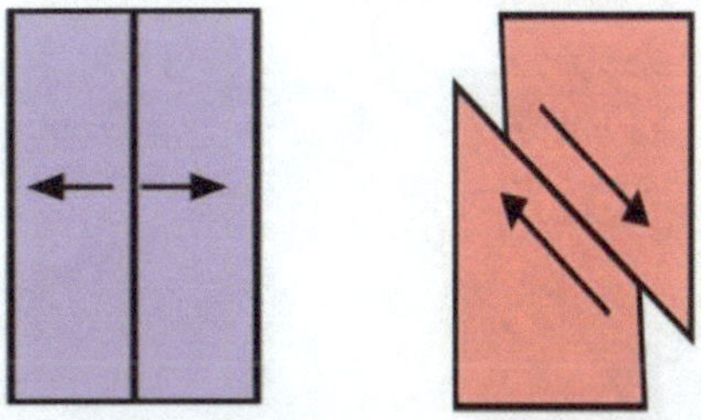

a. Open fracture in extensional domain

b. Shear fracture in compressive domain

Figure 13.11: Rock breakage and shearing

Open fractures (Figure 13.11a) occur where the vertical load is much higher than the horizontal, whereas shear fracture occurs when the vertical load has been affected by horizontal stress components (Figure 13.11). This can be represented in a Mohr's diagram, where the extensional domain occurs to the left before the 0-point where the vertical axis represents the torque and cohesion of the rock (Figure 13.12).

Rock Test and Mohr's Diagram

The force applied to a rock may result in confining, compression, tension or shearing the rock; such a rock is said to be acted upon by stresses. Mohr's diagram provides a way of plotting S_1 and S_2 and using it to understand the general behaviour of a rock when stresses, or rotating drilling bit, are applied to it. To determine the Mohr's curve, the stresses of the rock are measured in the laboratory and plotted on the X and Y axis to obtain what is known as Mohr's diagram (already introduced in *Chapter Nine* with respect to subsurface stresses and hydraulic fracture). Rock strength test help to determine the type of drill bit to be used and to predict other drilling parameters. One way to achieve that is the use confining stresses to plot the Mohr's diagram.

As explained previously, when a rock sample is confined in the laboratory, such that the stresses acting around the rock in the x and z axes are equal (let's say $\sigma 2=\sigma 3$), then a major force can be applied directly to it in the y axis (let's call this $\sigma 1$) and record at which point it breaks or shears. Then the confining pressure is varied to obtain various $\sigma 3$ and $\sigma 1$. When $\sigma 3$ and $\sigma 1$ are plotted on the horizontal axis, and the centre ($\sigma 3 + \sigma 1 / 2$) is used to plot a circle whose radius is $\sigma 3 - \sigma 1 / 2$, similar to the diagram in Figure 13.12 is obtained, which is the Mohr's diagram.

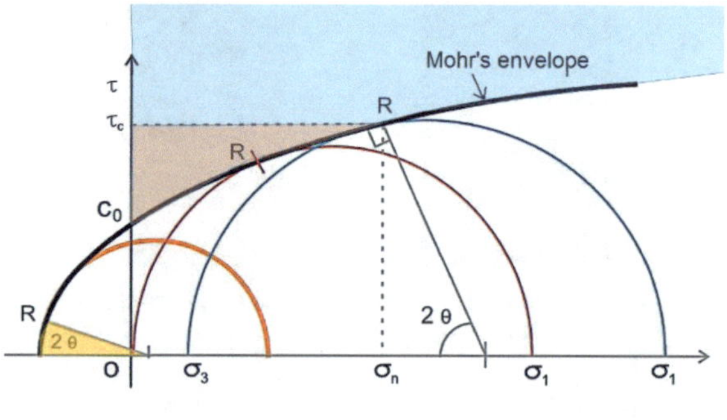

Figure 13.12: Mohr's diagram showing the Mohr's envelope (and different domains with varying rock behaviours)

To throw more light on what has been discussed, a lot more can be said of the rock properties and the impact on drilling, from this diagram. When the upper limit is drawn by a line that follows the tangent of each circle, this Mohr's envelop crosses a vertical y axis known as the *shearing component* (or torque, τ). τ gives an indication of how the rock will fail, or be broken while drilling it. The point the envelope bisects the τ axis gives the cohesion of the rock, *C* (noted as C_0 in diagram). Any torque value higher than this point, or above this envelope, would have resulted in the failure of the rock. Any value within, below the envelope, shows when the rock would not have not failed. Above the Mohr's envelope is the failure domain of the rock tested; below it is the domain in which the rock has not failed.

A more brittle material will have a straight envelope, whereas a ductile material tends to have a curved envelope as in the example in Figure 13.12. The angle 2θ between the radius, from the centre of the circle to the tangent (point R) line of the envelope on the circle, is related to the *internal friction* of the material, given by other equations. It is the measure of the ability of the rock unit to withstand shear stress. Today computer analyses are used to simplify the use of stress relations. As stated above, the Mohr's diagram concept is applied in understanding rock resistance and stresses. Stress in the subsurface is important to drilling in many ways: one way is in deciding the appropriate drilling bit and weight on bit while drilling, and in following the path with less resistance (Chester and Logan, 1986. Burland, 1990; Sibson, 1998, 2003).

Stresses in the Subsurface

To further explain the concept of stresses and how they affect rock behaviour and drilling, there is need to look at each of the stress components in more details. The major or principal stresses acting on a rock are:

S_1: maximum principal stress acting directly (say on the x-axis)

S_2: second or intermediate principal stress (acting at the other axis, say y-axis)

S_3: minimum principal stress (acting on the last cardinal axis, say z-axis).

As in a solid in the laboratory, the force applied vertically by the load of the Earth creates one normal stress (S_1) and two tangential stresses (S_2 and S_3) acting say horizontally. In Figure 13.13, the directions of the stress acting directly on the planes are called normal principal stress. Often S_1 is greater than S_2 which is greater than S_3. They are commonly called vertical (or overburden stress S_1) and horizontal stresses (S_2-intermediate and S_3-minimum). In Figure 13.13, S_2 and S_3 could be interchanged on the horizontal axes, depending on which axis has the higher value (S_2).

Similar to the way these stresses are used to understand rock properties in the laboratory, they can also be applied to drilling.

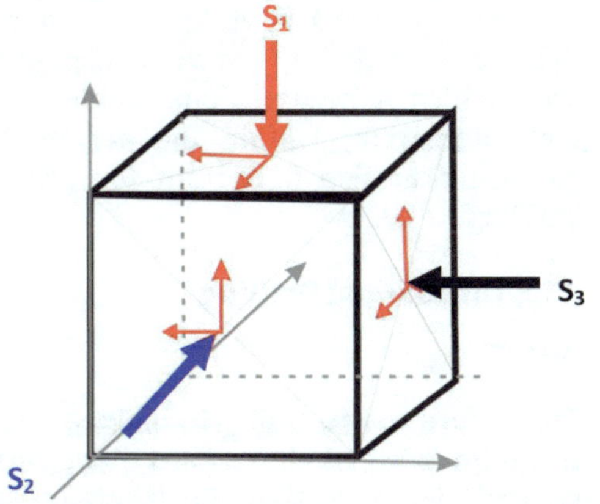

Figure 13.13: Principal stresses diagram showing the vertical and horizontal stresses (Compare with σ1, σ2 and σ3 explained in Mohr's construction in Figure 13.12).

Drilling in the direction of the minimum principal stress (S_3) in a horizontal well will make it easier. But apart from directly affecting the strength and drillability of a rock formation, stresses in the subsurface also control the pressure in-situ, which is one of the highest challenges of drilling (Brace, 1978; Breckels et al., 1982; Westbrook et al., 1994). A combination of the stresses existing in-situ and the additional pressure by the rotating drill bit culminate around the well bore. The principles of geomechanics are used to understand this, through calculations and concepts to be discussed in the subsequent sections.

The pore pressure relationship with rock load, leading to the development of the Terzaghi equation, is closely related to the behaviour of stresses in the subsurface. It relates the lateral dynamics of fluid and grain properties to vertical stress:

Vertical stress (OBG_v) = *Fluid pressure* (P_{fluid}) + *Effective Stress* (E_s)

(This equation has been explained in Chapter Twelve, on Pore Pressure Prediction and Interpretation)

Effect of Temperature

Temperature affects the drilling of a formation. Subsurface temperature is taken into consideration while drilling, casing and during production, because they directly affect drilling and casing tools and the fluid type, as well as possible chemical reactions. Rocks containing colder fluids fracture or break faster while drilling through them, as the rocks will be less 'ductile'. Knowing the original borehole temperature is important for establishing a geothermal gradient and for mitigating drilling challenges. The challenges with high-temperature and high-pressure wells, known as HT/HP wells, are currently of major concern to the industry. At certain depths, rocks behave differently, likewise the fluid they contain and the functioning of the drilling equipment downhole.

Temperature tools are attached to drilling stems and the temperature they measure are affected by the drilling processes. While measured Bottom Hole Temperature (BHT) is unreliable due to disturbed formation before measurement, extrapolating the temperature to the initial time before circulation can be achieved by a correction using the *Horner Plot* (Figure 13.14).

The Horner plot is a cross-plot of *build-up* or *drawdown temperature* on the vertical axis against a time coefficient on the horizontal axis (as shown in Figure 13.14). Several of the estimations are plotted to give a good interpretation. It works by estimating the effect of circulation or cooling and retrograding to remove the effect. Ideally, waiting for several days after stopping circulating would start restoring the hole to its initial temperature.

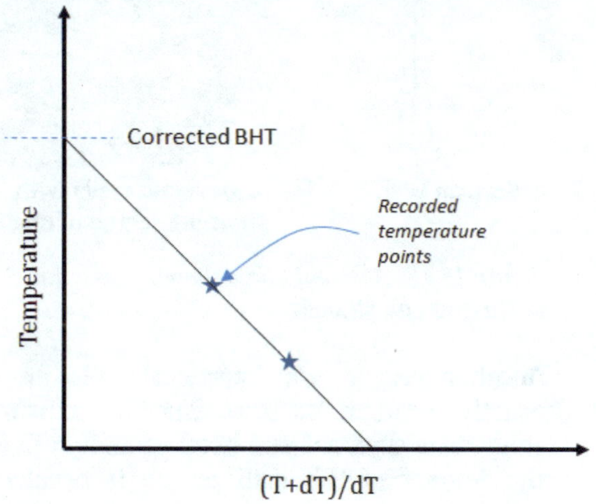

Figure 13.14: Simplified Horner Correction Plot, giving the original (corrected) Bottom Hole Temperature (BHT)

In Figure 13.14, the corrected BHT is obtained by extrapolating measured temperatures to a time before end of drilling and circulations. This Horner plot requires at least two temperature measurements at different depths and circulation time. T is the time of logging since stop of drilling and circulation

Vertical versus Horizontal Well

A vertical well has a simple trajectory in which the Measured Depth (MD) and the True Vertical Depth (TVD) are the same. The MD, TVD, Azimuth, Inclination, Vertical Section, Dogleg, and other survey variables remain the same downwards, from top to bottom. However in reality, no well is truly perfectly vertical. There would be slight changes in the intended vertical penetration. Due to varying subsurface conditions, there is need to carry out directional drilling (Anon, 1989; Barret and Lyon, 1988; Cooper, 1994). Some holes are intentionally drilled at different angles so as to be able to reach the targeted reservoir.

A vertical well actually does not have zero (0) deviation all through, due to the nature of the tool and the drilling process; it has a minimal deviation of about 1 to 2 degrees, usually less than 3^0 as shown in Figure 13.15.

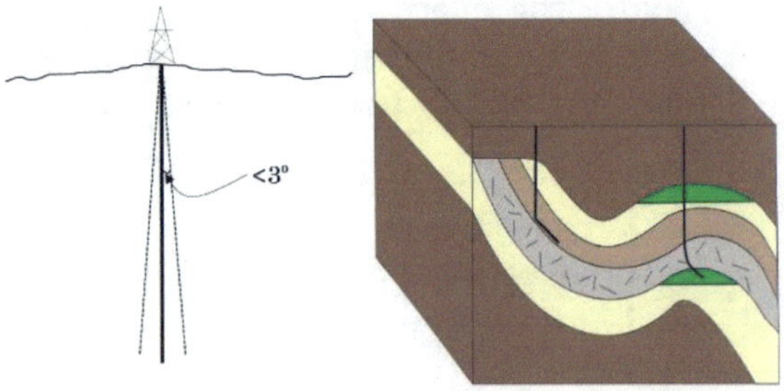

a. Vertical Well

b. Deviated wells with different degree of dogleg

Figure 13.15: A straight hole showing less than 3^0 inclinations on all sides

Another reason why vertical wells are not exactly straight is the varying subsurface conditions such as local heterogeneity affecting the impact of the drill bit as it breaks the formation.

Horizontal wells are deviated much more than 3^0 and often the deviation is intentional. Some deviated wells move away from the vertical smoothly while others are more crooked; however there is a maximum possible change of degrees with depth (known as *dogleg severity*) after which it becomes impossible to case the well. Remaining within the permissible hole curvature is also important for drilling tools stability.

13.5 Directional Drilling

Basics

This refers to the ability to make a well depart from its vertical downward penetration and control its direction horizontally or near horizontal below the surface. It is achieved by using a wedge object that deviates the trajectory of the well. One of such objects is known as a **whipstock**. A *whipstock* is a steel ramp shaped wedge inserted into the borehole, usually at the base on the cement plug, to enable wellbore deviation by deflecting the bit at an angle away from the vertical (Figure 13.16).

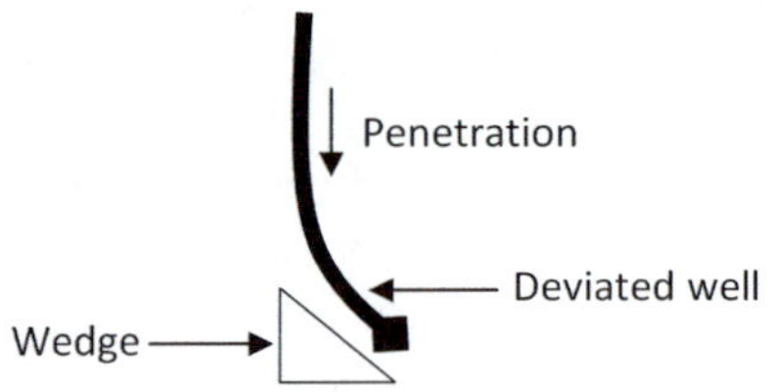

Figure 13.16: Sketch showing how a wedge object deviates a well

Today, directional drilling has advanced methods to change drilling direction (using a **mud motor** for example, discussed below), moving ahead from the very early methods (Eastman, 1954; Bourgoyne et al., 1986; Edlund, 1987; Eddision and Symons, 1990). The calculation and application of desired trajectory is achieved through modern techniques. With this ability to redirect a well, some may look straight while others may take the form of an s-shape as shown in Figure 13.17.

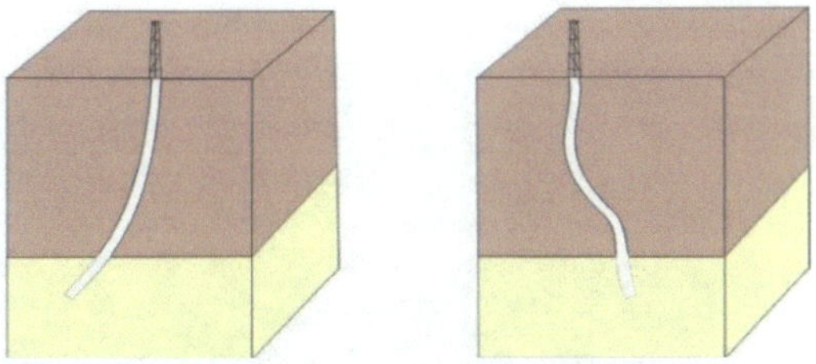

a. Curvilinear b. S-shaped deviation

Figure 13.17: Examples of deviated wells

Directional drilling is necessary in drilling non-vertical wells, especially for long drains. The well tracks are measured or surveyed (using Measurement While Drilling - MWD methods) and plotted on charts which help the driller and geoscientist to follow the well trajectory.

Why Drill Directional or Deviated Wells?

1. **Relief well:** As mentioned earlier, when there is a problem downhole involving pressure build up, another well is drilled to reach the porous layer with the overpressure, to dissipate the pressure, from an offset distance. Such a well has to be deviated, so as to arrive at the same target but from a different surface location (ref. Figure 11.12). A similar procedure is used for geothermic (heat) production from existing wells (where a renewable energy technology is used to produce energy from the deep lithosphere), and for fluid injection in enhanced recovery of hydrocarbons (described in a previous section).

2. **Clusters:** To reach different reservoir targets using multiple wells, but one manifold; there is need for multiple deviation downhole. This is known as multi-well platforms or clusters. A production manifold is an equipment, usually at the sea bottom offshore, installed close to pipelines for the collection of produced oils. Wells will have to be drilled such that well heads are close to the manifold, so as to direct produced fluids to collection flowlines.

In Figure 13.18 a., the map view shows the reservoir targets but the vertical layering of the units is not obvious; the wells deviate from the centre of the cluster to the outer black circles, reaching reservoir (green) with example intersected on the left by cross section A-B;. In b., oil pools in different reservoir targets are reached by the wells. In c., only lenses of two reservoirs intersected by the cross section C-D are shown.

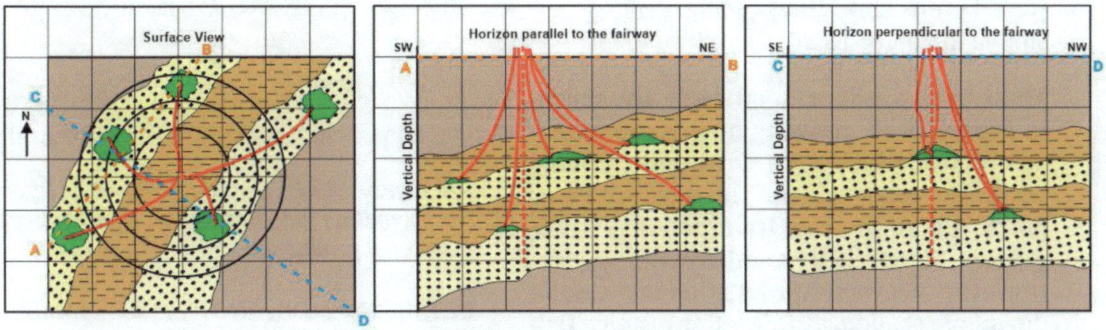

a. Map View b. Section A-B view parallel to fairway c. Section C-D view perpendicular to fairway

Figure 13.18: Multiple deviated wells of a cluster targeting different reservoirs

3. **Sidetracking:** example, for better downhole drainage or failed fishing. A fish in this context is a tool lost downhole. Wells are drilled away as sidetracks to the main well in order to drain one or more reservoirs horizontally. This is similar to a cluster, but different as the interest on deviation occurs from one already existing well, and targeting the drains. It can also be used when drilling can no longer continue, because a tool is lost downhole and cannot be *fished* (retrieved). The well may be plugged at that depth (seal by injecting some cement), then resume drilling and start to kick off, side-

ways, from a certain depth. That is *sidetracking.*

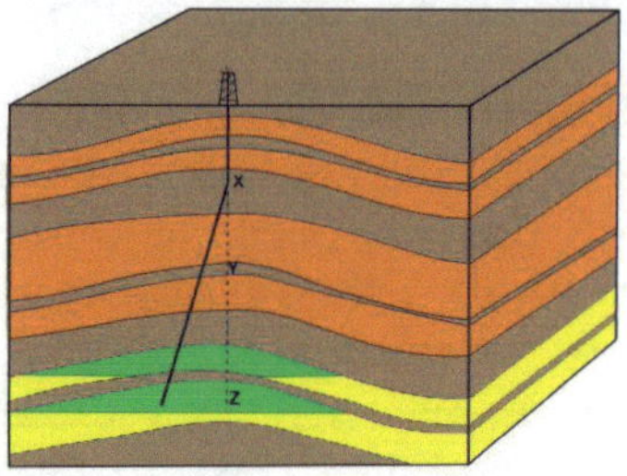

Figure 13.19: Sidetracking of a vertical well

In Figure 13.19, the well was drilled from top to Y earlier (intended get to Z), before there was a problem leading to plugging the well with cement from Y to X. This was followed by sidetracking from X to the same reservoir and depth at Z on the left. The original proposed trajectory would have followed the dotted line from Y to Z.

4. **Geological barriers:** Geological structures and bodies such as salt diapirs, faults, dipping beds and domes are reached by deviated wells where they are the target (when they have accumulated hydrocarbons) or avoided where they are a barrier to the reserve. A well could be a barrier to another well, due to spacing requirements between wells. Other causes of deviation, such as inaccessible site and need to remain onshore (just before shoreline), while drilling, are shown in Figure 13.20.

   Wells from onshore to offshore are shown in Figure 13.20, showing how a mountainous region, which does not permit drilling vertically, was avoided. The next well was positioned to produce from multiple stratigraphic targets, and another with a sidetrack that allows for production in a case of downhole challenges (like after losing a **fish** – a drilling tool). This is followed by an injector well for enhanced oil recovery, and a near-shore well targeted to produce offshore pools from a convenient onshore location. An offshore well was then used to target production from more than one pool further down slope.

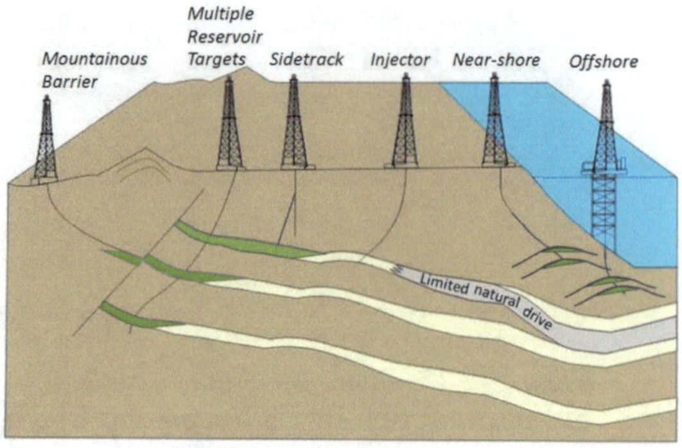

Figure 13.20: Well from onshore to offshore showing different reasons for deviation (modified from Beebe, 1961)

5. **Horizontal fracking:** In unconventional resources, fracking and reaching deposits such as tar sand or shale oil require horizontal wells. Fracking has limited distance of propagation, requiring that drilled wells approach the area of interest as much as possible. This is achieved by horizontal or lateral wells; fracking on vertical well has limited reach compared to horizontal wells (Figure 13.21).

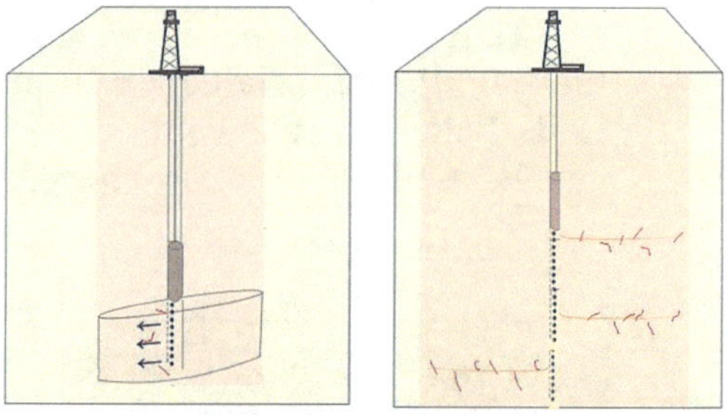

a. Vertical fracking   b. Horizontal fracking

Figure 13.21: Horizontal deviation laterally into formation for fracking purposes in unconventional production

6. **Horizontal drain:** This is common for long narrow horizontal reservoirs such as turbiditic sandstones. They occur as channels or lobes that extend laterally, requiring that the drain section of the well is horizontal. This often requires Geosteering and multiple development injectors and producers (injection and production wells). As shown in Figure 13.22, horizontal drain is different from extended reach well. In a., the horizontal drain is within the reservoir while in b. the horizontal path started way before reaching the reservoir.

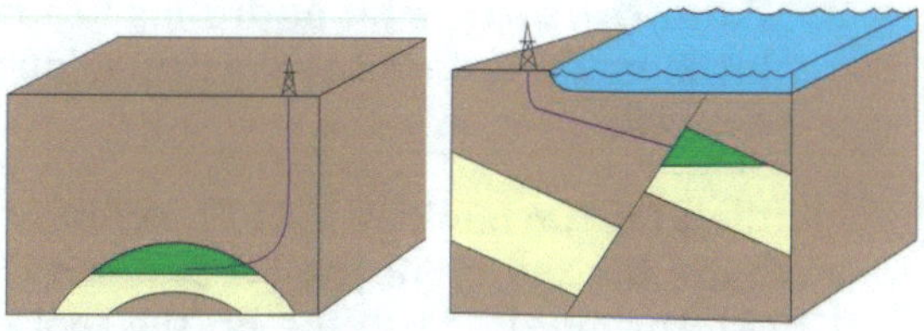

Figure 13.22: Wells for horizontal drain of a reservoir and a deviated well to reach distant reservoir

Directional Drilling Techniques

Directional drilling employ a combination of techniques. Below are some of the tools used in deviating a well and how they work is necessary:

1. **Stub**: a simple stub or wedge with a slant is placed at the bottom of the well after drilling some distance down. When the bit is lowered above it, it tilts the head in such a way that a deviation is formed while drilling further (ref. Figure 13.16).
2. **Mud jetting**: a nozzle close to the drill bit is made to jet mud in one direction, so that the bit moves in the opposite direction, leading to deviation of the well.
3. **Bent drive sub**: the bit is attached to a flexible unit attached to the drilling string. Bending the unit, known as bent drive sub, deviates the direction of the well.
4. **Mud motor**: this allows the flow of mud within a motor system to add different pressures at different parts of the motor, thereby orienting the drilling head.
5. **Rotary steerable systems**: an example is the Power Drive system, which is a modern technique in which the drill bit has a mechanized tool that drives the drill head to different directions in other to achieve directional drilling. There are the "Push-the-bit" and "Point-the-bit" systems. The push-the-bit requires additional pressure to create a deviation, while the point-the-bit teleguides the face of the tool. They are more precise and the desired direction is monitored from a screen on the surface.

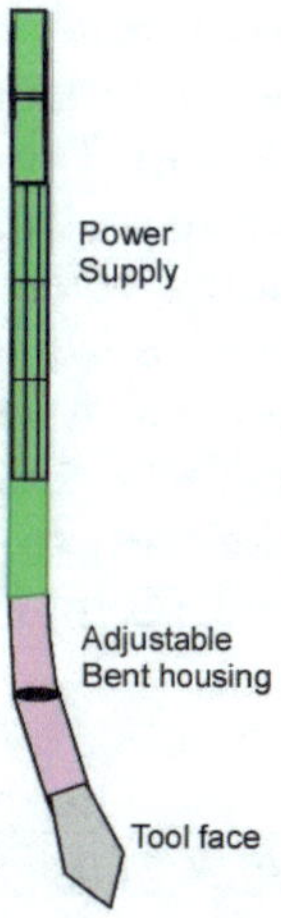

Figure 13.23: Downhole motor and bent housing

A representative downhole system that can be powered as a motor and bent at the adjustable unit is shown in Figure 13.23. The rechargeable battery in it is stored on the upper part of the string, while the adjustable bent or bendable unit is closer to the end of the bit.

13.6 Measurement While Drilling

Measurement while drilling (MWD) is the art of taking information as well as sending down information to direct the well. MWD could be said to be a logging that focuses on the direction of drilling. This is because a log is a record of a given parameter with depth or with time across a geologic unit. That the logging is taking place while drilling introduces the concept of Logging While Drilling (LWD).

Any test carried out while drilling is measurement while drilling, but in the industry, MWD is used to class those measurements which pertain to the drilling itself, while logging pertains to those related to formation evaluation. MWD and LWD help to localize the position of the drill bit and direct it to the target. How well the directional drilling system reaches its target gives its ***precision*** (mean absolute deviation or repeatability) and ***accuracy*** (closeness to the actual desired target).

Sensors shown in the tool diagram of Figure 13.24 include, from left to right, those for measurement of gamma ray, resistivity, inclination and rate of penetration, respectively. This tool is fitted as part of the Bottom Hole Assembly (BHA) for drilling and logging the well.

Before logging begins, it is important to know the offsets of the sensors in the tool (compare Figure 13.24 and Table 13.2). Sensor offset indicates how up on the tool a given log sensor will start reading, when the bit is at the bottom. By implication, a zone is logged only after having drilled past it (except for logging tools "at bit"). In Table 13.2, the distance of the sensor (on the tool) to the bit is measured; as an example, the GR tool is 33.24 m (offset) from the bit.

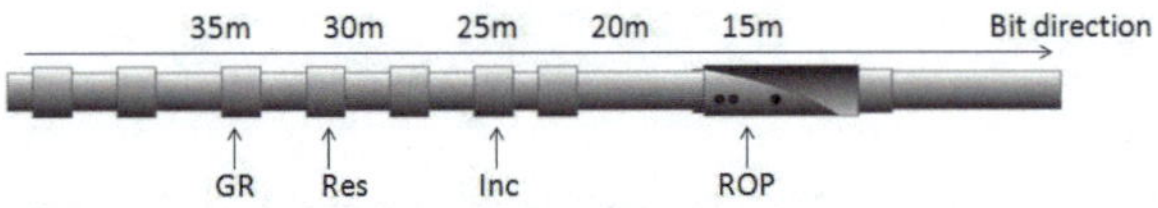

Figure 13.24: Tool diagram and sensors showing different offsets from bit

The understanding of this '***lateness***' in arrival of information is important for well safety (reaction time, where a log value is depended upon for safety). It is also important to consider a ***rat hole***, which is an additional hole to allow the logging tool reach intended total depth (TD) for each section. The quality control before the interpretation of data may require a ***relog***, involving logging a section twice so the interpreter can properly ***splice*** (cut) and ***merge*** (attached them). This effectively links an upper logging phase to a lower one by correlating the small section logged twice.

Table 13.2: Example of a sensor offset recording

Tool Name		Sensor	Distance to Bit
GYRO	:	ROP	15.11 m
TELEMETRY	:	Inc	25.88 m
ARC	:	Resistivity	31.19 m
ARC	:	GR	33.24 m

MWD records geometrical information such as the azimuthal survey and inclination. It also records the drilling dynamics such as the downhole torque and the drag. The geometrical information on direction is achieved by two directional drilling technologies (which are discussed further in the next section):

1. **Push the bit:** added pressure forces the tool to bend in between or close to the drill bit, which then makes the drilling head to deviate.
2. **Point the bit:** the bit or drilling head of the bit is made to face the particular direction desired.

Multilateral Technology (MLT) combines the push the bit and point the bit systems (Figure 13.25). During the push the bit the tool is said to be in *sliding mode* and during the point the bit it is in *rotational mode.* Rotation occurs during normal drilling while sliding is drilling without rotation, especially while building angle. There are two ways to communicate and redirect the bit:

- The **power drive**: information transmitted electronically redirects the bit face (it is also known as rotary steerable systems, RSS)
- The **mud motor**: mud flow through the spiralled string changes the direction to which the drill face points

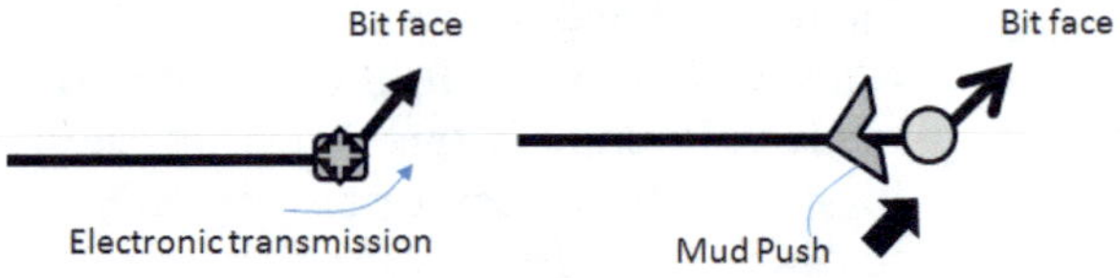

a. Point the bit b. Push the bit

Figure 13.25: Illustration of Point the bit and Push the bit systems

Principles of Magnetometer and Gyroscope

Directional results are communicated to the surface tool and the orientation of the well is recorded. This is achieved by using *magnetometers* (as with a compass) or *gyroscopes* (for a given known direction). Magnetometers work by sending information on the direction of the bit with respect to the magnetic (or geographic) North field at a particular location. Gyroscope gives an indication of how much the well path has deviated from a given direction, as in *traverse mapping* where the degree of deviation from an initial direction is used to trace and retrace a direction.

Trajectory and Survey

A **well trajectory** is the path the drilling tool follows from the surface location or from a current depth point to the required target. The **target** is the required destination of the drill tool. Well trajectory is controlled and represented using:

a. *The drilled length known as "measured depth".* This is measured using physical tools such as the drill string depth, known by adding the total number of drill pipes used.
b. *The inclination given in degrees and giving the vertical deviation.* This is measured using inclinometers.
c. *The azimuth, also in degrees but showing the horizontal plan.* This is measured using magnetometers (as with compass) or gyroscopes (for a given known direction).

In the representation of Figure 13.26, the Azi in red is measured clockwise while the Incl in blue is measured away from the vertical. The resulting coordinates are represented with a universally known reference system such as the Universal Transverse Mercator (UTM) coordinates. The depth, inclination and azimuth measurements, often presented in ASCII and LAS file formats, are known as the **survey**.

Figure 13.26: Representation of survey showing the measured and true vertical depths referenced to the Rotary Table (RT), Inclination (Incl) and Azimuth (Azi)

Following the Well

A successful well follow-up is achieved by constantly surveying the well using the magnetic method, for example. This is plotted by the wellsite and operations geologists in a survey sheet, as shown in Table 13.3. The survey gives the deviation of the well away from the vertical, and is especially useful in deviated wells. A survey may be taken at the end of each drilling stand or drilling pipe.

The survey in Table 13.3 gives the well head coordinates and other information necessary to identify and position the well, such as field name, Universal Well Identification (UWI), and Coordinate Reference System (e.g. UTM). These are placed in the survey header and make up part of the survey report.

Table 13.3: Illustration of Survey Sheet (compare with Table 11.1)

Parameter	MD	TVD	Incl	Azim	Vert. Sect.	Northings	Eastings	Latitude	Longitude	DLS
Unit	m	m	deg	deg	m	m	m	DMS	DMS	°/30 m
1										
2										

(DMS = Degrees, Minutes and Seconds; Depths referenced to RT or Rotary Table)

With the availability of survey data based on planned well path, the deviation curve can be plotted on a *well trajectory sheet*. This is updated with each actual survey measured while drilling, and the geologist is able to follow the well, comparing the actual trajectory with the prognosed. This will help to identify errors or avoid consequences of poorly directed or geosteered well (see Figure 11.11 for *well profile* showing trajectory in map and section view). Some consequences of poor trajectory follow-up in a geosteered well are:

- Missing the target
- Collision
- Delayed drilling
- Equipment damage

13.7 Geosteering

Geosteering is the control of a well trajectory in the desired path as it is drilled downhole. Geosteered well is a special case of directional drilling where the basic deviation is not the only objective, but also a careful steering of the hole both horizontally and laterally (as in a horizontal meandering reservoir channel for example). The drill bit may be steered: moved up or down, left or right, downhole. This is important for draining long horizontal reservoirs, especially in turbiditic channel reservoir in deep offshore sediments. Geosteering and directional drilling is riddled with many terminologies, most of which will be defined below.

Geosteering Principles

Most wells leave the **surface location** downwards from a vertical position and then deviate (Figure 13.27). That point from where a well deviates from the vertical is known as **kick-off-point (KOP)**. That angle of deviation when measured from the vertical is known as **inclination**. Laterally, a similar moving away from the present direction is measured azimuthally; the **azimuth** is the angular direction from the North. The angle between the initial azimuth and the target direction is known as **lead angle**. A change in direction of the well is known as **dog-leg**. The straight section of the hole is known as the **slant**. If the hole is curved, the radius at that point is known as the **radius of curvature**.

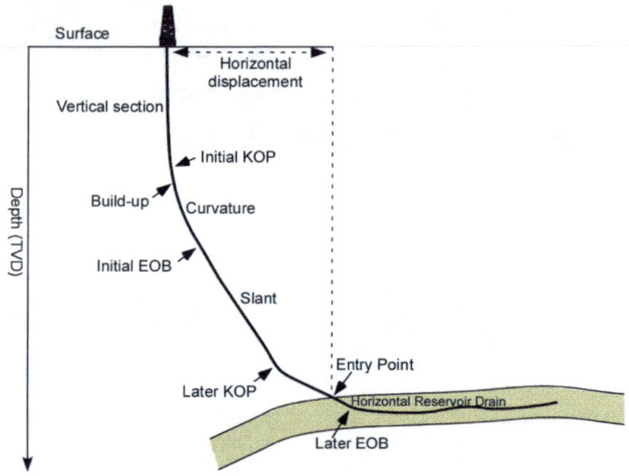

Figure 13.27: Well path of a directional well with Geosteering within a reservoir drain

The ***displacement*** or *horizontal displacement*, as shown in Figure 13.27, is the horizontal distance between the point of initial vertical drilling on the surface and the entry point in the reservoir. That is the horizontal distance the well has move away from the well head vertical axis. The term ***vertical section*** or ***VSec*** is used for the measure of the varying displacements of the well, for different sections away from the vertical axis, directly towards the intended TD at a particular azimuth (Figure 13.28). It is actually the horizontal distance from the central point of the wellbore to a point of interest on the well trajectory, along the given azimuth.

In Figure 13.28, Well path **Y** in **a**. has its maximum displacement at the Total Depth. In **b.** an idealistic vertical meandering of a well **X** is used to show possible smaller displacement at well TD compared to vertical section (distance from the well central axis at the time) somewhere along the well path. This would occur if the true vertical depth and the target reservoir is reached earlier with more deviation than initially planned.

An increase in inclination along the well path is known as **build-up**. Every 10 or 30 m, or every 100 feet, the change in degree gives a gradient known as the **build-up rate**.

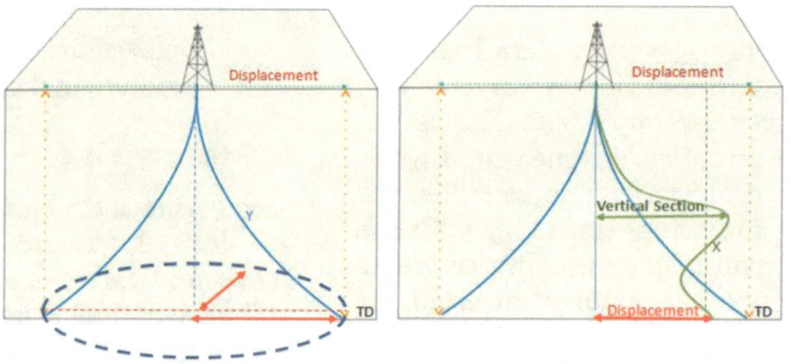

a. Well Displacement at TD b. Vertical sections along well path

Figure 13.28: Final horizontal displacement and varying vertical sections along well path

When the inclination stops building up, it is known as **end of build-up** (EOB). A mere decrease in inclination is known as **drop-off** and drop-off gradient is given in deg/10 m, deg/30 m or deg/100 ft. Where the deviation is so high that the well path is in a horizontal or near horizontal hole, the path is known as a **drain,** which is often used to produce hydrocarbon from long horizontal reservoirs.

Horizontal Reservoir Drain

When drilling within a drain, within a horizontal reservoir (in which there is an intention to run a casing, a liner or especially to leave an open hole) for production or injection, there is need to ensure the bit does not pass the reservoir and enter the shale (or non-reservoir unit) for the final depth. In the sketch below (Figure 13.29a), the thick line represent the well which left the "drain" at the point with the red star. This is what the drilling team needs to avoid, because the shale will introduce dirty sediments into the drain.

The Geosteering team will take regular survey (at the end of every drill stand of say 30 to 40 m long). The survey should be in line with the estimated vertical and lateral deviation of the reservoir, if in a sand channel for example (see Figure 13.29b). In addition to this, the geosciences and LWD team will study the log obtained while drilling to ensure that the responses correspond to what is expected in sand / reservoir.

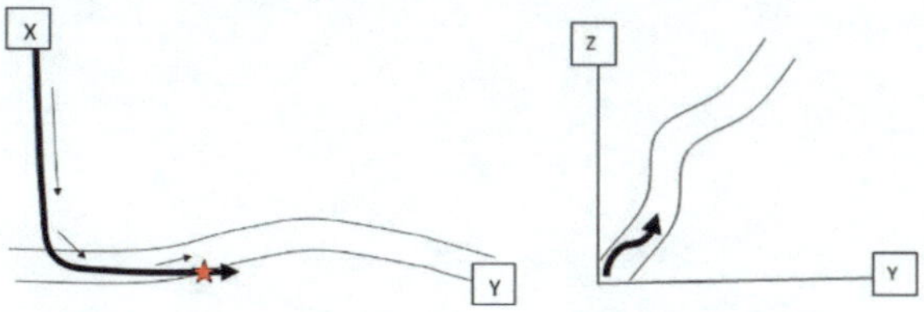

a. Vertical view of a sand drain

b. Lateral view of a sand drain

Figure 13.29: Geosteering within a meandering sand channel reservoir

Once within the drain, because of the need to prevent leaving (rather than identify) the reservoir, a tool that reads / looks ahead some distance to the next lithology or boundary is desired. Most Geosteering logs are azimuthal, meaning they look in all radial directions while the bit rotates.

Figure 13.30 shows a log 'Y' that responds when leaving the sand body, that is, it detects responses when closer to the sand boundary 'X'. The arrow below shows the drilling trajectory and the drill bit at Point 1, from where it will follow the drain through Point 2 to Point 3.

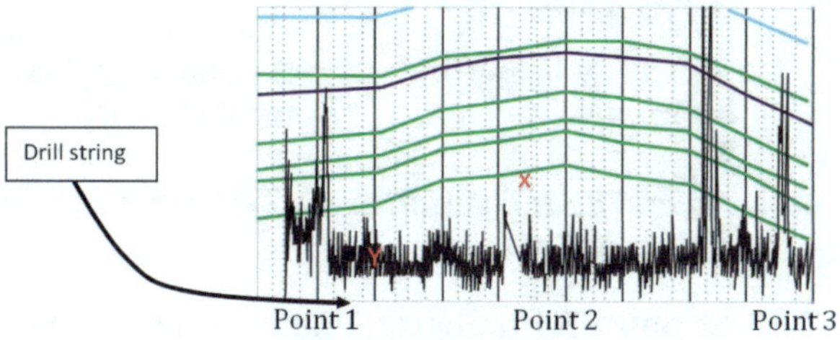

Figure 13.30: Geological section along well path

Even if the strata were not visible (which are indicated with green horizons in Figure 13.30), it can be deduced that at Point 1, the log shows that the drill bit is close to the sand-shale (higher log response) boundary. So the Geosteering team decides to *drop-off* and it is Ok at Point 2. Depending on the scales, there might be need to *build-up* a little at Point 2 so as not to leave the sand downwards. At Point 3 the bit is again very close to the top unit. The path will have to be deviated downward at Point 3, else it will go out of the drain (as in Figure 13.29a) to tag the shale above. Geosteering logs could be image, resistivity or GR logs, and their principles are similar.

Trigonometry in Geosteering

To move from a vertical to a horizontal position, a simple calculation is used to obtain what vertical length (TVD) would be reached using a given inclination angle (θ). The TVD is important because the reservoir prognoses remain the same in TVD but the measured depth depends on what angle of inclination while drilling. Figure 13.31 is an example, assuming point X to be the entry point of a target reservoir drain, at the given TVD.

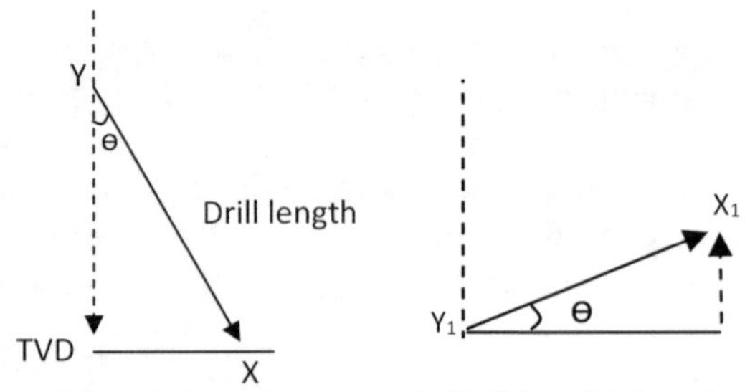

a. Build up below the horizontal

b. Build up higher than the horizontal

Figure 13.31: Calculation for TVD and inclination for a given drilled length

At point Y (in Figure 13.31a), using trigonometry,

$$\text{Cos}\ \theta = \text{TVD} / \text{Drilled length}$$

Therefore,

$$\text{TVD} = \text{Cos}\ \theta * \text{Drilled length}$$

If the directional driller is to build upward above the horizontal, the drilled length will be from Y_1 to X_1 (as shown in Figure 13.31b). It is an upward drilling direction; TVD reduces.

At point Y_1, using trigonometry,

$$\text{Sin}\ \theta = \text{TVD} / \text{Drilled length}$$

$$\text{TVD} = \text{Sin}\ \theta * \text{Drilled Length}$$

This TVD in Figure 13.31b is the vertical depth drilled upwards, to be subtracted from the initial TVD to get the final TVD.
There are many other ways to achieve this calculation, just as there are many more cases of Geosteering trigonometry related problems and solutions. Consider Figure 13.31b, the measured inclination after drilling downwards is actually $90^0 + \theta$. A cosine function is used to relate the above examples to a general formula.

Depending on whether there is need to add or subtract θ in the equation, a rearrangement of the formula makes it possible to multiply the cosine of ($90^0 \pm \theta$) and drilled length to always find the TVD.
This would mean that if the length needed to reach a reservoir entry point from a given depth is known, then it would be possible to determine at what angle to drill. Today modern software solutions do the calculation and the user will not need to work this out manually most times. Machine learning and computer-assisted calculations are improving the determination of the angle at which to drill, for desired directions and targets, in real-time.

EXERCISES

1. a) List 4 safety issues that should be considered before planning any well.
   b) List 4 factors that should be taken into consideration when presenting different stages of operation on a well program.

2. a) What is spudding? How it is carried out.
   b) What is a riser? Explain its most important functions during drilling.

3. a) Describe the structure of a petroleum well from top to bottom.
   b) What is a well architecture? Describe the processes that design it.

4. a) What are the functions of the drilling fluid mud?
   b) List the two main types of drilling fluid.
   c) What is a bottom hole assembly?

5. a) What is the purpose of cementing a well?
   b) Describe the two types of cementing job.
   c) How is the quality of the cementing checked?

6. What is Measurement While Drilling (MWD)? Describe the 'Push the bit' and 'Point the bit' directional drilling technologies.

7. Describe the necessity to drill directional or deviated wells in the following:
   a) Relief wells
   b) Clusters
   c) Sidetracking
   d) Geological barriers

8. What are the basic tools used to deviate a well and how do they work? Differentiate between Measured Depth (MD) and True Vertical Depth (TVD).

9. a) What is a well trajectory, and what are its controlling factors?
   b) What is a well survey?

10. What is Geosteering? What are the consequences of poor trajectory follow-up during a geosteered well? Define the following Geosteering principles:
    i. kick-off-point (KOP)
    ii. inclination
    iii. azimuth
    iv. lead angle
    v. dog-leg
    vi. slant
    vii. radius of curvature
    viii. build-up and build-up rate
    ix. end of build-up (EOB)
    x. drop-off and drop-off gradient
    xi. drain

REFERENCES & FURTHER READING

Anon, 1989. Horizontal Systems Growing more Specialised, Parts I and II: *Offshore*, V. 49. p. 36 – 43.

Atkinson, B. K., 1987. *Fracture Mechanics of Rocks*. (ed.), Academic Press, London.

Bourgoyne Jr., A. T., K. K. Millheim, M. E. Chenevert, and F. S. Young Jr., 1986. Applied Drilling Engineering. Richardson, Tx, USA: Society of Petroleum Engineering.

Barret, S. L. and R. G. Lyon, 1988. The Navigation Drilling System Proves Effective in Drilling Horizontal Wells in the Java Sea. *SPE 17238, IADC/SPE Drilling Conference*, Richardson, Dallas, Tex., p. 503 – 510.

Beebe, B. W., 1961. Drilling the Exploratory Well, in G. B. Moody, ed., *Petroleum Exploration Handbook*, McGraw-Hill Book Co., New York, p. 17-36.

Breckels, I. M. and H. A. M. Van Eekelen, 1982. Relationship between Horizontal Stress and Depth in Sedimentary Basins. *J. Pet. Technol.*, 34, p. 2191–2199.

Burland, J. B., 1990. On the Compressibility and Shear Strength of Natural Clays. *Geotechnique*, v. 40, p. 329–378.

Chester, F. M. and J. M. Logan, 1986. Implications for Mechanical Properties of Brittle Faults from Observations of the Punchbowl fault zone, California. *Pageoph*, v. 124. no.s. 1/2. p. 79–107.

Cooper, G. A., 1994. Directional Drilling: *Scientific American*, V. 270 (5), p. 82 – 87.

Eastman, H. J., 1954. Lateral Drainhole Drilling: *The Petroleum Engineer*, V. 26 (12), p. B57 – B73.

Economides, M. J., L. T. Watters, and S. Dunn-Norman, 1997. Petroleum Well Construction, John Wiley & Sons. Duncan, Oklahoma: Halliburton.

Eddision, A. and J. Symons, 1990. Downhole Adjustable Gauge Stabilizer Improves Drilling Efficiency in Directional Wells. *SPE 20454, 65th Annual Technical Conference and Exhibition*, New Orleans, LA., Richardson, Tex., p. 509 – 516.

Edlund, P. A., 1987. Application of Recently Developed Medium Curvature Horizontal Drillings Technology in the Strawberry Trend Area. SPE 6470, *IADC/SPE Drilling Conference*, Richardson, Dallas, Tex., p. 1057 – 1063.

Gatlin, C., 1960. Petroleum Engineering, Drilling and Well Completions. Englewood Clifs, N.J., USA: Prentice-Hall.

Haskell, N., S. Nissen, M. Hughes, J. Grindhaug, S. Dhanani, R. Heath, J. Kantorowicz, L. Antrim, M. Cubanski, R. Nataraj, M. Schilly and S. Wigger, 1999. Delineation of Geologic Drilling Hazards Using 3-D Seismic Attributes. *The Leading Edge*, p. 373–381.

Hyne, N. J., 2001. *Nontechnical Guide to Petroleum Geology, Exploration, Drilling, and Production*, PennWell, 598 p.

Jaeger, J. C. and N. G. W. Cook, 1979. *Fundamentals of Rock Mechanics*, 3rd ed., New York, Chapman and Hall, 513 p.

Sibson, R. H., 1998. Brittle Failure Mode Plots for Compressional and Extensional Tectonic Regimes. *Journal of Structural Geology*, v. 14, p. 25–50.

Sibson, R. H., 2003. Brittle-Failure Controls on Maximum Sustainable Overpressure in Different Tectonic Regimes. *AAPG Bulletin*, v. 87, no. 6, p. 901–908.

Westbrook, G. K., B. Carson, R. J. Musgrave et al., 1994. Proc. ODP, *Initial Reports*, 146 *(Adapted from Natural Resources Canada).*

CHAPTER FOURTEEN

MUDLOGGING AND GAS MEASUREMENT

14.0 Mudlogging and Safety

Mudlogging is the assessment of subsurface and drilling conditions by the collection and analysis of data from drilling fluid (mud), cutting samples and drilling parameters. It involves the creation of a detailed record of information that comes to the surface from a well while drilling, especially rock cuttings and gas, and the monitoring of rig site activities. These provide clues regarding the specific nature of rocks and fluids in the subsurface, and on safety. To be discussed in this chapter are:

- The roles of mudlogging
- Monitoring of drilling parameters
- Sample collection and analysis
- Laboratory tests, petrography and geochemistry
- Gas measurement, interpretation and applications
- Comprehensive reporting with master log
- Quality control and safety

The mudlogging process is even more vital for rig safety. Gas measurement is not only useful for assessing petroleum results, but also for well safety. Monitoring of sensors installed on the rig and assessment of drilling parameters can aid in spotting potential dangers, such as high levels of poisonous gases and well kicks.

14.1 Roles of Mudlogging

Though mudlogging is essentially contracted out to specialized engineering and geoscience firms (usually third-party mudlogging companies), the geologist working on the well or rig site needs to understand this side of the business. The mudlogging team is made up of geologists and data engineers. The company geologist follows up the mudlogging team and as such should understand what goes on in the unit. He should check the various sensors measuring drilling parameters which the **mud logger** places on the rig. This should be compared with that of the driller, where possible, as part of the check mechanism.

In addition to estimating and recording the depths and mud properties, the duties of the Mudlogging Unit (ML Unit) are:

- Monitoring of Drilling Parameters and Process
- Drill Cuttings Collection and Description
- Laboratory Measurements
- Gas Monitoring
- Daily, Weekly and Monthly Reports (see Appendix I)

The mudlogging unit monitors almost every procedure related to drilling (Purcell, 1977; Whittaker, 1991). The mud properties are determined by the **fluid supervisor** and the **mud engineer**. The drilling parameters are measured by sensors on the rig and determined by the drillers. The mudlogger monitors mud properties and drilling parameters, recording them at given depths all through the period of the drilling (see Table 14.1).

The Wellsite Geologist (WSG) supervises such recordings by mudloggers and transmits the data to the base in town. When the depth is plotted against the ROP for example, rate of penetration at any depth could be used to infer lithology changes, drilling breaks (where the formation hardness suddenly changes), and other drilling challenges.

Table 14.1: Sample recording of mud properties and drilling parameters

Depth (m) MD/RT	Depth (m) TVD/RT	Mud Properties	Drilling Parameters
1500	*1300*	Type: NABM MW: 1.19 sg FV: 145 s/qt Filtrate: 3.0 O/W: 70/30	WOB: 7.51 Ton RPM: 71 rpm Flow: 2498 l/mn SPP: 127.5 bar DrillTRQ: 12.39Kft-lb

FV signifies Funnel Viscosity; the other abbreviations are given among drilling parameters in the next section

The inclination and azimuth of the well trajectory are as well be recorded, such that estimation of TVD in deviated wells is often automatic. This is because the vertical height can be computed from trigonometry *(as demonstrated in Chapter Thirteen, on Drilling and Geosteering).*

14.2 Monitoring of Drilling Parameters

Mudloggers monitor the drilling activity and send in a daily drilling activity report to the WSG. Standard monitoring, depending on how rapid the drilling process is, is usually performed by a two-man crew working 12-hour shifts. A Data Engineer or Analyst may also be there to work with them. They track every activity going on at the rig floor and maintain a daily drilling activity log known as *Daily Drilling Activity* log or DDA. The DDA log records the up and down movement of the top drive, thereby recording the hook height and plots these alongside comments from the mudlogger. With these, experienced oil field personnel can retrace exactly what went on all day on the rig. The WSG reviews this document and sends it to the Operations department in town.

An example of a drilling parameter, which is the ROP, has been mentioned above and is very important in normal drilling conditions. Torque and SPP becomes important when there are challenges. The WOB and RPM as well as the HK height are controlled by the drillers. Figure 14.1 is an example of how ROP may plot in a given formation.

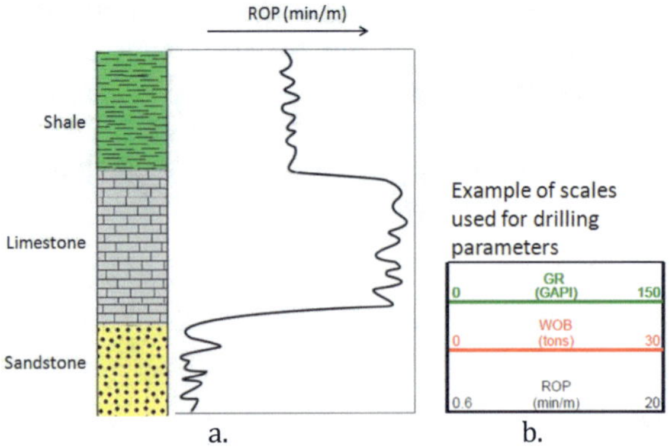

a. Rate of Penetration (ROP)
b . Example scales for GR, WOB and ROP.

14.1: Examples of drilling parameters

In Figure 14.1, Weight On Bit (WOB) is measured in tons from about 0 to 30 while ROP is in minutes per meter from about 0.6 to 20 in the example scale above. ROP though could be higher or lower. ROP may also be measured in meter per hour (m/hr). Gamma ray (with Gamma ray API unit - GAPI) may be plotted in the same column with drilling parameters for comparison.

The drilling parameters, most of which are recorded by the various rig sensors, are:

- Torque (TRQ)
- Rate of Penetration (ROP) of the bit
- Rotations Per Minute (RPM)
- Weight On Bit (WOB)
- Hook Height (HK Height)
- Stand Pipe Pressure (SPP)
- Flow (FLW) values of the Flow Pumps
- Mud weight In (MWin) and Out (MWout)
- Temperature In (TmpIn) and Out (TmpOut)
- Gas (C_1 to C_5 and other downhole gases) and Total Gas (TG)
- Other drilling parameters

In Figure 14.2, higher ROP is observed in shale lithology, which progressively reduces in siltstone and sandstone. This is because (when measured in minutes per meter) lower ROP means faster drilling (in the sandstone lithology). When measured in meters per hour, it is the reverse; higher ROP would then mean faster drilling (in the sandstone relative to the shale). Depth is very important, as all measurements are tied to it. Drill cuttings description as well as gas readings are as well recorded in the sheet. *Using the examples in Table 14.1 and Figure 14.1, an attempt could be made to plot and fill the columns in the Mud Log sheet in Figure 14.2. This is only a part of the mud logger's duties.*

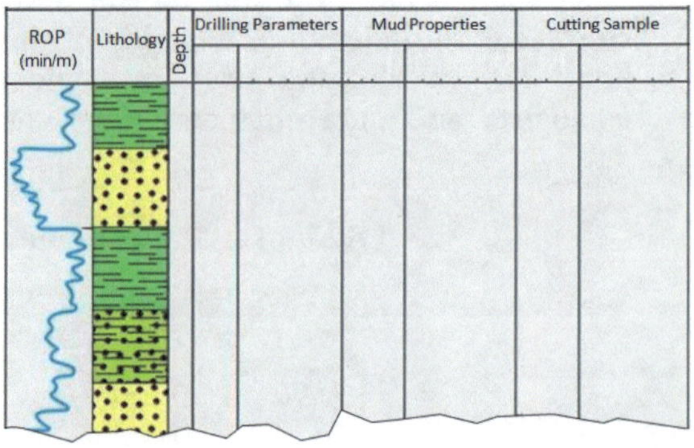

Figure 14.2: Simplified mud log sheet showing how the columns may be filled on a computer screen or in a document manually

As will be discussed later in this chapter, the mudlogging process involves the provision of useful information associated with the gas measurement, such as:

- Gas calibration information (plots and checks)
- Gas events (peaks with their percentages)

These help the mudlogging unit to monitor and report indicators of smooth or problematic drilling processes.

The mudlogging team may provide the WSG a time and depth ASCII file each day, which when plotted shows these drilling parameters variation with time and with depth respectively. The ASCII is the *American Standard Code for Information Interchange*; Log ASCII Standard (known as LAS) is the file-format most used in the oil and gas industry for transmission, use and storage of well log data.

After the drilling team have set up the sensors, before the start of routine duties, the WSG carries out an audit of these systems on the rig floor and those in the mudlogging cabin itself. Then he prepares a report to show if every item is set up properly and functionally. The WSG also gives a *Statement of Quality Management* (SQM) report to the Operations Geologists or Head of Operations at the base. The WSG closely monitors the operations and activities of the mudlogging contractor by way of close supervision. Data and equipment records are kept in collaboration with the mudlogging team. Validation of proper execution of contractual terms and payment to the mudlogging service contractor is based on this.

14.3 Sample Collection and Description

Most of the drill cuttings are collected from the **shale shakers**. This is the point where drilling mud circulating at depths is collected at the surface (Wittaker, 1991). Mud coming to the surface usually brings up rock cuttings which are collected and evaluated by the mudlogging team (Low, 1951; March, 1968; Swanson, 1981). Samples are collected every 10 m in non-reservoir formation (e.g. shale) and 5 m (in the reservoirs) or as may be requested by the Drilling Program or Well Location Report. They are properly treated and bagged, with samples left out for analyses. If the ROP (in m/hr) is very high it will become impossible to safely catch samples in time. This is because the required distance (5 or 10 m) is already drilled before the **sample catcher** can go to the shakers, prepare and bag the sample, and then return. If this is the case the sampling interval should be increased, if ROP must be maintained or cannot be reduced. "Double bagging" of samples (where sample from same unit is bagged twice to meet up with uncollected demanded sample) is a waste of time as it produces invalid data.

The first part of the analysis is to identify what types of sample they are. They could be normal or differently shaped cavings and pack-off. ***Cavings***, such as *chunks* or *splintered* shaped, are left in the hole longer than normal cuttings because they are larger. They may be recovered during well circulation when drilling is momentarily stopped to clean the well. They may indicate instability in the shale section. Cavings come off due to cave-in into the formation; splintered shaped cuttings are sharp flattened fragments (Figure 14.3).

Figure 14.3: Sample types showing splintered cavings (pen cover used for scale)

Pack-off means the cutting was literally packed off from the well wall. This may be an indication of *overpressure* or *underbalance*. What they exactly signify depends on the shape, the lithology, drilling mud and pressure conditions. Samples are considered normal if their look and sizes correspond to what is expected for the lithology and type of bit used to drill. Small threads which are cutting impact on rocks may appear on normal samples and should not be described as part of the sample analyses.

Methods of Sample Description

Sands and ***shale*** are relatively easy to describe. Attention is focussed on sand/sandstone, especially in the vicinity of the target reservoir. For the description, it is usual to first mention the **mineral** for sandy lithology, then its **colour, hardness, sorting, texture,** presence of **mineral** and **matrix**, before the **porosity** and finally whether there is hydrocarbon **(HC) presence** or not. Below are examples of ***Sample Description;*** that of the sandstone is overemphasised to highlight nuances in description.

SANDSTONE (70%): quartz, light olive grey, friable to moderately hard, fine to coarse grained, predominantly medium grained, moderately sorted, sub-angular to sub-rounded, sub-spherical to spherical, silty, trace of argillaceous matrix, strong silica cement, trace of peletal glauconite, fair intergranular visual porosity. Abundant dull yellow fluorescence.

CLAYSTONE (20%): grey to light grey, firm to moderately hard, earthy,
sub-blocky to sub-fissile, non-calcareous.

SILTSTONE (10%): dark brown, locally grey, friable, earthy, blocky, grading to very fine sandstone.

Table 14.2: Example of rock properties and observations reported in cuttings description

ROCK	ROCK PROPERTIES			CUTTINGS SAMPLE			
	COLOUR	**HARD-NESS**	**TEX-TURE**	**SORTING**	**POROS-ITY**	**CAVING**	**HC Shows**
Sandstone	white, brown, yellow, grey	Variable, soft to hard	Fine to medium grain	Poorly sorted (prove-nance: proximal)	5 - 25 %		Fluo-res-cence
Shale	Grey, brown, black, green	Soft to medium	Fine grain, varying degrees of fissility		10 - 30 %	Splintered (underbal-ance drilling with lower mud weight)	
Limestone	Grey, black	Medium	Granular		5 - 20 %		
Dolomite	White, grey	Medium to hard	Granular		1 - 5 %		

When describing the hardness, it is important to mention if it is a mineral hardness within the sample or that of the rock fragments. Table 14.2 combines the report of rock properties as well as other indications of well condition from drill cuttings. An estimation of percentage is shown in Figure 14.4.

Description of drill cuttings is often riddled with repetition and contradiction. Attempt is made to refine the above description by being more succinct. An example of contradictions is to say the sand is fine to coarse, then again that it is medium grained. It is rather common to have fine to medium grained or medium to coarse grains. If a sample has the same grain size, then it is most likely well sorted. Do not expect poor sorting and then the same grain size.

Here is an example of ***Hydrocarbon Shows***, described after doing a fluorescence test on the above sample. Again it is slightly overemphasised to show the nuances in show description.

Strong Show: bright yellow spotty fluorescence; instant streaming milky white cut; instant diffuse milky white crush cut; bright thick white residual ring fluorescence; thin straw coloured residual film. No odour.

The mudlogging team usually has a similar guide for describing **Shale, Siltstone** and **Carbonate**; the guides are often pasted on walls in the mudlogging unit and also in the WSG's office or work folders. In areas with carbonate, it is ensured that the proper methods of distinguishing dolomite and limestone, and other carbonates, under the microscope are applied (Archie, 1952; Dunham, 1962; Folk, 1974; Tucker and Wright, 1990). Limestone is composed of calcium carbonate, sometimes with accessory rock materials such as quartz, clay and pyrite.

Limestone and dolomite are similar in terms of their mineralogical characteristics, with dolomite formed by the alteration of limestone by the action of magnesium-rich water to form a sedimentary rock of calcium-magnesium carbonate. For noting the percentages of shale and sandstone, or other lithologies, below is a guide (using colour differences) in Figure 14.4. An accuracy of 5% is reported but in practice a 10% precision is acceptable.

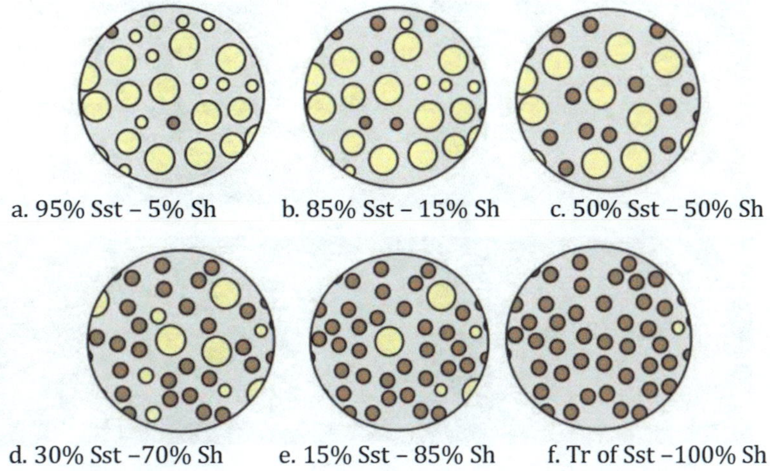

Figure 14.4: Illustration of percentages of sand and shale in samples of drill cuttings

Limestone versus Dolomite

- *Limestone (> 90% calcite):*
  With 10% HCl, instant and violent/floating component that moves around and dissolves in minutes; leaves acid frothy
- *Dolomitic limestone (50-90% calcite, 10-50% dolomite):*
  Immediate but moderate reaction which remains continuous for a while, and moves from top to bottom of acid
- *Calcitic dolomite (10-50% calcite, 50-90% dolomite):*
  Slow and weak reaction that accelerates to continuous reaction after few minutes, with some bubbling at the bottom
- *Dolomite (> 90% dolomite):*
  Slow and hesitant bubbles which evolve one at a time and may require that acid is warmed for reaction to proceed; leaves acid milky.

There may be need to:

- leave the sample for some time after applying HCl
- observe those samples that rises to the surface after some time
- observe the delayed reactions and less bubbling due to lower reactivity
- sometimes crush sample before applying the acid

Before a sample is described, it is prepared in trays and the depth noted. In the example in Figure 14.5a the depth label was placed on the sample collected at 1850 m (usually MD/RT). In Figure 14.5b, there is a bubbling when HCl was added to another sample.

Figure 14.5: Sample depth labelling and test for limestone and dolomite

Figure 14.6 is an illustration of the nuances in colour, grain size, cutting pattern of rock fragments, and the differences noted while describing samples. A trend may be noticed and sand grains may be identified. Note that colour alone may not be enough to identify the sand/sandstone and the shale. *Compare with the examples in Table 14.2.*

In Figure 14.6, the sandstone in a. is very sandy; sands appear as grains in cuttings. Note that the large cuttings in f., for example, do not mean large grains; they are fine-grained shale rock fragments. Percentages of sand versus shale should be apportioned to each of the samples from **a.** to **f.** in Figure 14.6. *Again, compare with the examples in Figure 14.4.* The properties to check and note for describing the samples are explained below.

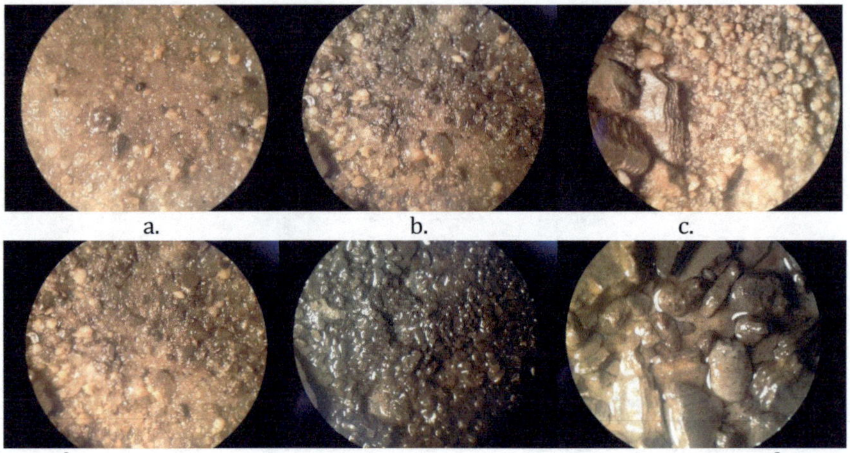

Figure 14.6: Sample of drill cuttings across sand-shale sequences observed under a 10x microscope

Presence of Mud Additives and Other Borehole Materials

Mud additives or other borehole materials should not be mistaken for lithological inclusions. Examples are **Nut plug** and **Mica**, which are Loss Control Materials (LCM). **Barite** is another mud additive used to increase the mud weight, a process known as **weighting-up**. Barite is orange brown and should not be mistaken for silt or very fine sand.

Cement

Cement often shows black specks, and is of a particular texture that shows glazed surfaces. In order not to mistake them for silty-, sandy-, or chalky-carbonate, apply some dilute HCl to it and observe if it changes colour. Cement tends to change colour to yellow or brown when acidified, or to pink when *phenolphthalein* is added to it.

Applied Dope to Pipe

The grease applied to the drilling pipes during connection, in order to reduce friction, known as ***dope***, may contaminate the cutting sample. Where this is not identified as greasy blobs, it can be isolated by fluorescence test. It gives a golden brown dull fluorescence.

Metal Shavings and Pipe Scaling

Corroded and non-corroded shavings of metal scaling from pipe may contaminate drill cuttings. One way to isolate them is with the use of magnet. Once identified, there is a need to take the necessary action, by consultation among the wellsite geologist, company man and drillers, to ensure the drill bit is not eating into the casing.

OBM Interference with Fluorescence

Oil Based Mud (OBM) may produce fluorescence similar to that of in-situ hydrocarbon. To avoid this, the sample is washed with base oil, then the sample scattered or crushed (not necessarily crushed but the grains separated) before testing. In fact, it is the reaction of grains (their interstitial fluids) that is of importance, and not that of the fluids staining them from the outside.

In Figure 14.7a, there are three horizontal lines of dishes on the spot tray (12 dishes altogether): the first line above has no samples; the second line in the middle has samples with no fluorescence while the last line below has sample with fluorescence. In b., although the samples in the left and right petri-dishes show fluorescence, partly due to contamination with Oil Based Mud (OBM), the fluorescence is seen to diffuse in the fluid away from the sample.

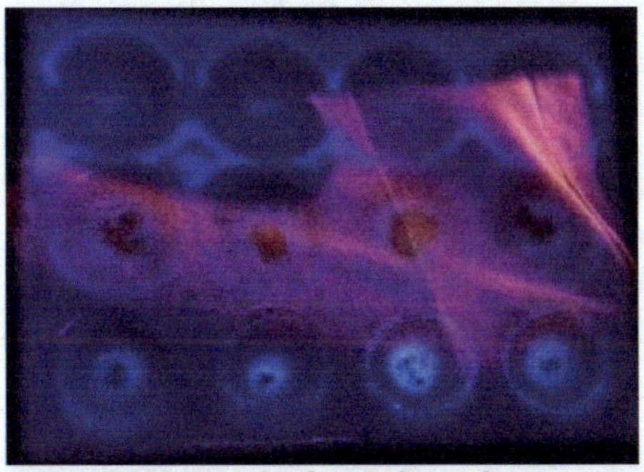

a. Fluorescence due to hydrocarbon shows

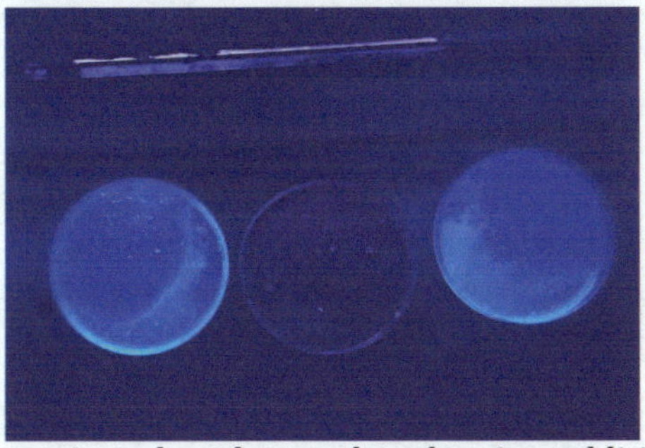

b. Contaminated rock samples showing additional fluorescence from oil based drilling fluid

Figure 14.7: Fluorescence in samples observed under a fluoroscope

To avoid confusing the fluorescence, obtain and retain a sample of the OBM used to drill, and the *base oil* used to wash the sample, for comparison with the nature of fluorescence observed on crushed / separated cutting sample on a *spot tray*. The colour of any fluorescing sandy reservoirs should be different when compared.

The mudlogging unit will provide description tables and charts to guide with the sample description and the chemical tests.

14.4 Laboratory Measurements

The Mudlogging Unit (ML Unit) also serves as a mini laboratory for carrying out chemical tests and measurements on samples. This may include porosity and density measurements, as well as others described below.

- **Fluorescence:** This indicates the presence of hydrocarbon. Isopropanol and methylcyclohexane are among the most commonly used for cut fluorescence analysis. Isopropanol is more recommended for HSE purposes, but methylcyclohexane seems to be more efficient. Strict restrictions and carefulness must be applied if methylcyclohexane is used. Trichloroethane is no longer in use due to its carcinogenic properties.

- **Calcimetry:** This indicates the presence of carbonate: limestone or dolomite. It is achieved by the use of standard Calcimeter, which gives the quantitative amount of limestone and dolomite contents. The values are plotted across shaly intervals to highlight the calcareous units.

- **Bulk Density:** This gives the density of the bulk sediment (rock matrix together with interstitial fluids in its pore spaces) drilled. It is measured by dipping the rock in a known volume of water without drying it, while measuring the volume of water displaced (thereby obtaining the volume of the rock). A weighing balance is used to determine the mass. Density is the ratio of the mass of the rock to its volume. The equipment setup for density measurement may be used to measure porosity. For porosity, an oven is used to dry the sample and the sample allowed to soak and absorb the fluid proportionally to the available pore spaces. *From the left:* rock sample, oven, weighing balance and graduated beakers placed on top of a lab locker, is shown in Figure 14.8.

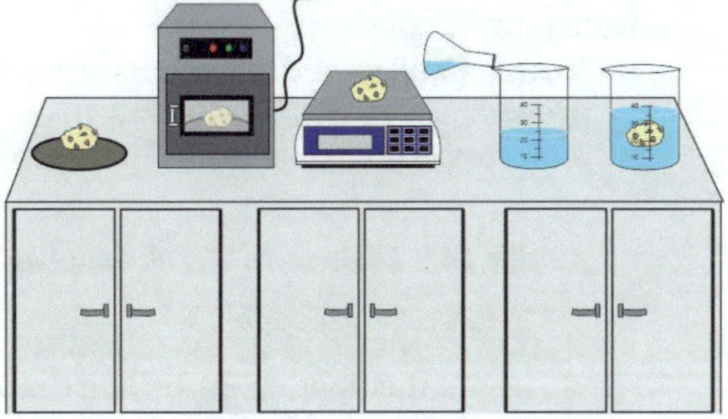

Figure 14.8: Equipment setup for (bulk) density and porosity measurement

A rudimentary way of calculating bulk density using samples is to use cuttings free of mud (washed), which are then allowed to soak in a representative water or brine (comparable with the salinity of the brine at depth from where the samples were collected). The density of the soaked samples are then determined. Density measurement of cutting sample or drilled rock layer is important in the shallow part of the reservoir where there are no density logs. Once density log has been acquired, a more coherent density value is obtained across the whole unit logged.

Petrography and Geochemistry

Petrography involves the use of petrological microscope to carry out observations in thin section. A thin section is a lean, flat slice of the rock under study, usually about 0.03 mm thick, attached to a glass slide (Figure 14.9). It is prepared in the laboratory to meet the basic standard for observation under a microscope.

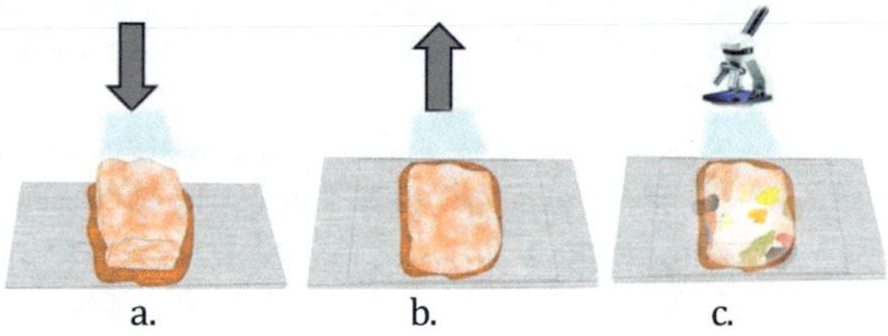

a. Place flat surface of the rock on a glass slide with epoxy and leave to dry
b. Cut off the rock from the glass slide leaving only a thin section
c. Polish and observe under the microscope for petrographic description

Figure 14.9: Preparation of thin section and observation under the microscope

Some of the information that can be obtained from petrography are:

- Sorting (well sorted sediments show that the provenance is distal whereas poor sorting indicates nearer or proximal provenance)
- Composition (organic matter, terrestrial remains, carbonate) indicates among others the rock type and depositional environment
- Clean clear sediments (such as sand) indicate a good reservoir

Sample description in the mudlogging laboratory, discussed above, is a form of petrographic analysis.

Geochemistry is applied in lab tests for fluorescence or the use of HCl in detecting the presence of carbonate. Other geochemical tests described in Chapter Ten, on *Unconventional Hydrocarbon Resources* can be applied in the mineralogy of cutting sample, likewise quantitative and qualitative analyses similar to those described in the chapter on Traps and Seal Assessment.

The geochemistry of rock samples from drill cutting is useful in knowing the chemical composition of the unit drilled. Source rock, reservoir and seal intervals are drilled through and as such information obtained show depositional, maturation, and hydrocarbon migration history. Fluids extracted from cores give indication of provenance and entrapment. Other information is relative to the compartmentalization in the reservoir and across the field.

An important geochemical analysis is the ***rock evaluation*** (the technique abbreviated as ***Rock-Eval***) involving the pyrolysis of small source rock sample in order to study the organic matter present in it. Pyrolysis is the decomposition of organic matter through heat in the absence of oxygen. In an exploration well, it can give the first direct information on the petroleum potential. The Total Organic Carbon (TOC) is obtained and its nature indicates the maturation and amount of hydrocarbons that can be generated.

Table 14.3: Limitations of rock sampling and analysis

Cuttings and Cores	Rock Sample at the Field
- Contamination by drilling fluid - Compacted by the process - Damaged while drilling - Not well representative - Affected during processing	- Weathered - Compacted - Disturbed during collection - Not well representative - Affected during processing

The Hydrocarbon Index (HI) values obtained from the Rock-Eval pyrolysis and TOC values are commonly used to approximate the hydrogen content and oxygen content of the kerogen, indicating the source rock quality in terms of its oil or gas generation potential.

Among the properties that may be determined are:

- Kerogen type (I, II or III)
- Total Organic Carbon (TOC) in %
- The Hydrocarbon Index (HI)
- Petroleum potential (S_2) in mg.HC/g of rock

Rock sample from drilling cuttings can be compared to cores and sampling from field analogues. Limitations exist in all cases and these are listed in Table 14.3. Data obtained from drilling cutting sample and from well core should therefore be used with care.

14.5 Gas Monitoring

The first step to gas monitoring while drilling is to record the gas peaks (events) and the total values (total gas) of gas captured from what is emanating from the hole (Dranchuk and Kolada, 1968; Whittaker and Sellens 1987; Wright, 1996). The mudloggers provide these gas measurements and inform the WSG of the gas events (Table 14.4).

Table 14.4: Example of gas events recording

DEPTH (m)	TIME	ACTIVITIES	Gas %
2180	02:53	Swapped to back-up	
2172	03:35	Cleared autoglass of mud and flushed line	
2187	04:56	Gas peak	
2291	05:03	Gas line injection test	
2391	05:37	Swapped gas line back to main	
2426	12:07	Reposition gas trap due to reduce flow	
2517	13:48	Drained water from autoglass	
2688	12:55	Flushed gas line	
2795	21:05	Drained water from drop out jar	
2818	21:33	Flushed gas line	

Gas Units and Estimate

The most common units for measuring gas are in parts per million (ppm) and percentage (%).

- 10,000 ppm of gas is equivalent to 1% gas

The mud weight of drilling fluid can be used to estimate the gas concentration, assuming that the change between when it was injected into the well and when it came out is only due to gas increase downhole.

To get an indication of the fraction or percentage of gas increase, the *density in* and *density out* of the drilling mud may be compared as below:

Change in gas concentration

= *(density in – density out) / density in*

It is the WSG duty to check with the mudlogging unit to ensure that gas calibration is OK (the total gas equipment is calibrated with methane). He should also ensure that there are backup systems (e.g. real time monitors and sensors) and that the gas line is clean and dry. He should keep an eye on the density out to know the percentage gas in mud (mud weight before and after). The report from the ML Unit is often passed through to the WSG; there should be a triangular communication pattern between the ML units/WSG, the drillers and the company man.

14.6 Gas Measurement

Gas Measurement While Drilling (GMWD) is used to identify and *characterize* the presence of fluid and *fluid type* in a given reservoir. It is also used to infer the possible *nature of the lithology* encountered while drilling. On the drilling side, it can help to identify *drilling breaks*, *underbalance* and early *kick detection*. The principle involves analysing gas results sampled as they arrive at the surface through the shakers and rig ***header box*** areas. The rig header box is a small box mounted on a screen of the *shaker* to receive and distribute drilling fluid from the flow line returning the fluid. GMWD is not reliable if used alone and must be collaborated with other data. Knowledge of the geology and the fluid expected is necessary for a good result. Different types of gases are measured; among the important gases in oil and gas industry are:

- *Primary dry gases:*
  Methane (C_1), Ethane (C_2)

- *Secondary wet gases:*
  Propane (C_3), normal Butane (nC_4)/ Iso-Butane (iC_4), normal Pentane (nC_5) and Iso-Pentane (iC_5)

There are other gases found alongside the hydrocarbon gases mentioned above, such as Nitrogen (N_2) which is inert and H_2S which is an acid gas. Normally, a gas is measured by the parts of its molecule in a million bases of other molecules of gas. When higher than 10,000 ppm, it is represented as percentage concentration. As mentioned earlier, measured 10,000 ppm is considered 1 percent of the gas because 10,000/1,000,000 = 0.01 or 1%. To measure escaping gas while drilling, there is need to use some form of trapping system to capture and channel it to the measuring device. This is achieved by the use of sensors which capture and send the gas in the drilling mud to the mudlogging unit for analyses. The Flame Ionization Detector (FID) is an example of a detector used in gas chromatography.

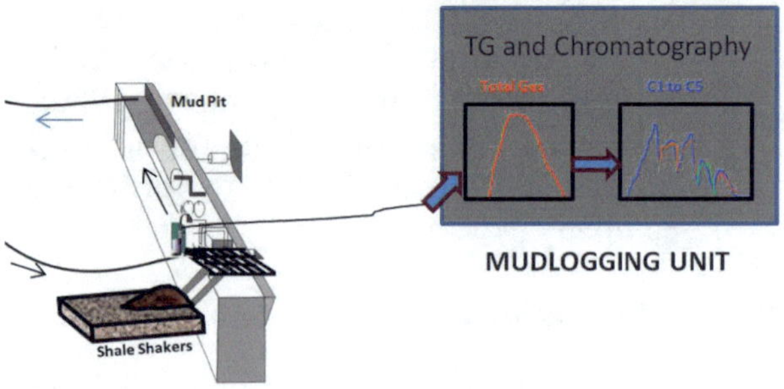

Figure 14.10: Gas measurement while drilling (GMWD) principle

In Figure 14.10, the GMWD setup shows the flow lines to the shakers, where a gas trap extracts the gas from the mud and sends it to the mud logging unit for analyses. The total gas is a measure of total methane equivalent in the gas reading.

Main Objectives of GMWD

Fluid Characterization

By identifying the fluid type getting to the surface, it is possible to deduce information such as heavy or light oil, *biogenic* (mainly C_1 to C_3) or *thermogenic* gas (mainly C_4 to C_5), biodegradation level, diffused gases and their evolution within a reservoir. It could indicate the contacts and transitions within hydrocarbons fluid pools. Changes in gas levels highlight porous and seal-tight zones.

Gas logging while drilling could provide information on:

- the presence of effective seals
- the vertical extension of waste zone
- the position of Oil or Gas Water Contacts

Lithological Information

Information that could be deduced from GMWD is the change in the tightness of the rock, the quality of the reservoir or sealing cap rock, and places that are more homogenous / heterogeneous and eventually their importance in serving as barriers/traps.

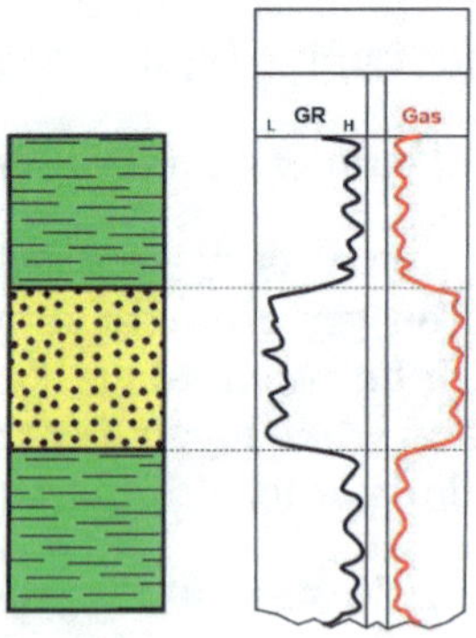

Figure 14.11: Lithological identification in a sand-shale sequence

In Figure 14.11, the GR log identifies the sand-shale sequence, while the gas plot shows porous sand completely saturated with gas.

Gas-Aided Drilling

Gas-aided drilling involves the identification of drilling breaks, underbalance and early kick detection from gas (Figure 14.12). Gas value can indicate a change in lithology, as porous sand or sandstone units will expel more gas. When there is an increasing trend in the background gas, it could indicate underbalance drilling (where the drilling mud weight is less than the equivalent mud weight of the formation pressure drilled).

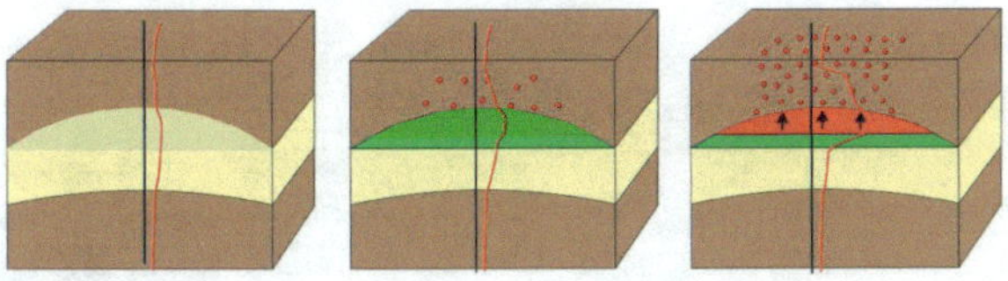

Figure 14.12: Principles of gas-aided drilling

In such instances where there is a sudden gas influx, it could indicate a kick (rapid rush of gas and formation fluid into the formation). Early detection of kick could help to avoid a blowout, which is the highest risk in a drilling rig.

The Gas System: Sensors and Analysers

To obtain gas with minimum contamination from downhole, gas lines are passed from the inlet or escape routes as they come out of the well mixed with drilling fluid (mud), through a sensor to a gas analyser. The *degasser*, which removes incoming air or gasses from the drilling mud, is one of such points where gas traps may be run to collect gas for analyses. The mud density in (injected into the well) and out (as it returns) all affects gas volume measurement as well as the temperature in and out. Gas sensors are also known as *gas trap*. There are two main types of ***gas traps***, which are described below.

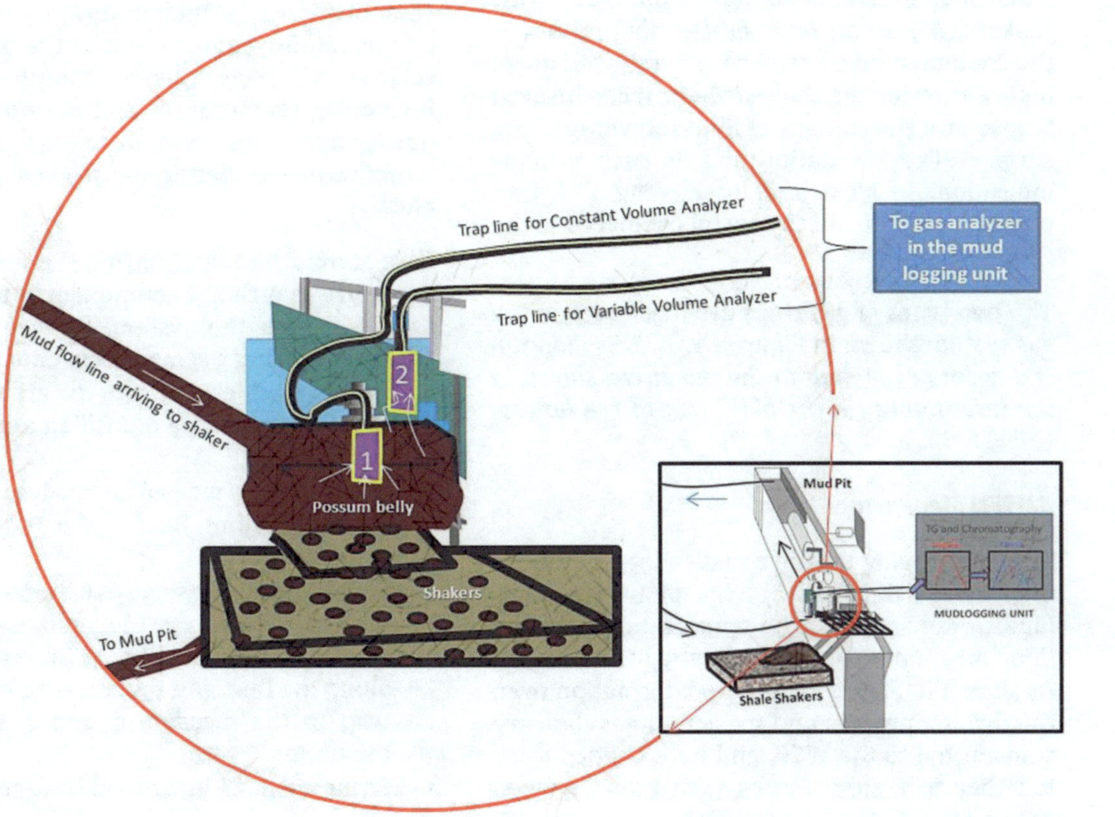

Figure 14.13: Gas system showing possum belly, gas traps and flow line to the (chromatograph) mudlogging unit (Compare with Figure 14.10)

Constant Volume Analyser (CVA)

The gas probe sucks a desired amount of returning drilling fluid (mud) and analyses it in a constant volume chamber. The result is reliable and the fluid detector is an all-in-one system, with the pump and gas detector in one component. The trap type is independent of the mud levels. The probe is fully immersed into the mud unit that transfers mud on the shale shaker. The mud and gas content moves into the gas trap where the gas is extracted. The gas trap has an agitation device, where there is air suction and extraction of gas from mud at the right temperatures. The volume and rate of air suction have to be monitored as well as the temperature. This pumping rate for the gas/air mixture needs to be constant to have a right measurement.

Variable Volume Analyser (VVA)

The gas sensor is placed just above the degasser or possum belly and needs to be adjusted if the mud rises or descends. The trap is hung at about four to five inches above the mud at the shakers. A *Possum belly* or *Distribution box* is the container where mud speed is slowed down before entering the shale shakers; it can be used to measure the amount of fluid returning to the surface after circulation and as such provide indications of kicks and losses. The VVA trap type is more difficult to control compared to the CVA.

The two types of gas traps are illustrated in the gas system shown in Figure 14.13. *This diagram is a zoom of the area in the red circle, shown in the inset figure (in the black box to the bottom right).*

GMWD Measuring Cabin

The mudlogging unit is usually close to the rig floor and to the shakers. This allows the mud-logger easy access to the sensors and to the rig floor for monitoring and adjusting the sensors. As shown in Figure 14.14, the information from the drilling process and measured gas data are transmitted to the WSG, and both parties work together to resolve issues with sensors when they arise.

From the probes, the gas moves to the measuring unit, where there is a measuring device. This is usually a chromatograph. There are different types of gas chromatographs. One of the most common is the **Flame Ionization Detector** (FID). It functions by a process known as ***elution***, which is the separation of the gas depending on their carbon components, due to the effect of temperature and pressure.

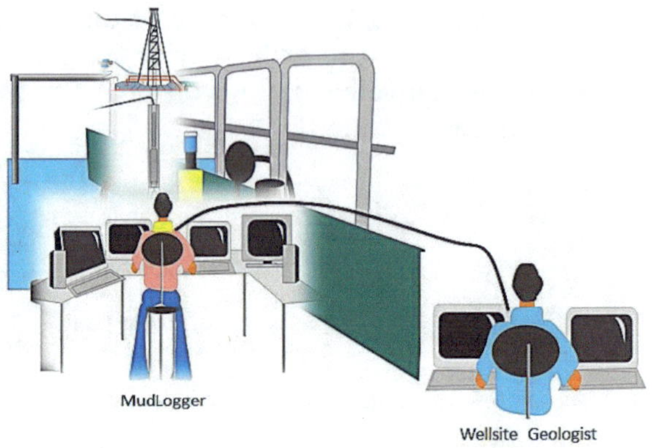

Figure 14.14: Mudlogger's cabin close to the rig floor and transmission of data to the wellsite geologist

FID provides a small flame in an enclosed chamber with a moderate pressure, which causes minute combustions of the gas variably, relative to their carbon components. With increasing temperature and pressure, there is quick but poor elution. Moderately lower temperature is better for a slow but proper elution.

Care should be taken during the measurement to ensure that the C_1 component is not too high as to saturate the system. Factors that affect efficiency during gas measurements are:

- Rate at which the gas is arriving (often dependent on rate of drilling and formation type)
- Quality of gas measuring devices
- Type of drilling fluid and its concentration (mud weight)
- Steady versus unsteady suction rates
- Transit time of gas (depends partly on the size of gas line, absence of sagged points along the line, and the distance from the gas trap to the measuring cabin); the shorter the better.
- Equipment set up and calibration

There are other factors such as temperature and pressure conditions, and the presence of other gases (recycled, swab, etc.) in the system that are important, as well as the mudlogger's competency. This is why there is need for a proper Quality Control (QC) of the gas data.

Detection

The FID chromatography measures both Total Gas (TG) and individual components of the gas from C_1 to C_5. Depending on the carbon content of the main petroleum gases, C_1 – C_5 for example, each would attach itself to an ***elution coil*** and get out at different times. That process of getting out at different times, known as **elution**, is poor if gas is flowing too fast or too slow (Figure 14.15). Therefore, there is need to capture the gas in systems that control the flow rate so that measurement could be consistent. Note that TG reading should always be equal or a little higher than the total of each of the reading of C_1 to C_5.

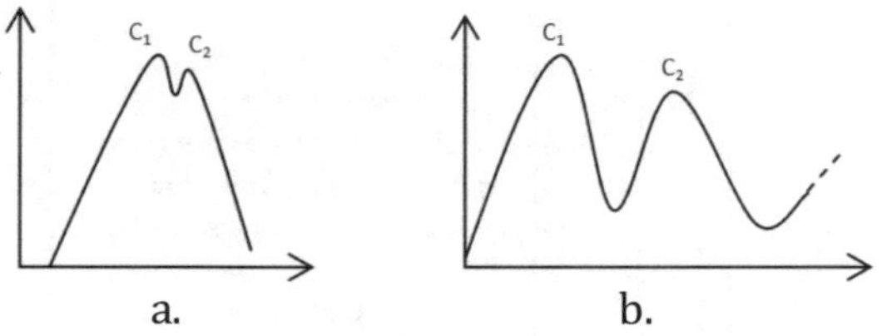

Figure 14.15: Principle of elution and C_1 - C_2 separation better in b. than in a.

The total of the chromatographic reading should not exceed that of the total gas (TG). That is one of the major and simple QC measures.

- **Total gas** (TG) measurement gives the total number of gas based on their carbon content or weight (methane equivalent)
- **Chromatography** (C_1-C_5...) measures the number of molecules of each gas.

For each measurement, a calibration is carried out to check how a given volume of known gas (often methane, known as C_1) would read off in the gas system. A gas system would be considered well calibrated if 5% methane injected reads the right value or very closely when injected in the system.

14.7 Comprehensive Reporting with Master log

At the end, a synthesis is made by log, providing all the information monitored while drilling. This document is known as the **master log** (Figure 14.16) which remains the main reference document for assessing the events that took place while drilling. While other documents such as daily geologic report may be updated, the master log should not be modified once delivered as it contains the understanding as it was during the drilling process. Ideally, the master log should record all the drilling parameters, which are not included in the simplified version in Figure 14.16.

The sample master log shown in Figure 14.16 includes the results of calcimetry (calc) and gas readings (C_1 in orange colour and C_2 in dark brown). Value of *inclination* (blue arrow at 2500 m MD) can be seen to be less than 90^0 measuring from the vertical black line. The *azimuth* (corresponding red arrow) may be estimated as about 220^0. The black triangular shapes on the depth column are *casing shoes* (a device attached to the bottom of the casing string surrounded by cement; it helps to guide the casing string downhole). Note the comment on the gas column, of the section where the gas trap was adjusted.

In Figure 14.16, the interpretations based on drill cuttings show the percentage lithologies (yellow for sandstone-Sst, brown for siltstone-Slt and green for shale-Sh) in multiple vertical columns while the lithology suggested by the gamma ray – GR (and to some extent the resistivity log - RT) is represented as only one lithology within the column. This final interpretation based on log (while having in mind the observation from drill cuttings) is used to decide casing depths, perforations, and production tubing that must be in a continuous length of sandstone or reservoir unit.

Note the gradual transition from the shale above, in Figure 14.16, through siltstone to a hundred percent sandstone based on the cutting sample. This is not observed in the interpretation based on the GR log. The gradual transition in the cutting sample may be due to mixing or contamination of the lithologies in-hole or at the shakers. The shakers are where drilling mud and cuttings are separated before the mud is again re-injected into the well. The places interpreted as Traces (Tr) may be interpreted as 5% of the given lithologies, showing the nuances and subjectivity of sample description. No limestone (Lst) was observed.

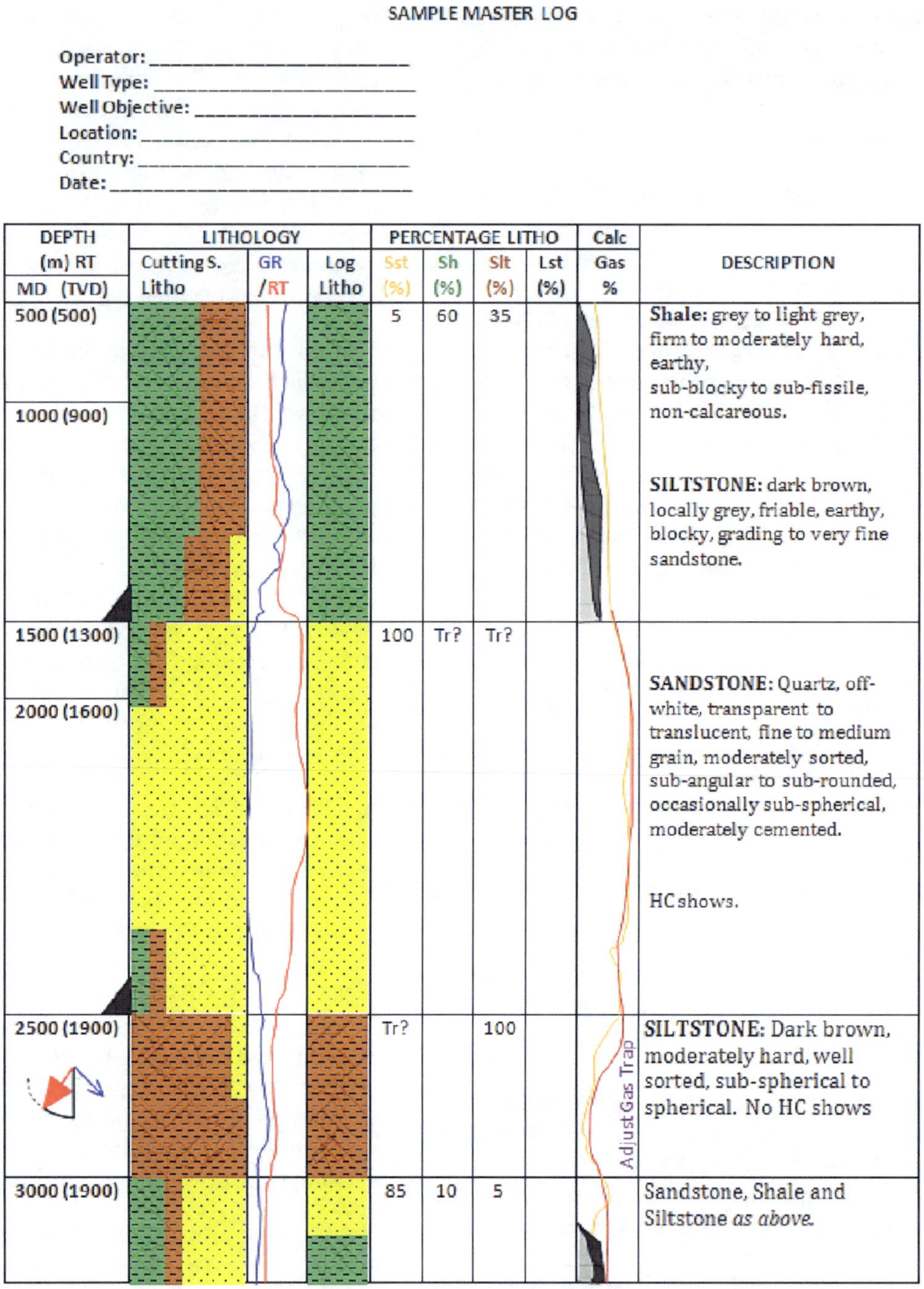

Figure 14.16: Simplified master log showing cuttings and log-based lithologic interpretation while drilling

14.8 Quality Control

The first step in analysing gas result is to identify possible sources of misleading data, such as peaks not related to actual gas measurement. There is need to identify and isolate ***trip gas, recycled gas, swab gas*** and other non-formation gases. Basic QC involves a *check of the depths* and use of *ratio plots to check* for consistency in data obtained. Remember that reading of total gas should be equal or slightly higher to total of individual gas readings.

Quality Control (QC)

It is necessary to check if the depth corresponds to other log responses, like porous unit showing higher gas value (in which case it is formation gas) or if it corresponds to a period when connection was made (connection gas) and so on.

- Gas values need to be compared with other drilling data and events.
- Do the different gases arrive so closely (that two components give information similar to their sum)? Has *separation threshold* been exceeded?
- A recycled gas may affect all readings, giving the same value increase on the background gas
- Gas contamination, leading to an increase in recorded value, may only affect one of the curves, as it is an external effect on the values.

There are two levels of QC: *Gas line* and *measuring system* checks involved in the quality control. These are discussed below.

Gas Line QC

Fluctuation in mud level could influence the gas reading, since the amount of gas captured in the trap will also fluctuate.

- Identify such fluctuation and isolate it from good data. Ensure that the gas line is clean.
- Carry out injection tests occasionally when in doubt. An injection test involves injecting certain known amount of gas at the trap and checking to see if the same amount (or very close to it) is read off in the cabin.
- Check that the ***lag depth*** is accurate; the depth associated to a gas reading is based on the lag depth (delay before fluid arrival to the surface)
- What possible Mud weight In and Mud weight Out could affect reading?
- What is the variation in the drilling parameters: pump flow, ROP and bit size? Is that affecting the amount of gas in the system?

Measuring (Chromatography) Checks

There are three parameters to QC on the chromatography. These are:

1. *C_1 Saturation:* The level at which C_1 saturates a given chromatograph depend on the specification of that particular equipment. This should be provided by the mudlogging company in charge of the gas measurement. Recorded level of C_1 should always be less than the given saturation limit.
2. *C_1 versus C_2 Separation:* Too much arrival of C_1 and C_2 closely spaced together is not good for proper reading in the chromatograph (see Figure 14.15). It makes it difficult to separate the values, such that one influences the other. Also the ratio of C_1 versus C_2 separation depends on the specification of the equipment and should be provided by the manufacturer (in the user manual). C_1 to C_2 ratio should always be less than the given specification. The cross-plot in Figure 14.17 may also be done for C_2 versus C_3.

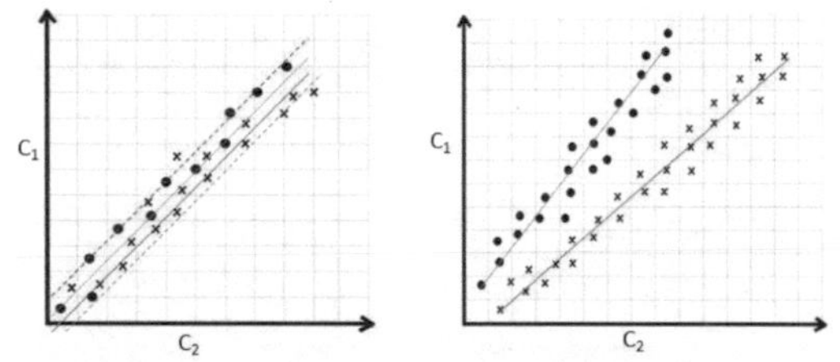

a. Poor separation b. Good separation

Figure 14.17: C_1 and C_2 separation checked in a cross plot

3. *Total gas* versus *sum of components:* Total gas (TG: essentially C_1 to C_5 of interest) reading should be equal or closely equal to the total of the individual reading of C_1, C_2...C_5 summed together. This is only logical, as the chromatography analyses gas passed on from the total gas measurement. This is known as *consistency* and is explained in Table 14.5. and Figure 14.18.

If this is not the case, it means that either there is:

- Error in calibration of the equipment (Total gas may be slightly higher, but not less than total of individual readings), or
- Presence of other gases (e.g. in zones rich in organic matter); in this case Total gas is slightly higher, which can be an interpretational indicator to the nature of formation.

Acceptable standard demands a **consistency**, such that,

$$TG \,/\, \text{sum } C_1 \text{ to } C_n = \pm 1$$

(mostly within 0.8 - 1.2)

This could be used as a calibration check, a pre-interpretation measure which helps to show that the data is consistent. It can however be used as an interpretative approach to check the presence of other gases as explained in Case 3 (Table 14.5)

$$TG / \sum C_1 \text{ to } C_5$$

The plot in Figure 14.18 shows acceptable case and cases indicating calibration issues or impact of geologic formation (compare with Table 14.5).

Table 14.5: Consistency check for total gas and chromatographic components

Scenarios	Total Gas / Sum (C_1 to iC_5)	Interpretation
Case 1	0.8 – 1.2	Acceptable Range (OK)
Case 2	< 0.8	Issues with calibration (Total Gas is less than the Sum)
Case 3	>1.2	Issues with calibration (Total Gas is greater than sum) or presence of other gases / organic matter not read by gas chromatogram

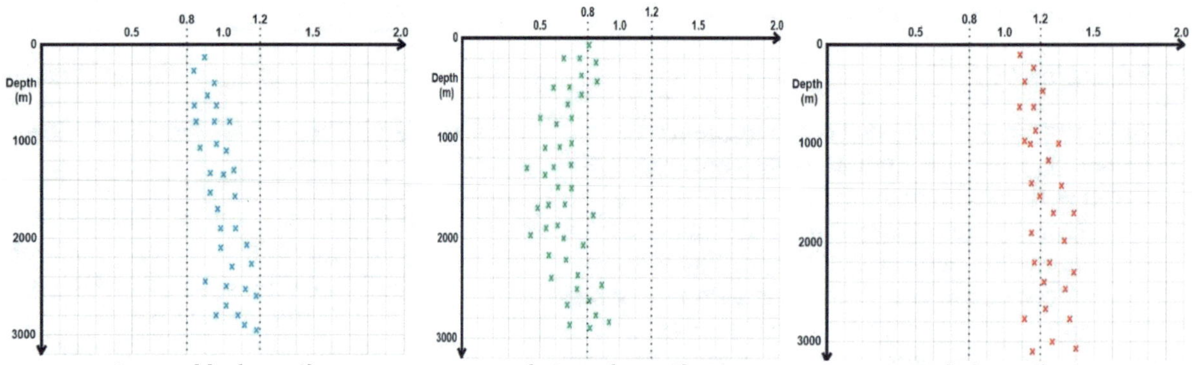

a. Acceptable (case 1) b. Low (case 2) c. High (case 3)

Figure 14.18: Consistency check plot: TG versus Sum of C_1 to C_5 (on the horizontal axis) plotted against depth (on the vertical axis)

14.9 Interpretation

Gas Sources and Types

Not all gases that are recorded in the system directly describe the formation; some are background gases or recycled gases. These could give an indication of what is going on while drilling; a change in background gas may mean more gas is being introduced or removed from the system. An important step before further interpretation of measured gas is to identify its source. The first interpretation starts by understanding what is read off on the screen, by the mudlogger and wellsite geologist, as it is detected on his monitor. The various sources and gas types are described below.

1. **Trip Gas (TrG)**: This gas occurs while tripping in or out and may not be observed when it was generated. Tripping is when the drill string is being run in hole (RIH) and when it is being pulled out of hole (POOH). Because tripping occurs for a long time, TrG is not tied to a depth.
2. **Swab Gas:** This is gas that may come in due to sucking of the formation while pulling out the drilling string. It may indicate an initiation of kick.
3. **Connection Gas (CG):** This occurs when drilling stops for pipe connection because there is a static condition and no additional pressure (ECD) from a rotating drill string or from pumping of mud. The mud pressure at this time is only mud weight dependent (mud weight is lower than ECD – Equivalent

Circulating Density). An increasing trend of CG indicates underbalance drilling and may lead to a kick.

4. **Circulation Gas:** Continuously circulated gas is a kind of recycled gas. A peculiar case is circulation when drilling stops, which brings up the cuttings at the bottom and all gas that has accumulated in the borehole since drilling stopped. Gas observed at the first lag time during circulation, after stop of drilling, is sometime termed the **"bottom's-up"** gas.
5. **Recycled Gas (RG):** This gas is the gas in the mud which is re-injected into the formation with drilling fluid. It would be higher if the degasser is not functioning properly or if the mud weight of the drilling fluid is relatively low. It may as well be introduced by the drill pipes from the surface. They add up to the background gas.
6. **Background Gas (BGG):** This is the gas level sustained over a period of time, which serves as the basis for evaluating new trend of increase or decrease in gas reading. The other gases listed above contribute to the background.
7. **Formation gas (FG):** This gas occurs mainly in the porous (reservoir) formation and presents as a sharp peak (above the background gas) when the rock formation is drilled. It comes up after one lag time (the lag time is the time it takes for a sample at depth to come to the surface), not necessarily when drilling has stopped and bottom's up gas value is due to delay and gas accumulation from other sources.

The gas recorded must take the lag time into consideration before associating it to depth or event; the gases identified in Figure 14.9 shows where they fit or are liberated from when lag time has been taken into consideration.

There are others such as ***Pump-off gas*** (influx due to pump being shut leading to temporal reduction in Equivalent Circulation Density, ECD), ***Circulation gas*** (influx at break of circulation, that is at the start of circulation, because accumulating gases in the borehole are transmitted to the surface at such time). One type of circulation gas is the *bottom's up*, which occurs when accumulated gas at the bottom of the well reaches the surface. There is also ***contaminated gas***, from mud additives, reactions, or other sources of contamination.

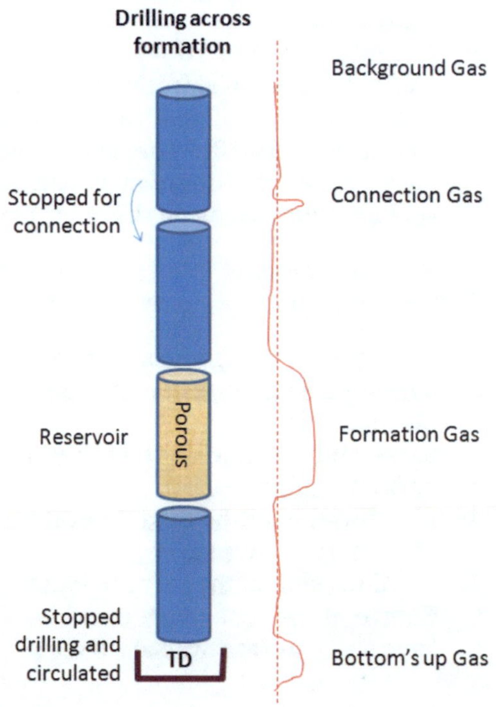

Figure 14.19: Gas shows: source and type identification

Most of these gases add up to the background gas which may interfere with the actual values of the formation gas, which is the gas in the reservoir of interest. There are no defined means of effectively separating these gases from the total gas or the background gases.

Logs and Assessment Curves

The Gas Quality Control described above may be plotted and presented as curves, to give useful information during interpretation. The three major curves, for the assessment of gas while drilling, would therefore follow as below.

1. *Saturation plot:* This is the limit after which methane gas saturates the gas measuring equipment, leading to underestimation of C_1. It varies for different given equipment. The upper cut-off may be set and methane or C_1 is plotted in ppm against depth, to notice where the value reaches that upper limit of saturation.
2. *Separation plot:* This plot tends to check the separation of C_1 and C_2 by plotting them in a cross plot (see Figure 14.17) or just against depth to check if they overlap (poor separa-

tion) or are distinct curves (good separation). *Figure 14.20a* of gas logs below for plotting C_1 to C_4 against depth show good separation.

3. *Consistency plot of TG / $\sum C_1$ to C_5:* This is a plot of total gas against the sum of the components to check if their ratio closes up to 1 (see Figure 14.18). The consistency plot gives an idea of calibration on one hand and geologic information on the other hand.

Some of the pieces of information that may be deduced from plotting gas readings are:

1. Porous units giving higher gas values
2. Varying curves for same unit due to multiple fluid type
3. Sharp shift from a curve indicating barrier or fluid contact
4. Identification of the fluid types from trends of primary and wet gases
5. A sudden shift of the curve to indicate a seal
6. Nature of the seal which could be deduced from sharp or transitional curves
7. Idea of lithology from nature of seal showing different transitions
8. Weak seal deduced from reading that suggest gas diffusion above reservoir

Gas diffusion may be seen as a weak dispersed non-coherent signal and could also be linked to observations highlighted in Figure 14.12 above on the principles of gas-aided drilling.

Dotted horizontal lines in the resistivity column of Figure 14.20 are Water Oil Contacts (WOCs) while those of the Neutron-Density column are Gas Oil Contacts (GOCs). These pieces of information are collaborated or further confirmed by the gas readings. Further interpretation involves separating information to those related to ***lithology*** and to ***fluid.*** Fluid evolution and differentiation is highlighted by further interpretation. Percentage C_1 versus TG in ppm may reveal hydrocarbon content, lithology and possible background gas. Other deductions are presence of organic matter, tight units and water.

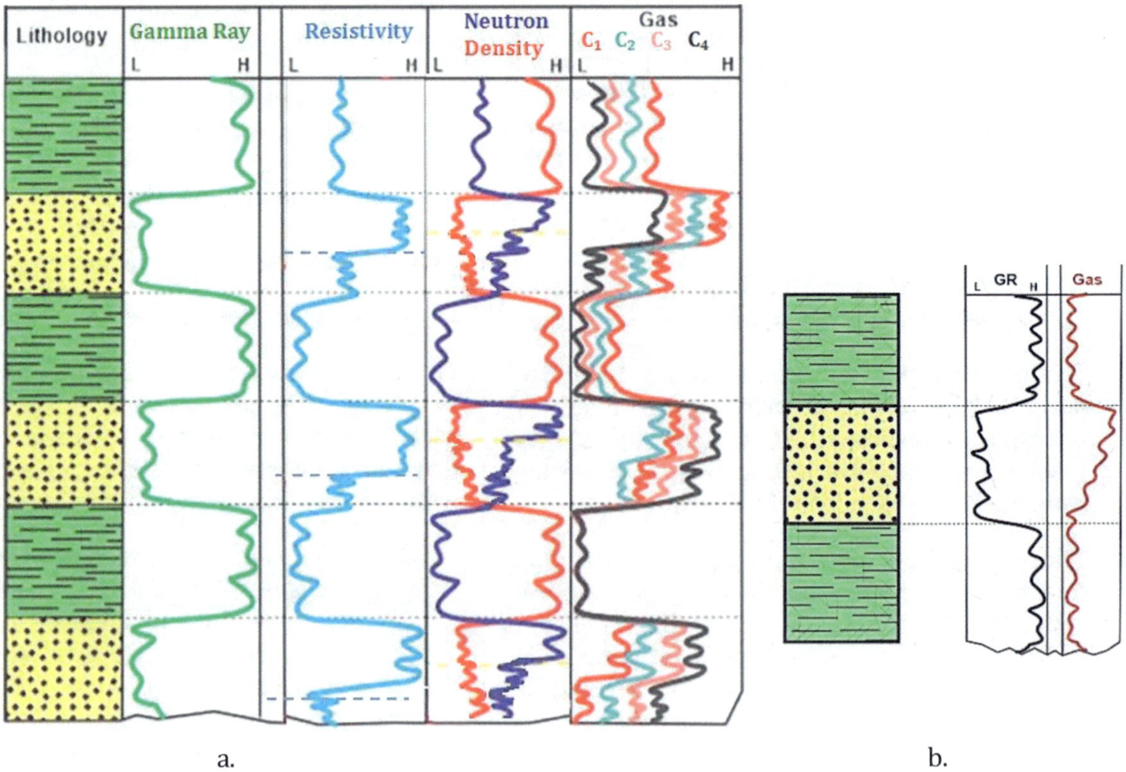

a. Use of gas data to collaborate other logs to identify reservoirs and contacts
b. Gas trend changing, indicating fluid evolution within same reservoir

Figure 14.20 Gas logs across formation showing plotted gas readings and their interpretation (L and H indicate low and high values respectively)

Wetness, Balance and Character

Although being replaced by more robust indicators of hydrocarbon types, there are three traditional gas ratios whose plots are still in use in the industry for classifying gas and oil types. These ratios were developed by ***Explog** in 1985 (now **BHI**).* They are also known as the *Haworth et al., 1985 correlatable ratios.* The three of them are:

- Wetness, W_h : $(C_2 + C_3 + iC_4 + nC_4 + C_5) / (C_1 + C_2 + C_3 + C_4 + C_5)$
- Balance, B_h : $(C_1 + C_2) / (C_3 + iC_4 + nC_4 + C_5)$
- Character, C_h: $(iC_4 + nC_4 + C_5) / (C_3)$

(The normal C_4 component is used where there are no iC_4 and nC_4. Character may be multiplied by 100 to get the answer in percentage).

The above could be entered into a working spreadsheet or wellsite analyses data or such software, and it computes the values.

- ***Wetness** values may be less for g*as and dry gas (less than 17.5) while heavier oil and oil show high wetness (greater than 17.5).
- ***Balance*** is less than wetness for residual oil or oil, whereas it is higher for dry gas and gas. So a good idea is to plot and find the difference between wetness and balance.
- ***Character*** is less than 0.5 if gas or coal and higher if oil. This is not always exact, so care should be taken in the interpretation and other data collaborated before reaching results.

In Table 14.6, there may be an initial column for geological formation and age as well as lithological column. Neutron Density log may be added, as well as sonic if available.

Table 14.6: Main Gas Log: QC and interpretation sheet

Depth Column	Drilling Parameters & GR	Lithology (Cutting and Log)	Resistivity Shallow & Deep	Gas Ratios TG / Sum C1 to C5	Wetness Balance Character	Total Gas & C1-C5	Shows	Mud Weight	Geological Description & Interpretations
m (Scale) →	ROP: 0.6-60 min/m GR: 0 – 150 API	% 1 - 100	0.2 – 2000 ohm-m	1 – 2 (0.8 – 1.2)	Wh: 0 – 50 Bh: 0 – 1500 Ch: 0 –10	% 0-100	% Gas, Fluorescence	S.g. 1 – 2	
50 100									
150 200									
250 300									
350 400									

Looking at the litholog or lithological column in Table 14.6, an attempt could be made to fill in the other data suggestively. Possible geological interpretations may be deduced from the data when the litholog is based on a careful cuttings description and compared with the main gas logs.

14.10 Application of GMWD

In summary, among the applications of results of GMWD interpretations are:

1. *Direct hydrocarbon indication*
   - High wet gas value could indicate a hydrocarbon presence
   - Heavier C3-C5 could indicate a reservoir with hydrocarbon
2. *Fluid characterization*
   - Oil, gas, water
   - Biodegradation
3. *Seal analyses and qualification*
   - Efficient versus inefficient seal
   - Gas diffusion versus leakage
4. *Fluid evolution and compartmentalization*
   - Vertical fluid evolution
   - Fluid contacts and transitions
5. *Rock type identifications*
   - Cap rock lithology versus reservoir lithology
   - Nature of reservoir: quality, tight zones

The mixer plot is a cross plot such as C_2/C_3 versus C_1/C_3 (example shown in Figure 14.21), a ratio of one value to the other, which is useful in understanding the fluid type.

As can be seen from the above, in addition to the usefulness as direct hydrocarbon indicator, gas measurement while drilling data can be used to build a petroleum model. To some extent, it can help in reservoir characterization, to highlight static and dynamic flow and barrier mechanisms.

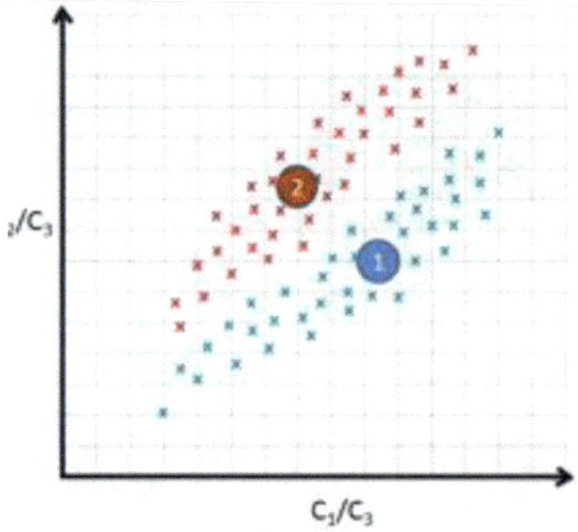

a. Cross plot showing two different hydrocarbon behaviours

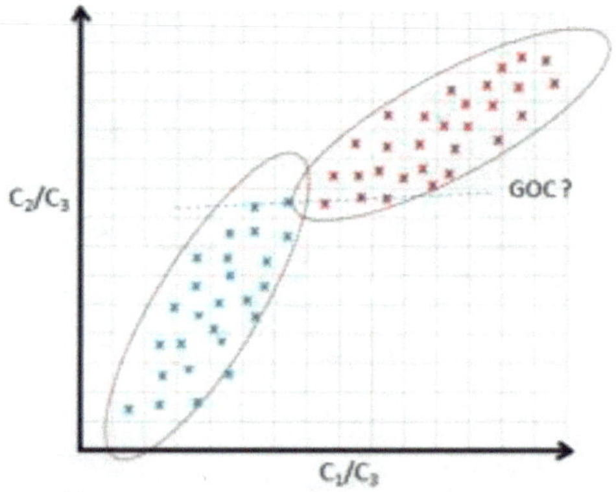

b. Possible fluid contact leading to two groups of associated points

Figure 14.21: Identification of fluid differentiation and contact using the mixer plot

EXERCISES

1. a) What are the duties of the mud logging unit?
   b) List the drilling parameters recorded by the various sensors during drilling
   c) How is drill cuttings collected, and at what rate are they collected?
   (d) What is the effect of high ROP on sample collection?

2. a) Describe the methods used to distinguish between limestone and dolomite
   b) Differentiate between dolomitic limestone and calcitic dolomite

3. Describe methods of identification of the following in drill cuttings
   a) Cement
   b) Applied dope to Pipe
   c) Metal shavings and pipe scaling
   d) OBM interference with fluorescence

4. Describe what the following parameters are used to determine in laboratory tests:
   a) Fluorescence
   b) Calcimetry
   c) Density

5. a) What are the limitations of rock sampling and analysis?
   b) Describe the importance of carrying out petrographic descriptions and geochemistry of rock samples.

6. Differentiate between biogenic and thermogenic gas. Why is there need to carry out GMWD? What are the different types of gases measured in GMWD?

7. a) What is the difference between the gas suction rate and the air suction rate from the trap in the gas line?
   b) Differentiate between the degasser, possum belly and header box. What are the differences in their usefulness when placing gas traps?
   c) What does it mean to say that a chromatograph has reached its saturation limit?

8. Explain the different types of gas shows: BGG, TrG, CG, RG, FG. How is trip gas, recycled gas, swab gas and other non-formation gases identified? Explain why this is important in GMWD QC.

9. What is the Mixer plot and how is it important in gas QC or interpretation?

10. a) How are multiple fluid types detected using GMWD interpretation?
    b) Explain the five major aspects of result sought for in GMWD interpretation – how is each of them deduced?
    c) How are the three gas ratios (wetness, balance and character) used in the identification of oil and gas? Provide the formula used to estimate each.

REFERENCES & FURTHER READING

Archie, G. E., 1952. Classification of Carbonate Reservoir Rocks and Petrophysical Considerations. AAPG Bull. 36 (2), p. 278-298.

Dunham, R. J., 1962. Classification of Carbonate Rocks According to Depositional Texture. In Classification of Carbonate Rocks, W.E. Ham ed., Memoir 1, 108-121. Tulsa: AAPG.

Dranchuk, P.M. and L. J. Kolada, 1968. Interpretation of Steady Linear Visco-Inertial Gas Flow Data, *J. Canadian Pet. Tech,* p. 36–40.

Exlog, 1979. Field Geologist's Training Guide: *An Introduction to Oilfield Geology, Mud Logging and Formation Evaluation*, Sacramento, CA: Exploration Logging, Incl, 301 p.

Folk, R. L., 1974. Petrology of Sedimentary Rocks. Hemphill Publishing, Austin, Texas.

Haworth, J.H., M. Sellens and A. Whittaker, 1985. Interpretation of Hydrocarbon Shows Using Light (C1-C5) Hydrocarbon Gases from Mud-Log Data. AAPG Bulletin 69 (8), p. 1305-1310.

Low, J. W., 1951. Examination of Well Cuttings. Quarterly of the Colorado School of Mines, 46 (4) 46.

March, J., 1968. Advanced Organic Chemistry: Reactions, Mechanisms and Structure. New York: McGraw-Hill Book Company. p. 533 – 534.

Noman, R. and Kalam, M. Z., 1990. Transition from Laminar to Non-Darcy Flow of Gases in Porous Media, in *Advances in Core Evaluation: Accuracy and Precision in Reserves Estimation*, ed. P.F. Worthington, p. 447–462.**Pixler, B.O. 1969**. Formation Evaluation by Analysis of Hydrocarbon Ratios. J Pet Technol 21 (6): 665-670.

Purcell, P.W., 1977. Mud Logging, Chapter 16, in Leroy, L.W., Leroy, D.O. and J.W. Raese, Editors, 1977. *Subsurface Geology,* Colorado School of Mines, *Golden*, 941 p.

Swanson, R. G., 1981. Sample Examination Manual: Tulsa, OK, AAPG Methods in Exploration Series No. 1, 35 p.

Tucker, M. E. And V. P. Wright, 1990. Carbonate Sedimentology. Blackwell Scientific Publication, Oxford, London, 482 p.

Whittaker, A., 1991. Mudlogging Handbook. Englewood Cliffs, NJ: Prentice Hall, 531 p.

Whittaker, A. and G. Sellens, 1987. Normalization of Gas Shows Improves Evaluation. Oil & Gas J. (20 April): 53.

Wright, A. C., 1996. Estimation of Gas/Oil Ratios and Detection of Unusual Formation Fluids from Mud Logging Gas Data. Paper CC presented at the 1996 SPWLA Annual Logging Symposium, 16–19 June.

PART FOUR

FORMATION EVALUATION AND RESERVOIR MANAGEMENT

This deals with petrophysics or formation evaluation and the roles of a petroleum company's asset team in the optimisation of an oil and gas field. A formation evaluation technique, known as logging, is used to record and interpret the intrinsic and induced properties of rocks encountered by a well. Logging may be carried out while drilling (LWD) or after drilling (Wireline). LWD and wireline interpretations are similar and play a key role in estimating porosity, saturation and oil in place.

Cores are also used to provide geological information. They are cut through a rock body in a cylindrical manner to extract a sample of the rock from the formation, as such they provide direct geological information. It is probably the only physical representation of the formation, providing information on petrophysical properties, state of the formation, rock mechanics and stresses, and hydrocarbons or other downhole fluids. Different coring tools are used for different formations. Smaller core samples used for basic acid test and chemical analysis, known as plugs, are cut perpendicular to the main core.

A well testing process is used to ascertain the flow conditions in the subsurface. Well test is part of a pre-completion process, which involves inducing and observing pressure changes in a drilled producing, observation or injection well. It provides useful information for completing the well, for optimizing productivity or injectivity, and for the overall field management. Effect of *skin* is considered, and once properly modelled, understood and improved, can be mitigated to enhance production.

Petroleum companies expect to make profit after the oil and gas have been discovered, produced and sold. Profit will depend on the oil price at the time of production, the cost of the well, or in general, on the total expenditure and expected income. Decline in fields and reservoirs is managed by reservoir revision or addition until abandonment. Abandoned fields can be reactivated to optimize production if new technologies make it viable or if increasing cost of crude oil makes less production profitable.

CHAPTER FIFTEEN

WIRELINE AND LOGGING WHILE DRILLING

15.0 Logging Principle

This chapter deals with the techniques used for formation evaluation, a process known as logging. This is achieved with the wireline or logging while drilling (LWD) methods, aimed at recording parameters over a given depth and time. Properties recorded are those related to radioactivity, density, sound, electrical, and the diameter of the well. LWD involves the use of a tool attached to the drill string to record intrinsic and induced properties of the formation while drilling. Wireline is done in the open hole after drilling, by passing the logging tool up and down the formation to be logged. Different aspects of Wireline and LWD to be discussed in this chapter are:

- General logging operations
- Depth and tool calibration
- The different tools and methods
- Fluid and pressure evaluation
- Porosity and permeability assessment
- Role of the wellsite geologist
- Quality control

A good log acquisition is vital for proper interpretation and utilisation for well development purposes. The usefulness of logs in petroleum results, estimation of water, oil or gas saturation, empirical assessment of log parameters, determining volumes for economic valuation, is only possible when the logs are properly acquired, quality-controlled and delivered to the petrophysics team for interpretation. *How this is achieved is the subject of this chapter.*

15.1 Logging Operations

In order to obtain a log or description of rock units and formation downhole, a tool is attached to a cable known as "wireline" which is lowered into the well after a drilling operation. When the tool is attached to a drill string, such that it records the properties while drilling is ongoing, the process is known as Logging While Drilling (LWD). The tool records intrinsic and induced rock properties as it is pulled across the units (Hingle, 1959; Alger, 1980; Hilchie, 1990). While natural radioactivity exists in certain rocks, others emit radioactive response because they have been excited by a radioactive tool.

Logs are able to record the properties of the fluid in the rock, directly as in well formation testers or indirectly as in the estimation of water or oil saturation from resistivity logs. The data obtained from logging can be used to derive other properties such as porosity and permeability. Wireline logging is also known as Openhole logging, because the hole is drilled and the string pulled out before it is carried out.

Openhole logging can either be:

1. Pipe conveyed (for pipe-based cabling)
2. Coiled tubing conveyed (for coil tubing-based cabling)
3. Tractor conveyed (for tractor-based cabling)

The tools are attached to a string lowered downhole in a cable, while logging downward and upward the hole (Figure 15.1).

To produce a wireline well log after the well is drilled, the hole is first cleaned by running in hole, circulating drilling mud and pulling the drilling equipment out of hole. A *sonde* or logging tool is lowered down the well (which is still filled with drilling mud) on a logging cable, to send signals to the formation, which comes back to the sensors on the tool. In LWD method, this is achieved while drilling, in which case the information is transmitted to the surface through the cable as well as recorded in the memory of the tool.

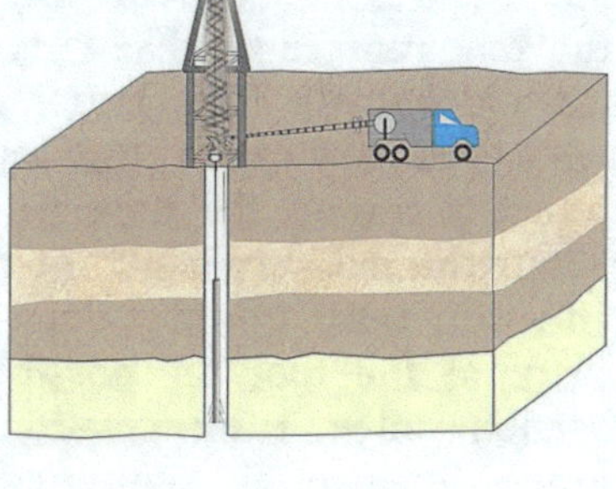

Figure 15.1: Principle of logging

Logging tools have a storage device from which data recorded while drilling may be downloaded when the tool is pulled out of hole. Recorded memory data is usually better than real-time readings. Logging may be carried out while running the tool in hole or pulling out of hole. Among the parameters sensed by the tools are the electrical, acoustical and radioactive properties of the rocks and their fluids, and sometimes the geometry of the wellbore.

Depth

Depth is the most important parameter logged. Other parameters are tied to it and cannot be properly interpreted if their depth is not known. There are two types of depth: **driller** and **logger** depths. The driller's depth is read out using the depths of the drilling pipes whereas the logger's depth is based on the logging tools or cable descended into the hole. Modern rigs have a sensor known as Depth Encoder, which records depth changes in real-time. Because a well is most stable during (wireline) logging, the *logger depth* is considered more precise. The logger depth is corrected for errors such as ***wheel error*** (*wear* and *slippage*) and for **stretch** *(stretching due to tension)*; a chart is used to correct the effect of stretching on logging cables. The final depth is tied to the gamma ray log, which is used to correlate the other logs.

A **Zero Point** is the point on the tool, descended downhole, from where the depth measurement starts; that is the point from where most depth references would be taken.

Calibration

Before a tool is brought to the rig site, it is calibrated at **base** (that is, at the office). Calibration ensures that the recorded value given by the tool represent what it is supposed to measure downhole. This initial calibration is known as **Primary calibration** and is meant to set a tool to respect the standard necessary to carry out the measurement. At the rig site, a **Verification calibration** is done to check the integrity of the tool for possible change in calibration after transportation. Another calibration known as **Sequence calibration** checks that the tool maintains its good stage of measurement before and after each job sequence.

Primary calibration (base) ⟶ *Verification calibration (rig site)*

15.2 Wireline Logging Tool and Methods

Logging tools look alike; often, what is seen is a long iron bar on which different sensors are attached. They form part of the bottom hole assembly. The types and mechanisms leading to the responses of different logging methods provided by wireline tools are described below.

Caliper

Some logging tools, especially those that measure lithology and density based information (*litho-density tools*), have a caliper that measures the diameter of the hole. The diameter of the hole can indicate lithology, since different lithologies behave in different manner. For example, mobile shale may *packoff* to form *cavings* while clean sand will leave *mud filtrate* on well walls to form *mudcake*. Caliper tools tend to be extensible, with an extender (which presses on the walls of the well). An extender contracts or extends, depending on varying diameter of the well they measure. This extensibility of the measuring tool may as well be achieved with a sensor, where the difference in hole size is rather received as signal impulses on the caliper tool's sensor.

To run caliper logs, it is necessary to know the original size of the hole (that is the drilling bit size), in order to relate increasing or reducing hole size to the nature of formation. Permeable formations will typically imbibe more mud filtrate during drilling, hence they will have a thicker mud cake. The caliper log can be used to locate thick filter cakes that form smaller diameter wellbores and identify permeable zones. This is why a sandstone interval may show smaller hole diameter than the shaly zones, as shown in Figure 15.2.

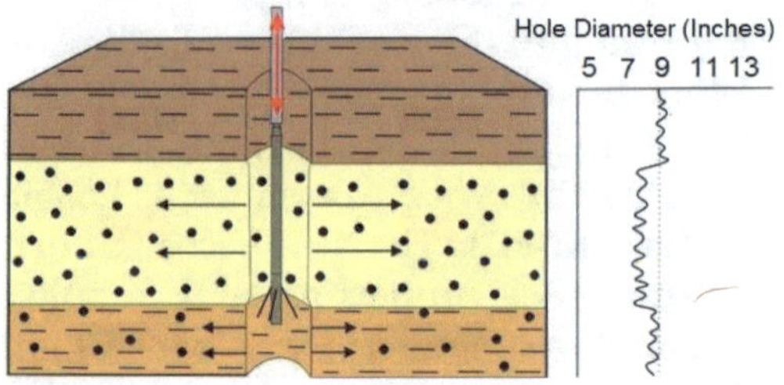

Figure 15.2: Caliper tool showing reduction of hole diameter across porous sandstone interval

In Figure 15.2, the reduction of hole diameter to 7 inches is due to formation of mud cake. Hole diameter is up to 9 inches in the upper and lower shale formation with less mud cake. Cavings or packoff in the shale intervals will further widen the hole. The red arrow in Figure 15.2 indicates possible measurement during upward and downward movements of the tool. Some calipers serve an additional role of stabilizing or centralizing well tools on which they are plugged, by latching or pressing on the walls around it – thereby centralizing the tool.

Since nature of image is related to hole size, and as contact with formation is necessary for certain well imaging tools, some calipers have been adapted to take ***micro-images*** of the wellbore. Modern *micro-images* provide true high-resolution colour image of wellbore, showing formation properties.

Nuclear

The nuclear method depends on the radioactive elements in nature such as Potassium, Thorium and Uranium. The particles emitted by these elements are measured in nuclear logs. Radioactive elements are found in organic matter (uranium for example) and reducing environments (such as bitumen and clays). Thorium is found in heavy mineral while potassium-rich clays are sometimes found in sandstones.

The three main nuclear methods are:

1. Gamma Ray
2. Neutron
3. Density

Gamma ray is the main lithological log, for distinguishing the reservoir and non-reservoir zones, in sand-shale successions. Gamma ray (GR) measurement is also used to estimate the ***Volume of shale (Vshale)*** of a rock, which indirectly relates to porosity and permeability of a rock *(This is further explained in Chapter Sixteen on Log Interpretation).* GR is sometimes run and interpreted in combination with density (Figure 15.3). The tools used in nuclear methods are often attached to the same Bottom Hole Assembly (BHA).

The density log measures an apparent density of the formation known as RhoB or Bulk Density, since it measures also the density of the overall package in the rock, including the fluids. The value of density log may reduce in sandstone similar to that of gamma ray, because porous sandstone is often less dense than compacted shale.

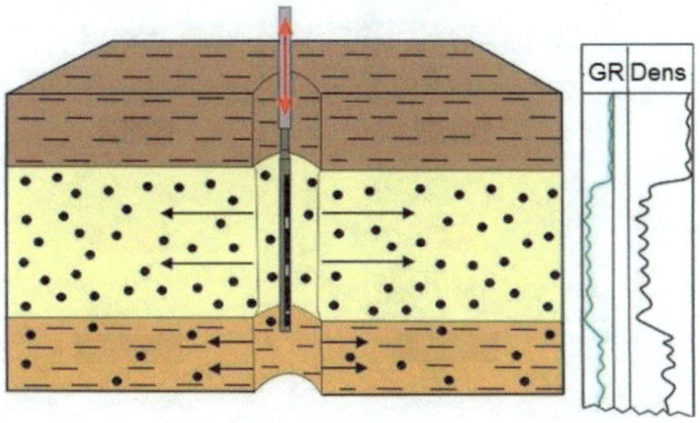

Figure 15.3: Density Log compared with gamma ray log in a sandstone-shale sequence

Neutron method measures the amount of neutron emitted (depending on hydrogen concentration) by radioactivity of the formation. The density, neutron and sonic/acoustic logs are known as ***lithology-porosity logs***.

Acoustic

An acoustic or sonic log uses the speed (or slowness) of sound to give the property of a formation. Slowness is the inverse of speed in logging terms. The principles of the acoustic log involve moving waves with different speeds in different rock types, which depends on the pore spaces, density and other properties of the rock. Acoustic tool usually has a transmitter and a receiver. The transmitter sends the sound waves while the receiver receives the response. The sonic log result is plotted as interval transit time (the amount of time for a wave to travel a certain distance), as a slow to fast interval transit time curve.

The principle of sonic recording is much like seismic measurement. From the principle of the acoustic log shown in Figure 15.4, the transmitter in blue will have a quicker transit time (T1) to the lower receiver compared to the arrival of T2 on the same receiver. Likewise the upper transmitter will have a shorter transit time to the upper receiver. By comparing these transit times, it is possible to deduce the different properties of the formation penetrated.

An important characteristic of acoustic tools is that their positioning in-hole has to be centralized, especially in large holes.

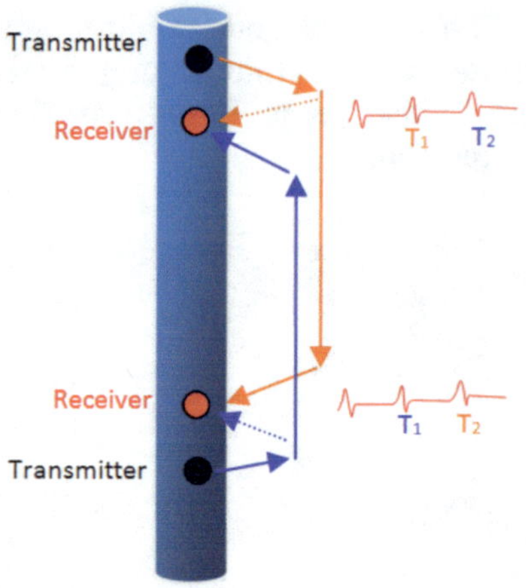

Figure 15.4: Sonic tool showing transmitters and receivers.

This is because it measures the speed of sound (which as well depends on distance to formation). Presence of fractures and dipping bed will lead to disturbances. A poor tool positioning will lead to what is known as **spiking**. This may be due to:

- Decentralization of tool after some time
- Breakout of formation during logging
- Cavings or caves in the walls of the hole

Resistivity and Spontaneous Potential (Electrical logs)

Electric logs work by sending electric current into the formation and reading the response. Spontaneous Potential (SP) measures small electric potential which becomes high in the presence of brine (salty water), in contrast with the conductivity of the drilling mud. It is used only with water-based or conductive drilling mud. SP measures the potential difference due to formation water and the drilling mud in the well. Its value is not high but average in shale due to low permeability. An oil-bearing reservoir will show reduced SP value (as in Figure 15.5) due to reduced presence of formation water or brine. On the other hand a resistivity log will show a high value due to increasing resistance of the reservoir (Figure 15.5).
In resistivity log, the resistance (or conductivity) of a rock unit is measured, by either *induction* or a *laterolog* measurement process. The conductivity is the inverse of resistance, and it is the ability of a material to allow current to flow in it. **Laterolog** measures the conductivity of formation by use of electrodes, as such it can work even in oil-based (non-conductive) mud.

Induction measures the resistivity of a formation by inducing electrical impulses on the drilling fluid, as such it cannot be used in non-conductive mud. It is adapted for water-based mud, where the reaction of the formation after 'induction' is measured.

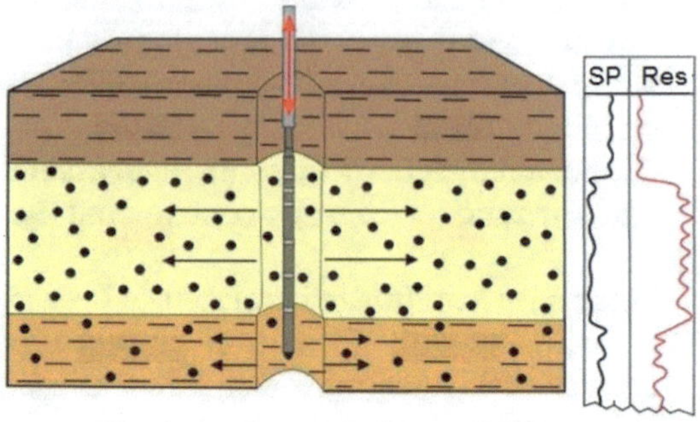

Figure 15.5: Resistivity logging and spontaneous potential test (the short white horizontal lines on the tool represent electrodes – transmitters and receivers)

Most critical uses of the resistivity logs are in the identification of fluid type and hydrocarbon contact. The first wireline log is an electrical log that measures resistivity. An electrical sonde with electrodes in contact with the rocks along the wellbore is lowered into the well. An electrical current is then passed through the rocks and the resistivity of the rocks (and their fluids) to that current is measured.

A resistivity measurement is either shallow, medium or deep, depending on depth of penetration from the wellbore into the formation. Two or more types of resistivities are often measured, with one shallow and one deep. While the shallow resistivity is used to know and eliminate the resistivity value of the drilling fluid, the deep resistivity sends electrical current far behind the *invaded zone* (zone permeated by the drilling fluid) to measure the true resistivity of the rock.

As such deep resistivity helps to determine the natural fluids in the pores of the rocks. Salt water conducts some electricity and has moderately low resistivity. Oil and gas however have high resistivity. An oil and gas reservoir will have a high resistivity, usually plotting to the right of log sheets.

Nuclear Magnetic Resonance (NMR)

The Nuclear Magnetic Resonance (NMR) tool involves a magnetic field and a radio receiver.

This is one of the modern logging tools which have Logging While Drilling (LWD) versions. It works by sending a disturbance which induces polarization of magnetic fields in rocks and leads to what is known as echo "decay". After the disturbance, the relaxation of reoriented magnetic signal, as it returns to the sensor, varies according to the nature of fluids in the formation, as such indicating pore spaces (porosity) and connecting spaces (permeability), as illustrated in Figure 15.6.

The tools sensor identifies signals based on the response of hydrogen nuclei present in a formation or formation water. Therefore, NMR can be used to determine the basic reservoir characteristics and fluid type. Some NMR tools can indirectly measure pore size of rock and the rate of the decay of the signal can be used to obtain the permeability even in high pressure - high temperature (HPHT) wells.

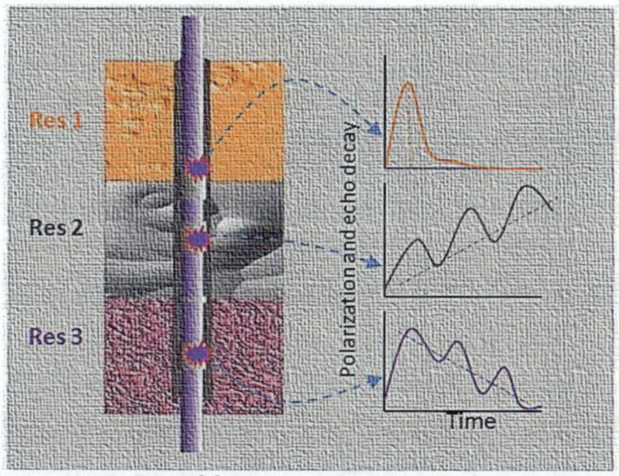

Reservoir 1: Possibly poor porosity at the onset and sharp transition to more effective porosity

Reservoir 2: Reducing bulk density leading to higher decay response

Reservoir 3: Increasing permeability leading to 'free fall' fluid response

Figure 15.6: Nuclear magnetic resonance showing polarization and decays for different reservoirs with time.

Cased-Hole Cement Evaluation

Most cased-hole measurements are wireline measurements carried out after a cementing operation and before drilling the next section of a well. There are two major uses, which are to verify:

- Casing to cement bond
- Cement to formation bond

Cement Bond Log (CBL) provides a representation of the integrity of the cement job, a measure of how much the cement adheres solidly to the outside of the casing. Variable *Density Log (VDL)* provides a representation of the acoustic waveform at a receiver of a sonic or ultrasonic measurement, across a cement in contact with formation wall, in which the amplitude is presented in shades of a grey scale to depict different degrees of attenuation.

The principle of the cement evaluation logs (such as CBL and VDL) has been explained in Chapter 13 with an illustrated example in Figures 13.10.

Sonic (acoustic) logs are also used to evaluate cased hole cement quality. In principle, where there is poor cementing, there will be more void or lighter density, such that sound signals will be slower in it. Sound signals will be faster in more solid cement.

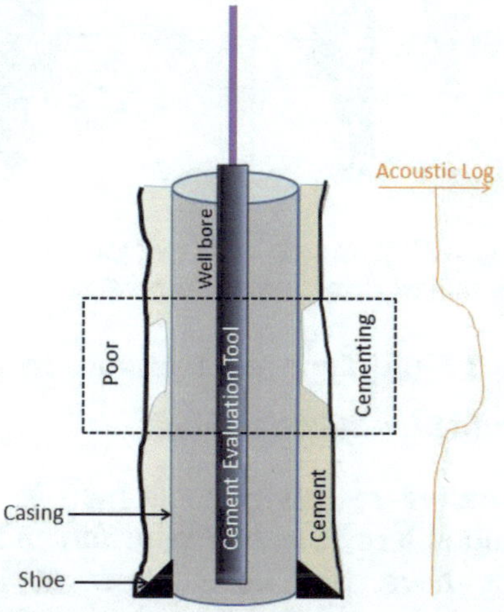

Figure 15.7: Acoustic based cement evaluation (behind casing) showing slower rate of decay of signal in zones with poor cement job

In the acoustic based cement evaluation, the rate of decay (or attenuation) of the resonance of the acoustic tool where there is poor cementing will be slower behind the casing, which will usually be filled with fluid (instead of solid cement). A good cement job will lead to a quicker decay / attenuation, meaning less decay time is recorded in the log (because of the good solid cement). The acoustic log will therefore plot to the right in regions of poor cement (as shown in Figure 15.7). A modern cement evaluation log is the *Segmented Bond Log (SBL)*, which shows more attenuation with more bonds in cement.

Density Measurement

Density tool uses a small nuclear impulse to excite the natural radioactivity of the formation leading to a scatter, which gives the density of the formation as responses return to the detector. It is important for the source and detectors to be in contact with the formation for optimal result. This is achieved by sensors on 'extenders' usually attached to the tool. There are two types of extenders, the pad and the mandrel types (shown in Figure 15.8) that press on the formation wall during logging.

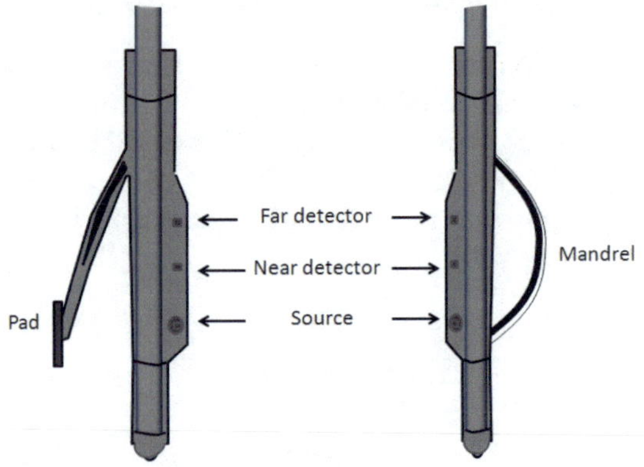

Figure 15.8: Density type tool showing the source, detectors and extenders

15.3 Fluid Type and Pressure from Wireline

Fluid Identification Tools

Wireline resistivity tools help to identify the fluid in a rock, as ***hydrocarbon*** and ***water*** does not have the same electrical conductivity. Logging tools that are particularly useful for fluid identification are the resistivity log and the neutron-density log combination (example shown in the next chapter: Figure 16.10). Modern wireline pressure test tools collect actual sample from formation and this is analysed on the surface to confirm fluid type. Wireline neutron and density tools mainly use nuclear source. Neutron tool emits neutron from a source and the emitted neutron collide with hydrogen atoms in the formation. Formations with high porosity filled with fluid, such as brine with hydrogen nucleus, will give higher neutron reading.

Pressure Test and Sampling Tool

There are wireline tools known as Well Formation Testers (WFTs), such as Repeat Formation Tester (RFT™) and Modular Formation Dynamics Tester (MDT™), which are used to directly measure formation pressures (Marin et al., 2005).

This principle of pressure measurement using a Well Formation Tester has been discussed in Chapter 12 (with an illustrated example in Figure 12.18). A gauge known as probe is placed on the formation, then fluid pressure is measured (Figure 15.9). The fluid may also be sampled and eventually brought to the surface with the logging tool. This is a more precise way of identifying formation pressure.

The direct pressure data from pretest, when plotted against depth, can give the fluid gradient in the formation and different fluid types show different gradients.
In Figure 15.9, a minimum of three points are needed to obtain a pressure-depth gradient and give an idea of the fluid.

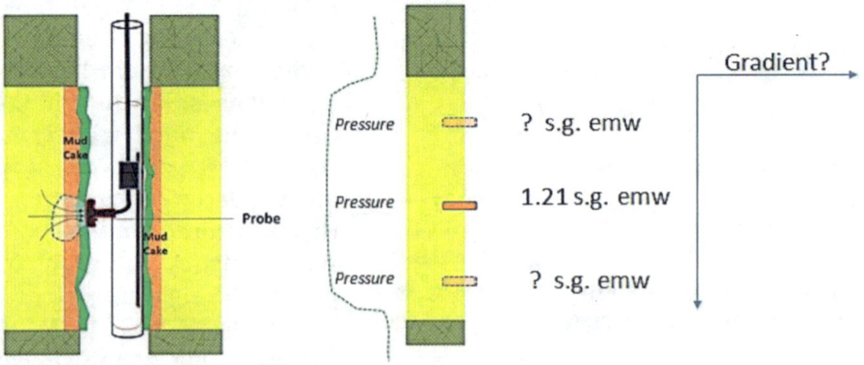

Figure 15.9: Well Formation Tester showing pressure measurements (converted to Equivalent Mud Weights) in a sandstone reservoir

Some pressure logging tools also have a down-hole chamber for collecting fluid sample from the formation, which would be analysed on surface to identify the fluid type. Pressure data can help to identify the:

- pressure boundaries across the drilled formation
- Fluid compartmentalization within the reservoirs
- Reservoir quality and possible depletion

15.4 Porosity and Permeability from Wireline tools

Porosity and permeability data are very important in the oil industry, as it determines what amount of hydrocarbon fluid can be produced. Porosity and permeability can be measured in the laboratory for sediments or cores that have been brought to the surface or taken from outcrops. In the subsurface, however, drilling tools are unable to directly and accurately measure permeability. Logging tools however give indication of an increasing or decreasing trend; for example, lower porosity and permeability are expected for a compacted (high-density) formation compared to the less dense uncompacted sediments (Borst, 1982; Dewhurst et al., 1998; Hartmann and Beaumont, 1999; Yang et al., 2004).

As mentioned above, Nuclear Magnetic Resource (NMR) can be used to measure porosity indirectly. WFT tools can also give an idea of fluid mobility (ability to flow), by measuring pressure and temperature. Although temperature is not directly associated with porosity and permeability, it can give indication of *mobility of fluid* in a given lithology. Mobility is similar to permeability, but it is measured at field scale and depends on viscosity.

The main logging tools for estimating porosity and permeability are:

- sonic
- density
- neutron

Sonic tools can be used to evaluate porosity because the compressional velocity of sound in fluid or void is less than in solid matrix; acoustic energy take longer to go from the transmitter to the receiver in more porous rocks.

Often some form of calibration or comparison / relationship is used to estimate a given value of porosity from logs. Porosity of rocks (e.g. sandstone and limestone) can be estimated from the ***density logs***, as shown in the plot (Figure 15.10). This relationship may not be exactly linear due to secondary porosity resulting from *dissolution* or fractures. This should be taken into consideration in computation of porosity from density logs.

Neutron log is also a good log for estimating porosity, but the nature of fluid makes neutron logs not to directly indicate porosity, where the pore space is not filled by water. The density–neutron log combination, or use of other porosity log (e.g. sonic) simultaneously, is preferred for porosity estimation from cross-plots (more on this in *Chapter Sixteen on Log Interpretation).*

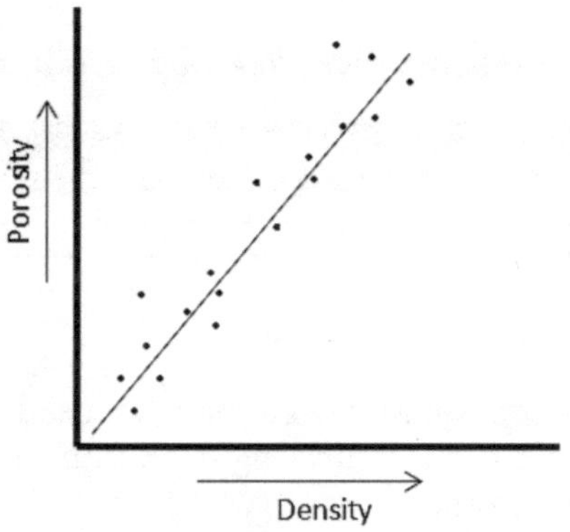

Figure 15.10: Illustration of a proportional relationship between porosity and density of rocks

A similar plot relates porosity to permeability, such that for a given basin or field, a relationship may be used to estimate permeability from permeability. Again, such linear relationship between porosity and permeability does not apply for rocks with secondary porosity (Traugott and Heppard, 1994; Walder and Nur, 1984).

Just as neutron-porosity, density logs are also affected by fluid types or fluid content of a rock. This is because most logging tools measure formation bulk density, which is a function of matrix density, porosity, and fluids contained in the pore space (Pickett, 1972; Borst, 1982; Alberty, 1994; Dewhurst et al., 1998); Formation bulk density measured by the log must be corrected for borehole irregularities before use in porosity determination.

15.5 Witnessing of a Wireline Logging

During a wireline operation, the Wellsite Geologist (WSG) should stay in the logging unit, beside the Wireline Engineer, to ensure that wireline logging operation is well performed. They both follow the operation of the tool, collaborating with drillers to Run In Hole (RIH) and to Pull Out of Hole (POOH) of the logging tool. Tool failures are reported as lost time, which may affect the billing of the logging firm to the oil companies.

Some of the duties of a Wellsite Geologist include:

- Communicate acquisition program to the wireline engineer
- Follow every step of the acquisition program
- Ensure that cable tangling and injury are avoided
- Ascertain the right correlation for reservoirs and their depths
- Confirm that sensors are functioning properly
- Check repeatability by repeat tests (when POOH for example)
- Perform Quality Assurance and Control (QA/QC) of each data recorded
- Prepare and update the Log Quality Control (LQC) Sheet
- Keep communication line open with the *base* (*office in the town*) and issue a report at the end.

More on this in Appendix I.

Other information that may be included in the report of Table 15.1 are:

- Tool characteristics
- Nature of hole (open or cased hole)
- Circulation duration and time and
- Log quality comments.

Table 15.1: Sample wireline logging report table

Wireline Type	Run Number	Time Drilling completion	Last Circulation	Total Lost Time	Logged Interval	Total Operation Time	Max Temperature	Mud Type	Rotary Table Elevation	Seabed Depth	Total Operation Time

A separate spreadsheet may be used to plot each operation, indicating evolution of logging time with depth.

As Logging While Drilling (LWD) becomes more conventional, the LWD versions of most logging tools are prioritized over the wireline equivalent.

15.6 Logging While Drilling (LWD)

All forms of logging are similar and most wireline tools are also available as LWD (Table 15.2); the principles are the same, but the wireline and LWD techniques vary. *There might therefore be few mentions of principles already discussed under wireline, in this section.* Emphases in the following sections are on the impact of (and on) drilling, while logging, compared to wireline where the hole has been left open for a while before logging. In addition to cost saving, what makes LWD different from wireline is the benefit of taking real-time decision or action that affect drilling.

Telemetry denotes the remote transmission of downhole data, with the advantage of obtaining information on drilling mechanics and formation evaluation through pressure waves in the mud to the sensors. It is a kind of direct or automatic logging through the wellbore to the surface, which helps in locating changes in geologic strata and assessing integrity of the well interior. Modern *measurement while drilling* tools incorporate telemetry in their system, sending signals through electronic pulses, amplitudes responses or wirelessly as electromagnetic waves.

Again, depth is the first important factor to consider during any logging process. It is one parameter that Measurement While Drilling (MWD), Wireline and Logging While Drilling (LWD), share. The depth of interest is not only the vertical depth (depth of measurement), but also the lateral depth into the formation (depth of investigation).

Table 15.2: LWD tools / methods and their applications

S/N	LWD Logging Tools/Methods	Applications
1	Resistivity	• Correlation of lithology / stratigraphy / hydrocarbon zones • Determining specific resistivity of strata • Estimating effective porosity of reservoir
2	Image/Photolog	• Searching for lost drilling tools (fish) • Checking for debris in the well • Pinpointing large pore spaces / dipping beds / geosteering
3	Lateralog	• Delineating lithology • Determining shale porosity • Indicating compaction
4	Sigma	• Determine water saturation • Inferring salinity of connate water • Checking reservoir invasion
5	Neutron	• Estimation of porosity • Inferring fluid type • Delineating lithology
6	Density	• Estimation of porosity / pore pressure prediction • Determination of gas-bearing formations / fluid type • Estimation of percentage of evaporites
7	Sonic	• Determining depth & thickness of porous zones • Estimating porosity / seismic calibration / cement bond • Identifying fractured zones / pore pressure prediction
8	Spectroscopy	• Inferring fracture direction • Identification of percentage of shale • Distinguishing between carbonaceous and shale layers
19	Magnetic Resonance	• Determining porosity & permeability of a formation • Determining pore size distribution • Determining pore fluids (gas, oil & water)
10	Formation Pressure	• Modelling of reservoir dynamic pressure / well safety • Reservoir characterization / compartmentalisation • Distinguishing fracture and matrix flows

The depth of investigation depends on the:

1. Strength of the signal
2. Parameter measured
3. Signal emitted
4. Tool spacing
5. Acquisition condition

In LWD, the Rate of Penetration (ROP) of the drilling bit and logging tool may be too fast for proper measurement, or to reach the intended *depth of investigation.* Proper measurement requires that the tool spend some time at a given depth, which is not possible if drilling is too fast. That is why it is necessary to control tool speed in order to obtain good logs. There is also need to adapt the mud *flow rate*, as a very high flow rate may disrupt the information reaching the tool sensor (especially for the real-time data). As mentioned above, apart from the real-time data displayed during LWD measurement, a memory data stores more accurate data and is downloaded when the tool is brought to the surface.

One way of assessing the effectiveness of the intended result of LWD is checking the *resolution*, the *depth of investigation* and the *log quality* control after logging. **Physical resolution** is the smallest increment which a tool can measure and this defines the *separability* and *recoverability* of the data. Each dataset has to be separate from the other and enough time is required for it to be recovered before the next one is sent to the recorder. **Axial resolution** is the measurement per rotation and this could give the number of points assessed depending on chosen speed of rotation and descent of the tool (Figure 15.11).

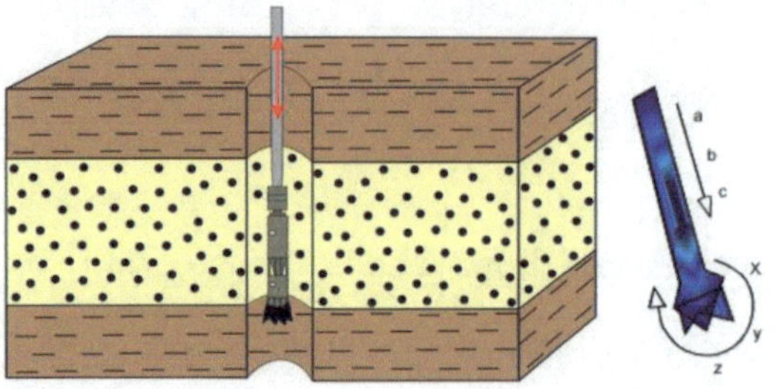

a. b.

a. LWD tool attached to the drilling Bottom Hole Assembly (BHA)

b. Physical and axial resolution

Figure 15.11: Sketch showing LWD logging and logging resolution depending on rate of penetration.

In Figure 15.11 b. while drilling past point a, b to c in direction of the arrow, the LWD tool may measure data increments at points a, b, or c, or may not be able to detect any increment even if parameters are changing. That depends on its ***physical resolution***. The number of points it can measure as it rotates from x, y to z in direction of the arrow depends on its ***axial resolution***.

Despite a careful monitoring of logging procedures, intrinsic or external situation (such as borehole conditions) could result in problems with logs. Some of the problems that could result when logging are:

- Fracture effect (this makes drilling fluid to flow into the formation, leading to invasion, or the irregularity of the formation may result in disturbed data)
- Low frequency and attenuation
- Horn in resistivity (spike as tool crosses a bedding boundary)
- Spiralling (wavy response on log, common in caliper log and other logs affected by spirals in borehole caused by drilling bit on certain formations)

There is always the need to correct logs for ***mud invasion***, ***casing*** (where the upper section of the well has been cased-off before drilling ahead) and for ***environmental*** factors such as attenuation of the signal due to impact of drilling and hole-size. Neutron Log, for example, would need to be corrected for:

- Hole-size (changes in each section or phase)
- Stand-off (distance of the tool to the well wall)
- Mud cake
- Salinity of drilling mud
- Salinity of formation water
- Density of mud
- Temperature and pressure

An example of how to correct a log is illustrated in Figure 15.12 with a density log, where the tool may be closer or farther away from the wall. Another log known as ΔRho (i.e. change in ρ; where Rho or symbol ρ designate density) checks this ***'stand-off'*** and the differences on it are used to correct the value given by the affected density log. Values are added where there is higher *stand-off* (tool less in contact with the wall) and appropriated where there is lower standoff (contact with the wall) respectively. Sometimes the log areas with stand-off are removed completely when doing computations for parameters such as Volume of Clay (VClay) or other cross-plots that relate to actual petrophysical contents.

Bad hole affects logs such as density log; a caliper log along with a ΔRho log is used to highlight and correct bad hole sections (Figure 15.12). Bad hole sections needs to be removed in logs for empirical calculations. Mud cakes formation makes a hole smaller in diameter while cave-in makes it larger.

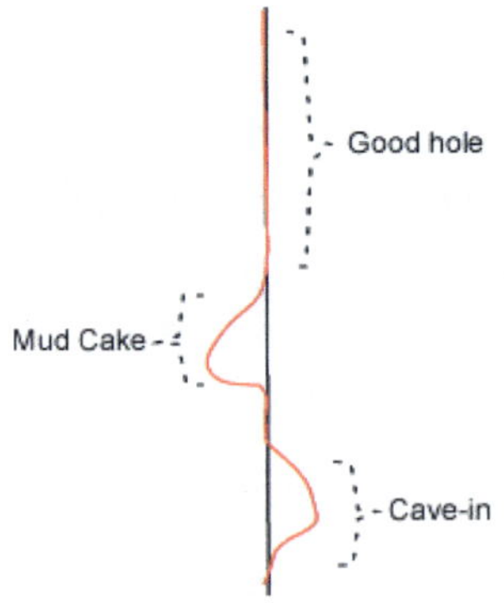

Figure 15.12: Sketch of Caliper log showing bad hole - cave-in and mud cake

Some aspects of well operations affect logging, especially any process that affects depth. Proper tracking of the depth at which logging tools measure formation properties is important. *Driller's depths* (as estimated by drillers) needs to be compared to *logger's depth* (as estimated by loggers). Drillers place a device known as *depth encoder* on the *drawworks* (machine on which the rope/line used for lowering and pulling out drilling pipes or string is coiled) to measure depth during drilling and the signal is transmitted to the loggers (mud-logging and LWD team).

The length of the bottom hole assembly (BHA), drill pipes and logging tools, lowered into the borehole is used to estimate total depth at each stage of the drilling. Sometimes depth encoders on the drawworks may give inaccurate results and need to be compared with other sensors and depth estimation methods. The depths obtained by each party using the different methods are tallied, compared and matched after the descent of each drilling or logging stand. During logging, *correlation pass* (*a test explained later in this chapter*) may be used to ensure that logging sensors are at the intended position downhole, based on the matched depth.

15.8 LWD Logging Tools and Real-Time Benefits

Sonic While Drilling

The principle of LWD sonic log is similar to that of wireline, where waveform arrivals increase with the transmitter-receiver spacing. With different spacing, it is possible to determine the nature of the formation, based on the change of arrivals at the different receivers (Figure 15.13). The difference in time of arrival with increasing separation or distance to the transmitter is known as the **move out**. This is given in **microseconds per foot** (**μsec/ft**) and represents the **slowness** of the waveform through the formation being evaluated.

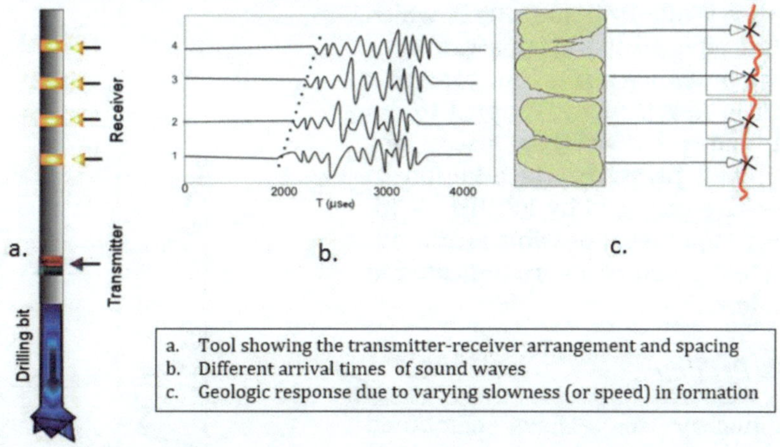

Figure 15.13: Principle of the LWD sonic log

With an LWD sonic tool, the acquisition is made before significant fluid invasion into the formation and alteration of the rock properties occur. The advantages are in quicker drilling decisions and that, with the provision of formation interval transit time, seismic-to-well calibration can be updated in real-time. One of the limitations, though, is the possible distortion of signals from drilling noise.

Nuclear Magnetic Resonance (NMR) While Drilling

The principle of the NMR tool for LWD is similar to that of wireline, as explained in the previous section on wireline. The amplitude and shape of decay, of an echo, give trends that indicate the petrophysical properties of the formation while drilling. The decay train depends on *porosity*, *nature of fluid* it contains and the *pore distribution* linking the mobile or bound fluids.

The trends shown by decay in real-time are the *surface trend* (in which fluid to pore wall interaction can indicate of pore size); *bulk trend* (in which fluid molecules interaction can indicate viscosity); *diffusion trend* (in which the relaxation of polarized magnetic field may indicate pressure and temperature as well as fluid contents). Obtaining some of this information may be useful while drilling. Knowledge of porosity and fluid type are useful in determining where to take pressure tests and carry out fluid sampling.

Formation Pressure While Drilling

LWD pressure measurement provides a near real time *formation pressure* and *mobility* during a temporal stop while drilling. There is no need to pull out of hole before lowering the tool as in wireline. The probe spring out and plug to the formation wall, measures the pressure, then drilling continues. Circulation with drilling fluid is maintained to avoid sticking of the drilling bit in weak formation.

Formation Pressure While Drilling (FPWD) logs can help in geo-decisions and in interpretation of fluid gradients while drilling. It makes it possible to evaluate compartmentalization of formation earlier than in wireline. Coupled with Annular Pressure While Drilling (APWD, which measures stand pipe pressure), this can optimize drilling by providing a real time reservoir information. The APWD tool is coupled to the drilling BHA and provides data useful for understanding well pressures (in addition to formation pressure provided by FPWD). With fluid sampling, now also possible with advanced FPWD tools, fluid types are indicated in LWD pressure logs.

Nuclear While Drilling

Some LWD nuclear tools have combined sensors for porosity, lithology, sigma and Gamma ray. They investigate the same rock body such that the various parameters can be compared for the same zone. ***Porosity*** is investigated using *neutron counts* while the ***lithology*** is investigated by spectroscopy. ***Sigma*** is a parameter obtained by a thermal stimulation and capture of the response of the formation. ***Density*** is based on the same principle as in standard neutron-density formation evaluation.

The ***Neutron-Gamma-Density*** combination gives a deep measurement in LWD and provides reliable data. **The density** measurement gives not only the average density of the formation (RHOB), but also the density in the different (4) quadrants, i.e. Bottom, Left, Upper, and Right, measured while the tool rotates. The density tool also images this, giving an image density that can indicate some properties of the rock, such as **compaction** and extrapolated **porosity**. These can guide on the validation of predicted pore-pressure and on other useful information while drilling (Matthews and Kelly, 1967, Prince et al., 2005).

Some nuclear logging tools do not use in-built nuclear source, nor is a nuclear material introduced before measurement. The tool interacts with the formation to generate a small nuclear signal that is measured. A sourceless neutron log is based on the principle that the faster the die-away or decay after a rock formation (with water) is excited to release nuclei, the higher the response that will be captured. Likewise t**he sourceless gamma ray** tool which uses the principle to capture gamma-rays after the neutron burst,. Gamma ray value is higher in shale and lower in sandstone as shown in Figure 15.14.

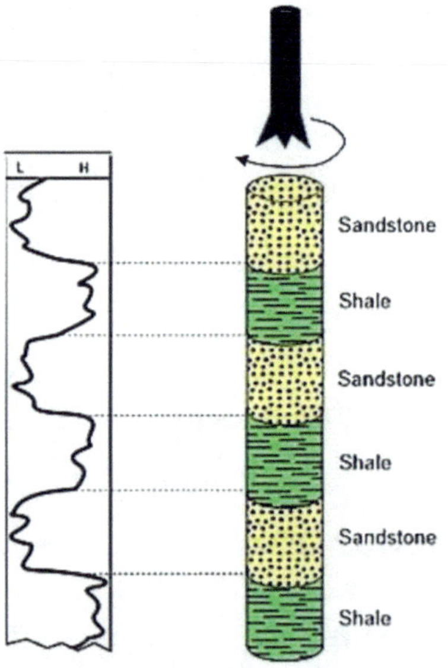

Figure 15.14: LWD gamma Ray while drilling providing immediate information on lithology in a sandstone-shale sequence

Resistivity While Drilling

The use of LWD resistivity tool is a robust approach that is very useful for identifying hydrocarbon zones. Almost all wells today are drilled with a resistivity (as well as gamma ray) LWD tool in the BHA. Resistivity tools can provide reliable boundary characterization due to their capacity to carry out directional measurements while drilling, hence the use in target identification and tracking solutions. Resistivity logs also provide borehole imaging in real-time.

As the tool rotates, it records an image showing either that the drilling head is within the desired formation or veering away to another formation with a different resistivity (such as from sandstone reservoir sand to shale). This can give a way-ahead indicator and steering guide in directional drilling and serve as necessary warning to avoid leaving the reservoir drain. LWD resistivity tools are used extensively for geosteering.

15.9 LWD Quality Control (QC)

Standard processes are used to ensure proper logging and check the quality of LWD logs (Pickett, 1972; Adams, 2005; Rendeiro et al., 2005). Few of the necessary steps and corrections, related to action of drilling or drilling situation such as movement of rig at sea, are:

- ***Tide and Heave:*** impact on depth measurement while logging due to the movements of the water body offshore. Drillers attempt a heave compensation to stabilize vessel and tools while logging offshore.
- ***Correlation pass:*** initial logging across an interval with known depth as reference.
- ***Slip and Stick Test:*** up and down movement of drilling string to ensure it does not stick just before logging. Sudden release after sticking is known as slip, which will give false depth due to sudden release of stretch on the tool.
- ***Tool Tension:*** the tool may first be lowered to settle at the bottom of the well, till there is no tension, then lifted slowly while logging to limit stretching.

Most correlation passes are done using the gamma ray log to ensure the right depth during formation pressure testing while drilling (Figure 15.15). LWD may be carried out while pulling the tool out of hole (after an initial descent with logs providing correlation points). The Gamma Ray (GR) log will highlight the sandstone and shale, for example, as such helps to place the pressure logging tool at the right or intended depth of the porous sandstone reservoir. This ensures that pretest is attempted on the desired reservoir target.

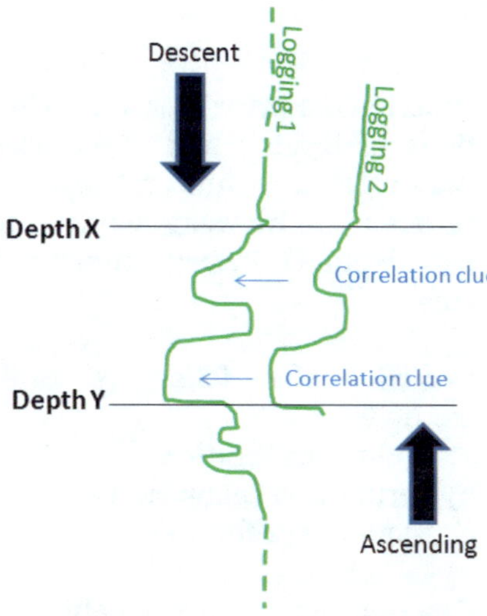

Figure 15.15: Illustration of correlation pass showing calibration of depth based on correlation clues or events on the log

To be recorded in the Log Quality Control (QC) table are:

- Job no°
- Tool n°
- Log type
- Date of tool calibration
- Date job was performed
- Oil or gas field and country
- Scale of reading and sampling rate
- Time after circulation before start of logging

Factors to check as part of the QC by the company geologist include:

a. How reliably and readily does the contracted logging firm provide requested information?
b. The repeatability of the logging response (during logging down and up)
c. Depth accuracy and resolution
d. Is the tool sticking?
e. What is the tension on the string?

The geologist on site will have to check the contractors' reports to ensure the completeness of heading and general presentation before sending the Log Quality Control (LQC) report to *base* (*office in the town*). Some basic consistency (Society of Professional Well Log Analysts, 1975) is expected in such reports. Calibration reports are attached in the form of excel plot or .pdf file (Portable Document Format).

EXERCISES

1. Describe the principles and field operations of logging and differentiate between driller depth and logger depth. Differentiate between depth of interest and depth of investigation while logging.

2. Describe the following calibration methods:
   a) Primary calibration
   b) Verification calibration
   c) Sequence calibration

3. Describe the principles behind the use of the following wireline logging methods:
   a) Caliper
   b) Nuclear (Gamma Ray, Neutron, Density)
   c) Acoustic
   d) Resistivity
   e) Laterolog
   f) Nuclear Magnetic Resonance (NMR)

   What other tools can you mention and compare with their LWD versions? What are the necessary precautions to be taken in order to obtain good logs during LWD?

4. Which logging tools can be used to estimate porosity and permeability? Compare and contrast the laterolog and induction resistivity logging. Use a figure to illustrate your response.

5. What is Spiralling or Spiralisation of logs? What causes it and how do you correct for it? What is horn in Resistivity Logging? What causes it and how do you correct for it?

6. Explain the functioning of Segmented Bond Log with a schematic illustration. Differentiate between Pipe, Coiled tubing and Tractor conveyed Wireline logging.

7. How does depth sensors work – can you list some examples? Differentiate between Telemetry and Sigma logging; what does the Sigma signify as a log response?

8. As a wellsite geologist, what are the necessary steps you would take to ensure that the wireline versus LWD logging operation is well performed?

9. What is correction for Heave and Tide? Differentiate between approach test point and stick test.

10. What is Stand Pipe Pressure and how is it important while drilling? Differentiate between APWD and FPWD – show a schematic to illustrate how APWD is measured.

REFERENCES & FURTHER READING

Adams, S.J., 2005. Quantifying Petrophysical Uncertainties. Paper SPE-93125, *SPE Asia Pacific Oil and Gas Conference and Exhibition.*

Anderson, B.I., T.S. Collett, R.E. Lewis, and I. Dubourg, 2005. Using Open Hole and Cased Hole Resistivity Logs to Monitor Gas Hydrate Dissociation During a Thermal Test in the Mallik 5l-38 Research Well, Mackenzie Delta, Canada. Paper SS, *46th SPWLA Annual Logging Symposium.*

Borst, R. L. 1982. Some Effects of Compaction and Geological Time on the Pore Parameters of Argillaceous Rocks. *Sedimentology*, v. 29, p. 291–298.

Dewhurst, D. N., A. C. Aplin, J. P. Sarda and Y. L. Yang, 1998. Compaction-Driven Evolution of Poroperm In Natural Mudstones: An Experimental Study. *Journal of Geophysical Research*, v. 103, p. 651–661.

Hartmann, D.J. and E. A. Beaumont, 1999. Predicting Reservoir System Quality and Performance, in E. A. Beaumont and N. H. Foster, eds., *Treatise of Petroleum Geology – Handbook of Petroleum Geology: Exploring for Oil and Gas Traps.* Tulsa, AAPG, p. 9-154.

Hilchie, D. W., 1990. *Wireline: A History of the Well Logging and Perforating Business in the Oil Fields.* Boulder, Colorado. p. 200.

Hingle, A. T., 1959. The Use of Logs in Exploration Problems. *Soc. Exploration Geophysicists,* 29th Mtg. (Los Angeles).

Marin, Y., Jacobson, A., and J. Codera, 2005. A New Generation of Wireline Formation Tester. Paper M, *46th SPWLA Annual Logging Symposium.*

Matthews, W. R. and J. Kelly, 1967, How to Predict Formation Pressure and Fracture Gradient. *The Oil & Gas Journal*, February 20 Issue, p. 92–106.

Pickett, G. R., 1972. Practical Formation Evaluation. Golden Colorado, *G. R. Pickett, Inc.*

Prince, C.M., Dixon, M.W., and L.L. Haynes, 2005. The Use of High-Resolution Core Imagery in Reservoir Characterization: An Example from Unlithified Miocene Turbidites. Paper SCA2005-02, *Society of Core Analysts Annual International Symposium.*

Rendeiro, C., Passey, Q., and H. Yin, 2005. The Conundrum of Formation Evaluation in High-Angle / Horizontal Wells – Observations and Recommendations. Paper SPE-96898, *SPE Annual Technical Conference and Exhibition.*

Society of Professional Well Log Analysts, 1975. Glossary of Terms and Expressions Used in Well Logging. Houston. *Soc. Professional Well Log Analysts,* 74 p.

Traugott, M. O., and P. D. Heppard, 1994. Prediction of Pore Pressure Before and After Drilling – Taking the Risk out of Drilling Overpressured Prospects. *AAPG Hedberg Research Conference*, v. 70, p. 215–246.

Walder, J. and A. Nur, 1984. Porosity Reduction and Crustal Pore Pressure Development. *Journal of Geophysical Research. 89*, p. 11539–11548.

Yang, Y. L., A. C. Aplin and S. R. Larter, 2004. Quantitative Assessment of Mudstone Lithology Using Geophysical Wireline Logs: Application of Artificial Neural Networks. *Petroleum Geoscience*, v. 10, p. 141–151.

CHAPTER SIXTEEN

LOG INTERPRETATION

16.0 Aspects of Log Interpretation

Having looked at several logs in the previous chapter, this chapter deals with their interpretation. After obtaining log data, its usefulness depends on the ability to interpret and optimise subsurface information. Log interpretation involves the analysis, classification and computation of related parameters in order to understand geological information in the subsurface. There might be need to extrapolate or derive other information from the initially measured downhole properties, leading to a better knowledge of the characteristics sought for in a formation.

Log interpretation may form the basis for regional basin studies and local field correlation, for the understanding of stratigraphy and petroleum results. Final oil pay zones are defined by logs; the realizable production and economics are based on proper log interpretation. This requires the need to understand:

- Principles of log interpretation
- Understanding of log responses
- Porosity and fluid saturation
- Quick look interpretation
- Further software-aided interpretation
- Integration of other data to logs
- Log interpretation deliverables

The basic principles of interpretation are the same for wireline or Logging While Drilling (LWD). There are however differences in quality and data type and in the necessary Quality Control (QC) before interpretation. Wireline gets final data once acquired, but because LWD has a Real-Time (RT) and a Memory (MR) data, there are two phases of the interpretation. The first phase is the real-time interpretation of displayed recording and while drilling. The second phase is after drilling, when the memory data has been downloaded; as such, LWD also has a post drilling data for interpretation. Their interpretation may be achieved during intermittent stop in drilling, when the drill string is pulled out of hole (e.g. for casing or cementing of a well section), before running in hole again to continue drilling. Memory data is more reliable and is used once a drilling phase is over, whereas

16.1 Log Interpretation and Basic Definitions

The understanding of log types and definition of some petrophysical terms are important in log interpretation (*definition of terms are available in the glossary of the Society of Professional Well Log Analysts, 1975*). Some of them (including resistivity, spontaneous potential, gamma ray and bulk density), which has been discussed previously, are revisited here, before referring to them in the course of log interpretation. The first list below is not necessarily that of properties that are interpreted from the logs, but few log types that may be interpreted to obtain the properties:

- **Resistivity:** the resistance offered by a given rock unit to current; it is the inverse of the conductivity of that rock.
- **Spontaneous Potential (SP)**: a measure of the natural electrical potential (charges) when formation water (which may be connate, i.e. trapped in the rock during its formation) is made to interact with water-based mud (which is a conductive drilling fluid) in the presence of clays (shaly rock) with free ions.
- **Gamma Ray (GR):** a measure of the natural or GR radioactivity of a rock formation, which can be indicative of lithology in a sand-shale sequence.
- **Bulk density:** the density of a rock unit and the fluid it contains.

The second list below are that of some properties estimated from logs, sometimes by indirect measurements or by using cross-relations and calibration within a given field or basin:

- **Porosity:** the amount of pore space in a unit volume of a rock or formation; which is the total volume that will be occupied by fluids if present

- **Permeability:** the ability of a fluid to flow through a rock unit, which depends on the connected pore spaces and fissures.

- **Saturation:** the fraction of pore space occupied by water, oil or gas, which gives the *water, oil* or *gas saturations* respectively.

- **Vclay** and **Vshale:** Vclay is the amount of clay mineral in a rock unit. This is different from the Vshale (which is the amount of shale in a rock unit). Vclay can be estimated from Vshale if the amount of clay per given volume of shale has already been established for a given formation.

Log interpretation may be *visual, empirical* or *relational*. Visual interpretation involves the initial identification of reservoir and non-reservoir sections, fluid responses and log quality, by simply looking at selected logs. The empirical interpretation is worked out by computing values recorded or measurable from the log. The relational interpretation is deduced by comparing log values or trends using charts, known as **nomograms (**also called **nomograph)** or **abaques** and **cross-plots**. These are graphical calculating diagrams, where a result is deduced by plotting the relationship among variable data from the logs (Schlumberger, 1968, 1979).

As such, an important approach to log interpretation is the use of **cross-plots** and **nomograms** to see the relationships among data and data points. Such cross-relations charts may be used to determine unknown parameter dependent on two or three known log values. The functioning of a simple chart is shown in Figure 16.1. In this illustration, the three lines: a (red), b (blue) and c (green) may represent parameters of interest such as **resistivity**, **salinity** and **temperature**. Assuming **i** is unknown; if the log interpreter gets a value of 3 for one parameter that depends on another parameter **a** (not **b** or **c**), then the value **3** may be plotted up to parameter **a** (at point x) and then **i** may be read-off on the vertical axis. Alternatively, two values may be plotted from the two axis to meet a line (where different lines represent different unknown parameters).

With the right cross-plot, the resistivity of the mud filtrate $\mathbf{R_{mf}}$ and that of the mud cake $\mathbf{R_{mc}}$ may be used to find **salinity**. Resistivity of formation fluid is dependent on salinity. The equivalent mud weight (EMW) of brine and hydrostatic pressure (when not fresh water) is also dependent on salinity.

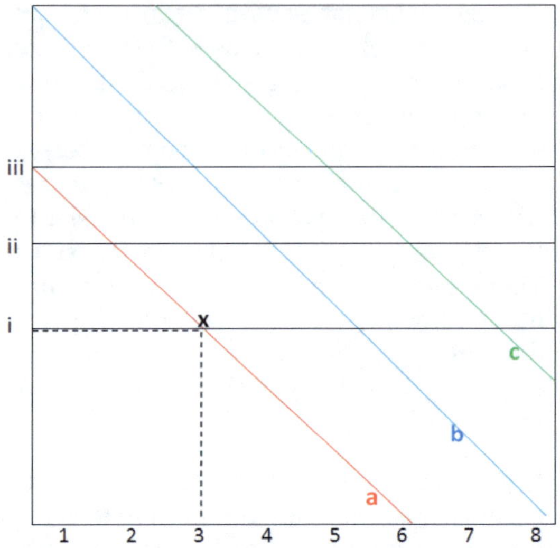

Figure 16.1: Illustration of the working of a three-parameter cross-plot

Log Responses

To interpret logs, the first step is to know the various log responses, and the next is to learn to integrate them in the prediction of lithology before going on to interpret other petrophysical properties of the unit. Once this is done, the interpreter can go on to estimate the fluid properties, gross and net pay zones and their thickness. From this, oil in place and recoverable hydrocarbons can be estimated (Hilchie, 1978; Glover, 2013).

Figure 16.2a. shows spontaneous (self-potential) log response in porous zones. In Figure 16.2b, higher resistivity in the limestone and sandstone is due to additional impact of hydrocarbon. In b. resistivity variation (compare limestone with standstone) is due to oil bearing reservoirs masked by impact of lithology (litho). The limestone matrix is less resistive to electricity than the sandstone matrix, giving higher resistivity response in the sandstone for similar oil saturation. *Resistivity (Res) is measured in ohm-metre or Ω-m while Spontaneous Potential (SP) is measured in millivolts or mV.*

Varying oil saturation; either high, medium or low (as shown in Figure 16.2a) is not easy to identify in practice with visual qualitative interpretation, due to the role of other factors such as lithology (Figure 16.2b). Resistivity values may already be affected by lithology, thereby creating bias in the impact of oil saturation. Neutron-density log is used to differentiate the impact of fluid type (oil versus gas) on logs. ***Shale*** often has:

- Low resistivity
- High gamma ray
- Low neutron

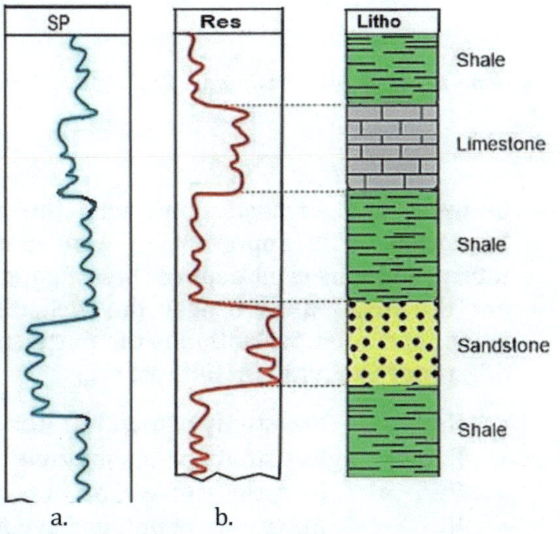

Figure 16.2: Impact of oil saturation and lithology on electrical logs response

The resistivity of a porous rock partly depends on the fluid (either water or hydrocarbon) it contains. Since saline water (brine) usually found in oil field is conductive, ***sandstone*** will show low resistivity if water-bearing or high resistivity if permeated with hydrocarbon. Several log interpretation software and graphic aids (these are cross-plots, nomograms, or diagrams known as *abaque*) exist, but require that the interpreter knows certain parameters for the zone being interpreted. Example of these parameters include:

- Depth
- Temperature
- Resistivity of Mud
- Salinity

Both depth and temperature are obtained from the recording in the logging tool. Using specific *nomograms* or graphic cross-plots, the interpreter could read-off porosity from certain log. An example of a cross-plot from which porosity may be estimated is the one relating neutron, density and sonic logs.

The aim of log interpretation is to obtain ***qualitative*** and ***quantitative*** information from the logs. Initial qualitative interpretation involves the identification of the reservoir, their contacts and fluid types, while the quantitative deals with determination of *empirical values* of porosity, fluid saturation and oil or gas in place. Logs do not directly give the rock or fluid properties from the formation (except where actual fluid samples are collected while logging). They provide the parameters from which the rock or fluid property can be estimated. This is achieved by combining several logs.

Composite Log

When several related logs are combined on a single plot sheet, it is known as *composite log*. After drilling, the composite log combining some important logs that have been *Quality Controlled (QCed)* is presented as a partly interpreted Final Log and scaled to a page size. Composite logs provide an integrated approach to log interpretation. A composite log may among others include:

1. Gamma ray / Caliper / SP
2. Resistivity (shallow, intermediate and deep)
3. Neutron Density
4. Pressure test (pretest) points
5. Other well information such as well architecture and test types.

A composite log with basic interpretation may include saturations and pay zones. In Figure 16.3, the *header* of a composite log is shown with log units, indicating how the information may be combined to provide interpretation on formation and fluid composition of units of interest. Examples of the logs in this example have been shown previously in this book. *An attempt could be made to add curves showing how they may appear in the LWD composite log, taking a clue from the curves of similar logs shown in Figure 16.4.*

A composite log may provide the stratigraphy of rocks encountered in the subsurface, the lithology as well as the geologic descriptions of cutting sample (Compare with Figure 16.4 showing actual log curves).

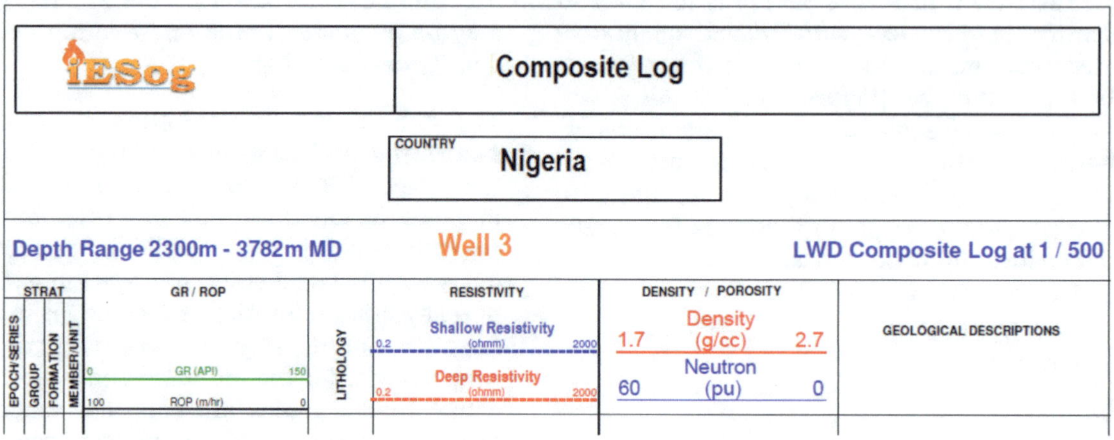

Figure 16.3: Example of Composite logs header, made up of important logs such as Gamma Ray, Rate of Penetration, Resistivity, Neutron and Density

In the Logging While Drilling (LWD) composite log of Figure 16.3, the service company name or logo (iESog) is shown at the top left. *Gamma Ray* (GR) and *Rate of Penetration* (ROP) curves are shown in the first log column. *Shallow* and *Deep* Resistivities are shown in the second. Density and Porosity logs are shown in the third column. The unit for gamma ray log is that of the American Petroleum Institute (API). The neutron log is recorded in porosity units *(pu).* That of density is in gram per cubic centimetre (g/cc).

Invaded Zone versus Actual Formation

The *depth of investigation* achieved and reported on a log varies for different tools and according to the parameter being measured. Closest to the wellbore is the **flushed zone**, where there is a maximum of drilling fluid injection and disturbance in the formation. This is followed by the **invaded zone**, where wellbore fluid has managed to permeate the formation. There is a **transition zone**, which will have an intermediate resistivity, between the *invaded* and the *uninvaded* zone. The main interest is on the **uninvaded zone**, which is deeper into (and has the resistivity of) the formation.

A tool that has a shallow depth of investigation may be reading the zones permeated by drilling fluid and not the actual formation. A major difference to note before interpreting a log is between the invaded zone and the actual formation. It is important to ensure one is interpreting the right depth of investigation and not the invaded zone near the well, that is, when formation evaluation is the purpose. The different resistivities of interest are:

- Rmf = Resistivity of mud filtrate
- Rxo = Resistivity of flushed zone
- Rw = Resistivity of formation water
- Rt = Resistivity of uninvaded zone

At least two types of resistivity tools are run while logging, to measure the ***deep*** to ***shallow resistivities*** of formation. The separation between the shallow and the deep resistivity indicates the resistivity difference between the drilling mud (which may invade the formation slightly) and the resistivity of the formation (deep resistivity). Sometimes up to three (or more resistivities, with increasing depth into the formation) are recorded by log tools, including the shallow, medium and deep resistivities. Differences among the three help in the proper interpretation of the rock bodies and formation fluid.

In Figure 16.4, the depth column is between the gamma ray and the resistivity log columns. Depth is given in meters (m). The shallow resistivity log shows a low resistivity across the entire log response (due to a water-based mud) while the deep resistivity reads higher resistance in the hydrocarbon bearing section of the formation (it reads a lower trend of resistivity in the water bearing section). The resistance in the formation is not very high in the sandstone section (suggested by the gamma ray), as would have been expected for a fully saturated hydrocarbon bearing reservoir (see dotted red line showing possible higher resistivity).

On the other hand, the dotted blue line shows that deep and shallow resistivities may overlap in the fully water-saturated zone, where they both read low value of water response whether in the invaded zone or deeper into formation. There are also situations where both read high resistivity in the hydrocarbon bearing reservoir zone.

The log on the right in Figure 16.4 measures the ***Photoelectric Factor (PEF).*** The *PEF* log indicates the *photoelectric absorption* factor (Pe), which is related to atomic number (Z) raised to a power of 3.6 ($Z^{3.6}$). It is measured by some modern formation density tools. *Pe* is relatively low in sandstone and high in clay or iron-rich formation. This makes it a very good lithology tool independent of porosity.

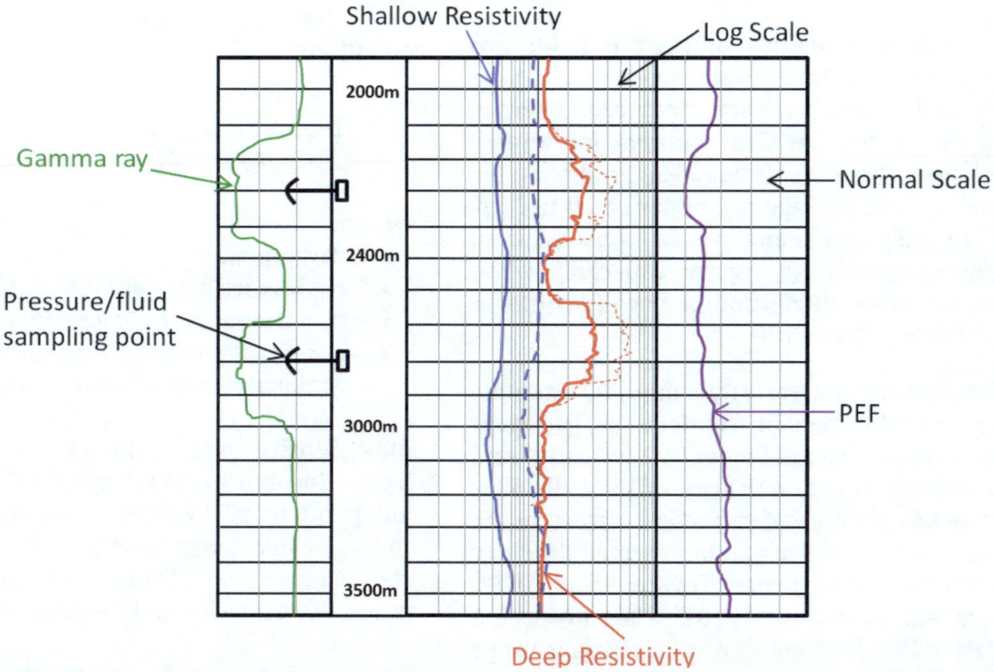

Figure 16.4: Shallow and deep resistivity curves plotted on a log scale and other logs plotted on a normal scale

However because of its shallow depth of investigation, it also gives information on the mud cake and the invaded zone. High iron-bearing filtrate in the mud cake, or the presence of barite, makes it to as well read high value in sandstone or other porous reservoir where a high mud cake is formed quickly. This makes it difficult to interpret the log in such situation.

The ***invaded zone*** will give pressures and resistivity values often higher than that of the formation drilled because of the high density mud weight and oil based mud (if the case) used during drilling. The ***mud cake*** formed during drilling and after the drilling fluids have filtered into the formation have a low permeability and alters petrophysical parameters around the wellbore. The resistivity of this shallow zone is isolated during log interpretation; it is the *shallow resistivity.*

Fluid Type from Pressure Test

In the Figure 16.4, only one pressure point was taken within each reservoir. Normally, there is need for a minimum of three points in order to establish a gradient, where the fluid type is unknown. In this case, since fluid samples were collected in the pressure test, the nature of fluid has been confirmed. There might be no need to take several pressure points (used to establish a gradient for identifying fluid type) in each reservoir.

16.2. Porosity and Permeability from Log

Porosity

One of the most important information sought during log interpretation is the *effective porosity* (connected pore spaces). This is because (effective) porosity is necessary for estimating (effective) fluid saturation and ultimately hydrocarbon in place. Porosity may be estimated from logs or measured directly from cores in existing wells.

The average porosities of mature fields are known, meaning that several wells would have been drilled and porosity measured. Average porosity ranges are also common for known lithologies. Reservoir lithologies, sandstone for example, have average porosities of 10 to 30% (0.1 to 0.3) depending on its consolidation; higher porosities may occur, especially when there has been dissolution or vugs (cavities in the rock) as in limestone.

In the absence of a porosity values, for example in a new basin / field or frontier zone, geoscientists simulate close values of porosity and permeability based on nearest offset wells. This may be based on the compaction history of the basin, seismic profiles and interpreted lithology or sedimentary sequences (Brace, 1968, 1978; Coates and Dumanoir, 1973). They propose a tentative lithology column, suggesting a range of porosity for the intervals to be encountered in the well.

Porosity is obtained from three important logs:

- ***Density log*** – dense material has less free space in it, so a porous material will show lower density.
- ***Neutron log*** – neutron logging measures the energy loss as the emitted neutrons collide with hydrogen atoms in the formation. Formation with high porosity filled with fluid, such as brine with hydrogen nucleus, will give higher neutron reading.
- ***Sonic log*** – this is because sound travels faster in dense materials, so more porous media will be less dense and has higher travel or transit time.

Formulas exist for estimating porosities from the empirical values of the above logs, but cross-plots are used more regularly where values vary over depths (and you cannot possibly do a calculation for each point). Any of the two logs above can be combined to estimate porosity more accurately, each with limitations. A direct computation of porosity is achieved from the bulk density. As mentioned previously, formation bulk density is a function of matrix density, porosity, and fluids contained in the pore space (Alger, 1980; Borst, 1982; Alberty, 1994; Dewhurst et al., 1998). Porosity from bulk density can be calculated using this equation:

$$\Phi = \frac{\rho_{ma} - \rho_b}{\rho_{ma} - \rho_f}$$

where:

- Φ = porosity
- ρ_{ma} = matrix density (see Table 16.1)
- ρ_b = formation bulk density (log value)
- ρ_f = density of the fluid saturating the rock at the borehole – usually mud filtrate

Bulk density may be measured and imputed from a log, but matrix density, which is density independent of the fluid contained in the pore space, is necessary for the application of the above equation. Table 16.1 gives a usable range, which varies in different basins.

Table 16.1 Example of Average matrix densities of some lithologies (modified from Hartmann and Beaumont, 1999)

Lithology	Density (g/cc)
Sandstone	±2.60
Limestone	±2.70
Dolomite	±2.80
Anhydrite	±2.90
Salt	±2.00

Neutron-density log combination may provide both the density and neutron values that may be used in a cross-plot to obtain porosity when the lithology is known. This is illustrated in Figure 16.5, where the yellow, blue and orange lines are the lithology lines, and in-between them are the calibration of porosity. The point where the vertical and the horizontal axes meet on the lines, corresponding to the lithology, gives the porosity value.

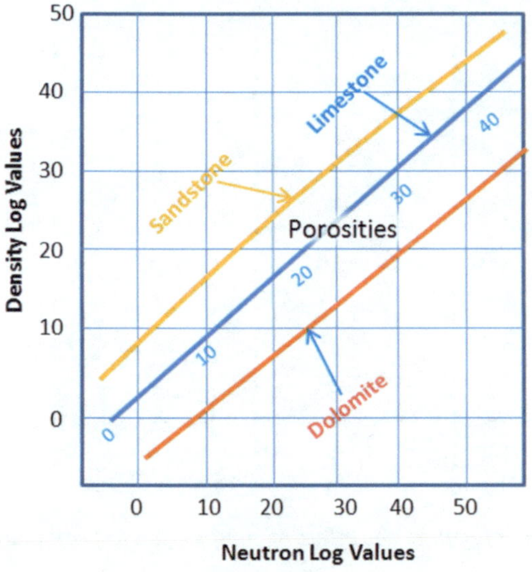

Figure 16.5: Example of a cross-plot for porosity determination from neutron-density logs

To use the cross-plot in Figure 16.5, the values obtained at the depths of interest from the density and neutron logs are plotted up to the reservoir *lithology line* in the cross-plot (sandstone, limestone or dolomite) to read-off the different calibrated porosities (0 to 40%). The cross-plot may also be used to determine the lithology if the porosity and neutron-density values are known. In this example, only the calibration of the limestone porosity line was shown.

Porosity from Vshale / Vclay

Where there are available logs, porosity is estimated from the assessed cleanliness of the (sandstone) reservoir or shaliness of the non-reservoir lithology, or directly based on log (e.g. gamma ray, neutron-density) values. Porosity may also be estimated from ***Vshale/Vclay*** (estimated volume of shale / clay) computed from gamma ray log. Volume of shale (Vshale) is used in evaluating how clean a reservoir is; the concept is that the more the clay, the less the cleanliness of the reservoir sandstone and the less the porosity. *Vshale* from Gamma Ray (GR), at any point of measurement, is given as:

$$\mathbf{Vsh_{GR} = (GR_{meassured} - GR_{min}) / (GR_{max} - GR_{min})}$$

Where:

$\mathbf{Vsh_{GR}}$ = Volume of shale from GR

$\mathbf{GR_{measured}}$ = Measured Gamma Reading (at a given point of interest)

$\mathbf{GR_{min}}$ = Minimum Gamma Reading (sand baseline)

$\mathbf{GR_{max}}$ = Maximum Gamma Reading (shale baseline)

Vshale from gamma ray can be calibrated locally, such that a certain percentage of it, say 70%, represents the Volume of Clay (Vclay) or actual clay content in that lithology. This is the case in some Niger Delta fields. Vclay is also used for estimating the sealing potential of caprocks and fault zones.

Calibrations of Vclay within a basin or a field could indicate what Vclay corresponds to a given porosity; the relationship between Vclay and Porosity is then plotted or a constant is used to obtain porosity when multiplied with Vclay values. *How this principle is used in fault seal analysis, where clay content indicates the sealing or non-sealing capacity (depending on the 'porosity') of fault zone, has been discussed in Chapter Nine on Traps And Seal Assessment.*

Permeability

Porosity is often related to ***permeability***, which is a measure of the ability of fluid to flow within a rock unit. It depends on interconnectivity of pore spaces or fissures in a rock. It determines the capacity of the rock to produce fluid (hydrocarbon of interest) from a well rock formation. Both porosity and permeability are measured from *rock cores* where available. Reservoirs are cored when they are first drilled during exploration.

Permeability measurement has been discussed in *Chapter Nine on Traps And Seal Assessment).* Mobility results from pressure tests and from well test (discussed in *Chapter Eighteen on Well Tests, Completion, Production and Management*) are also useful for assessing permeability in terms of fluid flow within the formation and production.

Pore Pressure from Logs

While interpreting log data from existing wells, the operations geologist may be interested in predicting the pore pressure for drilling the next nearby well.

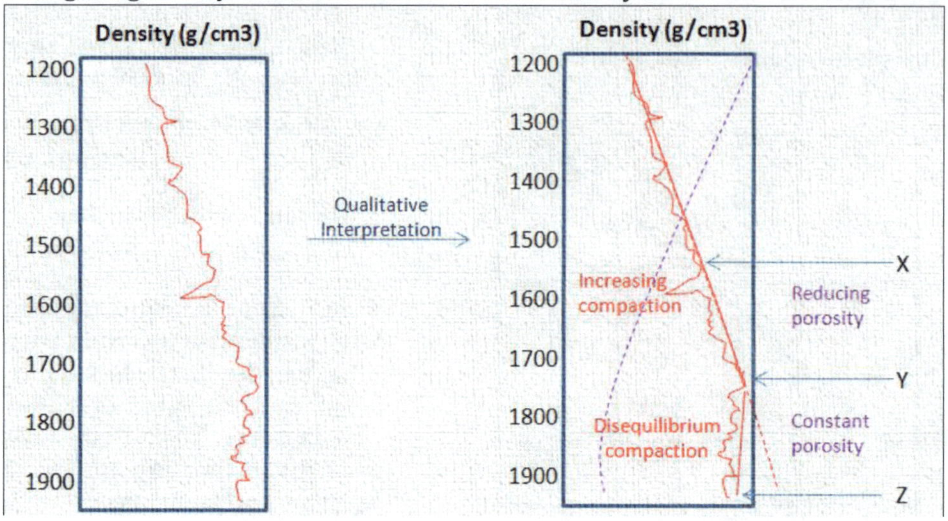

Figure 16.6: Indirect use of porosity data from density log to determine the effect of compaction in pore pressure prediction

Density and porosity logs are applied in in *pore pressure prediction* (Matthews and Kelly, 1967; Eaton, 1975; Traugott, 1996, 1997), discussed extensively in Chapter 12 on *Pore Pressure Prediction and Interpretation.*

This is illustrated in Figure 16.6, where the density increases all through **X** and deviates from the trend (dotted red line) at **Y** and stops increasing till **Z** when there is no longer normal compaction. This is the zone of overpressure and possible changes in safe drilling window for this zone must be considered for planning the well. Abnormal compaction or disequilibrium compaction leads to overpressure generation.

Saturation from Logs

The petrophysics team in an oil company is interested in the hydrocarbon in place and its evaluation starts by estimating **fluid saturation** for the available pore spaces or porosity of a given reservoir. Saturation is the fraction of pore volume filled by fluid (usually water, oil or gas) within a rock unit. Fluid saturation is measured at the laboratory or it can be indirectly derived from resistivity logs (*explained below*). Once porosity has been estimated, the next step is to determine the fluid saturation.

One of the major objectives of log interpretation is to know what type of hydrocarbon is in the reservoir (fluid type may be determined from resistivity and neutron density logs or directly from fluid sampling; ***the saturation*** may be determined from resistivity, as demonstrated in section 16.3 on Quick Look Interpretation). Saturation is measured in fraction of 1, or in percentage; water saturation, therefore, is the fraction of the pore volume of pore filled with formation water, just as oil or gas saturation is the fraction filled with oil or gas respectively. Because **water** and hydrocarbon are the dominant fluid in the subsurface, it is considered that:

Hydrocarbon Saturation
+ Water saturation
= 1

Oil + Gas + Water = 1

This is a major principle in the estimation of oil in place. Any of the above fluid making up to 1 could be zero. All that is needed is to know the water saturation and then the hydrocarbon saturation and eventually volume in place can be determined.

PVT from Logs

The reservoir team is concerned about the different fluid phases in the reservoir, and how these affect actual volumes that will be recovered at the surface. They sometimes have to rely on the temperature and pressure recorded by logging tools, such as Wireline Formation Testers (WFTs). Some WFT fluid sampling tools also has a detachable PVT bottles that are used to collect samples downhole while maintaining borehole pressure - temperature - volume conditions, until it gets to the surface. *Such samples may also be collected during a 'mini-well test' such as the drill stem test (DST).*

The fluid is then flashed at the surface *separator* or in the laboratory; it is exposed to gradual changes in (atmospheric) temperature and pressure and allowed to separate into its various components. The volumes of water, oil and gas, and especially the temperatures and pressures where this occurs (e.g. bubble point, due point), is used to plot a PVT phase diagram that indicates what the fluid behaviours will be during production. The fluid phase (gaseous or liquid) behaviour affects the actual volume of resources that will be recovered in standard conditions. PVT phase diagrams uses pressure and temperature (related to depth) to predict fluid phase changes, making it possible to predict Gas to Liquid Ratio (GLR) or Gas-Oil-Ratio (GOR) at surface conditions for a given reservoir.

For a pure fluid, the PVT relation would look like the plot shown in Figure 16.7a. A phase interface indicating the ***dew point*** (when the fluid becomes liquid) and the ***bubble point*** (when the fluid becomes gaseous) will continue up to a ***critical point*** as shown in the figure. The *critical point* is where the fluid is neither gas nor liquid, or it is both, as they are in equilibrium. Below the critical point, the fluids coexist in such a way that it can become liquid or gaseous depending on changing volume (due to pressure or temperature changes).

For a pure fluid, the dew point and bubble points are always the same, that is why the phase interface is a single line up to the critical point. For a fluid with multiple composition, like hydrocarbon, differences in the components affect fluid phase behaviour such that there are several phase interfaces, leading to a phase envelope as shown in Figure 16.7b, which all meet at the critical point.

For the fluid mixture like hydrocarbon, as pressure drops during production from the reservoir to the surface, the initial components change from liquid to gaseous, and there is a mixture of 100% liquid less that exsolved gas (like 90% liquid and 10% gas) as shown in the Figure 16.7b. This continuous from the bubble point to the dew point, and vice versa. A reservoir fluid above the bubble point is classified as **undersaturated**, meaning that it does not have free gas.

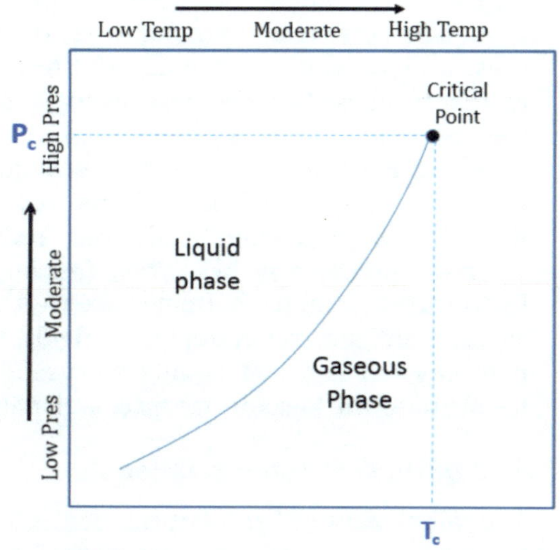

a. Pure fluid

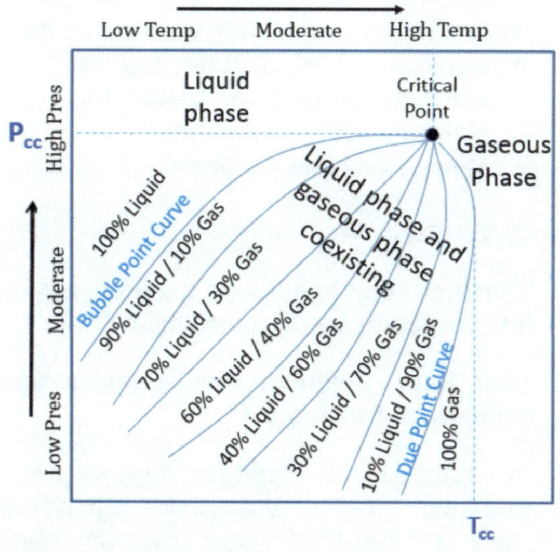

b. Multicomponent fluid

Figure 16.7: Phase diagram showing the controlling factors and tendency of fluid to have different phases

From the bubble point to the region below it, a reservoir fluid is classified as **saturated**, meaning that it has free gas. Below the bubble point, the fluid would exsolve gas. Beyond the *dew point* (to the right) is a full gas phase region, while beyond the bubble point (to the left) is the full liquid phase region. The **cricondenbar** is the limit pressure above which the two phases cannot coexist together, just as the **cricondentherm** is the limit temperature above which the two phases cannot coexist.

PVT diagrams have been established for known oil fields to relate fluid behaviour at depth with actual fluid content at surface. During oil production, care is taken to not lower reservoir pressure (by producing too quickly) to a point where the phase becomes gaseous while still in the reservoir. Water injection is used to increase or maintain pressure and consequently retain the liquid or oil phase in the reservoir. Even if gas production is desired, reducing reservoir pressure makes *lifting* (moving the hydrocarbon from the bottom of the well to the surface) difficult. In developing oil fields, initial petrophysical and PVT results are useful also for planning the injector-producer well pairs.

16.3 Quick Look Interpretation

The initial stage of log interpretation is known as *quick look*, which is done before proper interpretation by petrophysicists in the office using basic calculations and software assistance. The aim of qualitative interpretation is to answer some basic questions, such as:

1. What are the reservoir boundaries?
2. What are the fluid contacts?
3. What fluid types are there?

Qualitative

The three steps to follow in qualitative log interpretation are discussed below.

First Step: Delineate the Reservoir Sections from other Formations

In a sand – shale – tight (or silty) sequence for example, draw out boundaries separating the shale and the tight zones from the sand. In Figure 16.8, the sandstone reservoirs are clean and neatly cut-out from the shale above and below them.

With a gamma ray log, delineating reservoirs in a sandstone shale sequence is straight forward, especially where it is a clean (non-degraded) sandstone and there is no arkosic sand with high potassium feldspar. Arkosic sand shows relatively higher gamma ray and makes it difficult to separate reservoir sandstone from the non-reservoir intervals. Degraded sandstone, with shale or clay content, also shows relatively high or spiky gamma ray response.

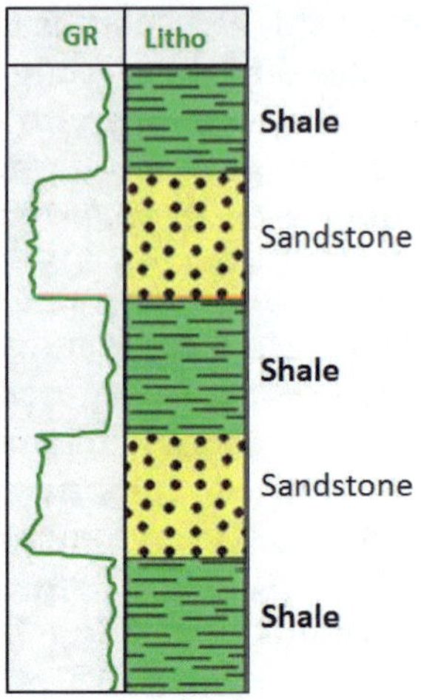

Figure 16.8: Gamma ray log showing delineation of sandstone reservoir in a sandstone-shale sequence

Second Step: Identify Hydrocarbon Section and the Water Leg / Zone

Hydrocarbon Zone

Hydrocarbon section will have the high resistivity while the water section will have relatively low resistivity. The contact between the two is drawn as the Water-Oil-Contact (WOC). Spontaneous potential (SP) logging is to some extent useful in identifying porous and salty water-bearing sections of the formation logged (Doll, 1948). It shows rather low value if the porous zone is oil-bearing (Figure 16.9). SP log shows relatively higher value in porous rock with brine composition. Once the hydrocarbon zone has been identified with the high resistivity, the neutron and density logs can be used to differentiate between the gas and the oil pool as shown in the next section.

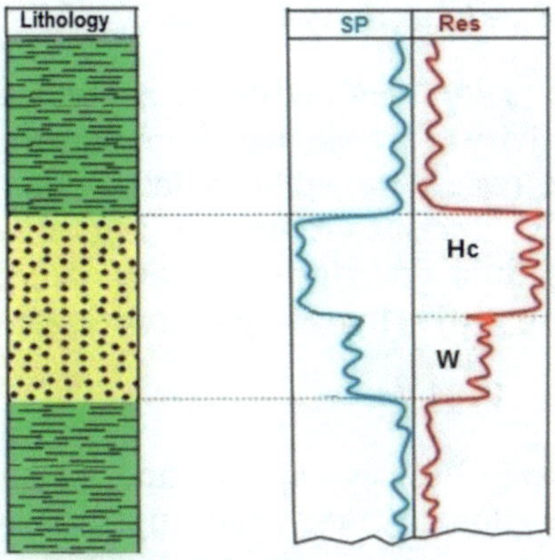

Figure 16.9: Delineation of fluid types and identification of water-oil-contact (WOC) using Spontaneous Potential and Resistivity logs

In Figure 16.9, the water-oil-contact (WOC) separates the water zone (W) from the hydrocarbon zone (Hc). Both the Spontaneous Potential (SP) and the Resistivity (Res) logs identify the WOC.

Water Leg

Identifying the hydrocarbon zone means isolating the water zone, which also has to be identified. In the sandy reservoir section, try to identify and isolate the "water leg" where the sand is filled or saturated with water only or nearly so (this will be useful later, *as shown in the later sections below*). In the clean waterleg sands, the deep and shallow resistivity will read the same or nearly so (if using water-based mud).

Notice in Figure 16.4 where the shallow and deep resistivities record low values. That is because at all depths of investigation in that interval, including the **invaded zone** and the **uninvaded zone** (deep formation), the formation is only water bearing and will read the lowest resistivity. If using oil-based mud, it is more complex but there will still be some observable difference compared to the oil zone.

Third Step: Identify Hydrocarbon Type in the Non-Water Wet Reservoir Zone

Differentiate the oil and gas using a neutron-density log combination (Figure 16.10). Looking at the neutron-density log, the separation is more when it is gas-bearing compared to when it is either oil- or water-bearing. Already water and hydrocarbon sections would have been separated using resistivity (and SP logs as explained above); so at this stage focus is on the separation between oil and gas in the hydrocarbon zone.

In Figure 16.10, the Low (L) and High (H) neutron-density Log readings are used to identify the hydrocarbon zone, within a sandstone lithology, as shown by the lithological column (Litho). Neutron-density log combination shows a *cross-over* (higher neutron and lower density values) in hydrocarbon zone, with higher separation of the logs in the upper gas cap compared to the oil pool. This is because density reduces in porous hydrocarbon-bearing (relative to water-bearing) reservoir while neutron response increases due to concentration of the hydrocarbon or high hydrogen content (with more mobile hydrogen in gas than in oil).

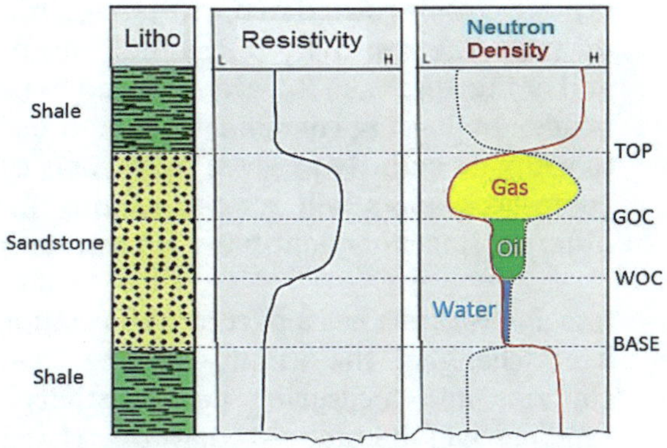

Figure 16.10: Neutron-density and resistivity logs combination used to distinguish fluid types

Quantitative

Once the height/thickness of useful reservoir section is delineated, quantitative analyses help to know the ***porosity***, the ***saturation*** of the fluid of interest, and with few other parameters, the ***total extractable*** or ***producible hydrocarbons*** in the reservoir can be estimated (Archie, 1942; Carothers, 1968; Crain, 1986). The permeability earlier on estimated or inferred from the porosity gives the *hydrocarbon movability*, which determines the producible fluid. As has been mentioned above, cross-plots and *abaques* exist for estimating porosity from resistivity and sonic logs.

Only the porosity in the net *completable* sands or reservoir is used for the estimation of producible hydrocarbons. The shaly sandstone or poor reservoir may be isolated with blanks (non-perforated 'screens') during production. That useful part of the sandstone is represented as percentage net-to-gross. Shaliness (any presence of shale degradation) of the reservoir and the other methods mentioned above (in the absence of *core* data) are used to rate the quality and decide on whether a given reservoir forms part of the ***net***. With the porosity and the resistivity values, it is possible to use empirical formula and determine the saturations.

The ***resistivity*** of a formation offers information relative to the fluid it contains, and is the basic log in use for most quick interpretation of fluid in the industry. Shallow Resistivity measurement **(*Rxo*)** is the resistivity of the invaded zone measured by one of the resistivity tools downhole. That measuring further to the non-invaded zone is the deep resistivity **(*Rt*). Rt** is more closely related to the formation resistivity **(R_o)**. The resistivity of the mud filtrate $\mathbf{R_{mf}}$ and of the Mud cake $\mathbf{R_{mc}}$ are recorded in the log header by logging companies and can be used to estimate **salinity** as given by various charts. Some cross-plots will require salinity for the different stages of quantitative interpretation.

Imagine that one has a pure sand formation and keep changing the salinity of the water it contains and measuring the resistivity each time, it will be noticed that the change of resistivity value obtained is directly dependent on the amount of salt added. It follows that a clean formation fully saturated with brine (saline water) will have its overall measured resistivity (R_o) value proportional to the resistivity of the brine it contains (R_w).

That is,

$R_o \propto R_w$

$R_o = F * R_w$ (where F is the constant of proportionality)

That constant of proportionality is known as *formation factor* (i.e. the aspect of resistivity change depending on the formation itself).

From the above,

$F = R_o/R_w$

Where:

F = Formation factor

R_o = Overall measured resistivity

R_w = Resistivity of brine

More salty conductive brine will mean less resistivity of a rock, which implies that rocks of lower resistivity will have less R_o and therefore less F. Rocks that are more porous will have more fluid or brine than less porous rock, so porosity (ø) is inversely proportional to F.

$F \propto 1 / ø$

Archie (1942) was able to find the constants of proportionality and come up with a formula. That basic formula given below is the key to the quick interpretation of a log, when the aim is to obtain the saturation and oil in place:

$$\mathbf{F = a / ø^m}$$

Where:

a is a constant

m is an exponent known as cementation factor.

(*a* and *m* varies in different basin and determined by empirical measurements).

When there is either oil or gas in the formation, then it is no longer that simple to determine the resistivity which then becomes a function of what percentage of hydrocarbon and of water is present. That is, what their various saturations are, a sum ($S_w + S_h = 1$) which should give 100% (Figure 16.11).

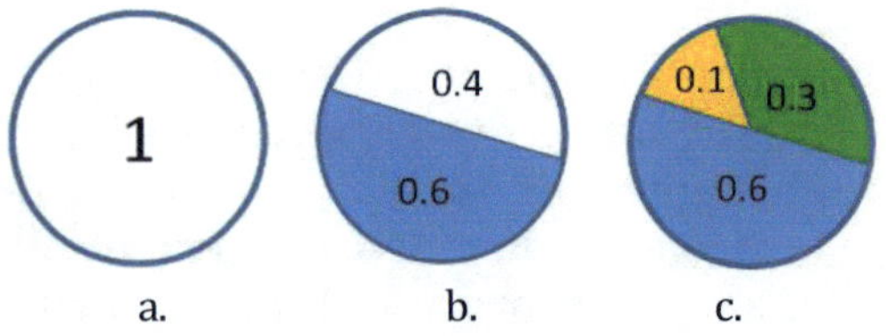

a. The volume of fluid in the entire pore space is a full fraction of 1 or 100%
b. Water fills 0.6 (60%) while hydrocarbon fluid fills 0.4 (40%)
c. The hydrocarbon is identified as oil filling 0.3 (30%) and gas filling 0.1 (10%)

Figure 16.11: Circles indicating pore space filled with different fractions of formation fluid leading to different water saturations of the formation

Therefore,

Water Saturation (S_w) = $(1 - S_h)$

Hydrocarbon saturation (S_h) = $(1 - S_w)$

Since hydrocarbons are not conductive, unlike water, the resistivity of the formation will depend on the above percentage saturations as given in the above relations. Archie's formula further relates the water saturation (S_w) to Formation factor (F) and the Resistivity of the water zone relative to the overall measured resistivity of the formation (R_t) measured by logs.

$$S_w^n = F * R_w/R_t$$

Where n is a constant known as *saturation exponent*, often taken to be 2 in most formations,

If n = 2,

Then,

$$S_w = \sqrt{F * R_w/R_t}$$

Having established earlier on that in a fully water saturated clean formation, $R_o = R_w$,

From $F = R_o/R_w$,

$F = 1$, in clean formation.

Therefore,

$$S_w = \sqrt{R_w / R_o}$$

The resistivity measured in the water level (with low resistivity) within the clean sandstone (with low gamma ray) can be read-off. The deep resistivity in the hydrocarbon saturated zone (comparing with neutron-density log for fluid type) can also be read-off. Then the quick look method explained above can be used to estimate water saturation.

The water saturation gives the oil or gas saturation. As shown above,

Hydrocarbon saturation (S_h) = $(1 - S_w)$

The hydrocarbon in place can be estimated by multiplying that fraction by the porosity, volume of rock body making up the reservoir, and other important petrophysical parameters. Most of this has been integrated in various industry software.

- Oil in place (OIP) =

  *Net reservoir fraction volume * Net pore volume * Net Hydrocarbon saturation*

  $= V_{res} * Phi\ (\emptyset) * S_h$

For the productivity of the formation, it is the effective porosity (PHIE) and effective water saturation (SWE) that is relevant. Effective or interconnected porosity and volumes are related to the permeability. The cleaner the sands, the more the interconnectivity of the pore spaces and the interconnected volumes. Clayey portions in dirty sands, and even smeared shaly traces at granular levels in sand, affect conductivities of a reservoir rock, requiring an interpretation that takes that into account.

16.4 Further Interpretation

Further interpretational solutions are provided by software using recursive and iterative procedures, a forward and backward computation to model actual oil in place while taking into account the results of the various logs and especially putting in consideration shaliness or Vclay content of the rock body. Archie's formula is not quite applicable to reservoirs where there is no clean sand for estimating R_w.

Petrophysical software allows the user to compute lithology, porosity, saturation and oil in place along the entire section logged, be it sandstone or shale, reservoir or seal. The final result looks like the illustration in Figure 16.12. This figure highlights the hydrocarbon section known as the ***pay zone*** (hatched in the figure). The fraction of the hatched space that represents hydrocarbon column is multiplied by the net reservoir fraction to get the net volume of oil in place (OIP) as shown in the equation in the previous section.

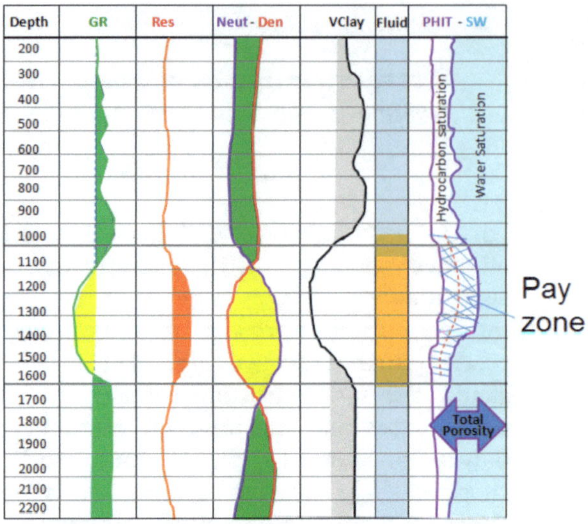

Figure 16.12: Illustration of log result column displaying the final results from petrophysical computation often aided by software

In Figure 16.12, the Gamma Ray (GR) log is shown in the column after the Depth. This is followed by the Resistivity (Res) and the Neutron-Density (Neut-Den) columns. Then the Volume of Clay (Vclay) and the Fluid type column. The Total Porosity (PHIT) and the Water Saturation (SW) are in the last column. This combination shows the various results of interest in petrophysical interpretation. These logs thereby identify the hydrocarbon section in the reservoir, the **pay zone,** and the fraction of the volume occupied by hydrocarbon fluid (oil, gas) and water. The total saturation should be equivalent to the total porosity to make up the full fraction of 1 or 100%.

At the end of the process, a petrophysical result table is provided (as shown in Table 16.2). This table includes the reservoir intervals of interest, the NET sandstone and the ratio of the net sandstone to the GROSS reservoir rock – known as the net to gross (NTG). It also shows the Total Porosity (PHIT) and the effective porosity (PHIET) as well as the Total Water Saturation (SWT) and Net Water Saturation (SWE) which is important for knowing the producible oil. The Net producible hydrocarbon is the difference between the SWE and 1.

The table gives the total of the reservoir petrophysical results thereby giving the total pay from the well; that is the total oil zones. This can be used for updating the economics, the profits expected from the well. The economics is simply obtained by multiplying the cost per barrel of oil and the expected volume from the hydrocarbon pool. While the volume (Stock Tank Oil in Place or STOIP) has been worked out by computing areas of geometries and their height, only petrophysical results evaluates the quality of reservoir and gives the actual producible oil from that volume or pool.

Table 16.2: Petrophysical interpretation result table

Zone Name	Interval Top* (m MD)	Interval Base* (m MD)	Gross Interval (m)	Net Reservoir					
				NET [m]	NTG [m/m]	PHIT V/V	SWT V/V	PHIE V/V	SWE V/V
Res A									
Res B									
Res C									
Res D									
TOTAL									

This is why it is important to re-evaluate STOIP once a well is drilled and update reserve estimates for a field. Usually log interpretation result leads to a new interpretation and reservoir model update. Log interpretation may also be complemented by core and stratigraphic data, seismic check shots and aid in reservoir correlation.

Other logs or necessary information in Table 16.2 (compared to Figure 16.12) are the *net-to-gross (NTG), effective porosity (PHIE), effective water saturation (SWE)* and *net pay* or *Net,* given as empirical values. It is such empirical values that go into economic computation and financial results.

16.4.1 Integration of Core Data to Log Interpretation

Cored rock unit may be logged when out of the hole and the result should be similar to that obtained by logging the formation at depth. Log result from cores on the surface may be used to correct or match logs taken downhole. This is useful where depth mismatch is suspected. In Figure 16.13, a gamma ray log is taken across a cylindrical rock core.

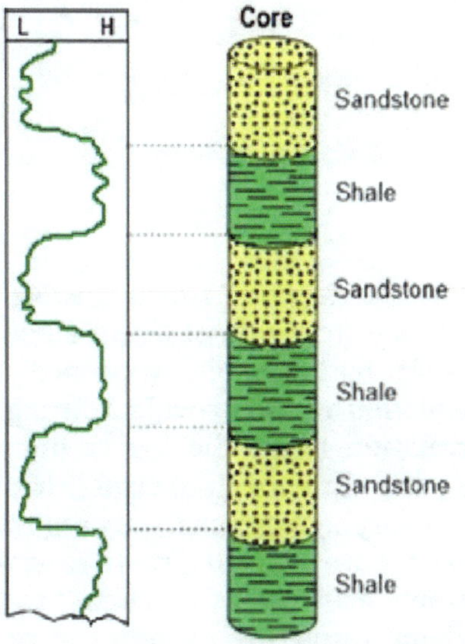

Figure 16.13: Illustration of core sample matched with gamma ray log showing low value (L) in sandstone and high values (H) in shale

Side cores, known as plugs, cut out from cylindrical rock units may be analysed in the laboratory to complement log results. Volume of Clay (Vclay) estimated from logs can be compared with total volume of clay minerals from quantitative mineralogical analyses of core data. Comparison between tangible core data and log results from important calibration information integrated in log interpretation.

16.4.2 Stratigraphic Association and Reservoir Correlation

In log interpretation, effort is made to group observed results to the history of the sediment described. This history is seen in terms of age, depositional environment and also in terms of the layering. One of the most useful applications of litho- and biostratigraphic study is in the correlation of units crossed while drilling. Microfossils are tied to log data for better and broader interpretation.

Integration of stratigraphic information to log is done at the initial stage of the basin studies. A stratigraphic column is placed alongside composite logs, providing a combination that answers certain questions about each section.

- Which sediment was below and which one above?
- How far away from the ocean was that section during deposition?
- Are there other reservoirs or formations to which these belong?
- Did these occur at the same time elsewhere?

When an interval is cored and compared with log, a better association is used to calibrate the interval; that is to establish their actual depths, age and other geological information of interest.

Looking at Figure 16.14, the integration of the logs in the correlation across geological strata helps to predict the lithology, good reservoirs, good seals, and better pay zones. Implication of the degraded zone shown to the right of the figure suggests that Archie's equation and quick look interpretation may not work. Fining up or down sequence in a log, especially in near shelf offshore environment, may help to establish the sequence stratigraphy which also helps in predicting where there may be good sandstone (like the lowstand systems tract; LST). Prediction may also be aided by biostratigraphic data, which are expected at the condensed section deposited during maximum flooding and seen on seismic as a marker known as the maximum flooding surface (MFS).

Tying a reservoir unit together at different places is a very important tool in the industry known as ***Reservoir correlation.*** It helps the geosciences department to extrapolate reservoir properties. Most calculations and estimations before drilling a new well are based on projected property similarities at offset well that encountered the same prospect.

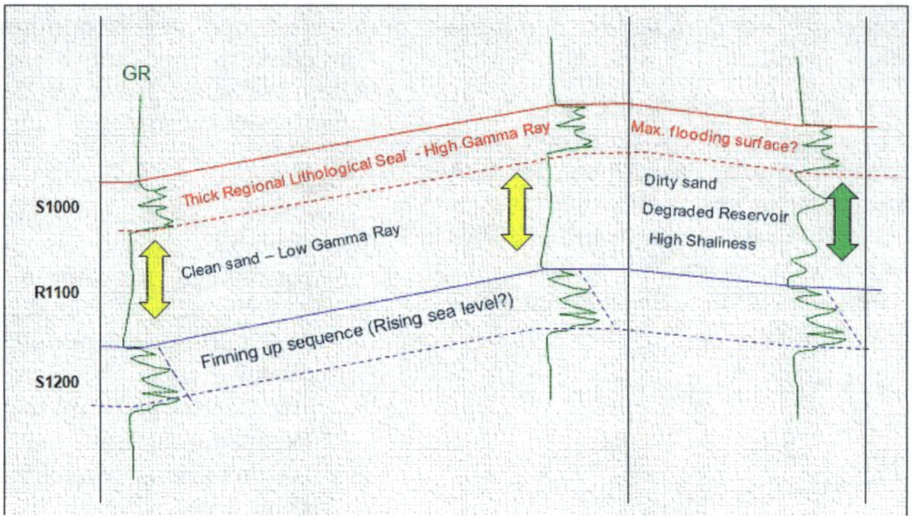

Figure 16.14: Log assisted correlation and stratigraphic association

Reservoirs forming similar thicknesses over a large surface or over the entire field development area can be correlated across most of the wells. Reservoirs with varying thickness and those that occur in a narrow space, either vertically or horizontally (like thin turbidite channel fairway), are difficult to correlated across nearby or offset wells even within the same field.

This is why correlation works better in fluvial continental widespread deposits than in single different paths deposited as turbidite complexes offshore with channels overlying each other. However, if a cross section were to be drawn within the channels themselves, following the meander, then it will be possible to correlate them at wells that encounter them at different places.

The aim of reservoir correlation is to identify reservoirs of interest in different wells at different locations, and to apply a given understanding to new developments. It also helps in the general understanding of layering and geologic history, thereby optimizing the knowledge necessary to develop the reserve. Seismic provides the most important correlation tool today in the industry; biostratigraphy is used to complement this once the well is drilled and samples collected.

Each region of the world has established biostratigraphic signatures used in stratigraphic and age correlation. In a single identified prodelta (farthest offshore region of a delta below the depth of effective wave erosion and relatively slow rate of fine-grained deposition) deposit for example, sequence stratigraphic correlation would provide a strong and reliable correlation when the log responses obtained are coherent enough to establish a depositional trend varying with changing sea level. Between two or more wells, the same reservoir tops and bottoms are tied as belonging to same levels, horizon, formation or series. Also, the farther wells are apart in the same oil field, the less their tendency of being correlatable.

Biostratigraphy, lithostratigraphy and sequence stratigraphy has been defined in Chapter Three on Stratigraphy. They are revisited here in the context of log interpretation.

Biostratigraphy uses microfossils observed in cuttings or cores to determine the age and sequence to which a unit belongs. It is used to tie the well to offset wells and to regional stratigraphy. It is also useful in correlation and dating of rock formations. For these to be effective:

- Sediment needs to have suffered less alterations
- Need to use a drill bit and drilling speed that does not destroy the fossils
- Preliminary lithostratigraphy and if possible sequence stratigraphy should have been done
- Geochemistry and petrography studies need to be performed on the samples

Lithostratigraphy uses the sedimentary signatures to identify the order of layering of a formation. When combined with log, stratigraphy could help predict what kind of unit (reservoir or seal types), would give a certain log response. *Sequence stratigraphy* is deduced for example, when observing a lithological grading from hemipelagic marine shale to sandy units deposited during lowstand, then through the highstand systems tract while the sea was rising and back again to the condensed sediment during maximum sea level.

Other forms of stratigraphy that could be useful in log interpretation, though less used, are chemostratigraphy and magnetostratigraphy. Chemostratigraphy is based on chemical indicators while magnetostragigraphy is based on changing polarities of substances due to changing magnetic fields with tectonic and structural reorganization.

16.5 Log Interpretation Deliverables

With association of seismic data, reservoir structure and geometry estimated from borehole images, dips and strikes, or other methods of evaluating reservoir extension, log interpretation gives useful information at the heart of oil and gas exploration and production. The ultimate aim of log interpretation is to estimate hydrocarbon saturation and **Hydrocarbon in place (HIP).**

In general, deliverables from log interpretation would include:

1. Fluid contacts (WOC, GWC)
2. Oil-Up-To (OUT) / Oil-Down-To (ODT) and Gas-Up-To (GUT) / Gas-Down-To (GDT) where the extension indicating fluid type on the log does not show the contact
3. Fluid saturations – water, oil and gas (S_W, S_o and S_g) respectively
4. Reservoir Net / Gross
5. Pay zones – Net Pay Gas and Net Pay Oil

For reservoir modelling, the log results in these deliverables are extrapolated from the well location to the whole field. Reservoir geologists and engineers play with facies and heterogeneity while populating the sandy versus shaly (or other lithology) regions according to the result of log data matched to seismic and other indicators of lithology. This provides the production index, expected total extractable oil and dynamic models / active communication that takes place during production.

The net pay oil and net pay gas is multiplied by their market price respectively to get total returns in millions of dollars (as shown in Figure 16.15). The cost of drilling/development and production is subtracted from the returns to assess the viability of the well at an appraisal stage.

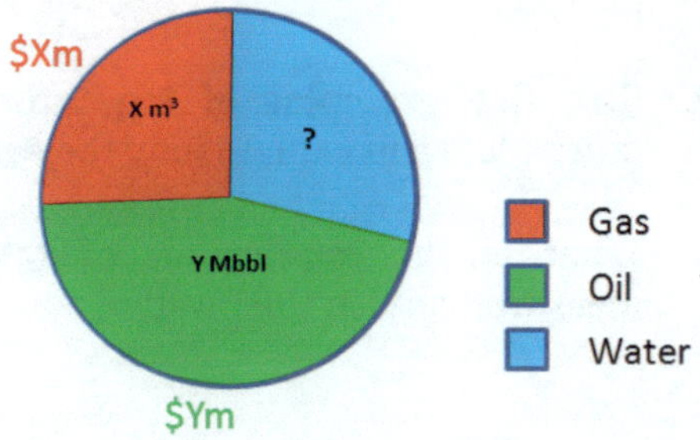

Figure 16.15: Extractable oil in place linked to volumes and economic results

The net pay of each of the wells targeting a reserve in a given field adds up to the total pay or income, just as the total cost of development and production for all of the wells is used to check the viability of an entire field. In practice, the profit has to far outweigh the expenditure for the oil company to pursue the project. For infill wells (later wells drilled in a producing field), most of the economics have been worked out and the aim is to maintain, optimize or revise production volumes.

EXERCISES

1. Define the following terms as they relate log interpretation:
   a) Porosity
   b) Permeability
   c) Saturation
   d) Resistivity
   e) Spontaneous Potential (SP)
   f) Gamma ray
   g) Vclay and Vshale
   h) Bulk density
   i) Vertical seismic profiling
   j) Interval transit time

2. What is a composite log? Give four examples of logs combination it contains.

3. Describe three logs used to determine porosity.

4. Give five examples of log interpretation cross-plots and explain how they are used.

5. What is the implication of taking well measurements in the invaded zone?

6. Explain the differences among shallow, intermediate and deep resistivity and how each of them complement the others in log interpretation.

7. Explain how the PVT phase diagrams function, showing example of domains and what fluid type will plot in them.

8. What behaviours are expected from the different log responses for a good reservoir sandstone, shale seal and fault zones? Are there sufficient information in a composite log to suggest the:
   - reservoir boundaries?
   - fluid contacts?
   - fluid types?

   Explain.

9. Illustrate the oil in place in an oil pool in a trapping structure, using a given formation factor and log responses.

10. Why is it important to integrate biostratigraphy and lithostratigraphy in log interpretation?

REFERENCES & FURTHER READING

Alberty, M. W., 1994. Standard Interpretation, Part 4 – Wireline Methods, in D. Morton-Thompson and A. M. Woods, eds., *Development Geology Reference Manual*: AAPG Methods in Exploration Series 10, p. 180–185.

Alger, R. P., 1980. Geological Use of Wireline Logs (p. 207-222) in G. D. Hobson, ed., Developments in Petroleum –2: London, *Applied Science Publishers, Ltd.,* 345p.

Archie, G.E.,1942. The Electrical Resistivity as an Aid in Determining Some Reservoir Characteristics, *Journal of Petroleum Technology.* Volume 5, No. 1.

Borst, R. L. 1982. Some Effects of Compaction and Geological Time on the Pore Parameters of Argillaceous Rocks. *Sedimentology*, v. 29, p. 291–298.

Brace, W. F., A. S. Orange and T. R. Madden, 1968. The Effect of Pressure on the Electrical resistivity of Water-saturated Crystalline Rocks. *Journal of Geophysical Research,* v. 70, p. 5669–5678.

Brace, W. F., 1978. A Note on Permeability Changes in Geologic Materials due to Stress. *Pure and Appl. Geophys.* v. 116, p. 603–614.

Carothers, J. E., 1968. A Statistical Study of the Formation Factor relation to Porosity: *The Log Analyst,* v. 9, p. 38-52.

Coates, G., and J. L. Dumanoir, 1973. A new Approach to Improve Log-derived Permeability: *Soc. Professional Well Log Analysts,* 14th Ann. *Logging Symp., Trans.,* paper R.

Crain, E.R., 1986. The Log Analysis Handbook (Quantitative Log Analysis Methods). Tulsa Oklahoma: Pennwell Corp. 684 p.

Dewhurst, D. N., A. C. Aplin, J. P. Sarda and Y. L. Yang, 1998. Compaction-Driven Evolution of Poroperm In Natural Mudstones: An Experimental Study. *Journal of Geophysical Research*, v. 103, p. 651–661.

Doll, H. G., 1948. The SP log, Theoretical Analysis and Principles of Interpretation: *Trans., AIME,* v. 179, p. 146-185.

Eaton, A.B., 1975. The Equation for Geopressure Prediction from Well Logs. *SPE 50th Annual Meeting*, Paper no. 5544.

Hartmann, D.J. and E. A. Beaumont, 1999. Predicting Reservoir System Quality and Performance, in E. A. Beaumont and N. H. Foster, eds., *Treatise of Petroleum Geology – Handbook of Petroleum Geology: Exploring for Oil and Gas Traps.* Tulsa, AAPG, p. 9-154.

Glover, P. 2013. Introduction to Petrophysics and Formation Evaluation. *MSc Course Notes.* University of Laval, Quebec, Canada.

Hilchie, D. W., 1978. Applied Openhole Log Interpretation: *Golden,* Colorado, D. W. Hilchie, Inc.

Matthews, W. R. and J. Kelly, 1967, How to Predict Formation Pressure and Fracture Gradient. *The Oil & Gas Journal*, February 20 Issue, p. 92–106.

Schlumberger, 1968. Log Interpretation/Charts: Houston, *Schlumberger Well Services, Inc.*

Schlumberger, 1979. Log Interpretation/Charts: Houston, *Schlumberger Well Services, Inc.*

Society of Professional Well Log Analysts, 1975, Glossary of Terms and Expressions used in Well logging: Houston, *Soc. Professional Well Log Analysts*, 74 p.

Traugott, M. O., 1996. The Pore Pressure Centroid Concept: Reducing Drilling Risks: *Abstract, Compaction and Overpressure Current Research,* 9-10 December, IFP, Paris.

Traugott, M. O., 1997. Pore/Fracture Pressure Determinations in Deep Water: *Deepwater Technology*, v. 4, no. 8, p. 68–70.

CHAPTER SEVENTEEN

CORING

17.0 Core Concept

The subsurface can be evaluated while drilling a well by analysing drilling parameters, sample cutting, well logging, pressure testing, fluid sampling, and by well testing after drilling. All these give very good and relatively accurate results. However, a core is the closest physical representation of the borehole and the formation, providing information on petrophysical properties, rock behaviour and stresses, and hydrocarbon content or other downhole fluids. Coring involves extracting rocks samples from the subsurface in order to provide geological information.

To be learnt in this chapter are:

- Types of core and coring tools
- How to prepare a core job
- Different stages of the coring process
- Challenges with coring operations and safety
- Core processing, logging and analyses
- Core data interpretation
- Use of field analogues and laboratory measurements
- Applications and well post-mortem

Pieces of rocks are collected as samples from core sections for microscopic description. A smaller core may be cut in a cylindrical tube, similar to the main core, but cut perpendicular to it. This is known as a core plug and the process is known as *core plugging*. Core plugs are used for basic acid test, permeability measurement, chemical analysis and other laboratory tests. Different coring tools are used for different formations, depending on the objective.

17.1 What is Coring?

Coring refers to the process of cutting through a rock body in a cylindrical manner, in order to extract a sample of the rock from the formation. A *core* is the sample of rock cut, which is removed from the wellbore during the drilling activity and after the core section, has been drilled through. Apart from the dill cuttings collected with drilling fluid by mud loggers, cores are the only physical representation of the formation, albeit in limited size. This makes a core one of the most important parameters in formation evaluation. Figure 17.1 shows different lithologies cut in a formation by a core.

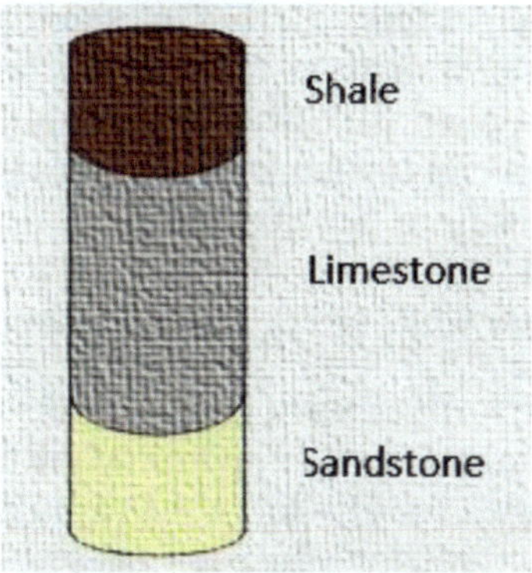

Figure 17.1: An illustration of a core showing three different lithologies

Despite its importance, most oil companies are now cutting down the cost of coring especially in oil fields with already established well parameters. In such cases coring is avoided except for exploration wells, new development wells or for specific petrophysical, seal or reservoir studies. A core provides:

- Petrophysical properties
- State of the formation (good or damaged)
- Rock mechanics and stresses
- Hydrocarbon (fluids) information

Depending on the coring directions, three types of core are recognized:

- *Vertical Cores:* Vertical cores are cut while drilling downhole in vertical wells
- *Horizontal Cores:* Horizontal cores are cut in horizontal wells, along reservoir well drain (for example)
- *Lateral Cores:* Lateral core refers to a cutting perpendicular to the direction of drilling. An example is the *sidewall core*.

Sidewall core is a smaller core, also known as a *plug*, cut into the well wall by drilling into it perpendicularly (sometimes by drilling into the main core after it has been brought to the surface).

An electronic device is used to orient a core head in a manner similar to geosteering. This makes it possible to core while a well is deviated. The process of coring and bringing it to the surface is known as *catching* the core. The wireline rotary method can catch several (up to 40 or more) sidewall core samples in one run. The challenge with such sidewall core is that it is taken from the sidewall which might have been invaded, flushed or previously damaged while drilling. Taking a sidewall core on a sample already in the surface helps to ensure that a good core plug is obtained. The sidewall coring tool in Figure 17.2 shows the protraction from where the formation is plugged and an extender on the other side to ensure good contact while taking the core sample.

There are two main types of sidewall coring:

- *Percussion Coring*: Involves firing a kind of open tube bullet in the formation to capture and retrieve the rock sample.
- *Rotary Coring*: Involves use of a rotary drill-bit (conveyed on wireline or pipe) to retrieve the rock sample after it has been drilled-and –severed.

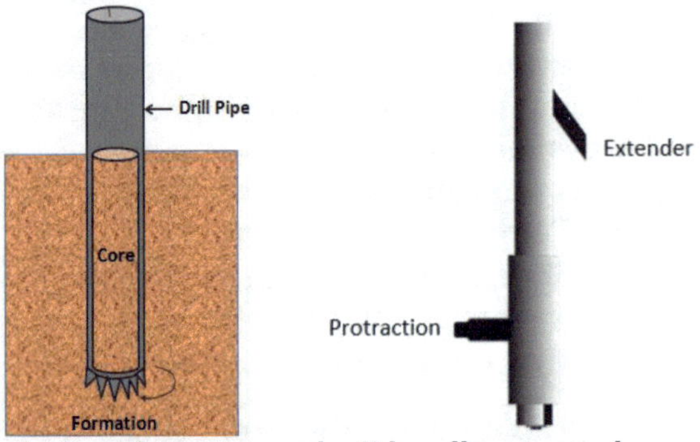

a. Coring a formation b. Sidewall coring tool

Figure 17.2: Coring principle and sidewall core

Rotary sidewall coring using wireline is a faster and less expensive way to take samples. After the well has been drilled, a sidewall coring device is lowered down the well until it is adjacent to the sample interval. A *percussion sidewall* coring tool, which may have up to 30 small core tubes called bullets with small explosive charges behind them, is used to plug the formation wall. This is done by detonating the explosives and the bullets are shot in the sidewall to take samples.

A wellsite geologist is supposed to monitor and witness the core catching process. The aim of witnessing a core job is to ensure a safe, best quality *usable* core, as well as maximizing *recovery* and *efficiency* of the coring process. This is more important for normal coring, compared to sidewall core.

- ***Recovery*** = core cut at depth per length recovered at surface
- ***Efficiency*** = length of core cut per length of coring barrel (length of barrel is maximum coreable length)
- ***Usability*** = amount of core that can be used for core analyses (after it has been taken to the laboratory)

17.2 Preparing a Core Job

The aim of good preparation is to obtain a good quality core that maximizes the amount recovered to the length of barrel run. The following factors should be considered when preparing a coring job:

i. Coring point
ii. Hole condition
iii. Coring assembly
iv. Bit type
v. Core catching technique

Preparing a *coring job* requires that the objective is properly assessed. The ***coring point*** is the depth of coring decided by the geologist based on stratigraphy and determined by observed changes in drilling parameters and gas data. A good knowledge of the ***hole condition*** is necessary for proper coring. The next step is to consider what tool will be used – the ***coring assembly***. It is important to consider the *mud condition* in order to avoid excessive flushing of fluid into the formation.

Other considerations will include ***bit type***, core ***catching*** technique and length of the barrel. These will depend on the lithology – softer formation will not support a very long barrel length (Clark and Shafto, 1987; Anderson and Azar, 1990; Maurer et al., 1990; Cohen et al., 1992). For the corehead selection, *Thermally Stable Diamond (TSD)* coreheads are better for less abrasive/abrasive formation while natural diamond cutters are most adapted for abrasive less fractured formations. The *Polycrystalline Diamond Compact (PDC)* cutters are the most adapted for larger formation types and they are becoming the most used drilling and corehead bits. TSD cutters are however better for very high temperatures, since they remains stable.

Table 17.1: Factors important in choosing coring bit types

S/N	Bit Type	Hardness/Fracture	Abrasiveness
1	TSD	Low to High	Low to moderate
2	PDC	Moderate	Moderate
3	Natural Diamond	Moderate	High
4	Impregnated Diamond	High	High

The oil company will have to make a choice of which coring firm to use, as coring job is usually contracted to specialized service firms. Some of the service firms or core companies that provide coring technology in the oil and gas industry are:

1. Schlumberger
2. Baker Hughes (Baker Atlas)
3. Halliburton
4. Corpro
5. Inteq
6. NOV

Most of the core companies offer similar technologies, with slight variation. Among the factors to consider while choosing a coring solution are the coring type required and technology. Modern coring tools should be able to achieve some of the listed challenges, with tools:

- for high torque coring
- for high temperature and pressure
- for different formations, lithology and fracture types
- adapted to be fitted with different bits or cutters
- adapted for different designs to meet coring challenges

Coring Tool Designs and Challenges

Does the core company integrate a *drill bit design* technology to the coring decision? For example, *aluminium* or *fiberglass* inner barrel which proves very useful in certain situations but not in others. There are several designs, which need to be used depending on objectives and subsurface conditions:

- ***A swivel design,*** which allows the inside core barrel to be stable while the outer one rotates. Some swivel designs have a seal to adequately isolate the core from the formation fluids.
- ***Sponge coring,*** which provides sponges to absorb the formation fluid from the core that can be analysed to determine fluid type, saturation, relative permeability and other information relating to capillarity.
- ***The wireline system,*** which allows the removal of the centre bit when the core point (depth of coring) is reached and then lowering the inner core tube into the core barrel with a wireline string. Once locked in, coring begins as the bit drills further. After coring, the inner core tube is pulled out with the core using a wireline string. Drilling resumes until another desired core point.

Choice of design will depend to some extent on objectives and challenges identified during coring job preparation (Collee et al., 1994; Skopec et al., 1995). Examples of major challenges are *core jam (core breaks and barrel gets stuck)* or *core block* (when the core refuses to enter the barrel). Another challenge is *disintegration* when coring in an *unconsolidated formation* (with lose sediments that may shift and not represent original arrangement in the formation*)*. *Core jamming* is one of the most challenging problems with coring, as it hinders the barrel from rotating when the core breaks.

Jam Mitigation

Right coring design and coring or drilling parameters may be adjusted to prevent core jamming. In case of core jam, an inner barrel is used to capture the core at the point of friction.

Unconsolidated Formation

In unconsolidated formation, a special enclosed catcher is used to secure and ensure that the core sample is not displaced (does not shake) after catching it.

Once these challenges have been met, the core has to be maintained in good conditions until it reaches the laboratory for analyses.

Core Orientation

A proper labelling system is necessary and packaging should be organized to maintain the identification of the core orientation. This has to be thought about during the preparation stages. Usually a cap is used to cover the two ends of the core. The cap used to cover the top and the bottom should have different colours, and the top and base of the core has to be labelled accordingly, as shown in Figure 17.3. The consequences of a costly mistake, where the top of the core is assumed to be the base, means the interpretation will invert the orientation of the formation bedding units.

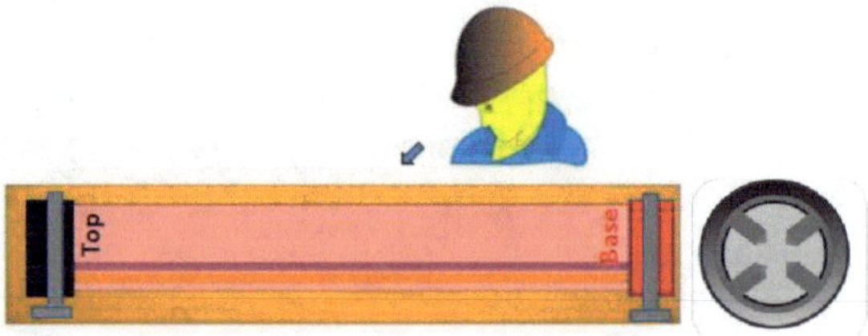

Figure 17.3: Core orientation and labelling

The wellsite geologist has to keep in mind the objective of the coring exercise and ensure the requirements are followed to reach those objectives. Capping and placeholders that keep the core off the ground will have to be provided even before coring starts. Figure 17.3 shows caps and placeholders (grips at the cap position) for holding and capping the core. Core caps (shown on the right) usually have projections or full closure to ensure core is held in place. The geologist has to determine the *coring point* (depth at which coring starts) and monitor lithological evolution to ensure that this depth is not missed.

17.3 Downhole Coring Tool

The downhole coring tool is made up of several parts, which may vary according to type of tool or inbuilt technology. One objective in choosing a downhole coring tool is its stability and adapted parts for a given core job. Important parts in most coring tools are discussed below.

Core Barrel and Core Head

The cylinder-like coring pipe is known as **core barrel**. Most *core barrels* will have a **core head** similar to a *drill bit*, just that it has a hole at its face through which the core can be caught while drilling (Figure 17.4). The PDC fitted corehead has a good rate of penetration (ROP) in soft to slightly hard rock but not in a fractured formation. It can be repaired and remodelled after wear to some extent. On the other hand, the TSD cannot be repaired and is expensive, but can work in hard fractured formation (as shown in Table 17.1).

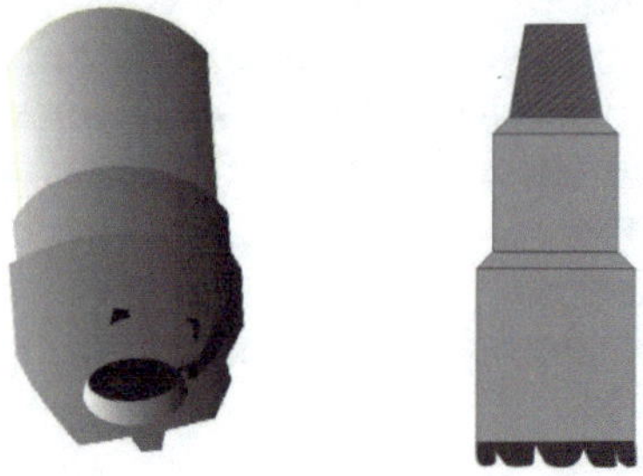

a. Corehead b. Normal drilling bit

Figure 17.4: Corehead showing a hole and a normal drill bit

Inside Liner or Sleeve

Another important make-up is an independent "inside" such as a **liner** or **sleeve** system that can remain stable/fixed while the outer core barrel is rotating as drilling progresses. This will work with a ***bearing*** or other *gyrating* mechanism that makes one item to rotate while the other is fixed. There are many inside types, some of which are made of either aluminium or steel. Some modern liners come as ***fiberglass***. The advantage of one type over the other depends on the use or purpose.

Stabilizer Groove

This is placed along the core barrel to ensure it remains centred in the hole (Figure 17.5). It helps to prevent core jamming, as core may jam when the barrel moves in a zigzag manner while rotating.

The process of grabbing the core sample after it is caught is known as **core catching** and the tool used for this purpose is known as a **catcher**. There are different types of catcher, among which are the *full closure* and the *partial closure*.

1. ***The full closure catcher*** completely cuts and closes in below while holding the core in place. This is good for soft formations.

2. ***The partial closure catcher*** only needs to grab or press on the core sample to keep it in place before the tool is lifted to cut the core.

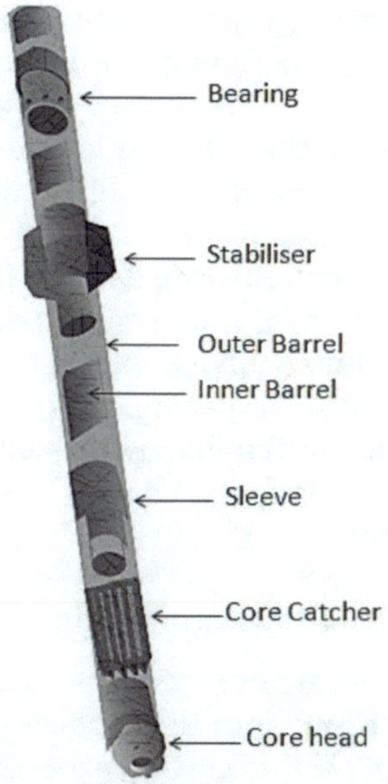

Figure 17.5: Coring tool

Modern catching process and wireline coring makes it possible for a core to be caught and removed before the *core assembly* is brought to the surface.

17.4 The Coring Process

The coring process involves major drilling operations, with personnel lifting and lowering pipes downhole as well as laying out cores. The cores are pre-processed and taken down in long cradles (Figure 17.6) to a safe zone from where they may be transported for full processing. A cradle is a carrier that aides movement, on which the core is placed. A core length may be long, contained in barrels, in several meters. The tip of the core barrel in a cradle is zoomed in red in Figure 17.6). Some cradles or temporal placeholders for cores have rollers for movement once on the floor.

The main coring process is divided into three phases, which are:

1. Trip-in
2. Cutting and catching phase
3. Trip-out.

The processes can overlap, it will however be discussed as separate phases for clarity below.

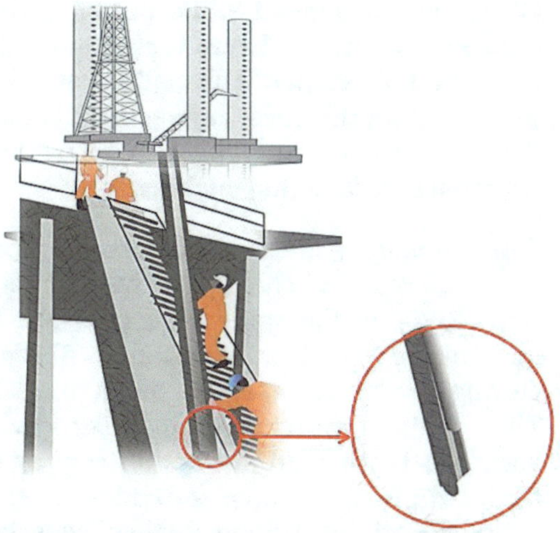

Figure 17.6: Coring job involving movement laying down of core in cradle from the rig to the pre-processing floor

Trip-In

This is the phase when the core assembly is lowered to the coring point. A good hole condition and careful passage through the *Kelly table*, through the *Blow Out Preventer (BOP)* and across possibly *gushy* or *cave in* section of the well is important. This is more challenging in deviated wells. Drilling fluid is used to clean core inner barrel while circulating. A process known as *reaming* (rotating the assembly) and *reciprocating* (up- and-down movement) as the core assembly descends is used to ensure smooth and slightly enlarged or rounded hole. Care should be taken not to damage the formation while reciprocating.

Then the corehead is gently touched on the bottom of the hole, at the coring point; a process known as *tagging bottom*. Just before reaching bottom, you ensure major circulation while tripping down, to detect when you are there without really drilling. This process is known as washing down. The depth might not be where you expect, so be careful at this stage. *Tagging* and *washing down* are normal procedures in drilling, like washing down to tag cement (after a cementing job).

Cutting and Catching

This is the phase when the core sample is actually cut at the base and captured in the barrel. The driller controls the rotation of the core string and the other drilling parameters, making sure that so much weight on the bit

(WOB) is not applied while pumping drilling fluid in a determined rate that should remain constant. Efforts should be made to avoid delay while cutting the core; connections should be done as quickly as possible as hole sticking and eventual core jamming might occur.

Core jamming is identified by careful monitoring of unexpected changes in rate of penetration (ROP), in the stand pipe pressure (SPP) and torque and flow of cuttings. Absence of cutting for example could mean *bit balling.* This is the formation of ball-like cover on corehead (with mud, formation clay or other fine sediments), which provides a smooth rotation without drilling further, especially if core jams at the core head.

Jamming (breakage) could occur at the base close to the core head and in the inner barrel. That in the inner barrel is easier to control and the core can be lifted if jamming is identified. Other forms of damage is *core milling* due to rotation of the inner sleeve and breakage resulting from slippage.

Trip-Out (POOH)

Another word for *trip-out* is *Pulling Out Of Hole (POOH),* just like in normal drilling procedure. In coring operation, it may involve the removal of the core barrel with the core sample. The most important parameter to watch out for at this time is the speed of tripping out. Too fast POOH could lead to an unrepairable damage to core as the petrophysical properties would be rearranged; core may even become deformed while tripping out (especially for soft formations). This is known as *trip damage.* An example is shown in Figure 17.7 of not only breakages, but also a shift in core shape due to deformation during movement.

Figure 17.7: Damaged core while tripping out

Well Safety While Coring

The major risks are *losing the core* downhole or having the *core barrel stuck.* Other challenges include experiencing either *mud losses* or *gain* while coring, their extremes could be major *lost circulation* (when some circulated muds no longer return to surface) or alternatively a *kick* (where there is a sudden inflow of formation fluid). Well kick could occur when coring if for example the tool is *stuck in hole* and the equivalent circulating density (ECD) is low; the well becomes underbalanced. Take the Coring Engineer's advice, but corroborate his suggestion with the geological knowledge. There is need to be familiar with solution to possible problems and well control operations in case any of these occur.

17.5 Handling and Transportation

It will be a pure waste of resources if cores are damaged when brought to the surface, after all the painstaking in the coring process. The most important aspect of handling is removing the core from the inner tubing. *Inner Tubing Separation (ITS)* involves:

- Carefully cutting the core (using a shear cutter or assembly) at the level of connection above the rig floor.
- Then, again carefully, laying down the core using a cradle.

It is a good practice to ensure that working area is not slippery and that it is well lit up.

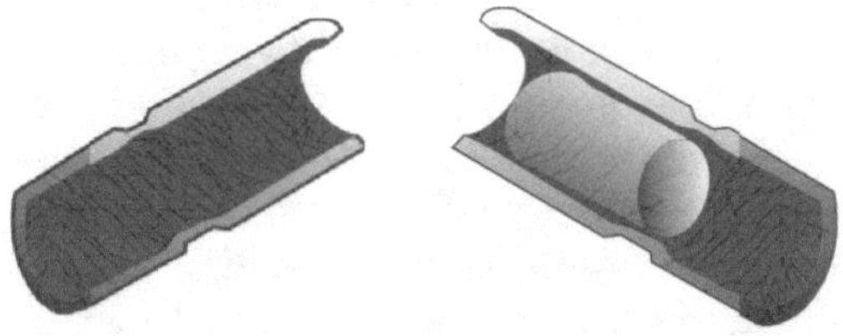

Figure 17.8: Core shown inside the inner tubing, as if the tubing is cut in half to show a core caught inside

The objective of the ITS should be to avoid core:

- Physical damage (denting, breakage and scratching).
- Internal petrophysical changes (such as compacting, realigning of grains).
- Rotation leading to flipped information (like not knowing the way up or down for the core).

The Wellsite Geologist (WSG) supervises the handling and processing until the packaging of the core. Depending on the objective, he might be involved at the initial state of the core analyses. There is need to give a minimum treatment to the core on site, apart from the basic labelling and sampling, especially for soft formation. Proper bars, foils, forms or other soft fillings are used to cushion and fit them into the boxes before transportation. Plans should be made at receiving end so that adequate hands and expertise are available to collect the cores. It can be so annoying when cores are damaged just before getting into the laboratory, after all the proper handing from the coring site.

17.6 Core Processing

Core handing mentioned above is part of core processing. It is discussed in this section in more details, separating the two aspects that involve basic pre-processing and further processing. Packaging and sealing for transportation are also part of pre-processing.

17.6.1 Basic Processing

Basic core processing concerns all that is done before the core is properly packaged, maintained securely and taken to a laboratory for further processing. These are:

- **Laying out** the core (often in rows) in cradles
- **Measuring** their original lengths (for record purposes and for estimating recovery and efficiency)
- **Cutting** them to required sizes (example, 1 m) and trimming off rough edges.
- **Pre-packaging** (capping and securing the core in a holder that makes it stable for parking.
- **Labelling/Marking** by using markers to write down the top and base, the lengths, depths and well name or other useful information.

Figure 17.9 is a cartoon to illustrate the cutting of core with ease, because the core has already been secured in an external sleeve or plastic, making it possible to manipulate it with flexibility. The core may be capped or recapped after cutting before organizing it for transportation or storage.

Figure 17.9: Personnel cutting cores to sizes

As stated earlier, proper labelling will help to avoid the core sections being mixed up, or turned in such a way that the upper and lower ends are reversed. Such reversion can lead to a major misinterpretation of formation. Even in physical examination-based interpretation, it could reorient fractures.

Further Processing

The aim of further processing is to keep the core in a condition that ensures its usefulness in the short and long-term and for proper transportation. It involves basic tests, paper work and preparation for advanced analyses. The different aspects of further processing are discussed below.

Core Treatment, Packaging and Preservation

These involve processes such as injection with foam or waxing. Cores are stabilized by the injection of foams. Soft sediment requires proper treatment and care should be taken with the packaging and preservation. In all this, Health, Safety & Environment (HSE) procedures should be kept in mind. Different countries and governments lay out different conditions for treatment and movement of subsurface data.

Core Sampling and Tests

After coring, some broken off parts are collected (or a small chisel is used to chip out small samples from the end of the cut sections). These are used for geological tests, description and thin section studies under the microscope. It is important to understand the layering of the beds, dips or strikes, to ensure the chips are representative. Core plugs are also used for further tests.

Core Plugging

As previously described, this involves cutting out a small cylindrical "plug" similar to the main core but cut perpendicular to it. It is a way of collecting samples for core sampling and studies, including basic acid test and chemical analysis. Core plug may be taken downhole, as described previously, or from retrieved core as shown in Figure 17.10; in this case the plugging tool is placed on the laid out core and 'shot' to collect the sample. A core for plugging may be held in a larger cylindrical grip or cap as shown in Figure 17.10b.

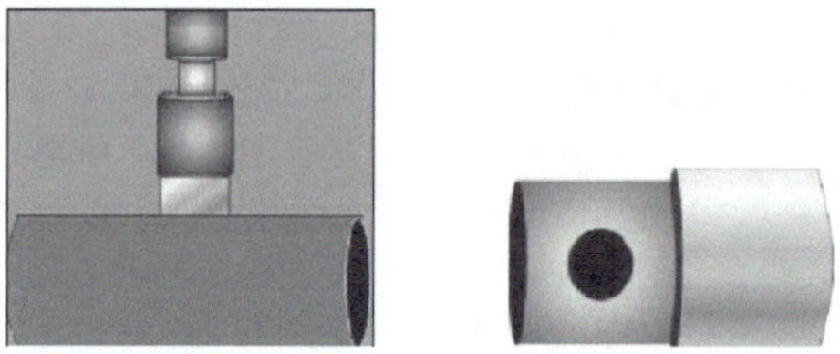

a. Core plugging   b. Hole from plugged core

Figure 17.10: Core plugging

Some of the uses of core plugs are:

- Analyses of water saturation (S_w)
- Analyses of Resistivity of water (R_w)
- Determination of Wettability Index (W_i)
- Determination of Relative Permeability (K_r)
- Assessment of Capillary entry Pressure (P_e).

Most of these parameters are relevant in petrophysical studies and description of well results.

17.6.2 Core Logging

Traditionally, core logging involves a physical description of the minerals and rock types observed, recording their thickness, hardness and the like. Just like in conventional outcrop logging, these features are used to establish lithofacies. Today, low-range gamma ray logger used with safety precaution may be passed over cores right at the rig site (as illustrated in Figure 17.11).

The aim of core logging and imaging is to establish:

- Lithological properties
- Sedimentary properties
- Structural features

From the above, it is possible to deduce the *reservoir quality* and *depositional environment.*

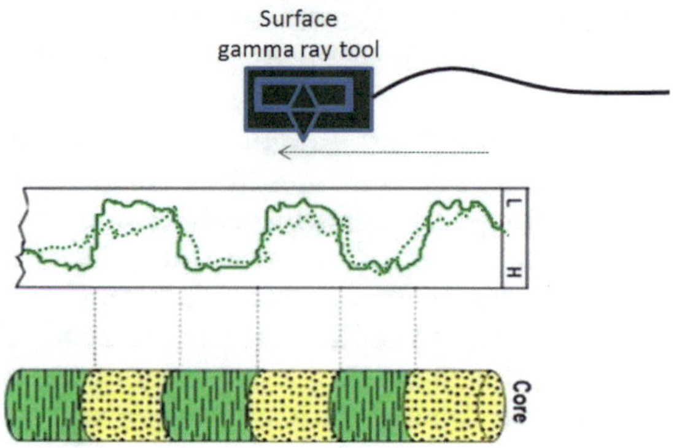

Figure 17.11: Surface gamma ray tool used on laid out core (result illustrated as dotted green line) compared to normal core gamma ray result (thick green line)

Logging a core electronically on the surface essentially gives only an idea of the high and low log values at several close-by points. Cores are used for the immediate assessment of a reservoir or seals encountered in a well. With cores from several parts of a field, it is possible to correlate and establish more regional conclusions on the petroleum system. Imagery, using medical-type scanners as well as log based core imaging, is also carried out once it arrives the laboratory (Prince, et al., 2005). Core logs are compared with images and field analogues in order to project and rescale the core result to the size of a petroleum basin. This upscaling is one of the necessary steps for putting every core result to full use in a larger basin scale.

17.6.3 Core Analyses

Cores are cleaned and prepared after acquisition for certain analyses, to provide good petrophysical data such as porosity, permeability and fluid saturation. A well-preserved core will maintain the fluid type within measurable pore space. *(The use of core plugs to determine S_w or R_w, W_i, K_r, and P_c has been mentioned above).* Where saturation is obtained from core analyses, it may help to complement information on fluid types and contacts. There are two levels of core analyses listed below.

Conventional Core Analyses (CCA)

- Petrography (thin sections)

- Porosity (measured as ratio of dry volume to that of wet volume)
- Permeability (measured in the laboratory by flowing nitrogen (N_2) through a core plug)
- Saturation (obtained directly from laboratory test or by log resistivity)

Advanced Core Analyses (ACA)

- Capillary entry pressure (from lab measurement of contact angle, estimation of pore-throat size and viscosity)
- Vclay (from quantitative chemical analyses to obtain clay mineral content)
- Wettability (from laboratory tests and microscopic observations)
- Biostratigraphy (microfossils of either animal or plant origin are observed under the microscope when the core is carefully crushed and washed)

17.7 Core Data Interpretation

Core data has a wide range of applications. What can be done depends on objectives and on available investment of resources. The need to use the right methods cannot be overemphasized. Data interpretation could be done in-house in the oil company or outsourced to a consultant or to the core company. Cores can be used for:

- Assessment of structural trends, fractures and local impact of tectonics
- Reservoir properties, compartmentalization and nature of fluids
- Seal evaluation, shaliness and sandiness of units, fault studies and Shale Gauge Ratio (SGR) determination
- Geochemical and petrophysical evaluation

For the oil company, the ultimate aim will be to determine the oil in place (OIP, which is the estimated amount of oil in a reservoir); the core gives the real data, actual lithology and general physical characteristics of rocks. It is often used to calibrate the results from well log formation evaluation.

17.7.1 Field Coring Work as Analogue

Despite the use of seismic acquisition, imaging and interpretation, field work and practical geological mapping of outcrops for analogy is still a norm in the oil industry. It gives geologists a practical contact for visualization of intrinsic reservoir properties. Rock properties observed in the field serve as analogues for actual reservoirs; extracted core samples in the field can be analysed for chemical and petrophysical properties and then compared with those obtained from wells downhole. Such properties which may be better understood from field analogue include:

- Physical reservoir architecture
- Porosity and permeability
- Microfractures and *banding* or layering
- Facies classification and heterogeneity
- Reservoir interconnectivity

It is important to always relate the reservoir properties of a well to what has been observed on outcrops, for further understanding of reservoirs and for the improvement of geoscience knowledge of rock formations.

In the field, directional coring can be adopted in order to evaluate hydraulic properties in different directions of both fault zone and wall. *In Chapter Nine (Figure 9.26: Directional coring of fault zones and walls) the coring directions and the orientations of the cored blocks or fault zones in the field were illustrated.*

In the field, hard units are cored using the coring machine while soft sediments can be cored using copper tubes.

Coring machines: a good field-coring machine should have an adaptable drill diameter, and have relatively light weight, deep core length and resiliency. However, the requirement of a special mix for the fuelling and the accompaniment of a water pressure tank makes it slightly cumbersome for use in rugged mountainous terrains.

Copper tube method: Copper-tubes, with sharpened tips, are adequate for the coring of the soft fault gouge by gradually tapping the tube at the base. As the tube forges its way into the muddy or incohesive gouge, it retains the core sample which is brought out onsite or in suitable conditions at the laboratory.

17.7.2 Laboratory Measurements of Permeability

Sample Preparation

Samples for permeability measurement may be cored to a cylindrical shape, (example, 2.0 cm) in diameter and where necessary trimmed to a length of between 1 and 4.5 cm (example in Figure 17.12a). The specimen is then dried at about 80°C in an oven to eliminate existing pore water before being placed into the fitting for the permeability measurement (Figure 17.12b). The sandy samples are dried for about 48 hours while the shaly samples are dried for a minimum of 72 hours. Some samples require that their edges be filed or coated with epoxy resin and left to dry before the measurement.

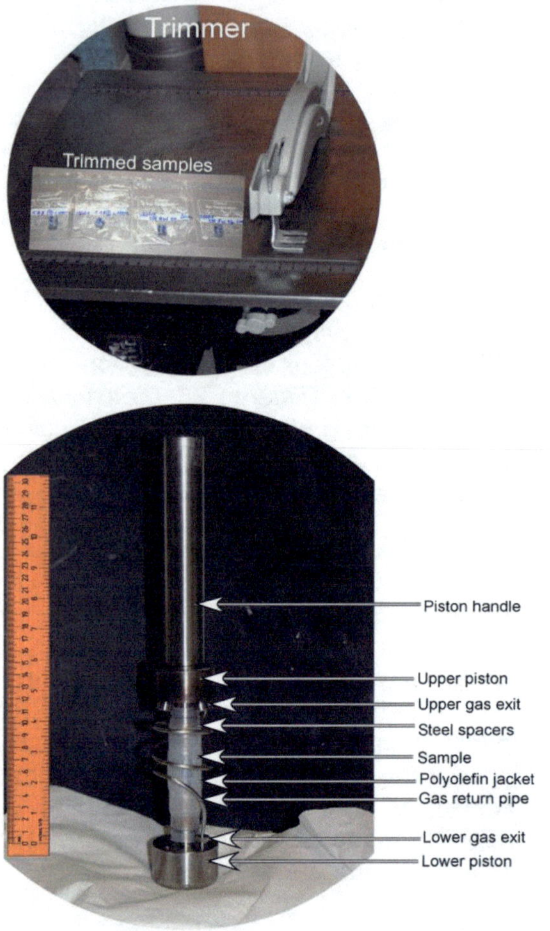

Figure 17.12: Sample preparation for permeability test (Nosike, 2009)

Precautions

To ensure that there is no leakage during confinement and to avoid bypass flow between the sides of the sample and the jacket during the experiment, each sample should be enclosed in up to three polyolefin jackets which may be contracted with heating to completely seal the sample fitting (as shown in Figure 17.13). A mesh filter (Figure 17.13a) should be placed over the specimen before placing it in contact with the steel spacers and the piston. These should be cleaned and replaced after a test and before the next fitting, to avoid particles in the air passage and specks that might have stuck to the mesh. The combinations (piston, spacers, gas exits and pipe) are also cleaned and hot air passed through them at high pressure to ensure moisture free conduits before each test (Figure 17.13b).

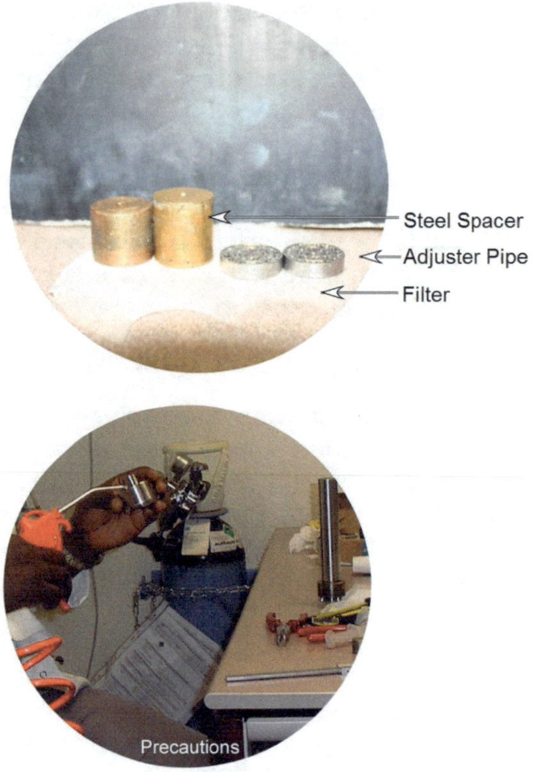

Figure 17.13: Precautions before sample fitting (compare with Figure 9.29: Equipment setup for permeability measurement) (Nosike, 2009)

Flow data analyses

When carrying out permeability tests, a table of the flow data should be made and the flow rates used to calculate the permeability. The viscosity of the fluid, often nitrogen gas, (is 16.62 x 10^{6} Pa.s), may be used and the units of the variables converted to the SI Unit to obtain permeability in m^2 (1 milliDarcy = 10^{-15} m^2). The test result at different constant pore pressures for each sample would be checked for consistency with Darcy law before the average gas permeabilities are calculated. A method known as *Klinkenberg effect correction* is used to obtain the absolute permeabilities for nitrogen liquid flow (Klinkenberg, 1941; Noman and Kalam, 1990; McPhee and Arthur, 1991).

17.8 Post-Mortem

At the end of logging and coring, with all the information obtained from a well, it is time to bring together most of what have been discussed in the previous chapters, relating them to the downhole reservoir objectives. When drilling has been achieved, it is relevant to consider if the subsurface conditions observed were those expected (Sneider et al., 1997). Using completed well results to compare with prognoses before drilling is known as **Well Post-Mortem**.

Evaluation methods from a well, for subsurface condition, of which results may be compared with prediction before drilling, include:

- Reservoir description (induced versus intrinsic properties)
- Pressure testing (formation)
- Sample cutting analyses (mud logging)
- Well Logging (LWD, MWD, Wireline, and Production logging)
- Fluid sampling (WFT with fluid module, GMWD and fluid type analysis)
- Analysing drilling parameters (Drilling break, etc.)

Depositional Environment from Core

Prediction of depositional environment is an important part of well planning. The environments usually range from onshore to offshore and the type of environment affects expected petroleum results. Core samples provide good information for confirming the environment of deposition, especially when observed sedimentary features are complemented with gamma ray log or other analyses such as medical-type imagery, as shown in Figure 17.14.

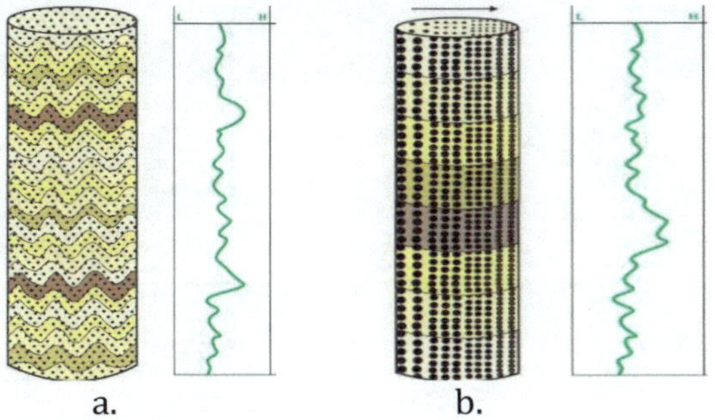

a. b.

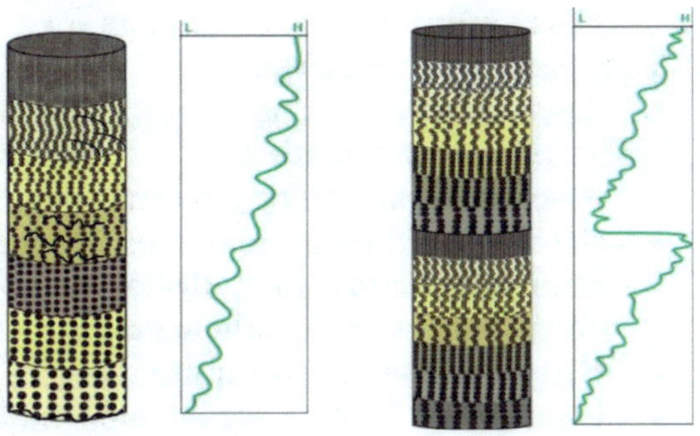

c. d

a. Aeolian dune: core sample showing consistent ripples of well-rounded fine sediments blown by wind
b. Alluvial fan: core sample showing graded, poorly-sorted deposition with mix of sand, silt and clay
c. Fluvial deposit: core showing localised ripples and erosional surface at the base with gravel or conglomerate, pebbles and stones, deposited by rivers.
d. Turbidite deposits: core sample showing Bouma sequence from repeating sea level rise and fall

Figure 17.14: Gamma ray log on core showing different facies, from terrestrial deposits to oceanic depositional environment (GR response: L=Low; H=High)

Reservoir properties are highly dependent on environment of deposition. When a well is drilled, the formations encountered are analysed to define these properties. In addition to core logging and interpretation, properties of the cored reservoir are evaluated to answer questions regarding to the well objectives, such as:

- Was the target reservoirs encountered?
- Are they clean sands?
- Are the reservoirs fractured?
- What type of reservoirs are they?
- What are the temperatures?
- What are the fluids or hydrocarbon types?
- How do the reservoirs correlate with known offset reservoirs?

Intrinsic Reservoir properties

Some of the evaluation processes of the rock properties and reservoir give information on intrinsic subsurface condition. These are properties that the rock exhibits and not those determined by interaction or induced properties. They are sometimes evaluated on a core. They include parameters such as:

- Clean reservoir rating (sorting, roundness and clay contamination)
- Porosity, permeability and saturation
- Net-to-gross (net sand or limestone reservoir section in a drilled formation versus non reservoir lithologies)
- Micro- and macro-fracturation
- Weathering and reservoir degradation (presence of trace minerals)

Reservoir Lithology

The nature of reservoir rocks is among the most important direct information provided by cores. The major property that makes a rock a reservoir is its capacity to store fluid or the introduction of secondary fluid storage spaces. Certain lithologies and changes provide porosity and permeability and enhance fluid storage; conventional reservoir lithologies include:

- Sandstone: with primary or secondary porosity
- Carbonates (limestone and dolomite): essentially secondary porosity
- Fractured basement (or other fractured rock types): serve as both migration pathway and storage

Shale Source Rock

Source rocks are less studied on core because most coring jobs take place in the reservoir. However wireline coring for mineral deposits and unconventional resources are carried out in shale. Oil shale for example is a non-conventional source of hydrocarbons cored to determine nature of deposit. Just as reservoir rock types have a role in the type of hydrocarbon it would store, source rock type is important on the kind of hydrocarbon that would be generated. The depositional settings of source rocks range from **terrestrial** to **marine**, and this determines the kind of hydrocarbon each type may generate. Core analyses provide information on the nature of shale source rock, its maturity and kerogen types.

EXERCISES

1. What is core plugging? How is this useful for core analyses, petrophysical study and petrography?

2. What do TSD and PDC represent? Mention other types of corehead, their specifications, advantages and disadvantages.

3. a) When should aluminium or fiberglass inner barrel be used? What other factor is important in choosing the core tool inner liner?

4. Explain each of these coring designs and advise on why each one should be used in a given condition:
   - A swivel design
   - Sponge coring
   - The wireline system

5. How does delay during pipe connections lead to hole-sticking and eventual core jamming? How is jamming identified?

6. a. What is a catcher? Differentiate between the various types.
   b. What is bit balling? In what kind of formation or lithology could it occur? What is core milling? Differentiate or associate this to breakage resulting from slippage.

7. a) Give details of how cores are packaged for transportation and preserved in the laboratory. What are examples of HSE concerns?
   b) Explain each of the tests carried out in conventional and advanced core analyses. What are the two methods of obtaining saturation from core?

8. What is the formula for estimating capillary entry pressure? Explain wettability and how to determine it from the laboratory. What does it signify for reservoir studies?

9. Why is there need to reconsider subsurface conditions after drilling? List four methods used for evaluating subsurface conditions from a well.

10. What are intrinsic reservoir properties? Give examples.

REFERENCES & FURTHER READING

Anderson, E.E., and J.J. Azar, 1990. PDC Bit Performance under Simulated Borehole Conditions, *SPE paper 20412*, presented at the SPE 65th Annual Technical Conference and Exhibition, New Orleans, Sept. 23-26.

Bouma, A. H., 1962. Sedimentology of some flysch deposits : a graphic approach of facies interpretation. Elsevier. 162p.

Clark, I.E., and G.R. Shafto, 1987. Core Drilling With SYNDAX3 PCD, *Industrial Diamond Review*, April.

Cohen, John H., Maurer, W. C., and P.A. Westcott, 1992. Design and Operation of a New Drilling Simulator, ASME PD Vol. 40, *Drilling Technology*.

Collee, P. E., Mallow, W. A. and S. R. Radford, 1994. Method and Composition for Preserving Core Sample Integrity Using an Encapsulating Material, *United States Patent No. 5*, 360, 074.

Klinkenberg, L., 1941. The permeability of porous media to liquids and gases. *Am. Pet. Inst. Drill. Prod. Pract.*, p. 200–213.

McPhee, C.A. and K. G. Arthur, 1991. Klinkenberg Permeability Measurements: Problems and Practical Solutions, in *Advances in Core Evaluation: Accuracy and Precision in Reserves Estimation*, ed. P.F. Worthington, p. 447–462.

Maurer, W.C., Anderson, E., Hood, M., Cooper, G., and N, Cook, 1990. Deep Drilling Basic Research," Report No. TR90-7 prepared by Maurer Engineering Inc. and the University of California at Berkeley, June

Noman, R. and Kalam, M.Z., 1990. Transition from Laminar to Non-Darcy Flow of Gases in Porous Media, in *Advances in Core Evaluation: Accuracy and Precision in Reserves Estimation*, ed. P.F. Worthington, p. 447–462.

Nosike, L., 2009. Relationship between Tectonics and Vertical Hydrocarbon Leakage: A Case Study of the Deep Offshore Niger Delta. *PhD Thesis*, University of Nice-Sophia Antipolis, France, 281 p.

Prince, C.M., Dixon, M.W., and L.L. Haynes, 2005. The Use of High-Resolution Core Imagery in Reservoir Characterization: An Example from Unlithified Miocene Turbidites. Paper SCA2005-02, *Society of Core Analysts Annual International Symposium*.

Skopec, R. A., Collee, P. E. and L. K. Shallenberger, 1995. High-Viscosity Gel Encapsulation for Core Preservation and Improved Reservoir Evaluation, *Journal of Petroleum Technology*, V. 47, no. 5, p. 399.

Sneider, R. M., J. S. Sneider, G. W. Bolger and J. W. Neasham, 1997. Comparison of seal capacity determinations: Conventional cores vs. cuttings, *in* R. C. Surdam, ed., Seals, traps and the petroleum system. *AAPG Memoir 67*, p. 1–12.

CHAPTER EIGHTEEN

WELL TESTS, COMPLETION, PRODUCTION AND MANAGEMENT

18.0 Reservoir Geology and Well Production

One of the aims of reservoir geology is to find the Hydrocarbon Initially In Place (HCIIP) and what volume it will represent when it gets to the surface. An overall appraisal study is done to guide on the drive mechanism (natural or artificial forces that displace hydrocarbons from the reservoir to the wellbore and up to the surface) that will be chosen for the production, and these can be either primary (natural) or secondary (artificial) recovery. An example of primary recovery is drive due to in-situ aquifer overpressure; an example of secondary recovery is gas or water injection. Sometimes advanced recovery methods, such as polymer injection and fracking, which change the rock properties, are known as tertiary recovery.

In primary recovery, the overall recovery efficiency depends on the efficiency of sweep of the areal extent, vertical extent and microscopic extent recoverable in the reservoir. This simply means that all the factors that control displacement in different directions are taken into consideration. It is stated as:

$$RE = AE * VE * ME$$

Where:

RE = Recovery efficiency
AE = Areal sweep efficiency
VE = Vertical sweep efficiency
ME = Microscopic sweep or Displacement efficiency

Recovery efficiency gives an idea of the sweep efficiency during production. The number of producer wells that will be needed for a field depends on the ratio of *production rate* at field plateau to the rate of hydrocarbons producible by each well. That is,

Number of producers
= Qo for field at plateau / Qo produced per well

Where:

Qo is the rate for producer

Secondary recovery, such as water injection, may increase the *recovery factor* and sweep more hydrocarbons to the producers, depending on the number of injector wells used. This may be decided based on the ratio of *water injection rate* needed at field plateau to the rate of water to be injected by a well. That is,

Number of injectors
= Qw for field at plateau / Qw injected per well

Where:

Qw is the rate for injector

Interference tests may be conducted to determine the location of the wells and to test faults connectivity. The *production index* gives a guide to the amount of hydrocarbons that may be produced by each well. Productivity index (PI) is the ability of a reservoir to deliver fluids to the wellbore of a given well, expressed as the volume of fluid produced (or production rate at surface) for a given *drawdown* (or pressure difference at the interface between the formation and the wellbore).

$$PI = Q / Drawdown$$

Where:

Q is production rate
Drawdown is the pressure drop

The drawdown may be expressed as,

$$Pe - Pwf$$

Where,

Pe is the pressure in the reservoir formation
Pwf is the pressure at the *well-reservoir interface* (known as *sandface*)

Then,

$$PI = Q / Pe - Pwf$$

The production index is obtained from well test, as such well test is a prerequisite for properly producing and managing a field. *These subjects are treated in this chapter.*

18.1 Well Test Principle

Well test is part of a process carried out after a well has been drilled, before *well completion* (*installation of the production equipment*). It is a downhole information acquisition mechanism that has the widest subsurface coverage away from the well, compared to core, log or well formation testers (Earlougher, 1977; Lee, 1982; Horne, 1995; Spivey and Lee, 2013). The principles behind a well test are:

- A fluid pressure change is observed naturally after a well is *shut-in* (closed) at the surface, or the change is induced (by pumping) and the pressure variation with time is read off on a gauge.
- The fluid pressure changes indicate the properties of (1) *the reservoir formation* and (2) *the well*.

Initiation of well test may be achieved when a shut-in well, stabilized by high drilling density / mud weight, is open and the wellbore pressure reduced, or when the well is induced to flow by other means (Figure 18.1).

The challenge with well test is in separating which of the pressure changes is due to the *wellbore* and which is due to the *reservoir formation*. Models have been created for expected results of well test, so that any match could indicate what is creating the change. There is no one conclusion for any observation, as such there is need for a careful interpretation and collaboration of other data. The main principle can be stated as:

Reservoir + Well = Fluid Pressure Variation

(The variation due to reservoir and that due to the well has to be separated).

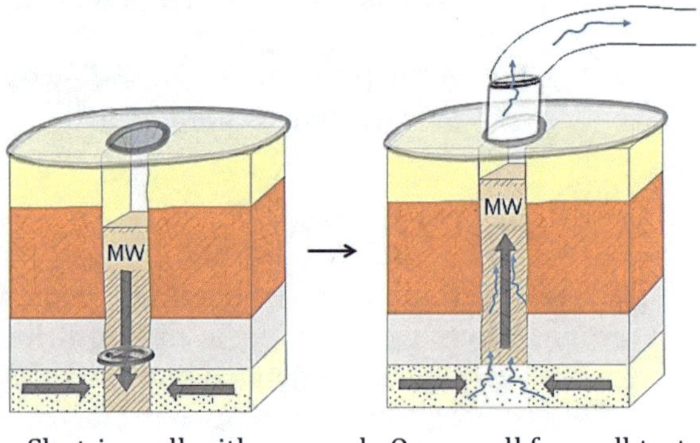

a. Shut-in well with a parker   b. Open well for well test

Figure 18.1: Open and closed wells showing the initiation of well test

Production Well Test

A **production well test** can be initiated by opening a closed well under pressure. That lowers the borehole pressure to atmospheric pressure, thereby releasing the fluid in it to flow out. This can also be achieved by a reduction of wellbore fluid pressure; reducing the mud weight fluid in the well (weighting down) leads to formation fluid inflow into the well from the formation and flow out to the surface. As the well flows to the surface, the fluid pressure is measured by gauges (Matthews and Russell, 1967; Lee et al., 2003).

The period during which there is pressure reduction in the well to initiate inflow from the formation is known as ***draw-down***. When *draw down* is stopped, inflow will lead to reestablishment of the initial pressure in the well and eventually an increase (known as ***build-up***) due to continuous inflow from formation. During well test, the well is left to flow for several hours and then shut-in so that it is allowed to go back to equilibrium (a process known as ***stabilisation***) as the formation fluid recedes.

Any changes after the well is shut-in, like a gradual flow back into the formation, show the intrinsic flow behaviours of the well and its surrounding formation. When a well behaviour is monitored after shut-in, it is known as **build-up well test**. Different well and formation type will show different *flow rates* into and out of the well, during *drawdown* and *building up*.

For example, if there is a nearby fault, then at some point the fluid reaching the well from the fault will affect flow, and change the trend. Likewise if there is a sand body or other structures in the formation that might affect flow across it to the well, it may be detected (Matthews et al., 1954; Gochnour and Slater, 1977; Holden, 1995; Wei et al., 1998; Beauheim et al., 2004). Whether a well will flow on its own or require pumping out is already an important result of a well test, showing whether the well will be produced naturally or will require a support of an injector. How long it flows naturally over hours or days gives an idea of how long the well will produce before needing support.

In a plot of *change in pressure* against *time*, the descending trend (drop-off) is monitored, and can give useful information on the necessary pressure-drop needed for production of the well – how long the well might produce and what improvement may be envisaged to enhance flow. This is similar to WFT (Wireline Formation Tester) principles, giving the mobility of the formation, but is occurring at a larger scale and it is the entire well that is probed (Horner, 1951; Jones et al., 1976; Towler, 2002; Thomas et al., 2005; Massonnat and Wigniolle, 2013). Draw down, build up and stabilisation in well testing are similar to that obtained in regular WFT (as shown in Chapter Twelve on Pore Pressure Prediction and Interpretation).

Production Index

The rate of production or flow out following drawdown is known as ***Production Index (PI),*** and is an important measurement to know the ***flowability*** of the well. It is perhaps the most important information from a well test. In this context,

PI = Rate of Production / Drawdown

The change in pressure during a well test can be divided into three phases:

1. The initial transient period when the well is still flowing constantly and yet to reach any interference boundary within the formation
2. The late transient period when the change begins to reflect the effect of boundaries or other interference within the formation
3. The steady state when the change is constant due to stabilization of the flow

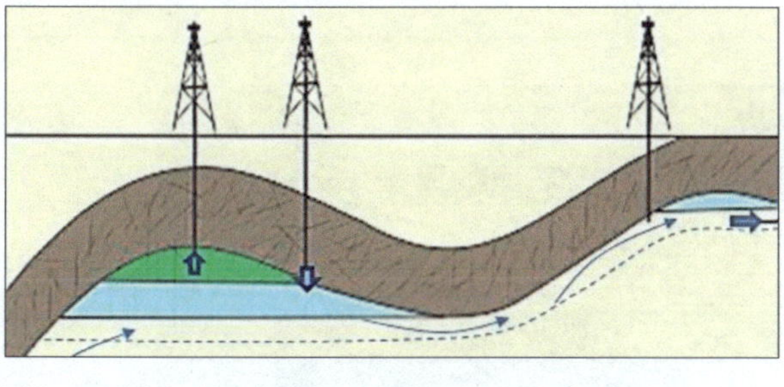

Figure 18.2: Production, injection and observation wells (from left to right respectively)

Well tests are done essentially before removing the rig used to drill the hole. However, testing ***rigless*** (i.e. without a rig) is possible especially if already producing in a well nearby and production equipment is available. Well tests can be carried out in a ***production***, ***observation*** or ***injection*** well (Figure 18.2). In each case, the knowledge obtained from the flow behaviour during the well test would help to achieve the objectives of the well.

Injectivity Test

When a well test is carried out in an injector, it is known as injectivity test. Water is pumped into the well at a constant rate until a stable pressure is reached, then the pump is turned off and the rate of pressure decrease at the gauge is measured. Injectivity test gives information on the permeability of the formation and the rate of flow into it, which is important for assessing supports to a producer. When injected fluid (pressure) is detected in the producer, it means the well is *interfering* or *communicating*.

Tracers (chemical substances used as markers) can be added to the fluid for detection or the pressure increase in the producer well is used to confirm that the wells are "talking to each other". Injectivity test also gives the *formation integrity*, which is the pressures that can be withstood by the rock matrix before fracturing. Where Leak-Off-Test (LOT) results are available, care is taken to not inject a well to the equivalent pressure to avoid fracturing the (top) seal. *(LOT is carried out to determine the pressure of fracture or the strength of a formation).*

18.2 Well Test Objectives and Types

As mentioned above, the objective of a well test is to obtain information about the well and the target reservoirs. This may include:

- Estimate volume of oil in place
- Check permeability of reservoir formation
- Build a geologic model in a field
- Reach some ***prognostics*** (explained in further details in the next section)

With a well test it is possible to:

- Characterize the wellbore condition and the reservoir properties
- Predict production volumes and duration (and enhanced recovery options)

- Investigate surrounding geological structures and bodies
- Validate exploration hypothesis about subsurface conditions and fluids

There are two major types or stages of tests, which are very similar but varying in their objectives. These are the:

Virgin Well Test

This is done the first time a well is drilled, often for exploration well or a new well in a development field. The aim is to confirm or invalidate geological concept developed during exploration and to quantify the production rate. Some important factors such as (1) **Productivity Index** (2) **Permeability** and (3) **Skin** are evaluated at this stage. Virgin well test would also help to establish the initial reservoir pressure, the minimum reservoir volume and the heterogeneities in the formation.

Follow-Up Well Test

This is carried out at a later stage of a well. Recent data are compared with previous data and the information obtained during the virgin well tests are revaluated for more accuracy and the impact of production. If new wells have been drilled, or if there has been a change affecting the subsurface pressures, then the well test results will show variation with the initial tests. The result may lead to implementation of new production strategies: e.g. new perforations or Enhanced Oil Recovery (EOR). Follow up well test would also help to identify changing fluid behaviour, changing skin effect, changing reservoir pressure as well as hydraulic connectivity across fault boundaries.

Mini-Well Test: Drill Stem Test (DST)

A well test carried out during drilling, or before ***Well completion***, is known as Drill Stem Test. DST uses the basic principle of pressure measurement to achieve well testing only for a section of the well, isolated with parkers (inflated pads in the form of rings), or in the open hole below the casing (uncased lower section of the well). The uncased **(or perforated)** section of a reservoir is open to the test. Just like with WFT, there will be a *build-up* period, during which the well is shut-in and the pressure in the well is allowed to evolve as it is being read off on the gauge. The DST is seen as a temporary well test to check if a well should be *drilled further*, *completed* or *abandoned*.

There are three acquisition levels and tools for a DST system:

- **Down hole tools:** like the packers which provide the upper and bottom seals
- **Seabed landing**: provide the hang-off point and the adjustable hangers (offshore)
- **Bottom hole gauge**: used for the actual pressure measurement

The tools are linked to a flow head where the flow is monitored, and with **valves** to allow the possibility to "kill" flow (i.e. stop flow) if necessary. The valve system is controlled on the surface through a **choke manifold** (high-pressure valves arrangement with piping and adjustable plugs used to control flow pressure). Once the fluid arrives at the surface, it goes through a **flow rate meter** to measure the flow per given time, to a **separator** (Figure 18.3)for the various fluid components (water, oil or gas), which are in turn passed to a **test meter** for analyses before it goes to the **burner**. The oil may be connected to a burner while the gas was flared in time past, but now piped for utility purposes.

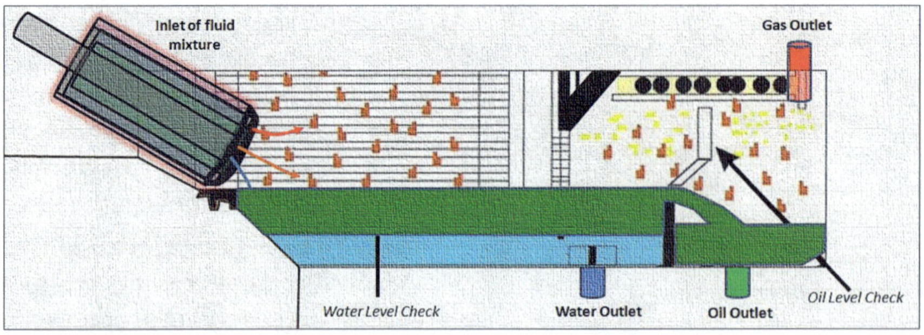

Figure 18.3: Basic configuration of a well test separator showing the outlets for different fluid components

18.3 Interpretation and Applications of Well Test

Major interpretational algorithms in well test follow the *derivatives* principle, which help to make *diagnostics* from the well test result (information on the well behaviour and on the surrounding formation). Based on the interpretation of the derivatives, ***formation damage, wellbore to well wall interface deterioration, near wellbore barriers*** hindering flow, among others can be evaluated for diagnoses. Though with limitations, this leads to a better understanding of how to improve the well (Chow 1952; Agarwal et al., 1970; Barker, 1988; Bourdet et al., 1989; Renard et al., 2009).

18.3.1 The Derivative and Diagnostics

The ***derivative*** is the result of an expression which expresses the additional change in pressure as a function of the original change in pressure per change in time. The derivative ΔP' is estimated as a variation in pressure (ΔP) over elapsed time (Δt) since the last change in flow rate. That is,

$$\Delta P' = \Delta P / \Delta \ln \Delta t$$

This allows for the diagnoses of differential pressure changes when represented on a log-log plot. When the derivative is plotted on the vertical axis against change in time, it looks like the plot in Figure 18.4. The pressure increase (ΔP) keeps increasing while the rate at which it increases (ΔP') starts to drop with time.

18.4: Plot of variation in pressure and derivative against elapsed time

The ***diagnostics*** refer to the different signatures that indicate what could be the situation down hole and across the formation. Three main examples of such signatures or diagnostics are:
(1) wellbore storage
(2) radial or linear flow
(3) mobility increase or reduction

There is need to ensure that the fluid is flowing within the formation, as a flow vertically or down along the interface between the well and the formation would be misleading. *Cross-flow*, from one well perforation interval to the other, across reservoirs, is a good example of such misleading situations. When flowing into the formation, the flow may be **radial**, **linear**, **semi-radial** or **semi-linear**.

Wellbore Storage (slope 1)

This is a state where the fluid is retained within the well. The slope of the derivative will be 1 (as shown in Figure 13.5), implying that the effect of pressure increase is proportional to changing time.

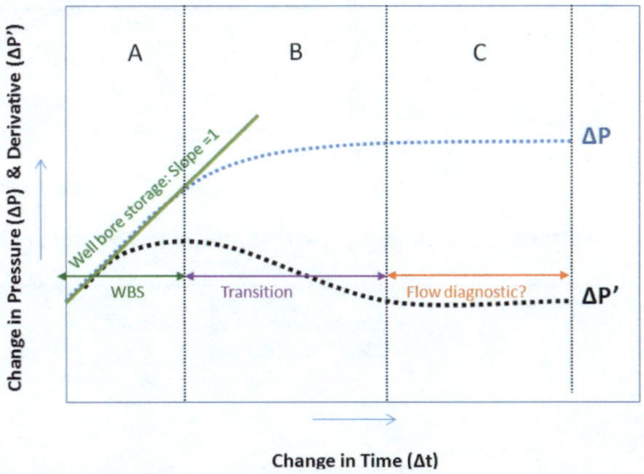

Figure 18.5: Plot of variation in pressure and derivative against elapsed time showing the wellbore storage (WBS) as a slope of 1

This is usually the case at the onset of well test until the initial barrier posed by the wellbore-reservoir interface is breached. How much a fluid will be retained within the well depends on how well it is in communication with the reservoir. It thereby gives an indication of constraining petrophysical factors such as permeability and mobility.

Beyond the wellbore storage (A) is a zone of transition (B) before stabilization (C) and a *horizontal*, an *upwards* or *downwards slope* on the derivative curve that gives the various diagnostics (radial or linear flow and mobility changes) described below.

Radial or Linear Flow

Most flows are first radial (normal circular outflow round a well in a homogenous formation) or linear (bounded on both sides like in a channel) before mobility increase or decrease.

There may however be a quick change in flow without the radial or linear flow being detected. Flow boundaries too close to the well may be harder to detect.
Radial flow:

Radial flow is the most natural flow, which occurs in a homogeneous formation. In a perfect radial flow, fluid will flow away from the well in a circular ring form as shown in Figure 18.6. This will show a **horizontal slope** in the derivative because there is a harmonious spread or loss of the fluid over time.

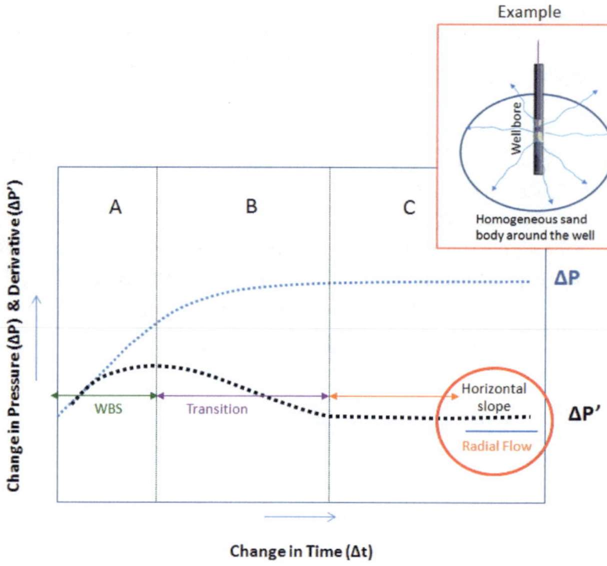

Figure 18.6: Radial flow (indicated in the red circle) showing a horizontal slope after stabilisation

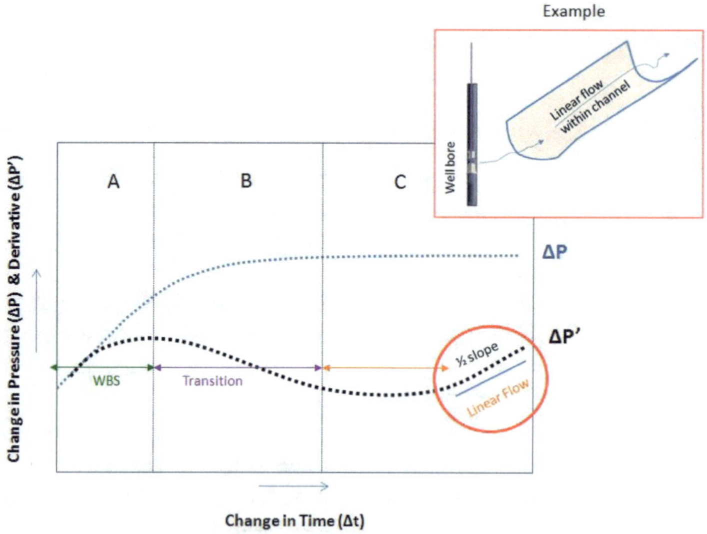

Figure 18.7: Linear flow (indicated in the red circle) showing ½ slope (without showing an obvious radial flow at the onset)

Linear flow

A linear flow is a special case of mobility reduction where there is a *no-flow boundary* on both sides. An example is a channel where the two sides are bounded by levees. This will show a **½ slope** in the derivative because there is no-flow boundary on both sides, leading to a preferential linear flow on opposite but linear directions. The rise in the derivative is about half what it would have been if there were no flow (recall that while the fluid is retained in the well there was a slope of 1 = Wellbore Storage, WBS). This is shown in Figure 18.7.

Mobility Increase or Reduction

Mobility Increase (negative slope)

An increase in mobility (during flow back) will mean that the change in pressure with change in time is negative (evolving negatively or regressing). It means that less fluid pressure is retained with time, due to more flow away from the well, leading to a "fall" in the derivative curve as shown in Figure 18.8. From this, it can be inferred that:

a. A petrophysical parameter of the formation such as Permeability (K) is increasing.
b. There is an extension and size of the reservoir within the formation is increasing (see inset image at the top right of Figure 18.8).
c. The reservoir interface is enlarged, like more contact of the sand body with the well

An example of mobility increase will occur in the widening of sand body, leading to increased flow. There is need to be careful as a change in fluid parameter during the well test, such as viscosity reduction, may enhance flow and lead to false indication of mobility increase (as if due to formation response).

Mobility Reduction (positive slope)

This is just the opposite of mobility increase. It will be indicated by a "rise" of the derivative curve as shown in Figure 18.9; it implies that the change in pressure with changing time is positive. It is increasing because pressure loss by mobility to the reservoir is low. An example is when the flow encounters a fault boundary, leading to reduction in flow across it. If the flow is completely stopped in that direction, a new direction may be preferred or enhanced by the fluid, leading to a drop in the rise and a second stabilization of the derivative. This is a case of a **no-flow boundary**.

It can be inferred that:

1. A petrophysical parameter such as Permeability (K) is reducing.
2. There is the presence of a no-flow-boundary, such as fault (see inset image at the top right of Figure 18.9).
3. The reservoir interface is reduced, like less contact of the sand body with the well.

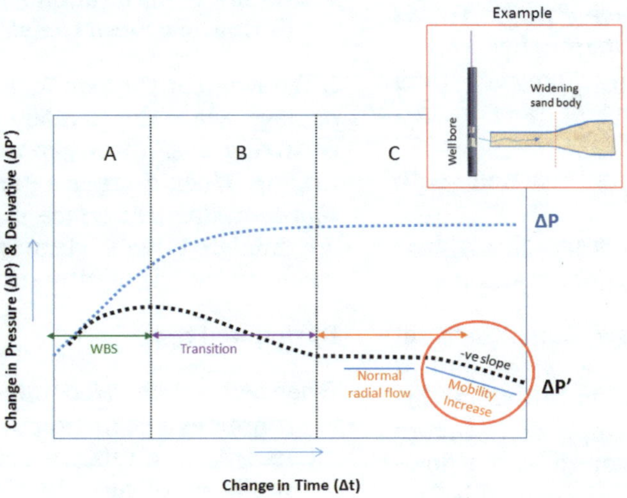

Figure 18.8: Mobility increase (indicated in the red circle) showing negative slope after an initial normal

radial flow

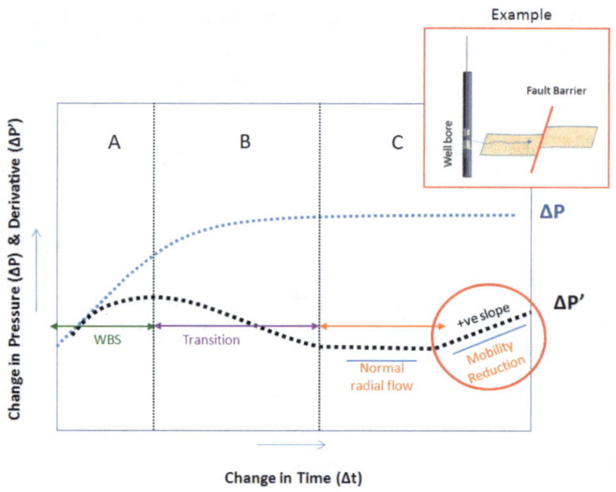

Figure 18.9: Mobility reduction (indicated in the red circle) showing positive slope after an initial normal radial flow

The distance to the fault or other causes of mobility reduction, which is the distance of the flow or of investigation at that time, can also be estimated. Mobility increase or decrease at a fault may indicate that the fault is leaking or sealing, respectively. A ***leaking fault*** would result in mobility increase while a ***sealing fault*** would result in mobility reduction. Again, there is need to be careful as a change in the fluid parameter, increase in viscosity for example, may reduce flow and lead to a false indication of mobility reduction (as if due to formation response).

18.3.2 Formation Damage

Formation damage around the wellbore is a major challenge in *completion of wells*. This is caused at *reservoir-to-well interface* by:

- Permeability reduction (induced local compaction or smearing of pore spaces with fine sediments or clay)
- Mineral precipitation or reaction with drilling fluid
- Mudcake formation over reservoir surface
- Completion process (e.g. poor gravel pack)

Damage increases the *flow impediment* at reservoir-to-well interface. A major aim of **well clean-up** after drilling is to reduce this impediment, at least temporally, before production. Duration of clean-up before flow-out is already an important information before well test. Well test results provide data for modelling wellbore interface and improving flow across it.

Wellbore Interface and the Skin Concept

The character of the interface between the wellbore and the reservoir is known as ***skin*** (Chow 1952; Agarwal et al., 1970). It is a dimensionless value estimated to assess the flow efficiency during well test. It gives the quality of communication between well and reservoir and is a function of several factors, among which are:

- Formation of mud-cake at interface between the well and the reservoir
- Degree of invasion of drilling fluid into the reservoir (invaded zone)
- Nature of perforation and gravel pack used during *lower well completion.*

It follows that the *skin factor* results from the damage with the tendency to inhibit flow, occurring near the wellbore during and after drilling. When there is a damage to the *formation-to-wellbore interface,* skin is high, whereas a stimulated well, cleaned out to enhance connectivity, has negative skin.

During well test:

When Skin > 0 (interface damage)

- It implies a reduction in flow into the well, leading to a relative pressure drop in the wellbore (Figure 18.10a), which will be measured and given by the derivative.

When Skin < 0 (well stimulation)

- It implies an increase in flow into the well, leading to a relative pressure increase in the wellbore(Figure 18.10b), which will be measured and given by the derivative.

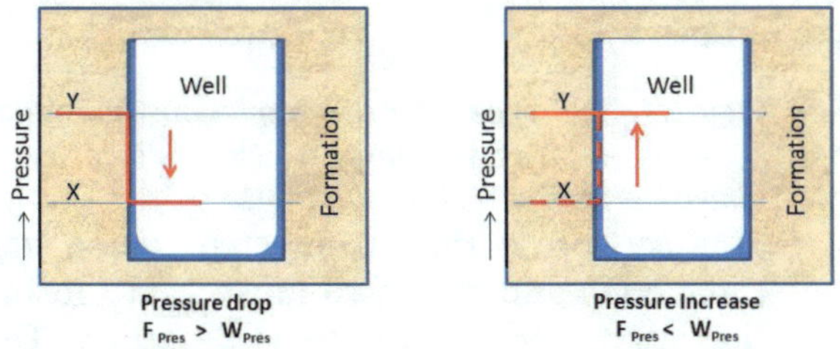

a. Positive Skin b. Negative Skin

Figure 18.10: Pressure changes in wellbore due to skin effect

In Figure 18.10a, the formation pressure **(Y)** is lower than the initial well pressure **(X)** as pressure drops within the wellbore due to pore interface during production or draw-down; in 10b, the initial well pressure **(X)** increases towards **(Y)** due to better flow into the well.

Stabilisation of derivative curve gives skin value, estimated as shown in Figure 18.11, which is usually between **- 7** and **>>100**; it may also be used to estimate the permeability of the near wellbore. The lowest value of skin is obtained by hydraulic fracturing or other effective methods of well stimulation. The highest skin is obtained when a well is completely plugged against the formation, leading to no-flow interface. Once able to understand and properly model the effect of skin, it is possible to improve (that is, reduce) it and enhance production.

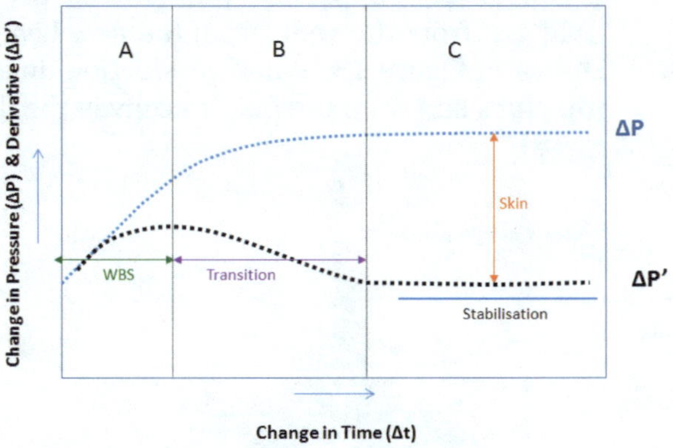

Figure 18.11: Estimation of skin from the difference between pressure change in a wellbore and the derivative at stabilisation

Examples of Enhanced Oil Recovery (EOR) methods which are based on skin reduction are:

1. Acidification
2. Hydraulic fracture
3. Solvent injection

Use of gravel pack behind casing and fracture are ways to improve the interface between the reservoir and the well - reducing **skin effect**.

18.4 Limitations of Well Test

There are several limitations of well test which have to be checked to ensure correct results and proper interpretation. Examples are:

- Multiple flow meters can present challenges with calibration.
- Without a proper control of flow rate on the choke, it is difficult to rightly estimate the impact of flow.
- The initial need to *clean-up* in order to start to recover fluid from formation may already alter in-situ conditions of reservoir to wellbore interface.
- While most flow principles are based on *Darcy's law*, the law is not very accurate at reservoir scale and conditions.

Despite the above, well testing remains a useful process in well development.

18.5 Well Completion and Stimulation

Well completion is the final stage in well development, after drilling and before full production or injection of the well (Gatlin, 1960; Bourgeois and Couillens, 1994; Hyne, 2001). It involves the preparation of the lower part of the well to allow proper communication with the reservoir formation and the installation of surface flow equipment. The first concern of ensuring a fluid inlet to the well is achieved by either:

1. Leaving the lower part of the hole open while casing off the upper part only
2. Running a special casing (and screens) that does not hang on the surface, known as liner, on only the lower section.

3. Perforating cased reservoir section of interest with down-hole guns

Well completion may be either normal or dual, as shown in Figure 18.12. In normal completion, the zone of interest is separated from the other intervals (using a parker for example) while a pipe or tubing is used to produce the well and direct fluids to the surface. Dual completion allows for production of different reservoir intervals at different times by more than one tubing.

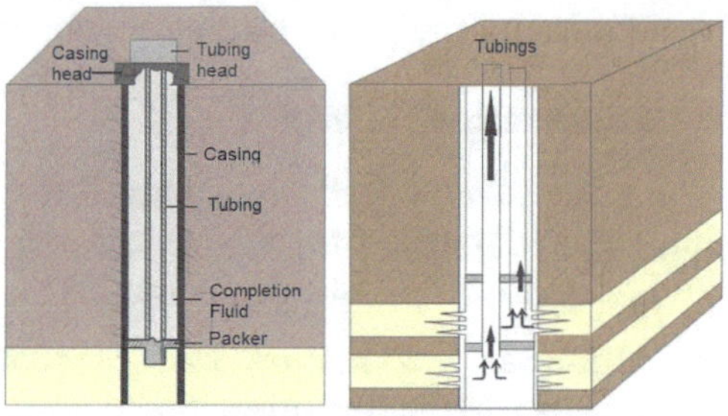

a. Normal completion b. Dual completion

Figure 18.12: Well completion showing single and multiple production from different reservoir intervals

Completion planning may start before the final Total Depth (TD) of the well is reached. Logs would have to be interpreted to ascertain the actual depths of clean sand and shale or other non-reservoir sections that should not interfere with completion. There are two major stages of completion: *Lower* and *Upper Completions.*

18.5.1 Lower Completion

This involves the setting of tools and apparatus in the lower productive or injection zone in the wellbore (usually the clean reservoir sands).

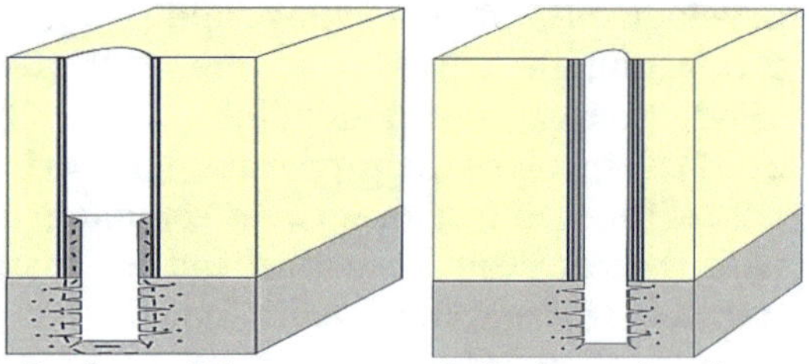

a. Perforated liner b. Perforated casing

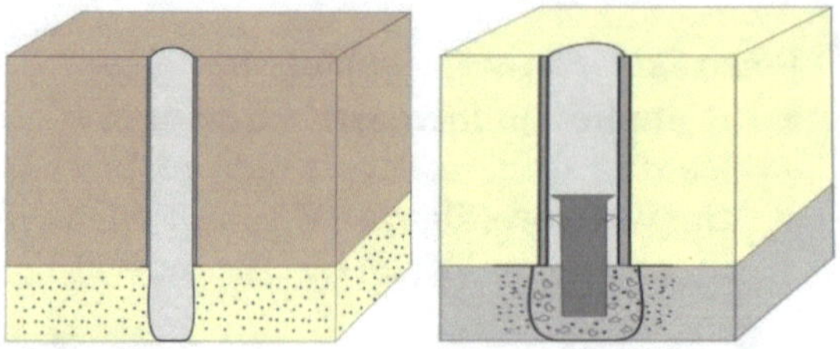

c. Open hole d. Gravel pack

Figure 18.13: Types of well completions (note that in a. the liner does not hang on the surface but just above the interval for production).

This section is then connected to the *upper completion* and tested to make every installation ready for production or injection. There are different types of Lower completion (Figure 18.13), such as the ***open-hole***, ***gravel-pack*** and ***liner*** or ***screen*** (shown in Figure 18.12). Where the completion is not in the open hole, either the casing or the liner is perforated. A *screen* is a pre-perforated set that can be combined with a liner to avoid the cost and consequences of perforation (such as formation damage).

18.5.2 Upper Completion

These are sets of apparatus installed on the ground level (onshore) or through the water surface to the seafloor (offshore) above production or injection well. They link to the wellbore and are useful for proper control of the flow from the well after the lower completion. These include the ***wellhead equipment***, the valves assembly known as ***Christmas tree***, the ***hanger*** for the tubing that transports the fluid to the temporary storage or processing facility, and many other ***pressure gauges*** and ***safety barriers***.

While lower completion controls fluid inlet to a producer well, upper completion controls the fluid exit from the well (from the flow head as shown in Figure 18.14 and production lines to the pipes laid from the field to convey the fluid away)

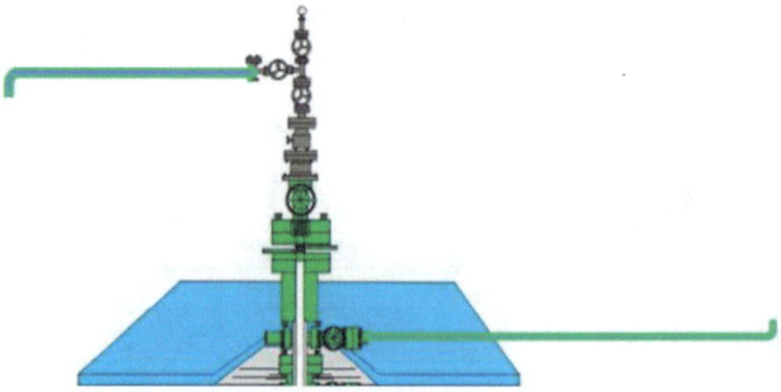

Figure 18.14: Flow head showing a collection of valves and gauges used to convey hydrocarbons and control hydrocarbon flow rate

18.5.3 Well Stimulation

This is the enhancing of flow across the *well-reservoir interface*, and eventually within the formation. The aim of stimulating a well is to ***reduce skin effect*** and enhance hydrocarbon flow into the well. This is achieved by chemical enhancement, acid or mechanical process such as reservoir fracturing.

Acidification

An adapted acid such as hydrochloric acid (***HCl**, for carbonate rock*) or hydrofluoric acid (***HF**, for tight sandstone*) may be used in the chemical process, but these are not without their downside (disadvantage) since acid can be a major hazard to both man and the environment.

Fracturing

Fracturing is achieved by a mix of ***proppant gel*** (like treated sand or ceramic fragments) and ***beads*** which remain in the cracks to maintain the open fractures and flow. Reservoir fracturing requires a prior study and understanding of the stress regime of the formation, to avoid fractures in unexpected direction which can lead to major pressure hazards. *Hydraulic fracturing* may cause seismic activity and as such regulations exist to either ban or control its application.

Swabbing

In a difficult situation, a well may be **swabbed** to initiate flow, especially after fracturing as a way of stimulating the well. ***Swab*** is the opposite of ***surge*** (where pressure is applied to flow into the well). In swabbing, the well is sucked by lifting a tight parker within the borehole, leading to a sudden inflow of formation fluid into the well as shown in Figure 18.15.

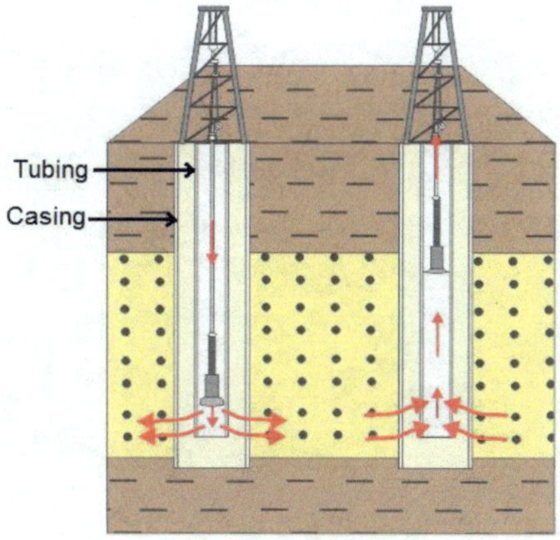

Figure 18.15: Well stimulation by swabbing –a surge is shown on the well on the left while a swab is shown on the right

This results when drilling mud from a well is reduced by swabbing, oil and gas flow into the well. It may be done after the well is completed to remove the last of the drilling mud / completion fluid or later to restore production in a producing well. A truck-mounted swabbing unit with a short mast is used to lower a swab tool down the tubing string on a wireline.

A swab job may also be carried out as a **work-over**, when it is done to restore production, in which case it is an intervention done on a well already producing when it develops a problem. A swab in that case is used to reinitiate flow or improve wellbore to formation interface. Meanwhile, a reminder that unintended swab is a major risk while drilling as it may lead to a ***kick***.

18.6 Field Production and Management

Well Pump and Lifting

A well may flow on its own for some time (natural depletion) before requiring support of an injector or abandoned. The lower completion foresees the different stages of well production and provides a ***downhole pump*** and in recent time an ***intelligent well completion (iWC)*** that may help to open a section of the well and close the other, increase or reduce production in a reservoir on a well, stop production temporary. It is a better solution improving what may be partially achieved by dual completion (Figure 18.12b) or a one-off parker installation to stop flow. Where a well cannot flow on its own, a pump is used to lift it to the surface into tubes that convey it away from the production site (Figure 18.16a).

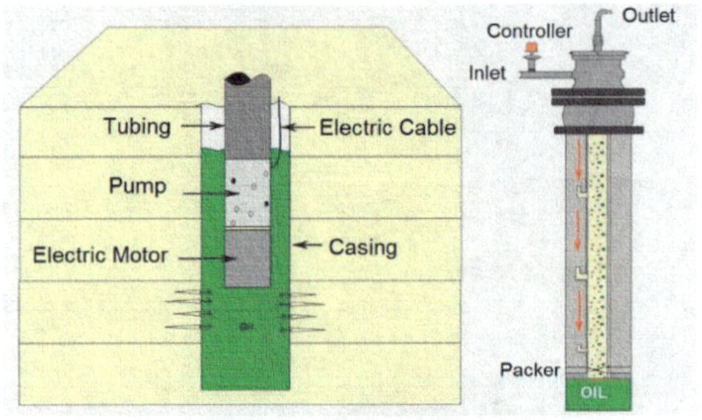

a. b.

a. An electric submersible pump downhole
b. A gas-lift well above oil separated by parker

Figure 18.16: Well fluid production using a p ump and a gas lift

In gas lift, the injection of high pressured gas reduces the density of in-well fluid and forces it to the surface through a separate production tubing or outer/inner barrel. Produced fluid would have to be separated to its components. When gas injection on the crest of a reservoir results in the production of the same amount of gas from a producer on the flanks, it is said to be ***gas recycling***. Towards the end of a field's life, gas injection may be stopped and previously injected gas is produced; the process of this production is known as gas ***blowdown***.

Some advantages of *intelligent well completion (IWC)* design are improved produced water disposal due to proactiveness, gas reinjection, oil recovery and check of water-cuts (incursion of water into production well). Intelligent well design is particularly useful when a producer is supported by an injector, and there is need to shut-off a lower flooded zone, to reduce water-cut and optimize hydrocarbon production in the upper zone. Such flooded zone may be identified by the use of Production Logging Tool (PLT) or 4D seismic monitoring result. IWC also helps to accelerate recovery when necessary and to decide on the need for infill wells. Oil recovery is better monitored and pressure maintenance or support at injection well may be improved based on monitoring impact at the producer well.

Separators

On production, reservoir fluid is separated to their various components and oil is stored in a temporary container. This may be a stock tank or a Floating Production Storage and Offloading (FPSO), if offshore. Gas may be piped to a utility, export or to a Liquefied Natural Gas (LNG) plant. There are many types of separators; the simplest and most common ones are based on **pressure** or **gravity** differentiation of water, oil and gas, as shown in Figure 18.17.

A two-phase separator may first separate the gas from the liquids before sending it to a free-water knockout (FWKO), which is used to separate the water from oil. A free-water knockout may fully separate the overall fluid into three phases. Most separators require multiple stages to get a clean fluid type, which may be achieved by multiple stage pressure separators as shown in Figure 17d below.

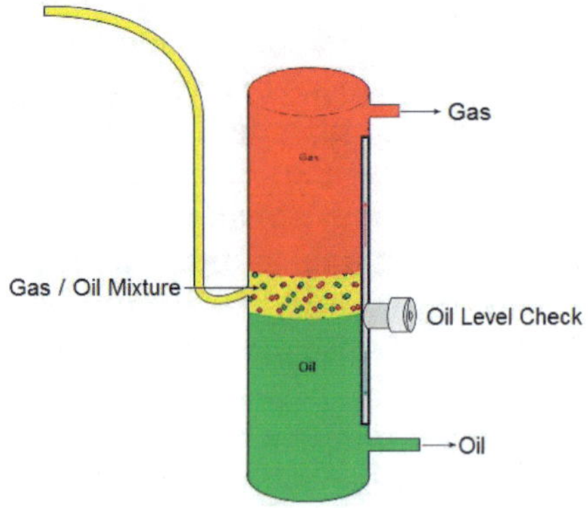

a. Vertical two phase separator

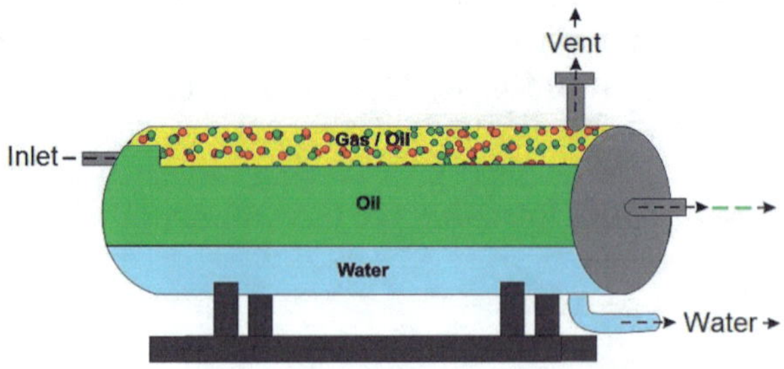

b. Horizontal free-water knockout (FWKO)

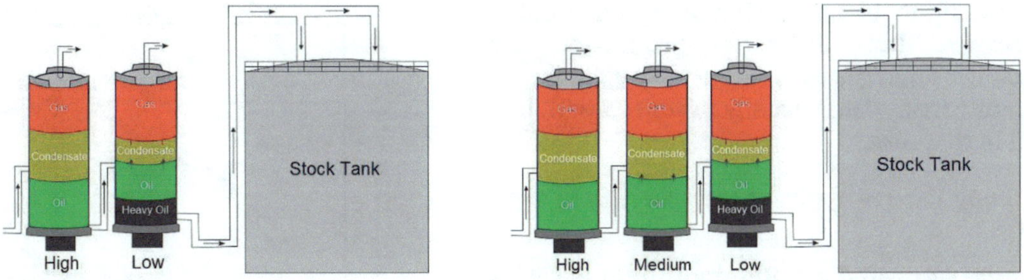

c. Two-stage pressure separator

d. Three-stage pressure separator

Figure 18.17: Principles of fluid separation illustrated in different types of separators

Reservoir Management

As production continues attempt is made to manage the field properly to produce economically as long as possible. A field is managed just like every other asset, so that it keeps producing results. A general vision on how the well will run is an important factor in ***reservoir management***. Effort is made to maintain the production rate and eventual decline. To have passed the *appraisal stage* in the first place, the *reserve* must be large enough to provide a long time production which will cover the *cost of delivery*.

The *delivery cost* is an addition of capital expenditure (CAPEX) and operating expenditure (OPEX). Total delivery cost must be far less than the oil price multiplied by the reserve, notwithstanding that some infrastructure costs added to CAPEX is moveable to other projects. When there is *decline*, OPEX will increase if there is an effort to maintain *production rate*; on the other hand OPEX can be reduced to make a well remain *viable* even with lower production.

This can be expressed as,

Delivery Cost = CAPEX + OPEX

And,

Delivery Cost << (Oil Price x Reserve)

for economic production.

Reservoir Fluid State Classification

The behaviour of the fluid as production continuous is an important parameter to take into consideration in planning field development and production (Buchwald and Thompson, 1973; Aziz and Settari, 1979). The reservoir fluid classes based on their state with respect to temperature and pressure are given by the phase diagrams. It shows the states of fluid (gaseous, liquid or critical) in a reservoir and the changes that occurs during production. During production, pressure and temperature reduces as the fluid comes to the surface, making it to take a new state. During the well test, sample fluid is collected and subjected to PVT (Pressure-Volume-Temperature) tests, which shows the critical temperature and the critical pressure, and the evolution of the fluid with changing temperature and pressure to represent what happens when the fluid is brought from reservoir conditions to the surface.

Based on the result of these tests, a phase diagram is plotted (as shown in Figure 18.18) to classify the reservoir hydrocarbon fluid and determine the conditions for its production. Some points in the curves that are important for the classification are:

- The **dew point** is temperature and pressure above which fluid phase is completely in gaseous state, and below which gaseous fluid phase begins to condense to water.
- The **bubble point** is the temperature and pressure above which reservoir fluid begins to form gases and below which it is in a liquid state.
- The **critical point** is the temperature and pressure at which the fluid is in a transient state between liquid and gas.

In this example of a PVT phase diagram shown in Figure 18.18, on the left side from the top, the fluid phase is only liquid while beyond the *dew point* to the right, only gas is produced. As shown by the arrow indicating liquid-to-gas ratio (Figure 18.18b), the amount of liquid to gas produced from the reservoirs reduces from left to right. If the initial pressure of the reservoir is above the *bubble point* curve, meaning it is an **undersaturated** fluid, a pressure draw down for production may exsolve some gas as it enters below the bubble point curve. Gas is usually produced with light oil in that case.

Among the reservoir hydrocarbon fluid types are:

Undersaturated oil: Reservoir fluid is said to be undersaturated if the initial reservoir pressure is above the *bubble point*, as shown (for Reservoir 1) in Figure 18.18a. Undersaturated fluids become gaseous in part as it is produced in the reservoir and gets to the *test separator* on the surface.

Saturated oil: The fluid is said to be saturated oil if the initial pressure is below the bubble point curve, as shown (for Reservoir 2) in Figure 18.18a.

Critical fluid: When the initial pressure at the reservoir is at or very close to the critical point, it is said to be critical fluid. Such fluid is in equilibrium between the gaseous and the liquid phase.

Wet Gas: A gas is wet if it is gaseous at reservoir conditions but produces some oil as it is produced through the wellbore to the surface, as shown (for Reservoir 3) in Figure 18.18b.

Dry Gas: A dry gas is gaseous in the reservoir and remains in the gaseous state even when it gets to the surface,(as shown (for Reservoir 4) in Figure 18.18b.

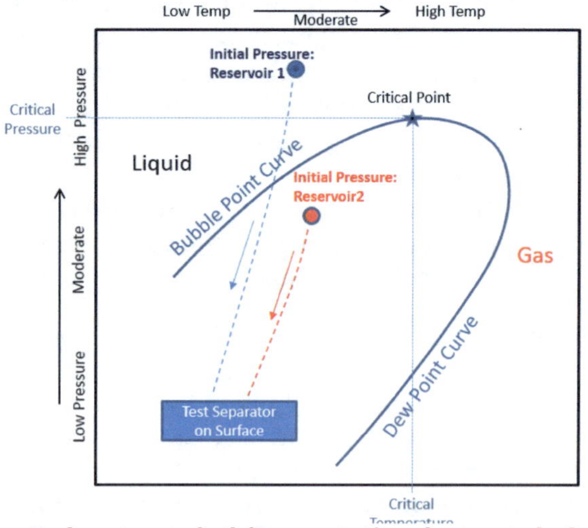

a. Undersaturated oil (Reservoir 1) and saturated oil (Reservoir 2)

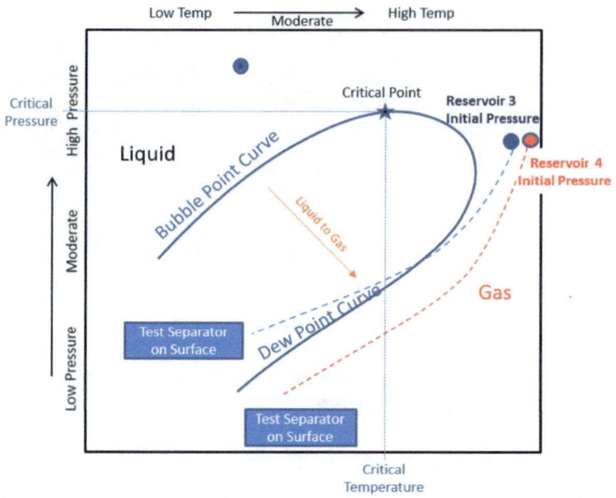

b. Dry (Reservoir 3) and wet Gas (Reservoir 4)

Figure 18.18: Fluid phase diagram showing saturation of oil and gas types

Condensate gas: A condensate gas has its pressure above the bubble point (as in undersaturated oil), but it is on the right of the critical point, such that the critical point is to the left and the reservoir temperature is higher than the critical temperature (as shown for Reservoir 5 in Figure 18.19a).

Other fluid classifications from PVT that may be shown in a phase diagram include **black oil** and **volatile oil**. Volatile oil, as shown (for Reservoir 6) in Figure 18.19b has its critical temperature close to the reservoir pressure. Its critical temperature and pressure is much higher in the phase diagram than that of black oil, as shown (for Reservoir 7) in Figure 18.19b. This means that black oil remains liquid even at a relatively lower pressure whereas volatile oil exsolves gas much quicker with lowering pressure.

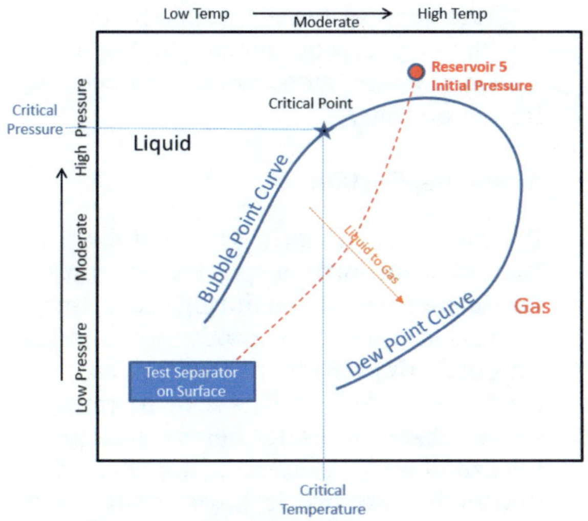

a. Condensate gas (Reservoir 5)

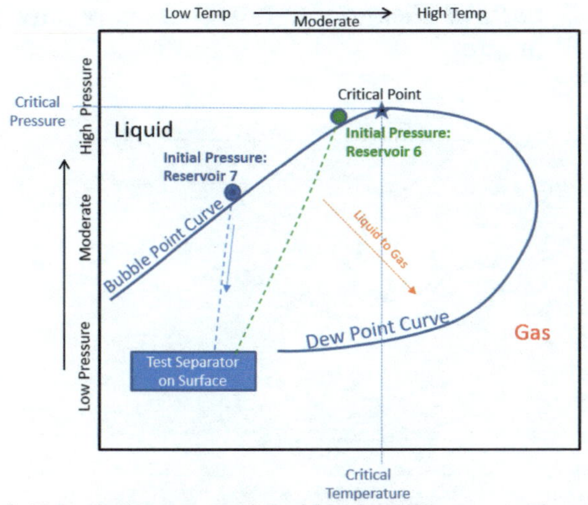

b. Volatile (Reservoir 6) and black oil (Reservoir 7)

Figure 18.19: Phase diagram showing condensate and oil types

18.6.1 Production, Delivery and Proceeds

Oil and gas companies expect to make profits after discovery, when the hydrocarbons are produced and taken to the market. The intermediate process of transportation, refining, storage and marketing is handled by the *downstream* sector of the business. The major oil and gas companies have a downstream unit that handles the processes from distribution to transformation of crude into other petroleum products.

In consideration for the viability of a well development is the oil price at that moment of production versus the cost of the well, or better: *total expenditures* versus *expected income*. It is a 100 million-dollar question. To evaluate the viability of a well or field development project or Field Development Plan (FDP), there is need to list the two sub-factors of investment versus turnover and note the difference to ascertain the proceeds. Some of the factors are listed in the Table 18.1, with the main controlling factor being the first in italics.

These are but a few details to consider, but at the end every oil and gas company does the arithmetic to see if development of the well is worth it.

18.1: Expenditures and return for consideration in a development project

SN	Factors of returns	Factors of expenditures
1	*Oil price*	*Well cost*
3	Barrels to be produced	Depth of well
4	Thickness of the reservoir	Location
5	Porosity and permeability	Number of days to drill
6	Formation and recovery factor	Facilities and their efficiency
7	Net-to-gross volume	Installations
8	Type of hydrocarbon (oil or gas)	Export and import duties

18.6.2 Field and Reservoir Management

The aim of reservoir management is to produce a reservoir at its *plateau*, then through decline until abandonment. Optimising production after the reservoir has started to decline is achieved by targeting unswept hydrocarbons in the interbedding, pinchouts, poor reservoirs and isolated bodies.

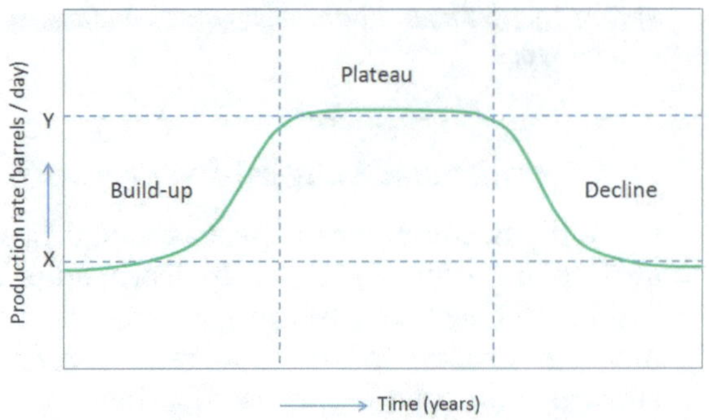

Figure 18.20: Production curve showing build-up, plateau and decline

The **plateau production** ***(Y** in Figure 18.20)* is reached after the initial **build-up** (**X** to **Y**) of production and the stabilisation level where production remain high and relatively constant with time. A field is said to **decline** (**Y** to **X**) when it goes below its plateau production. Overpressured reservoirs would produce without support for some time before decline. Decline may also occur because the oil pool is decreasing and more water coming into the former oil zone.

Managing Decline

Decline curves (Figure 18.21) show a reduction of hydrocarbon production with time. Optimising production during decline may only accelerate the rate of production and lead to eventual **depletion** or **water cut** in the producer well. A well is said to catch water when substantial water begins to be produced instead of oil. A reservoir is depleted when the downhole pressure is lower than the initial virgin condition and, often, such a reservoir no longer produces oil on its own (that is, by ***natural depletion***). It will then require an injector.

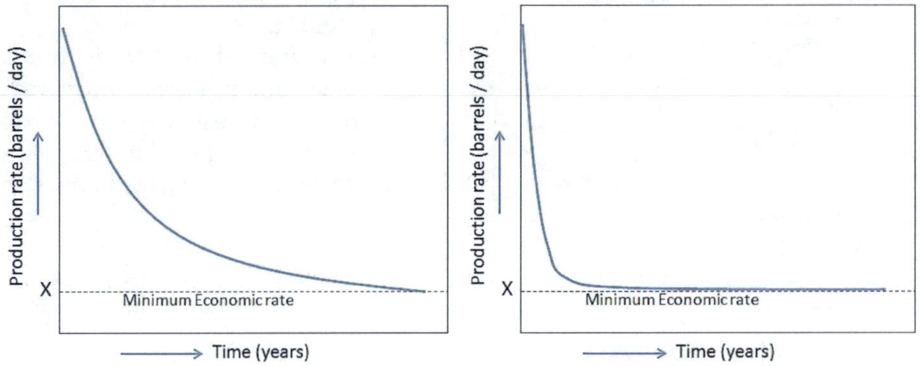

a. Normal decline b. Decline for fracture reservoir

Figure 18.21: General decline curves: normal and fractured reservoir

Infill wells, drilled after an initially approved number of wells during the *Field Development Plan (FDP),* may be producers targeting **unswept zone** or injector helping to sweep unrecovered STOIIP (Stock-Tank Oil Initially in Place) to the producer well. When water reaches the well, often the reservoir itself has been flooded up to the former oil level from a water injector or there is **water rise** from the **aquifer** as the overlying oil is produced.

Decline is usually gradual in normal oil production as shown in Figure 18.21a; however in fractured reservoirs where the oil is quickly produced, pressure drop and decline in oil production occur much faster as shown in Figure 18.21b. With decline, there is a minimum rate of production (**X** in Figure 18.21) at which the cost of production is no longer profitable.

In natural depletion of a well, water production (if this occurs) may decline linearly (dotted blue line in Figure 18.22a) for a long time because most aquifers are not quickly depleted. Underground water is unlimited and water production is only limited by reservoir size and permeability or structural boundaries delineating the targeted compartment. *Oil production* will decline gradually, depending on the amount overlying the aquifer in the trap (green line in Figure 18.22a). *Gas production* is quicker due to higher mobility and it will have a more rapid decline towards the end of the well life (red line in Figure 18.22a).

Due to the reduction in pressure, there may be *gas exsolution* (gas coming out of solution) from oil during rapid decline, leading to increase in gas-oil-ratio (GOR) near the end of production. A sharp rise in GOR during oil production may indicate that expanding gas cap has reached the well placed in the oil rim below it, thereby limiting oil production.

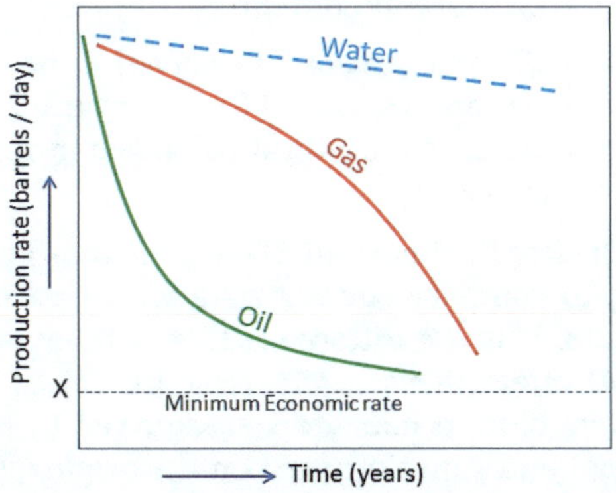

a. Decline curve for oil and gas in natural depletion

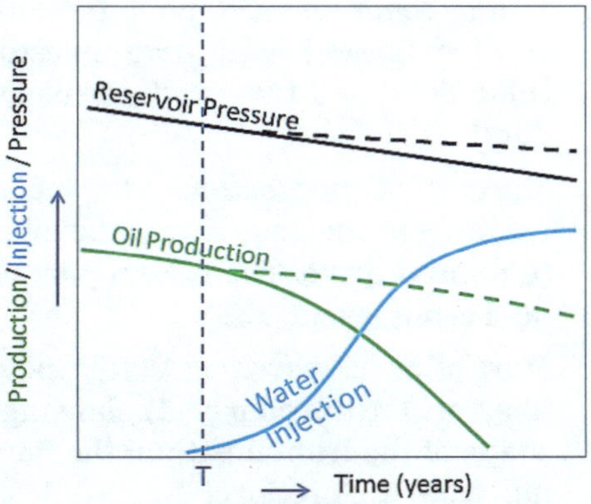

b. Change in reservoir pressure and oil production with water support

Figure 18.22: Decline curves for different reservoir fluids

Decline curves can be applied in reservoir management and help in decision making (history match and production forecast). One way to manage decline due to depressurization is to reduce production; another way is to drill an injector to support the well to boost pressure and sweep more oil. In Figure 18.22b, the beginning of water injection (at time **T**) resulted in higher production (dotted green line) and reservoir pressure was eventually stabilized or at least increased substantially.

In supported wells, water production may keep increasing due to injection but the oil production will reduce. A time comes when there is decline in oil production, despite increasing water injection, because the oil pool is being fully drained. Increasing pressure and water by injection does not always amount to more oil production. At such time the well may be abandoned or there may be more work to revise and upgrade field oil production performance.

In reservoir management, information of performance of a reservoir over the years can be obtained from:

- Production data (volume of fluid already produced over months or years; type of hydrocarbon: oil and/or gas and their produced ratios).
- Produced water (which shows how field's aquifer behaves during production).
- Permeability anisotropy of the oil pool (indicating direction of flow within the reservoir body).
- Fluid and pressure depletion (showing reducing fluid and consequently reduced pressure in reservoir).
- Fourth Dimension (4D) or time-lapse seismic (which shows other concerns of stability, alongside indicating hydrocarbon reserve).

Reservoir Revision and Addition

There are ways to maintain production from a field and adapt mechanisms to prolong the life of the reservoir. The two solutions to improving a declining field are:

1. Reservoir improvement (revision or addition)
2. Cost reduction (human, technical and material)

Adding a reservoir that was not foreseen in the layers of producing levels or revising / identifying more oil than foreseen in the same reservoir can help to maintain and even increase production in a declining field. When the life of a declining field is readapted (often for improvement), it is known as **reservoir revision**, and where the isolated reservoir bodies integrated in this process were not considered part of the initial field, it is called **reservoir addition**.

Cost reduction as a way of improving turnover on oil fields can have a downside if the development and field team are not carried along, or if *cost-cutting* touches on some *critical operations*. Critical operations are those that may have associated risks and downgraded situation due to financial constraints can lead to accidents. The aim should be to maintain the same quality of operations while adapting methods, timing and approach.

Where this is not handled properly, cuts in personnel allowance, welfare and even salary could create a poor working mood leading to failures and shutdown in production. Each minute that passes on a major oil field production installation counts and quickly runs into millions of dollars especially offshore. A single day loss due to failure-related Non-Productive-Time (NPT) could be equivalent to months of savings through meagre cost-reduction actions.

18.7 Reservoir monitoring

18.7.1 4D Monitoring

The industry has come afar, since the advent of 2D to 3D seismic. **4D seismic** refers to 3D seismic measurement of the same zone after a given time; the change over time adds the 4th Dimension. 4D indicates a change in reservoir behaviour due to changing fluid quality and quantity following production. For example, depletion / pressure loss could change the phase of a fluid from oil to gas, which is known as **gas exsolution**. 4D detectable changes are highlighted in seismic in the form of:

- ***Changing horizon:*** Markers appearing deeper or higher (due to changing speed of 2-way travel time arrival of seismic). Changes in speed occurs when the original reservoir property has changed, affecting the speed of waves during 4D acquisition.
- ***Push-up***: This is a particular case of *changing horizon* locally to reservoir, seen as push up in a fairway or channel on seismic section for example. Seismic responses will arrive quicker (since sound travels faster in denser material) in the presence of denser fluid due to water arrival. So emptying reservoir oil while it is being replaced by aquifer water during production will show a *"push-up"* in the seismic marker on a cross-section perpendicular to the sand body.
- ***Higher or low amplitude:*** Amplitude values obtained from seismic processing if changing over time indicates changes in the reservoir content.
- ***Different values of derived data:*** change in computed values from measured or production data indicate changes in reservoir behaviour.

The key to almost all 4D-seismic interpretation is to compare current data with past original data. It is the difference that is interpreted as **4D-effect**. This effect may be colour coded (like blue to indicate water or red to indicate gas) and superimposed on the original seismic as shown in Figure 18.23. As can be seen from the figure:

a. *Virgin condition:* Oil pool (green) above aquifer (green) with two injector wells (blue dots) and two producer wells (green dots)

b. *Start of oil production:* The incursion of water from the aquifer into the oil pool due to natural drive. The crestal part of the oil pool is not swept.

c. *Start of water injection:* Water moves from injector 1 to producer 1, showing a cone shape of the frontal path of the flow due to pull from the producer 1.

d. *Flow signature over time:* Water flow path shows limits of sand body, like a channel fairway limit. Injector 2 shows radial flow, probably not supporting producer 1. Producer 1 is depleted too quickly, leading to gas exsolution (red dots). *A-B / C-D are lines of cross-section shown in Figure 18.24.*

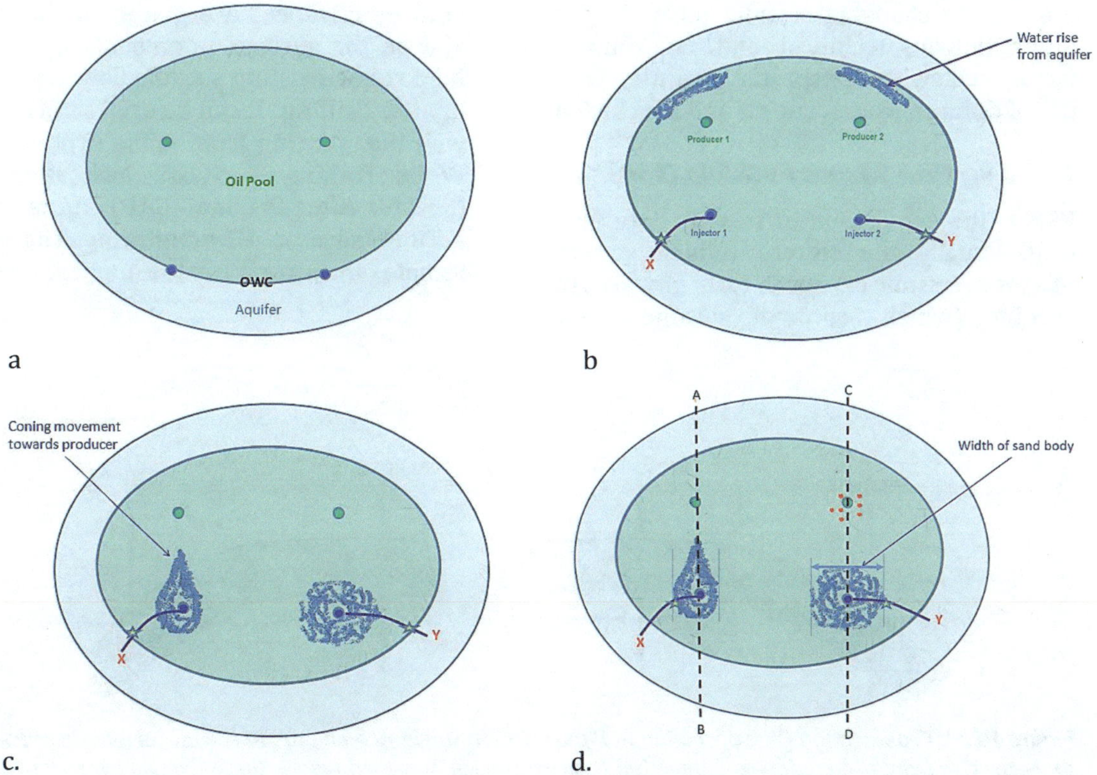

Figure 18. 23: 4D effect showing water rise, coning and gas exolution in anticlinal (domal) reservoirs

In Figure 18.23a, the injector (blue dots) targets the contact between the aquifer and the oil pool. However being a deviated or horizontal well, it runs from the well top (red X and Y in Figure 18.23b) through landing on the contact (green star) to final Total Depth (TD) within the oil pool.

With 4D effect showing the limits of the channel or reservoir fairway in Figure 18.23d, it becomes obvious that the injector only reached the reservoir further in (inner green star) at the channel levees or reservoir limits. 4D seismic monitoring can help to refine the *Architectural Elements (AEs)*, or geometries of the fairway, in narrow reservoirs with side boundaries, typical in a turbidite channel deposits.

These 4D effects can also be observed in sections. The cross-section represented by the line of section in Figure 18.23d is shown in Figure 18.24 a. and b. In Figure 18.23d, the increasing coning and peripheral aquifer rise suggests that water is rising also vertically as production continues, leading to a shallower WOC or OWC (Oil-Water-Contact) compared to the *initial Oil-Water-Contact (iOWC)* before production.

This water rise is not observed in Figure 18.24b (ignoring possible water rise due to natural depletion before injection), where the water is limited to the vicinity of the injection well. The well is not in communication with producer 2, explaining why there is no support and no sweep to the producer, leading to depletion and gas exsolution with increasing effort to produce. If the water rise and water flooding get to the producer, it is said that the well has caught water. A good asset team will slow down injection before then and probably find other ways to optimize the well, like drilling an infill well if there are still remaining unwept hydrocarbons.

A proper 4D interpretation would indicate reservoir behaviours and changing properties due to production activities. Among what may be indicated by 4D seismic are:

- Reducing fluid content
- Changing compaction / decompaction
- Increase or reduction in temperature
- Change in migration pattern / depletion

The challenge with 4D seismic is knowing which response in the later seismic acquisition is induced by production effect and those that are due to changing seismic technology and other human, technical and environmental factors. Likewise interpreter bias may lead to use of different codes, cut-off and conclusions.

18.7.2 Vertical Seismic Profiling (VSP)

VSP is the acquisition of seismic data from a well. Usually the **source** (which generates phases of seismic energy in form of waves) and **receiver** (which records the seismic waves on land or offshore) are placed in the borehole and on the surface, interchangeably. This can help refine existing seismic data while drilling or after drilling. It can be used at any stage of a well life, starting from while drilling (Seismic While Drilling – SWD), just after drilling (Seismic After Drilling - SAD) and later during a Time-lapse or 4D monitoring. The principle remains the same, only the application varies.

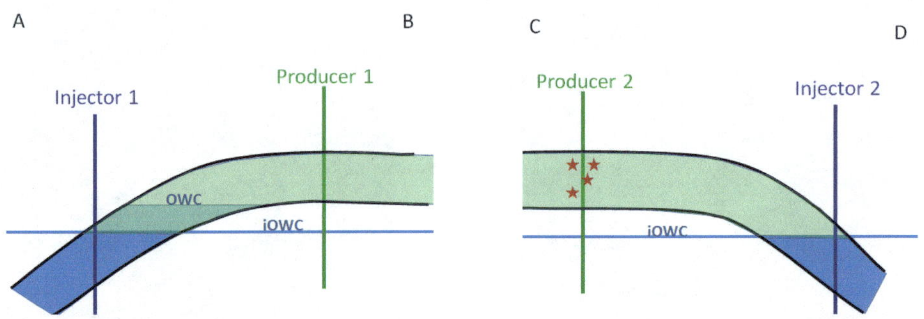

Figure 18.24: Cross section (line of section in Figure 18.23) showing a change in the initial oil-water-contact due to water rise from production and injection (gas exsolution occurred where there was no support) typical in a turbidite deposit with reservoir channels

As the name implies, VSP involves sending signals to move (somewhat) vertically from the surface to a point along the wellbore or vice versa. The source is either **fixed** or **mobile** when on the surface while the receiver is lowered into the hole by a wireline system. Because sources are sometimes directly above receivers, VSP can help to easily determine seismic ***interval transit time*** (the time it takes for a sound wave to go through a body) and the ***two-way time*** (the time it takes for the wave to go through the unit and return to surface or point of measurement).

Among the most common configurations for VSP are:

- ***Zero-offset:*** Source is placed directly above the receiver near the well head position.
- ***Offset VSP:*** The source is placed some distance away from the well head position and receivers are placed at several points along the wellbore.
- ***Walkaway:*** The receiver is at a fixed position along the wellbore, while the source is moved away at different distances from the wellhead position.
- ***Walkabove:*** This is used in a deviated/horizontal well, such that the sources and the receivers are directly above each other.

Figure of fixed and mobile VSP has been shown (Figure 6.18) in Chapter Six on Exploration Geology and Geophysics.

In some cases, drilling noise could serve as a source such that receivers are lined up along the surface away from the wellhead. This provides a form of seismic while drilling (SWD). Just as 4D seismic monitoring, the result from VSP is useful in assessing the continued viability (or otherwise) of a well.

18.7.3 Production Logging

Production logging is a kind of down-hole measurement carried out to measure important parameters and their evolution during production. Unlike wireline, LWD and WFT that attempt to measure **static** (or WFT that measure semi-dynamic/pressure condition in an open hole), well logging for production purposes helps to obtain **dynamic properties** of cased hole when the well is flowing. The aim is to answer questions such as:

- **What is the flow rate?** A production test method known as *Spinner survey* uses a flow meter to record profile of fluid movement in wells, measured by a spinner tool that detects any increase in flow velocity.

- **What are the boundaries** the fluid encounters as it flows within the formation? *Repeat Formation Test (RFT),* a pressure measurement that can also be used as production test, can help to obtain information on flow boundaries.
- **What is the effect of an offset well** on the present well? *Interference test,* a type of well test that carries out measurement between two wells, such as increasing flow in one well and measuring any changes in pressure on the other, can be used to answer this question.
- **Any fluid communication** between injector and producer well? *Markers* and *tracers* that can be injected on one well and detected on another well is used to check this. Oxygen activation logs or radioactive iodine tracer detector logs are used to identify the tracer introduced in one well at the other well, after a flow time to assess communication.

In addition to answering the above questions, *Production logs* may help to assign *oil production* or *water injection* to a particular reservoir interval. It helps to identify production quotas of producing zone and to discover flow problems, including leakages and water flow to producing reservoir. An unwanted situation such as vertical flow from one reservoir interval to another behind casing may also be detected. A **cross-flow** refers to a situation where fluid flows from one reservoir below or above, instead of getting into the well through completion perforations. It may occur due to passage ways more conducive from a higher pressure to the lower pressure reservoir, or in a situation where *skin* is high across the wellbore to formation interface.

Field Rehabilitation and Reactivation

The purpose of monitoring a reservoir is to determine its state, how to improve it, whether it is viable or not. ***Rehabilitation*** of an oilfield is the improvement of the production in a field that is already near **abandonment**. Abandoning a well involves sealing it to a point where it would not flow to the surface again and possible blocking flow into the reservoir (as shown in Figure 18.25).

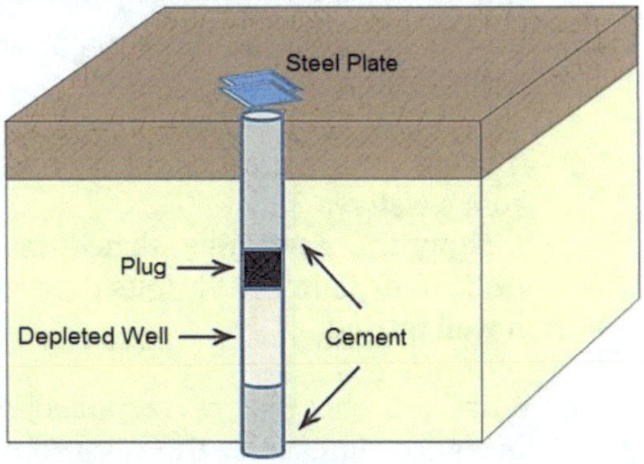

Figure 18.25: Plugged and abandoned well

A cement plug of several meters provides a good seal; cement may be used to fill the well or a plug, a strong parker, is placed above the reservoirs. Then a heavy steel plate placed above the well on the surface or seafloor. It is a standard best practice to also leave the surface vicinity of the well in the condition it was before the development of the well.

In reservoir monitoring, when production and 4D Seismic results are collaborated, it helps in decision related to field and reservoir management and may provide basis for restarting production of an abandoned or near abandoned field. Where the field has already been *abandoned* before production is resumed, the term is known as field ***Reactivation***. A field that has been shut-in or abandoned can be reactivated if new technologies lead to more production and make it viable or if the cost of oil and gas go high and become lucrative even with less production. Smaller companies obtain the field as a **marginal field** and restart exploration and production.

EXERCISES

1. a) Explain the principle of draw-down, build up and stabilization in well testing.
   b. The change in pressure during a pressure test can be divided in three. Use a diagram to describe them.

2. How is the flowability of a well determined? Describe the three major stages of well testing.

3. a) What information could be deduced from a well test?
   b. Show the derivative algorithms and use it to explain the various diagnostics in well testing.

b. What are the factors responsible for formation damage at the reservoir-well interface?

4. Describe the skin concept, and what basic EOR methods are used to reduce skin effect?

5. What is the aim of stimulating a well? How is it achieved?

6. How is 4D seismic used in well monitoring? Give examples.

7. a) What are the factors that determine the viability of well development?
   b) Why is there need to carefully manage a field or reservoir before abandonment?
   c) How can a declining field be improved?
   d. Why are wells abandoned?

8. Describe the two ways in which reservoir monitoring is carried out.

9. What are seismic first arrivals? How does VSP make calculations easy in determining transit and two way time?

10. a) Differentiate between reservoir revision and reservoir addition.
    b. Differentiate between Field rehabilitation and reactivation.
    c. What can make a fields that have been shut-in or abandoned to be reactivated?

REFERENCES & FURTHER READING

Aziz, K. and A. Settari, 1979. Petroleum Reservoir Simulation. London, *Applied Science*, 475p.

Agarwal RG, Al-Hussainy R and H.J.J. Ramey, 1970. An Investigation of Wellbore Storage and Skin Effect in Unsteady Liquid Flow. 1. *Analytical treatment.* SPE J 10, p. 279–290

Barker, J.A., 1988. A Generalized Radial flow Model for Hydraulic Tests in Fractured Rock. *Water Resources.* Res 24:1796–1804

Beauheim R.L., Roberts R.M. and J.D, Avis, 2004. Well Testing In fractured Media: Flow Dimensions and Diagnostic Plots. *J Hydraul Res* 42, p. 69–76

Bourdet D., Ayoub J.A. and Y.M. Pirard, 1989. Use of Pressure Derivative in Well-Test Interpretation. *SPE Reprint Ser* 4, p. 293–302

Bourgeois, M., and P. Couillens, 1994. Use of Well Test Analytical Solutions for Production Prediction. Presented at the *European Petroleum Conference,* London, 25-27 October. SPE-28899-MS.

Buchwald, R.W. and J.H. Thompson, 1973. *Gas Field Developmental Strategy from Exploration Well Test Data.* Dallas: Society of Petroleum Engineers of AIME.

Chow V.T., 1952. On The Determination of Transmissibility and Storage Coefficients from Pumping Test Data. *Trans Am Geophys Union,* 33, p. 397–404

Earlougher, R.C., 1977. *Supplement to Advances in Well Test Analysis.* New York City: Henry L. Doherty Memorial Fund of AIME.

Gatlin, C., 1960. Petroleum Engineering, Drilling and Well Completions. Englewood Clifs, N.J., USA: Prentice-Hall.

Gochnour, J.R. and G.E. Slater, 1977. *Well Test Analysis in Tight Gas Reservoirs.*

Holden, L., Madsen, R. and A., Skorstad, et al. 1995. Use of Well Test Data in Stochastic Reservoir Modelling. Presented at the *SPE Annual Technical Conference and Exhibition,* Dallas, 22-25 October. SPE-30591-MS.

Horne R., 1995. Modern Well Test Analysis, 2nd edn. *Petroway, Palo Alto,* CA

Horner, D.R. 1951. Pressure Build-up in Wells. Proc., 1951. Proc., Third World Petroleum Congress, The Hague, Sec. II, p. 503–523.

Jones, L.G., Blount, E.M., and O.H. Glaze, 1976. Use of Short Term Multiple Rate Flow Tests to Predict Performance of Wells Having Turbulence. Presented at the SPE Annual Fall Technical Conference and Exhibition, New Orleans, Louisiana, 3-6 October 1976. SPE-6133-MS.

Lee, J., J.B. Rollins, and J.P. Spivey, 2003. *Pressure Transient Testing.* Richardson, Texas: Society of Petroleum Engineers. *SPE Textbook Series*, Vol. 9. Vol. 9, p. 1–9.

Lee, J. 1982. Well testing. New York: Society of Petroleum Engineers of AIME. *SPE Textbook Series* Vol. 1, 159p.

Massonnat, G., and E. Wigniolle, 2013. Use of Well Test Interpretations to Constrain the Small Scale Petrophysical Anisotropy (kv/kh). Presented at the *SPE Reservoir Characterization and Simulation Conference and Exhibition*, Abu Dhabi, UAE, 16-18 September. SPE-166043-MS.

Matthews, C.S. and D.G. Russell, 1967. Pressure Buildup and Flow Tests in Wells, 1. Richardson, Texas: Monograph Series, SPE.

Matthews, C.S., Brons, F., and P. Hazelbrook, 1954. A Method for Determination of Average Pressure in a Bounded Reservoir. Trans., AIME, 201, p. 182–191.

Renard, P. & Glenz, D. & M. Mejias, 2009. Understanding Diagnostic Plots for Well-Test Interpretation, *Hydrogeology Journal*, 17, issue 3, p. 589-600.

Spivey, J.P., and J. Lee, 2013. Applied Well Test Interpretation. Richardson Texas: Society of Petroleum Engineers. *SPE Textbook Series,* Vol. 13. 386p

Thomas, O.O., Raghavan, R.S., and T.N. Dixon, 2005. Effect of Scaleup and Aggregation on the Use of Well Tests to Identify Geological Properties. *Reservoir Evaluation & Engineering* 8(03): 248 - 254. SPE-77452-PA.

Towler, B.F. 2002. Fundamental principles of reservoir engineering. Richardson, Tex: Henry L. Doherty Memorial Fund of AIME, *Society of Petroleum Engineers.*

Wei, L., Hadwin, J. and E. Chaput, et al., 1998. Discriminating Fracture Patterns in Fractured Reservoirs by Pressure Transient Tests. Presented at the *SPE Annual Technical Conference and Exhibition*, New Orleans, 27-30 September. SPE-49233-MS.

CONCLUSIONS

This book has hopefully taught you many, if not all, of what you need to start as a good geoscientist. Despite that, it is important to know that it is not just a question of the geoscientist's results, but economic, political and environmental issues that influence decisions on the development of prospects in the petroleum industry. Without limiting the importance of geoscience result, which is in itself paramount as a first priority, a good prospect could be *killed* if it is not discovered at the right time and place, and by the right persons. This is a truth you will do well to remember if you must progress positively with your career in the exploration and production business today.

APPENDIX I

WELLSITE DUTIES AND REPORTING

I.0 Company Representatives

The Oil company geoscience representative on the rig drilling site is known as *Wellsite geologist*. In a quarry field for unconventional hydrocarbons, this person may be known as *Site Geologist*. Colleagues at the site may just call him *geologist*. He works with the Company overall representative known as *Company man* and commonly called *Coman*. The geologist relates with the other main supervisors and reps on the rig site, such as Fluid supervisor, Mud engineer, and the drillers. The geologist at the rig site is charged mainly with the supervision of the Mud logging and the Logging while drilling or wireline units. Every geological input while drilling is referred to him while he reports back to the operations geologist in town or at base.

While many geoscientists in petroleum industry might not work as a wellsite geologist, understanding his duties will help in preparing one's work in the office in view of well operations. It will help staff in the asset management team, the reservoir geologists and engineers and the geophysicists to understand the data coming from the rig. For those who will work directly on the rig, or collaborate with rig personnel from base, the following section will be helpful in equipping them for the tasks ahead.

I.1 Before Going Onsite

Understand the rig site as much as possible, the particular rig and the challenges. There are rig-simulations, models and at least pictures and labels of the rig site on the internet. After reading as much as possible, it might be good to take a virtual tour or rig simulator, attend general training courses, which will save the personnel time on arrival at the oil field platform, before starting specific on-the-job training onboard.

Many oil field workers travel several kilometres, either directly to the rig or to be first being received at the geoscience unit of their employers. Take time to be acquainted with the itinerary.

When you arrive the first time to work in the subsurface or formation evaluation department of your company, get the contacts, especially phone numbers of appropriate useful persons such as the secretary and the operations coordinator. Then revise whatever document you find useful concerning the field you are going to work on. A meeting known as *Well Statement of Requirements (WSOR)* would have been held for any well to be drilled and the minutes are kept. The final WSOR report (FWSOR), also known as Well Plan or *Well Location Report (WLR*), will remain a useful reference while the drilling operation is ongoing. Make sure you have a copy.

The days of hands-on communication, letters, telegram are probably over. Telephone and internet email and video-reporting are normal in these days of videoconferencing. Maintaining a real-time visual contact with the rig, at base by the operations geologists and senior asset or development team, is beneficial, even though it might also have setbacks, some of which can be imagined. As a clue, know that overly supervising and micro-managing, not allowing the site geologists to have a mind and responsibility of their own, may impact on efficiency in the long term.

The issues to sort out before going to the rig include:

a. IT – login and passes
b. Flight tickets – to the rig and return plan to base
c. Allowances – at least initiate the process and conclude it later
d. Timesheet – your pay may depend on it
e. PPE – usually a hard hat, coverall, boots, hand gloves and a goggle.
f. HSSE–Health, Safety, Security and Environment; also be aware of local policies
g. Local contacts – Medics, logistics, the best restaurants in town for your time-off!
h. Make the Secretary or Operations assistant your friend!

You will eventually leave to the site by vehicle or by helicopter, depending on whether you go onshore or offshore. It is time to get ready to put to practice the training on survival (many of which may not come in handy except in an emergency). By now, you know your basic work, and the challenges.

I.2 Getting Used To Rig Site Rotation

For many, the challenges of isolation are many and not worth it. For some, the rig site provides uncountable advantages. Most rig site works are rotation-based, meaning you work for some time on the rig and then you are home on vacation for about an equal time. We already know the many challenges, but are there any advantages of working at the rig? It may depend on whether you are single or in a relationship. It depends on the country you are working in, your personal goals in life, what matters to you most and the quality of the rig and operations you are involved in. Time management becomes key. Sometimes it seems there is more time to spend with family with a regular few weeks of rotation, that is, compared to working every day in office without the chance of a regular time-off. And you have enough money from your rig allowances to spend!

Some people are regular *office staff* and go to the rig as work travel mission, with expense note fully paid. There are all forms of allowances added to your regular pay when you work on rig or offshore. In some countries or companies, you have Room-sharing allowance, Out-of-station allowance, Simultaneous operations allowance, and more, in addition to the regular Shift and Daily allowance. While it lasts, try to enjoy your stay on site and remember to keep in touch with home and make the best of your off-times. While fully a *rotational staff*, you do not come home every evening stressed with office work pressure. In the rig there are "heat" and "cold" moments when you are busy and when you have some time to cool off. And then when you are back home from rotation, depending on how well you plan with your family, you can always have quality vacation time together. Finally, only your own personal experience will tell; for many it is a relatively good experience that should not last too long.

I.3 Rig Site Reporting Responsibilities

As soon as you arrive at the rig site, start working on your reports immediately. Extract the important information and images from the Well Location Report or Drilling Program or Final WSOR to form a working document. The document will serve as a reference as the reports are updated from prognosis in the document to the actual results. Update the *Final Well Report (FWR)* or *Geological End of Well Report (GEWR)* each day as you generate the data in your *Daily Geological Report (DGR).*

I.3.1 Daily Reports

The daily reports, to be sent to town by the Wellsite Geologist (WSG), include:

1. **DGR:** a daily summary of geologic activity and results
2. **Master log:** shows a daily plot of the lithologies drilled among other information
3. **Mud logging reports:** provided by the Mud logging team on their tasks
4. **LWD/MWD reports:** log plot and LAS files of the log data provided by that team.
5. **Well Trajectory:** plots of the well path using the survey data

The WSG carries out a daily time analyses of drilling activity and follows up drilling activities, which he readily provides when called upon by the Operations Geologist (Ops Geol) through phone and emailing, and where possible video conferencing.

I.3.2 Weekly Reports

Weekly reports include:

1. Gas calibrations
2. Weekly safety briefs
3. Weekly reviews

These depend on the operation going on at the rig at the time.

I.3.3 End of Shift Reports

1. Handover: a short report prepared for the incoming replacement, to the "back-to-back" personnel coming while his counterpart on the rig is leaving. This will include the rig or operations history, current status, planned operations and contractors' information.
2. Backup data: backed-up information on the network or other devices, including folders, files and graphics.

Often, end of shift will involve logistics, room change, transport of samples, physical movements, and more, with manifests and related documents or reports to be archived.

I.3.4 End of Well Report

The *FWR* or *GEWR* is the key document provided after a well has been drilled. It is the actual version and response to the *Well Location report (WLR)*. Don't wait till the last day to update it. Add information to it daily and have a relatively final document at the end of the drilling. This may be forwarded to town where it will be revised for signing before it is sent to the authorities and archived.

A FWR may include:

- Well general information
- Drilled operation for each section
- Geology and petroleum results
- Pressure results verses prognoses
- Acquisitions result (Petrophysical logs, Gas, Core)
- Confirmed stratigraphy
- Well testing or abandonment

The FWR remains the main reference document for a well post-drilling.

I.3.4 WSG Morning Checklist

Perhaps the most challenging of WSG duties is the ability to be multitasked, to obtain quality check and report multiple data in a timely manner. Below is a checklist highlighting the data, processes of obtaining them and the procedures for reporting. A good practice is to quickly scan through the checklist each morning to ensure it is up to date. Then upload relevant files to a shared network or email them to base on request.

1. *Cutting Descriptions*

Go to the mud logging unit and describe the samples. A lithology description form is usually available at the unit, which should include:

- Sample depth
- Lithology percentages
- Lithology description and comments
- Calcimetry and acid tests
- Gas data and fluorescence test

2. *Save Real Time and Memory Files*

These may include:

a. *Mud logging files such as:*
    - Gas data file
    - Mud log plotted data in .pdf
    - Pipe connection
    - Calcimetry and fluorescence data
    - Mud logging Report
    - Drilling parameters

b. *LWD/MWD such as:*
    - Survey data
    - MWD morning report
    - GR and Resistivity LAS files
    - PDFs of logs

c. *Daily Drilling Report* from Company Man or Drilling Supervisor. Details may be discussed in the morning meeting / phone conference.

d. *Daily Fluid Report* from Fluid Supervisor or Mud Engineer. This will contain mud data. Example is Water-Based Mud (WBM) or Oil-Based Mud (OBM) / Non-Aqueous-Based Mud (NABM).

3. *Update Lithology and Other Logs (e.g. Composite Log)*

In our days this is done using software, such as HRH Gravitas®.

- Use the resistivity and gamma ray logs to update the interpretation of lithology from mud samples, and to refine cutting descriptions in the litholog.
- Print and convert the logs to portable digital file versions and save them in the network folder afterwards.

4. *Update Daily Geological Report*

You will need to update:

- *Operations summary*, from the company man's daily drilling report.
- *Lithology*, summarized from the litholog descriptions.
- *Gas data and their peaks*, compare the values with that of the mud logging team.
- *Average ROP*, estimate this by totalling ROPs over total time in *m / hr (or min/m).*
- *Formation tops,* confirmed from correlation and comparison with prognosed stratigraphic chart or offset well logs.

5. *Update Progress Chart (PC) or Prognoses Versus Actual (PVA) Chart*

An excel type sheet or software is used to update the daily progress of drilling where it is plotted graphically. Data is added each day to show length drilled, which extend the *actual* curve in the *Prognoses Versus Actual* plot. The PVA plot already should have the *prognosis* curve plotted on it. The information in the PVA may include:

- Prognosed versus actual depth.
- Lithology update.
- Prognosed mud weight versus actual mud weight.

6. Upload to Shared Network Base (e.g. PetroVault™ - Petrolink)

Ensure files are properly named, adding numbers or dates in the same formats for easy sorting. While at the rig site, avoid working directly on network folders as they could be slow; rather work on your local folder and then backup regularly. At the end, files are sent 'as is', or converted to ***.pdf***, ***.wrf***, or other display or web transfer friendly file format, then saved in a shared base where everyone involved could access it. Often this base is secure and access passwords provided to involved personnel.

From the above, it is seen that the WSG or site geologist is fundamental in terms of collecting, verifying and transmitting raw well data to base. That is the essential of his duties.

APPENDIX II

HYPOTHETICAL WELL

II.0 Defining the Hypothetical Well Project

Consider a hypothetical oil well in which there are many challenges and the geoscientist has to make input on geosciences-related issues. He or she needs to understand the working of a well drilling job to be able to achieve this. To define and develop a given well of interest, some pertinent questions to be asked are:

1. Are there any operational or environmental hazards?
2. How do we ensure good contractors and rig personnel relationship?
3. What are the well objectives?
4. What depths are the hydrocarbon targets?
5. What is the geology of the expected formations (lithology and pressures)?
6. What data would be acquired?
7. What is the drilling program?

The answers to these questions would have been sought long before the well is drilled. At the point of drilling, you will have certain documents in which the answers have been suggested.

*Let us name the hypothetical well as : **ABT 1X** (The suffix "X" means it is not yet started, once spudded (beginning to drill) it can be called just **ABT 1**.*

*We shall follow **ABT 1X** and look at each of these questions, basing the decisions on what has been seen in the chapters, sections and figures in the main text, in order to practically apply the lessons learnt.*

II.1 Are there any Operational or Environmental Hazards?

Risk and Hazards

The first objective when drilling a well is safety; finding a positive result is the second objective. The involvement of human life, regulatory and business constraints and the busy working conditions make it important to ensure that none of the various safety issues are ignored. In addition to operational hazards (well safety issues) are environmental hazards (harm indirectly to human or directly to non-human surroundings). These challenges must be considered before drilling the well. What are those hazards that could be directly related to the **ABT 1** operation?

Operational Hazards

a. ***Mud gains:*** if we drill underbalanced, pressured formation fluid may flow into the well, leading to a *kick*, which if not controlled will lead to a *blow-out*.
b. ***Mud losses:*** if we drill overbalanced, pressured drilling fluid in the well will flow into the formation. This may cause eruptions elsewhere or low well pressure when the fluid is lost. This would result in a backflow, fluid entering back into the well, leading to a kick and blow-out as well.
c. ***Hole instability:*** we may have large formation rock *packing-off* and falling into the well, or fragments of rocks *caving-in*, blocking the well and making logging and other operations problematic
d. ***Kick / blowout***: kick leading to a blow-out is the most dangerous situation; in case of fire it can compromise the whole rig installation and be fatal to personnel. However chances of blowout are low for the **ABT 1** because predicted pressures are not too high.
e. ***Well collision***: because there are other wells in the vicinity of **ABT 1**, and even within the cluster, we have to be careful with steering the well horizontally to avoid collision. *Deviation* data (*inclination*, *azimuth*, at each *depth*) is monitored.
f. ***Temperature and pressure:*** Too high temperature and pressure at a great depth may be problematic for the tools, but we are not very concerned here because we are not going too deep.
g. ***Poor hole cleaning***: This can lead to left over debris in the well and *tight spots* while RIH or POOH.
h. ***Bit balling:*** The head of the drill string may become moulded by constantly rubbing on sediments to form a ball-like head that no longer drills deeper despite weight on bit (WOB) and rotation per minute (RPM).

i. ***Differential sticking***: where parts of the well are free and the other parts sticking to the drill string, like in shaly units with swelling clay, we may lose out bit in-hole.
j. ***Faulting:*** faults could act as conduit if not sealing, allowing the fault zones to contain several materials such as mixed breccia, fluid and unpredictable contents in the subsurface. Drilling past them may cause few challenges to the operation, including possible losses or gains and sticking.

Environmental Hazards

Apart from the operational hazards, there may be risk factors for personnel on the rig, for the environment and the public afar. In the case of fluid coming to the surface, for example, there may be oil spill into the open sea and impact on aquatic life. Among environmental hazards that may be consequential to drilling **ABT 1** are:

a. Exposure to harmful substances such as chemicals.
b. Injury and damage while lifting material or carrying rig equipment.
c. *Shallow gas*, which may arrive to the surface before the installation of the BOP.
d. Harmful gases such as H_2S and even CH_4, which may also be flammable, could emanate from the well. Although chances of H_2S are minimal in **ABT 1**, due to non-sulphuric hydrocarbon presence in this basin.
e. Radioactive sources may be radiated during nuclear-based logging operation.
f. Hydrates, which occur near the seafloor could be an issue offshore, but was not the case in **ABT 1** well.

People must not be exposed to any of these risks; as such measures are taken to ensure they are prevented. Most national authorities provide rules and regulation against these hazards. Personnel are trained to avoid them. Keeping to the rules and ensuring strict Health, Safety and Environment (HSE) measures are absolutely important. Practical preventive solutions, which will ensure effectiveness, in the effort to curb most operational and environmental hazards include:

1. *Proper communication:* across the various units; language barriers should not be ignored.
2. *Adherence to procedures*: which should be subject to regular review and revision.
3. *Management of Change (MoC):* crew, operation and personnel change should be preceded by a handover, and there should be no flimsy change during delicate operations.
4. *Pre-phase and pre-job meeting:* should be done before every major stage of the operation, with participation of all those involved.
5. *Daily safety meeting:* HSE meeting and regular drills should be adapted to avoid and manage possible accidents.
6. *Overall Supervision:* of all operations and processes, such as monitoring *tripping speed* during RIH and POOH procedures.

II.2 Company Representatives, Contractors and Rig Personal Relationships

This is probably the most important factor in successfully drilling **ABT 1**. We are working with people from all walks of life, and no matter how well prepared, good drilling equipment and good reservoir prospects we have, it is humans who will carry out the operations. Working in a busy environment, with inherent challenges, demands that we work together. We need to adopt the "go-and-see-often (GASO)" policy. Some of the units in which we will find personnel to relate with are:

- Mud logging
- Directional drilling
- Mud Engineering
- Fluid Engineering
- Waste treatment
- Drilling crew
- Cattery and laundry

All personnel, especially the company representatives, should always care to know if there are any issues. Make sure to not ignore the basics, such as the *eating mess*, the *gym*, the *laundry* and the people that work there. How you relate with these people could determine your career progression and satisfaction. There is the human side. If you are on a rig, if you are the company representative, or the contractor, or one of the personnel, try to go by and see others (of course when they are not busy) and say hello and ask how the day is going. That alone could save a life!

II.3 What are the Well Objectives?

Wells drilled only for exploration purposes may be abandoned after *plugging*, which is after injection of cement to seal the well at bottom. That is, irrespective of the result. *Exploration wells* may be drilled with smaller drill bits, to create smaller holes, and as such not adapted for production of a major quantity of hydrocarbons. After appraisal of an oil field, wells are drilled for developing of oil, gas or other types of hydrocarbons (condensate, heavy oil, etc.) in an established oil field. These are either injector or production wells. Wells are drilled as *injection wells* when it is meant for injection of water or gas into it to produce other wells by. Those other wells from where the hydrocarbons are produced are known as *production wells*, drilled in fields in an oil company's asset.

So the objective of the majority of wells could be:

- Exploration
- Development
  - *-Injection*
  - *-Production*

The objective of ABT1X is ***Gas Production****.*

II.4 What Depths are the Hydrocarbon Targets?

The hydrocarbon level in a well is the Reservoir levels. Example is the sands or sandstone lithology with impregnated hydrocarbons. They are often named with suffixes as I, II, III, or as prefix such as A, B, C, or numbers 100, 200, 300 in line with known markers or seismic levels. Reservoirs occur in-between *seismic levels* or major level known as Markers (which could be named as say H100, H200, H300). The reservoir between H1200 and H1300 may be called R1200 (with top as H1200 and base H1300). The advantage of going in hundreds is that you could break it up to say R1250 (reservoir 1250) if an interval is identified between existing ones later.

ABT 1X has multiple reservoirs which are represented with roman numerals, with depths as below:

Reservoir		**Depth of Reservoir Tops**	**Thickness**
Reservoir I	:	*2950 mD / XXXX m TVDSS*	....
Reservoir II	:	*3155 mD / XXXX m TVDSS*	....
Reservoir III	:	*3450 mD / XXXX m TVDSS*	....

The thickness is obtained by subtracting the top depth from the bottom depth or base

Imagine that there is a consistent seal of 10 m shale between these reservoirs; could you fill in each of the reservoir thickness above? The actual vertical thickness filled with hydrocarbon, given in True Vertical Depth, TVD, is the *hydrocarbon height*.

We use XXXX m TVDSS to represent the true vertical depth referenced to the sea level (where the XXXX are actual numbers). A deviated well has a different Measured Depth (MD) compared to the True Vertical Depth (TVD) because the TVD is measured vertically and referenced to a point on or near the surface (example: Ground Level or Sea Level) whereas the MD is longer (depending on how deviated from vertical a well is). A depth referenced to (that is measured from) the mean sea level is represented as TVDSS, as in the above example. TVD/RT refers to depth measured from the rotary table.

II.5 What is the Geology of the Expected Formation?

ABT 1X well when drilled will encounter few geological formations before getting to the targets and probably after. Wells are drilled a little further after reaching the target (the targets are obviously reservoirs, like sandstone). This is to find a suitable non-reservoir depth (like shale) to land the casing, and provide enough room so the drill bit could pass slightly and allow logging of the reservoir behind it. The tools attached to the drill string are usually a few meters away from the bit, a little *offset* from the bit. Offsets or distances to the bit are usually recorded to help decide the additional depth, known as the "Rat hole", to be drilled from reservoir.

The prediction of these characteristics of these formations determines the drilling program and the Wireline or Logging while drilling acquisition to be made while drilling the well. The good prediction of the formation pressures is also necessary for well safety. A table, as below (Table II.1), will be given in the Well Plan or Well Location Report (WLR) or Final Statement of Requirements (FSOR), used to finalise the drilling program.

The Equivalent Mud Weight (EMW) is the density a column of mud running from the surface will require to exert the pore pressure at that depth. In the labelled row of Table I.1, 262 bar represents that pressure in the reservoir, whose EMW is 1.2s.g. The well must be drilled with actual mud column slightly higher than this, to avoid a kick. To evaluate each of the formations or units encountered while drilling **ABT 1X**, and compare how the prognosis matches with the actual, the geologist will have to interpret among others:

- Drill cuttings, which are samples cut from formation while drilling and collected from the shale shakers.
- Lithological descriptions, produced by describing the cuttings sample and comparing it with logging data and responses from drilling parameters.
- Mud logging lab results, such as calcimetry, density measurement and fluorescence tests.
- Actual pore pressure, from Well Formation Tester results obtained by logging.
- Drilling parameters, such as drilling breaks (change in Rate of Penetration – ROP) assuming weight on bit (WOB) is constant.

Other information would come from the company's third parties or contractors (examples are actual mud weight used, changes resulting from losses and gains while drilling, and more).

In **ABT 1X**, we shall use real time resistivity (at bit) evolution to attempt to monitor real time pressure. This is because in brine filled pores, there will be a reduction in resistivity if there are more fluids trapped inside, as in a compacted formation with trapped overpressure (due to undercompaction, where the salty water was not expelled during rapid sedimentation). Such reduction with be gradual with depth, not sudden. Also resistivity increases considerably with influx of gas, and **ABT 1X** is a gas well with risk of possible gas kick (that is, major influx of gas into the well if we drill in an *underbalance* condition, with lower mud weight compared to formation pressure). A relative increase in reservoir formation is expected, but a sudden increase in the shale section may indicate danger.

II.6 What Data Would be Acquired?

Sometimes data acquisition, especially petrophysical data, is the main purpose of an exploration and even development wells. Such data help to gain insight before planning further exploration projects or to refine existing development and improve field results.

Different types of data are acquired from a well, and these include:

- Depth (TVD, TVDSS, TVD RT)
- Petrophysical Logs (e.g. gamma ray, resistivity and neutron-density)
- Fluid (Water, Oil, Gas)
- Temperature & Pressure (WFT pressures, borehole temperature)
- Stress (formation integrity tests, leak off tests)

Table II.1. Geology of the formations encountered by a hypothetical well

Depth (m) TVD/MSL	Geological Age / Epoch	Formation	Dominant Lithology	Pore pressure	EMW s.g.
0 - 650	Pleistocene	Formation A	Silt		
650 - 1910	Pliocene	Formation B	Shale		
1910 - 2150	Miocene	Formation C	Sandstone	262 bar	1.20
2950 - 3155	Oligocene	Formation D	Shale		
3155 - 3450	Eocene	Formation E	Sandstone		

Table II.2: Data acquisition program for ***ABT 1X***

Depth (m) TVD/MSL	Drilling Section/Csg	Petrophysical Logs	Fluid Sampling	Pressure	Stress
0 - 650	26 inches CP	Shallow seismic check shot	None	NCT plot	NCT equiv.
650- 1910	23 inches	Resistivity, Gamma Ray, Neutron-Density	Gas While Drilling	NCT plot	NCT equiv.
1910– 2150	16 inches	Resistivity, Gamma Ray, Neutron-Density	Gas While Drilling	NCT plot	FIT
2950 - 3155	12 ¼ inches	Resistivity, Gamma Ray, Neutron-Density and Cement Bond Log	Gas While Drilling	WFT pressure test	LOT
3155 - 3450	8 ½ inches	Resistivity, Gamma Ray, Neutron-Density, Casing Collar Locator, and Sonic	WFT and fluid sampling	WFT pressure test	None

Csg = Casing. CP = Conductor Pipe. NCT = Normal compaction trend. FIT = Formation Integrity Test. LOT = Leak Off Test

Depth is probably the most important data, upon which other data are based. Petrophysical logs measure the sound (sonic log), the speed of waves (VSP log), the resistivity or conductivity (resistivity log), the radioactivity (gamma ray log), the presence of porosity (neutron porosity), and more.

Porosity is detected by neutron log because most of the pore spaces are filled with hydrogen-related fluid.

Table II.2 shows the data acquisition program for **ABT 1X.**

The conductor pipe section is the upper part of a well, dug by pushing the special casing (or conductor pipe) down the unlithified sediments down to a refusal depth or until it meets resistance. At that point, a drill bit is introduced into the CP to start drilling.

Why did we choose these logs for **ABT 1X**?

- GR for defining the top of a formation, especially as our formation is switching from sandy to shaly units.
- GR-Resistivity combination is used to define top markers and fluid contacts. Resistivity increase could also indicate a sudden gas influx, and kick. This is especially important as this is a gas well.
- Sonic is used to derive and evaluate porosity, as sound travels faster in less porous or denser media.
- Cement Bond Log and Casing Collar Locator (CBL-CCL) scanner in cased hole will locate casing collar and cement-casing bonding.
- Neutron-density is used for reservoir evaluation, also for contacts and identifying fluid type and their extension, and for further petrophysical studies.

The above logs will be used to derive other important parameters after drilling, such as Vclay and to some extent permeability, the latter of which could also be determined at field scale after well testing.

II.7 What is the Drilling Program?

The drilling program is the main document, with other sub-programs such as casing, cementing and mud programs, used to drill the well. It is the most comprehensive document for the drilling operation. It contains every information about well management, HSE, well type, details for drilling each section, BOP tests, and well completion. It is similar to the equivalent document, Well Location Report or Final Well Statement of Requirements, used by the geologist on site to follow the well.

Drilling involves essentially digging a hole, placing a metallic casing within and around it and cementing for stability. Drilling operations are often described with driller's jargons. Do not try to ensure that the jargons are in their right tenses, singular or plural, or referring to the past, present or future. Most of them will refer to present in reports, but understood to say it the way it was while operation was or will be ongoing. Most of them are abbreviated in reports, and may vary slightly with an oil well operator or drilling company, but the general usage remains the same. Below are the terms to be familiar with in an operational drilling procedure. We shall mention them as we look at the critical stages in drilling **ABT 1X**.

A Pre-job safety meeting (PJSM) is held before every major operation. During the PJSM, the drilling operation is discussed and planned for efficient and safe execution. For **ABT 1**, there will be a daily morning meeting by 07h00 and evening by 19h00 to briefly review:

- Safety
- Logistics
- Operations
- Feedback from each unit

As drilling of the hypothetical well starts, personnel took the major step to drill into the ground, known as **Spud (spudding).** To do this, there is a **Rig up (R/U),** meaning driving to install the drilling equipment ready to drill. The pipes and tools used for drilling and attached to the **top drive** (drilling equipment that moves up and down pulling or pushing the tools with it) form an overall known as **Bottom Hole Assembly (BHA).** The BHA often rotates in a fluid that aids drilling, known as either **Water-Based Mud (WBM)** or **Non-Aqueous-Based Mud (NABM)/Oil-Based Mud (OBM),** depending on whether the fluid is more of oil or only water-based.

During drilling, the equipment placed above the wellhead, to prevent sudden gush out of overpressured fluid, is known as **Blow Out Preventer (BOP)**. After a normal drilling or other down-hole process, the BHA is **Pull Up (P/U)**. To maintain a smooth friction free descent, there is a **Wash Down (W/D)** involving the injection of fluid while going down with the drilling tool, that is, while **Run In Hole (RIH)**. When not in use, the tools are **Lay Down (L/D)**. If there is need to use them for the next stage, then they are **Make Up (M/U)**, that is prepared and taken up again. Eventually, the aim of drillers is to drill to the intended final depth, known as final **Total Depth (TD),** without having any tool **Lost In Hole (LIH)**. After drilling each section, they Pull **Out of Hole (POOH)**. Once drilling is over, the drilling equipment will again be uninstalled, known as **Rig down (R/D)**.

*The drilling program used to drill **ABT 1** is made available for every key person at the rig as a reference document.* It contains:

a. Casing sub-program
b. Cementing sub-program
c. Mud sub-program
d. Completion sub-program

a. Casing sub-Program

Because longer holes can easily loose stability or collapse, there is a careful selection and decision on the length of each section depth and casing pipes, and the depths to which they will be cemented. At great depths of wells, where pressures could be very high, it is not enough to have a steel casing to ensure it will be strong enough to withstand the in-situ conditions in the subsurface over the duration of the well after drilling.

Casings are tested for both physical stress and chemical reactions. The sub-program provides the characteristics of the casing and their adaptability to the current well.

Among the factors considered before choosing a casing for **ABT 1** are:

- *Tension:* ability to withstand own weight if suspended
- *Burst:* ability to not be burst open, even if filled with liquid oil, gas or water at the highest expected pressure.
- *Collapse:* ability not to be broken if falling from a height equivalent to the maximum it may be lifted to during the well operation.

In high pressure and high temperature condition, special or high resistant casings are used that meet the above conditions.
The casing sub-program outlines the different casing lengths, and how they will be joined to reach the length of each well section. Where a *casing liner* (one that does not get to the surface) instead of normal casing is used, its characteristics are also given. The most important of the physical properties and casing operation details are:

- Casing diameter (often given in inches)
- Casing lengths

- Casing shoe depth and thickness
- Casing characteristics such as tension, burst and collapse
- Casing capacities

The casing sub-program covers the tests that might be carried out to ensure the casing withstands the expected pressure. *In **ABT 1**, it was ensured that the Maximum Expected Wellhead Pressure (MEWHP) and Well Full of Gas (WFG) tests were carried out.*

b. Cementing sub-Program

Cementing sub-program gives the types of cement that will be used, and how it will be mixed (to form the ***slurry***) with different densities or specific gravities; these are used as either ***lead Slurry*** (injected first while cementing) or ***tail Slurry*** (injected last while cementing). Since the well is filled with drilling fluid/mud before cementing, a substance known as ***spacer*** is placed between the cement and mud so that they are immiscible, before pumping the cement downward and up behind the casing.

The cementing sup-program gives information on the details of the cement preparation, like *gas tight slurry*, the possible *top-up jobs* or *secondary cementing*, the *plugging* and *abandoning* cementing processes. The specifications of the cement stinger (cementing tool) to be used are also given in the sub-program.

*In our **ABT 1** well, the casing will be cemented up to the surface with lead slurry of 1.50 sg and tail slurry of 1.80 sg up to 50 m above shoe. The spacers are 20 m³ Vis Pill and will be pumped as spacer slurries:*

- *Lead slurry: 130 m^3 slurry from depth AAA m to BBB m (Ground level/surface).*
- *Tail slurry: 40 m^3 slurry from XXXX m to YYYY m (50 m above XX" casing shoe.)*

A biocide and H_2S scavenger are added in the displacement fluid to prevent microbial activity. The cement will be displaced 4 m below float collar.

At completion of the cement operation, the pressure in the stinger is ***bleed-off*** and there is a check for backflow (to avoid some cement remaining in the system).

c. Mud sub-Program

The mud or drilling fluid properties and the mud weights to use at each section is detailed in the Mud Sub-program. It will include procedures from *mud transfers* into *pits* and the list of checks to make during drilling. While drilling, the mud engineer and fluid supervisor provides an *on the fly* short note giving the history of actual mud properties in use and a description of those to be used at the next stage. The note covers the ongoing operation, with updates of possible changes made based on feedback from the well.

The full mud sub-program provided the objectives of using the drilling fluid in the well, which include:

- Hole cleaning
- Hole cooling
- Hole gauging
- Controlling lost circulation
- Controlling shallow hazard
- Controlling kick

The mud type is given, with the agents used to treat it.

*In **ABT 1**, the mud types and agents used are:*

- Brine/seawater (available in abundance while drilling offshore or near shore. In ABT 1, it was used for the *top hole*, i.e. while *jetting* the uppermost *conductor pipe*)
- HiVis sweeps (high viscosity mud to clean hole while drilling)
- PAD Mud (special mud to reduce fluid loss, with adapted mud weight and pH)
- LCM (*Lost Circulation Materials introduced in mud while drilling to control losses)*
- NABM (*Non-Aqueous-Based Mud)* or OBM *(Oil-Based Mud)*

In addition to the above, the mud sub-program includes:

- Viscosities of the cleaning agents, such as *high viscosity (hi-vis)* sweeps
- Other mixing agents such as biopolymer, gels, caustic soda, calcium carbonate, sodium chloride and graphite
- The expected pH of the mixture.
- Cost and logistics of putting the mud in place

It will provide the ***kill mud*** formulation, which will be used to control the well in case of a kick.

In ABT 1, there was also a contingency plan, in case operations go differently from the original plan. This includes rescue in case of accident and emergency.

d. Completion sub-Program

After drilling, there is a completion operation which may involve:

- Well perforation for production: the casing or liner is perforated and used as the flow vents for producing the hydrocarbons from the reservoir.
- Well test: the well is allowed to flow (if overpressured) or made to flow (by reducing borehole pressure) and the flow rate measured to obtain the natural characteristics of the reservoir.
- Well plugging and abandonment: where the well is not meant for production, cement is injected into it and abandoned. Sometime temporarily, or permanently if it has already served its purpose.

Often, the well site geologist's involvement is reduced at this time but the reservoir geologist may be present to take part in the tests.

***ABT 1** is drilled for production purpose. The well will be tested after drilling. We shall consider plugging and abandonment if the well is not positive.*

The hypothetical well will follow a trajectory that can be represented in the lattice shown in Figure II.1.

Expand and fill the table of summary (Table II.3) below, tentatively as you plan and execute the well.

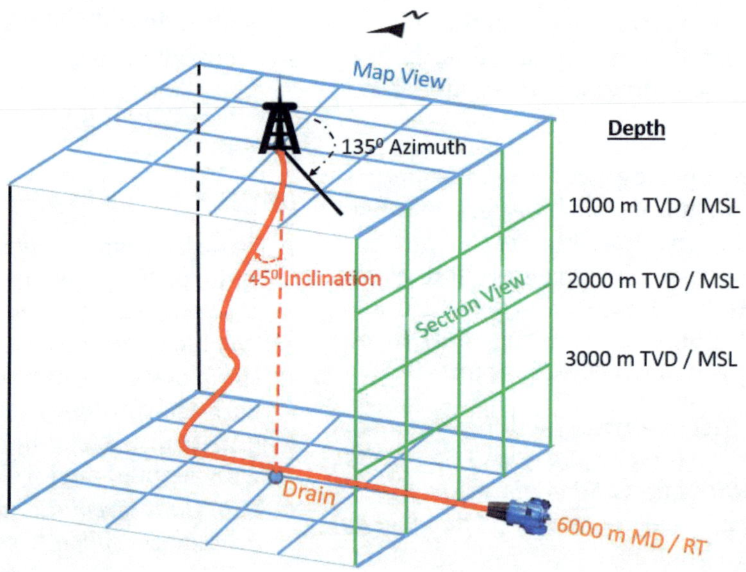

Figure II.1 Drilling Lattice

The physical model of this lattice was first constructed by the author to ease illustration during the 2020 iESog Annual Geoscience Training. During the demo, trainees tie a flexible cable representing a well path, to follow the trajectory from the wellhead to the drain section, plotting accurate azimuths and inclinations at each lattice junction.

SUMMARY: HYPOTHETICAL WELL PLANNING AND EXECUTION

WELL OBJECTIVE: Exploration - Appraisal - Development - Production?

HSE CONCERNS : Operational – Environmental Hazards?

Table II.3: Program for the execution of a hypothetical well

DEPTH / REF AGE / FORMATION	LITHOLOGY	LOGS / DRILLING PARAMETERS	PORE PRESSURE PREDICTION	WELL PLAN	TRAJECTORY	*CASING*	*CEMENTING*	*MUD*	FLUORESCENCE TEST	CUTTING DESCRIPTION	PETROLEUM RESULT	*COMPLETION*	COMMENTS
	SILT												
	SHALE												
	SANDSTONE												
	SHALE												
	SANDSTONE												

APPENDIX III

CONTRACTS, AUDITS AND LOGISTICS

III.0 Oil Licence and Mining Contracts

Most geoscientists working in the industry will come across a contract, or be involved in its implementation, along the line in their career. When not directly involved, the geoscientist is yet executing various technical works daily and using technical documents that enact the contractual agreements. The geoscience personnel help to achieve one contractual clause at every stage of their work. The geo-scientists working as a company representative on the rig or relating with service companies need to know the basic terms in the smaller agreements.

The various types of oil licenses determine the major contractual agreements between the parties. The main agreements are mentioned below, before discussing the auditing or quality control, which ensures that terms are duly respected during execution of the contracts. The *operator* is the person or organization that puts the exploration and production into reality. A major part of the exploration and appraisal may be handled by any of the parties in the agreement; the actual drilling is handled by the operator oil company. The operatorship may also be relegated to a Technical Adviser party in the agreement.

There are several types of Oil and Gas exploration and production agreements, among which are:

- Concession
- Joint Venture / Operations Agreement (JV/JOA)
- Production Sharing Contract / Agreement (PSC/PSA)
- Risk Sharing Contract (RSC)
- Service Contract (SC)
- Marginal Field Concession (MFC)

The geoscientist will do well to understand the workings of each of these contract types. In practice, most oil companies work together and share the risks in an oil venture by having a percentage share in the oil mining or oil license block. All parties are involved in the check, with regular partnership meetings, while the field operator takes the lead.

III.1 Interface between Major Players

People involved or who play a role in the oil agreement, driving oil and gas supply, are often referred to as players. Most players are major oil company and governmental authorities that own an oil mining or operating license. It includes technical advisers and any sponsor or each person that has a major stake. At the execution stage, the oil service companies perform the actual outsourced tasks and are paid for each service they render, making up the total cost of exploration and production. The success of the collaboration between the oil company and its services firms affects the overall success of the contracts. Hence the various meetings, call for tenders, auditing and quality management. Final payments and deductions for services rendered are made only after audits and quality checks.

Often, a special department is appointed in an oil company to be charged with Non-Operated Assets (NOA), which covers business or aspect of the business that involves other partner and non-operated fields.

III.2 WSOR, CFT, AUDIT and SQM

The *operations geologist* and the *asset* or *development geologists* and *geophysicists*, alongside with the drillers, check and follow the contracts and audit plans to ensure that expected tasks are delivered by the service companies or contractors. The geologist should understand the aspects of any contract that affect geoscience operations and should plan out the sequence each task will occur on the rig. The wireline acquisition, for example, based on contractual agreement with the wireline company signed before drilling, will require that some acquisitions are "Firm" while others will be "Contingent". The former being firm involves logs that must be done as an evaluation process, irrespective of the result of the well. The later described as contingent depends on the result of the well, like finding hydrocarbon or not. If the well finds hydrocarbon, the logging takes place, else it is cancelled. The agreement and budgeting have to be concluded before drilling and the costs stated in the

FWSOR or WLR. This means that payments would be revised after drilling.

Below are the essential processes involved in executing the contracts.

- ***FWSOR:*** *Final Statement Of Requirements*
- ***CFT:*** *Call For Tender*
- ***Audit/ QC and QA:*** *Service contractor audits*
- ***SQM :*** *Service Quality Management*
- ***LQC:*** *Log Quality Control*

The Well Statement of Requirements (WSOR) is the preparation process for drilling a well, usually one to three (WSOR-0, WSOR-1 and WSOR-3). It starts once there is a prospective well identified by the exploration team. There are several reviews and the WSOR3 becomes the Final SOR (FWSOR), which may serve or is used to prepare the *Well Location Report (WLR)*. Once the decision to drill is firm, there is a ***CFT*** to service companies. There are discussions on best practices, best service and cost effectiveness, through which choices are made on who provides the logging, wireline, coring, and other services which are not done directly by the operator oil company. At each stage of the processes execution by the service firms, there are ***Quality Control (QC)*** measures.

III.3 ML Audit and Quality Assurance (QA)

Before the mud logging process, the geoscience representative on the rig will audit the equipment, sensors and stocks and sign if everything is OK. Any issue flagged at this point is reported to base in the audit document and is resolved before closing the case. The mud loggers are among the first team of personnel to arrive on the rig for geoscience service, so it is important to know that all they have installed are functional. That will determine the proper functioning of the other geoscience activities to follow.

A ***mud logging audit*** is carried out with a "Yes"/ "No" form to be signed by involved parties at the end. 'Yes' to okayed tools or installation and 'No' where there are issues, with a space to add comments. The comments may specify what should be done. The audit will include:

A Summary Page

- Objectives, observations and witnesses
- Conclusions and recommendation

The Body of the Document

- Rig type and sections
- Sensors and rig installations
- Mud logging unit and crew
- Hydrocarbon (Hc) and non-Hc detector and degassers
- Data acquisition and monitoring
- Computer systems, software and backups

Monitoring and HSE Awareness

- Safety related gadgets and sensors
- Safety apparatus such as fire extinguishers
- Functionalities of the unit with respect to safety
- Specific PPE for the unit such as gas masks

III.4 Wireline & LWD Audit and QA

It is similar to a Mud logging audit process, and will be carried out with a "Yes" / "No" form to be signed by both parties at the end. It will include:

Summary Page

- Objectives, observations and witnesses
- Conclusions and recommendations

Body of Document

- Logging unit and crew
- Tool house
- Explosives and radioactive materials
- Operation hazards measures
- Computer systems, software and backups

Monitoring and HSE Awareness

- Safety related gadgets and sensors
- Safety apparatus such as fire extinguishers
- Functionalities of the unit with respect to safety
- Specific PPE for the unit such as radioactive shields

III.5 SQM and LQC

While the Audit is the beginning of the Quality Control measures, the ***Service Quality Management (SQM)*** covers all the other practices that ensure the quality of services delivered by the contractor firms. These include regular *Performance Evaluations* through random service quality checks and meetings. Once such control is the ***Log Quality Control (LQC)*** performed at the end of each logging cycle, between the Wellsite geologist and the service contractors (essentially ML and WL/LWD/DD).

The Directional Driller (DD) is often in the same unit with the LWD team. Performance evaluation can be rigorous, since each aspect of the services evaluated and scored with specific marks in percentages. The report of an LWD operation run, in the form of ***LQC***, will include:

- Well designation
- Mud types
- Operating time versus total drilling time
- Data lost versus data repeated
- Drilled length versus logged length
- Lost Time (LT) and drilling hours
- Average ROP, ECD, Temperature

The report will be signed by the representatives on both sides, after including general data such as names of personnel involved, LWD run types, logging types and directional or MWD data. SQM and LQC are the aftermath of the runs and services offered; it is a revalidation of procedures and result which will check the trust established during audit.

III.6 Logistics

One aspect of service delivery that is often taken for granted by personnel is logistics on the rig site. While the wellsite geologist is not directly responsible for logistics, it turns out that he has to supervise the proper packaging, loading and shipping of geological samples. This is especially for new company subsidiary where exploration is in progress, or oil has just been discovered, as such there are yet no established efficient logistics department that coordinates activities between the rig site and the base. For shipping of sample and collection equipment, the WSG is in the best position to understand the content of the boxes and direct logistics accordingly.

A *manifest* must be established for every item to be shipped and the boxes properly labelled. Challenges with logistics depend on the location, the type of rig and country regulations. Regulations decide what may or may not be transported. The geologist has to be keen; he or she will, for example, not ignore a rock core painstakingly obtained to be destroyed en route. The WSG will care to know how and by what means it is transported. Likewise when there is need to reserve tickets and obtain a PPE for a work mission, it is in the personnel's interest to follow it all through. It is like management training, a start into becoming responsible in the labour market, not as an individual, but as part of a whole, functioning within a versatile organisation. In the real world, someone has to assume the responsibility and take the actions necessary for progress.

AFTERWORD

We are winding up fossil fuel exploration, but that transition seems to take longer, especially in developing oil host countries. These countries are yet struggling with energy challenges and a mix with petroleum is inevitable. The exploration and production fields with already existing installations are yet to be fully mined. While some major oil companies sell off blocks and equity in their assets in host countries, there are uncountable assets worth billions of dollar in oil exploration sites that cannot be dismantled, but probably transferred to a new workforce.

It is time for oil producing host nations to get ready in case of sudden abandonment and for the increasing need to ensure a proper energy mix in this critical time. There is need to be concerned in our time, to ensure that physical assets integrity and infrastructural maintenance and monitoring, and procedural appropriation is maintained.

Today, geologists and geophysicists need to grasp the overall aspect of petroleum exploration and production – from the basics of petroleum geoscience, through the various E & P disciplines to well operations and petrophysics. It is no longer tenable to present as either geologist of geophysicist, but we all become geoscientists as we integrate the various disciplines in the geoscience and reservoir engineering domains. This is no time to emphasise scales, tedious calculations that confuse beginners, which take too long and lead to distractions; it is time to go to the core and enrobe the overall processes quickly, getting young candidates ready to work. When details and semantics are emphasised, readers who are presented with sophisticated technologies lose grip of the fundamental knowledge that suffice to produce oil and gas from petroleum fields. It is a mop-up exercise for the oil and gas industry.

In a time of energy transition, in a declining phase for fossil fuel, in the face of changing resource value and media waves, as in a seemingly sinking boat, what is required are the best men to keep it afloat. Despite this urgency, the catastrophic approach must be avoided and geoscientists have to own their responsibility and do the needful. Energy demand in the world is increasing and some persons have to wear this cap of necessity. Though there has been a reduction and non-replacement of some staff in the mainstream petroleum sector, man power demands continue to be a challenge in the industry. A recruitment of a new type of workforce, who maintains the expected energy mix during the global crew change, is ongoing at the same time. It is not a choice, it is a necessity.

This comprehensive work has been provided to fill in the gaps, as a fall-back option for that new workforce needed in a transition era, for an evolving industry. These skills will also find use in other domains of geoscience during this quest for artificial intelligence, digitalisation and diversification of resources. Exporting oil technology to mining and processing, to refining of by-products and renewable energy, is the ultimate aim.

INDEX

ALSO BY THE AUTHOR

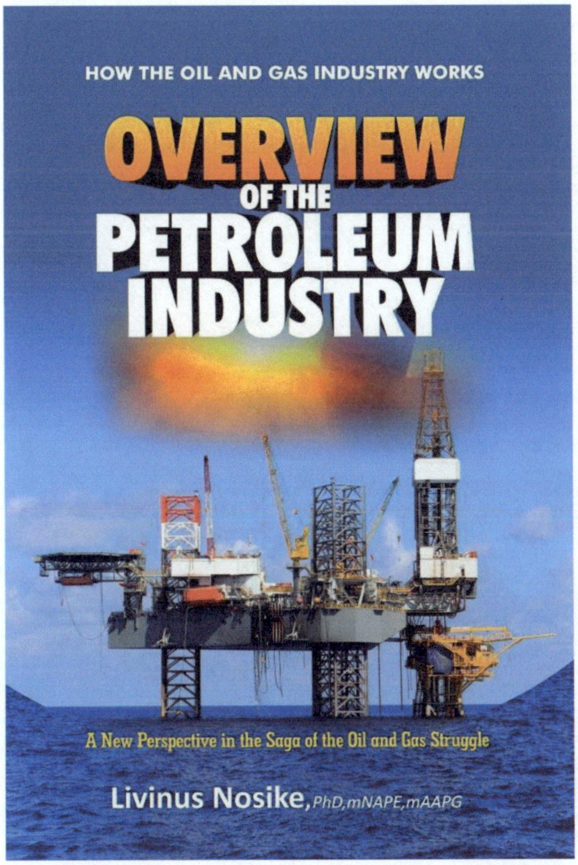

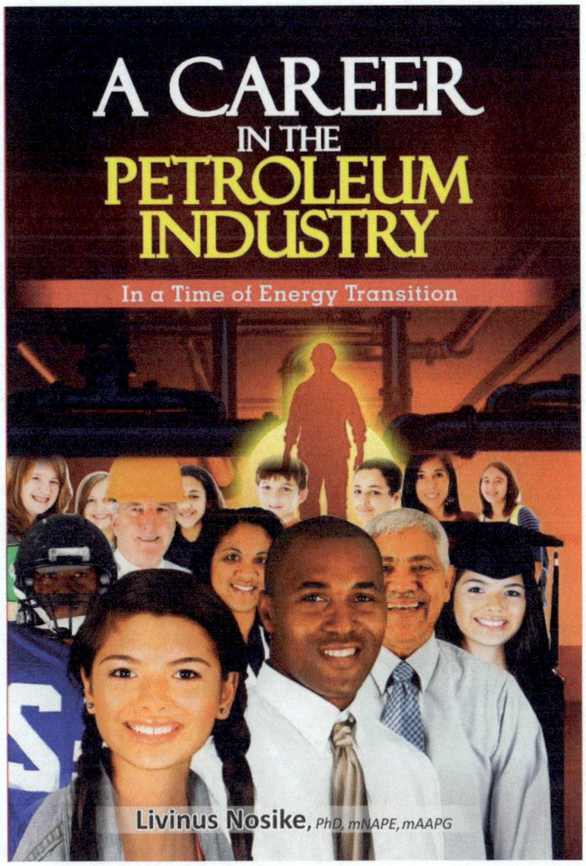

For ***Training on Exploration and Production Geoscience****, Contact:*

INTEGRATED ELVEE SERVICES (IES) OIL AND GAS (OG) DIVISION

iESog

MD & Chief Consultant: Dr Livinus Nosike

Head Office: 146 Ihe Street, Off Agbani Road, Oneday Layout, Enugu

www.iesog.com / contact@iesog.com / services@iesog.com

+234(0)7033209571, +234(0)7037628734

Made in the USA
Monee, IL
05 November 2026

67804733R00219